Communications in Computer and Information Science 931

Commenced Publication in 2007
Founding and Former Series Editors:
Phoebe Chen, Alfredo Cuzzocrea, Xiaoyong Du, Orhun Kara, Ting Liu,
Dominik Ślęzak, and Xiaokang Yang

More information about this series at http://www.springer.com/series/7899

Jong Hyuk Park · Hong Shen
Yunsick Sung · Hui Tian (Eds.)

Parallel and Distributed Computing, Applications and Technologies

19th International Conference, PDCAT 2018 Jeju Island, South Korea, August 20–22, 2018 Revised Selected Papers

Editors
Jong Hyuk Park
Department of Computer Science and Engineering
Seoul National University of Science and Technology
Seoul, Korea (Republic of)

Yunsick Sung
Department of Multimedia Engineering
Dongguk University
Seoul, Korea (Republic of)

Hong Shen
School of Computer Science
University of Adelaide
Adelaide, SA, Australia

Hui Tian
School of ICT
Griffith University
Gold Coast, Australia

ISSN 1865-0929 ISSN 1865-0937 (electronic)
Communications in Computer and Information Science
ISBN 978-981-13-5906-4 ISBN 978-981-13-5907-1 (eBook)
https://doi.org/10.1007/978-981-13-5907-1

Library of Congress Control Number: 2018966510

This Springer imprint is published by the registered company Springer Nature Singapore Pte Ltd.
The registered company address is: 152 Beach Road, #21-01/04 Gateway East, Singapore 189721, Singapore

Message from the PDCAT 2018 General Chairs

The International Conference on Parallel and Distributed Computing, Applications, and Technologies (PDCAT) is a major forum for scientists, engineers, and practitioners throughout the world to present their latest research, results, ideas, developments, and applications in all areas of parallel and distributed computing. Beginning in Hong Kong in 2000, PDCAT 2018 was held in Jeju, Korea, after 18 years of a successful journey through various countries/regions including Taiwan, Japan, China, Singapore, Australia, New Zealand, and Korea. For the 19th event we invited new and unpublished papers.

The conference papers included in the proceedings cover the following topics: PDCAT of Networking and Architectures, Software Systems and Technologies, Algorithms and Applications, and Security and Privacy. Accepted and presented papers highlight new trends and challenges of parallel and distributed computing, applications, and technologies. We hope readers will find these contributions useful and inspiring for their future research.

Our special thanks go to the program chairs: Houcine Hassan (Universitat Politecnica de Valencia, Spain), Hui Tian (Griffith University, Australia), Yunsick Sung (Dongguk University, South Korea), and all the Program Committee members and reviewers for their valuable efforts in the review process that helped us to guarantee the highest quality of the selected papers for the conference.

December 2018

James Park
Yi Pan
Young-Sik Jeong
Hong Shen
Qun Jin

Organization

Honorary Chair

Doo-soon Park	SoonChunHyang University, South Korea

General Chairs

James Park	SeoulTech, South Korea
Yi Pan	Georgia State University, USA
Young-Sik Jeong	Dongguk University, South Korea
Hong Shen	University of Adelaide, Australia
Qun Jin	Waseda University, Japan

Program Chairs

Houcine Hassan	Universitat Politecnica de Valencia, Spain
Hui Tian	Griffith University, Australia
Yunsick Sung	Dongguk University, South Korea

Workshop Chairs

Daewon Lee	Seokyeong University, South Korea
Ka Lok Man	Xi'an Jiaotong-Liverpool University, China

Publicity Chairs

Byoungwook Kim	Dongguk University, South Korea
Kyung-Soo Lim	ETRI, South Korea
Neil Y. Yen	The University of Aizu, Japan
Rossi Kamal	IEEE Seoul Chapter, South Korea
Yingpeng Sang	Sun Yat-sen University, China

Local Committee

Jungho Kang (Chair)	Baewha Women's University, Korea

Program Committee

Ahmed El Oualkadi	Abdelmalek Essaadi University, Morocco
Alexiei Dingli	University of Malta, Malta
Andrzej M. Goscinski	Deakin University, Austrailia
Cheonshik Kim	Sejong University, South Korea

Cho-Chin Lin	National Ilan University, Taiwan
Chuan-Ming Liu	National Taipei University of Technology, Taiwan
Depei Qian	Beihang University, China
Dorairaj Prabu	Broadcom Corporation, India
El-Sayed M. El-Alfy	King Fahd University of Petroleum and Minerals, Saudi Arabia
EunYoung Lee	Dongduck Woman's University, South Korea
Haibo Zhang	University of Otago, New Zealand
Hirohi Ishikawa	Tokyo Metropolitan University, Japan
Hui Tian	Griffith University, Australia
I-Cheng Chang	National Dong Hwa University, Taiwan
Jaeshup Oh	Sookmyung Women's University, South Korea
Javier Martínez Torres	Centro Universitario de la Defensa, Spain
Jun Wu	Beijing Jiaotong University, China
Jun Yan	University of Wollongong, Australia
Kaijun Ren	National University of Defense Technology, China
Kapetanios Epaminondas	University of Westminster, UK
Kenli Li	Hunan University, China
Li Tao	Xi'an University of Posts and Telecommunications, China
Manu Malek	Institute of Electrical and Electronics Engineers Inc., USA
Manuel Dominguez-Morales	University of Seville, Spain
Maytham Safar	Kuwait University, Kuwait
Michaeal Sheng	Macquarie University, Australia
Nitaigour Premchand Mahalik	California State University, USA
Prasan Kumar Sahoo	Chang Gung University, Taiwan
Rajkummar Buyya	University of Melbourne, Australia
Ren-Song Ko	National Chung Cheng University, Taiwan
Sheng-Shih Wang	Minghsin University of Science and Technology, Taiwan
Soon M. Chung	Wright State University, USA
Sun-Yuan Hsieh	National Cheng Kung University, Taiwan
Teofilo Gonzalez	University of California Santa Barbara, USA
Ye Tian	University of Science and Technology of China
Yidong Li	Beijing Jiaotong University, China
Yu Zhang	Los Alamos National Laboratory, USA
Yu-Chen Hu	Providence University, China
Zhiyi Huang	University of Otago, New Zealand

Contents

Algorithms and Applications

Software Systems and Technologies

Security and Privacy

Networking and Architectures

Spectrum-Centric Differential Privacy for Hypergraph Spectral Clustering

Xiaochun Wang, Yidong Li(✉), Yi Jin, and Wei Wang

School of Computer and Information Technology, Beijing Jiaotong University, Beijing, China
{16120425,ydli,yjin,wangwei1}@bjtu.edu.cn

Abstract. In real world, most of complex networks can be represented by hypergraphs. Hypergraph spectral clustering has drawn wide attention in recent years due to it is more suitable to describe high-order information between objects. In this paper, we focus on the spectrum of hypergraph-based complex networks, propose a spectrum-centric differential privacy scheme by using stochastic matrix perturbation. The main idea is to project the hypergraph Laplacian matrix into a low dimensional space and perturb the eigenvector with random noise. We present a differential privacy mechanism for hypergraph clustering based on the exponential mechanism. We evaluated the computational efficiency and data utility of the existing methods on both synthetic datasets and real datasets, and the experimental results reveal that our proposed mechanism achieves a better performance in both data utility and efficiency.

Keywords: Differential privacy · Information security · Hypergraph Spectral clustering

1 Introduction

In real world, in order to analyze and model the complex system which is go beyond simple pairwise interactions, researchers introduce a framework for clustering and community detection using hypergraph-based representations. Unfortunately, complex networks often refuse to publish their hypergraph data due to privacy concerns. A natural way to bridge the gap is to publish the anonymized ones. However, such anonymization is vulnerable to privacy attacks [1].

In recent years, differential privacy [2] has increasingly become a widely used privacy preserving mechanism in data publishing. In 2014, Google RAPPOR [3] provides a crowdsourcing collection framework based on random response mechanism and satisfies differential privacy. However, the basic goal of this method is frequency statistic. For complex heterogeneous information networks, it can be applied to the task of degree counting, but it is difficult to meet the requirements of other research tasks, such as spectral clustering. In 2014, Wu [4] proposed a privacy preserving method for graph spectral clustering, which ensures that the existence of a single user is insensitive to the spectral analysis task through perturbing the spectrum of graph, but the method is not suitable for hypergraph analysis task.

J. H. Park et al. (Eds.): PDCAT 2018, CCIS 931, pp. 3–14, 2019.
https://doi.org/10.1007/978-981-13-5907-1_1

In this paper, we apply the traditional differential privacy method based on random response and Laplace mechanism to the task of hypergraph spectral clustering. In view of the shortcomings of the existing methods, we propose a new method based on the exponential mechanism, aim to ensure data privacy at the same time maximizing the utility of hypergraph spectrum. We develop a scheme for publishing hypergraph-based complex network data with differential privacy. The scheme should satisfies two requirements. First, the released randomized graph should keep spectrum not much changed or at least can be reconstructed from the randomized graph. Second, the scheme should achieve the desired privacy guarantees. The main contributions of our work as follows:

- First, we develop a hypergraph spectrum privacy preserving method based on differential privacy for the hypergraph spectral clustering task.
- Second, we give the theoretically proof of global sensitivity calculation for eigenvalues and eigenvectors query.
- Finally, we conduct a detailed theoretical analysis and experimental evaluation of the proposed method from three aspects including privacy, data utility and algorithm scalability, and the experimental results show that our method performant better than the existing traditional methods.

The remaining parts of the paper are organized as follows. Section 2 reviews the related works. Section 3 introduces some basic notions of our approach and the existing protocols to achieve differential privacy. In Sect. 4, we present our spectrum-centric privacy preserving approach. Experimental results are shown in Sect. 5. We conclude this paper and discuss our future work in Sect. 6.

2 Related Works

The research on the release of graph under differential privacy is first started with the release of simple graph. According to the preserved objects, we divide the existing works into node differential privacy [5, 6] and edge differential privacy [7]. In complex networks, spectral clustering is used as a common dimension-reduction tool in various data mining tasks [8]. The general properties of real networks has gained much attention due to the proliferation of network data. Most analysis [9–11] has been confined to real-space characteristics, e.g., degree sequences, shortest connecting paths, and clustering coefficients. In this paper, we investigate this problem by focusing on the spectrum of networks since it has been shown the spectrum has close relation with many graph characteristics and can provide global measures for some network properties.

A few schemes have been developed to approximate eigenvectors and eigenvalues of matrices in a differential private manner [12, 13]. Their main idea is to perturb the original matrix by adding random noise and then publish the perturbed matrix. The key limitation of this approach is the high cost in both computation and storage space. In last year, Ahmed et al. [14] proposed a more efficient approach by random matrix projection which reduces the dimensionality of the published matrix. Wu et al. [4] directly perturbs the eigenvector of the original data by Laplacian noise.

While the above works have achieved privacy in simple graph spectrum publishing, researchers have struggled to develop suitable adaptations of these techniques on complex networks. A principal difficulty naturally arise: the adaptation of differential privacy from graph spectrum to hypergraph spectrum.

3 Models and Existing Protocols

Let V denote the set of vertices (objects), and let E be a family of subsets of V such that $\cup_{e \in E} = V$. Then we call $G = (V; E)$ a hypergraph with the vertex set V and the hyperedge set E. Hypergraph is a generalization of graph, a hyperedge containing just two vertices is a simple graph edge. The connection relationship of hypergraph G can be represented as a $|V| \times |E|$ related matrix H, with $H_{i,j} = 1$ if there is a link between node v_i and hyperedge e_j, in other words, node v_i contains in hyperedge e_j, and $H_{i,j} = 0$ otherwise.

$$\mathrm{H}(v, e) = \begin{cases} 1; & v \in e \\ 0; & otherwise \end{cases} \tag{1}$$

The Laplacian matrix for unweighted and undirected hypergraph can be represented as $L = HH^T$. L is a semi-positive definite symmetric matrix with $L_{i,j} = c$ if there is c common hyperedges between node v_i and node v_j. In function (2), d_i is the degree of node v_i.

$$L_{i,j} = \begin{cases} c, i \neq j \\ d_i, i = j \end{cases} \tag{2}$$

We divide the existing differential privacy protocols into following three types according to their perturb mechanism:

(a) Random response-based unitary encoding [3]. An input v is encoded as a length-d bit vector, with only the bit corresponding to v set to 1. Here two key parameters in perturbation are p, the probability that 1 remains 1 after perturbation, and q, the probability that 0 is perturbed into 1. Depending on parameters choices, we have two protocols, UE and OUE [15].

(b) Laplace mechanism [16]. Given any function $f : \mathrm{D} \rightarrow R^d$, expression $A(D)$ of output satisfies the following equation is satisfied ϵ-differential privacy.

$$A(D) = f(D) + \left(L\left(\frac{\Delta f}{\varepsilon}\right)\right)^d \tag{3}$$

In the function (3), $\mathrm{L}(\Delta f/\epsilon)$ is Laplace noise function, the size of the noise added on the function A is related to Δf and ϵ. The outputs probability of function f in adjacent datasets A and B are almost the same under Laplace noise perturbing.

(c) Exponential mechanism [16]. Given a utility function $q : (D^n \times R) \rightarrow R$ for a database instance D, the mechanism A gives ϵ-differential privacy if it satisfies the following condition:

$$A(D, q) = \{r : |P_r[r \in R] \propto \exp(\epsilon q(D, r)/2GS(q))\}. \tag{4}$$

3.1 Unitary Encode-Based Hypergraph Clustering

In this paper, we apply the unitary encode method to the hypergraph publication. In order to adapt to the hypergraph related matrix, we adjust the definition of unitary encode and derive the sequence encode (SE). Here we further explain the SE mechanism. We first get each node's hyperedge sequences $seq(v_i)$, Table 2 shows the hyperedge sequence derived from the hypergraph related matrix in Table 1. Then, we adopt a three-step operation to perturb the encoding process.

$$seq(v_i) = \{j | v_i \in e_j, i \leq |V|, j \leq |E|\}$$

Table 1. Hypergraph related matrix

Related matrix $\boldsymbol{H}$	e_1	e_2	e_3	e_4
v_1	1	0	0	0
v_2	1	1	0	0
v_3	1	1	1	0
v_4	0	0	0	1
v_5	0	0	1	0
v_6	0	0	1	0
v_7	0	0	0	0

Step 1: Encoding. Encode each value j in $seq(v_i)$ into a bit string, a length $|E|$ binary vector where only the j-th position is 1.

$$UE(j) = [0, \ldots, 1, \ldots, 0]$$

Step 2: Perturbing. Perturb(B) outputs B′ as follows:

$$\Pr[B'[i] = 1] = \begin{cases} p, B[i] = 1 \\ q, B[i] = 0 \end{cases} \quad p = \frac{e^{\varepsilon/2}}{e^{\varepsilon/2} + 1}, q = \frac{1}{e^{\varepsilon/2} + 1} \tag{5}$$

Step 3: Union.

$$SE(seq(v_i)) = \bigcup\nolimits_{j \in seq(v_i)} UE'(j) \tag{6}$$

Table 2. Hyperedge sequence

Node	v_1	v_2	v_3	v_4	v_5	v_6	v_7
Sequence	$\{1\}$	$\{1,2\}$	$\{1,2,3\}$	$\{4\}$	$\{3\}$	$\{3\}$	$\{0\}$

For example of node v_2, we have $seq(v_2) = \{1,2\}$, the true sequence encoding of v_2 is as follows:

$$\begin{aligned} UE(1) &= [1\,0\,0\,0], UE(2) = [0\,1\,0\,0], \\ SE(seq(v_2)) &= UE(1) \cup UE(2) = [1\,1\,0\,0]. \end{aligned}$$

After perturbation, we may have:

$$\begin{aligned} UE'(1) &= [1\,0\,0\,0], UE'(2) = [0\,0\,1\,0], \\ SE(seq'(v_2)) &= UE'(1) \cup UE'(2) = [1\,0\,1\,0]. \end{aligned}$$

3.2 Laplace Perturbed Hypergraph Spectrum Algorithm

In this section, considering the spectrum of the hypergraph, we adjust the traditional Laplace mechanism-based differential privacy and apply it to hypergraph data, thus we propose the Laplace perturbed hypergraph spectrum algorithm LPHS.

In Algorithm 1, we output the top-k eigenvalues $\lambda^{(k)} = (\lambda_1, \lambda_2, \ldots, \lambda_k)$ and the corresponding eigenvectors $u^{(k)} = (u_1, u_2, \ldots, u_k)$ under ϵ-differential privacy with Laplacian mechanism [16]. We first derive the sensitivities for the eigenvalues and eigenvectors in Results 1, 2. We then calibrate Laplace noise to the eigenvalues and eigenvectors based on the derived sensitivities and privacy parameter. Because the perturbed eigenvectors will no longer be orthogonalized to each other, we finally do a post-process to normalize and orthogonalize the perturbed eigenvectors.

Result 1. Given a Hypergraph H with its Laplacian matrix L and rank R, the global sensitivity [4] of each eigenvalue is $GS_{\lambda_i}(H) = \|P_L\|_2, \; i \in [1,n]$; According to Hoffman-Wielandt theorem [18], the global sensitivity of the first k (k > 1) eigenvalues vector $\lambda^{(k)} = (\lambda_1, \lambda_2, \ldots, \lambda_k)$ is

$$GS_{\lambda^{(k)}}(H) = \sqrt{k(2R-1)}. \tag{7}$$

Result 2. Given a Hypergraph H with its Laplacian matrix L and rank R, the global sensitivity of each eigenvector $u_i (i > 1)$ is

$$GS_{u_i}(H) = \frac{\sqrt{n}\|P_L\|_2}{\min(\lambda_i - \lambda_{i-1}, \lambda_i - \lambda_{i+1})}. \tag{8}$$

Specifically, the sensitivities of the first and last eigenvector are respectively $GS_{u_1}(H) = \frac{\sqrt{n}\|P_L\|_2}{\lambda_1 - \lambda_2}$, $GS_{u_n}(H) = \frac{\sqrt{n}\|P_L\|_2}{\lambda_{n-1} - \lambda_n}$.

Algorithm 1. Laplace perturbed hypergraph spectrum algorithm LPHS

Input: Hypergraph Laplacian matrix L, dimension parameter k, privacy parameter ϵ;
Output: The first noisy k eigenvalues $\tilde{\lambda}^{(k)} = (\tilde{\lambda}_1, \tilde{\lambda}_2, \ldots, \tilde{\lambda}_k)$ and corresponding noisy eigenvectors $\tilde{u}^{(k)} = (\tilde{u}_1, \tilde{u}_2, \ldots, \tilde{u}_k)$
1: Decomposition A to obtain the top-k eigenvalues $\lambda^{(k)} = (\lambda_1, \lambda_2, \ldots, \lambda_k)$ and the corresponding eigenvectors $u^{(k)} = (u_1, u_2, \ldots, u_k)$;
2: Distribute ϵ into $\epsilon_0, \ldots, \epsilon_k$
3: Add Laplace noise to $\lambda^{(k)}$ with ϵ_0 based on $GS_{\lambda^{(k)}}(H)$ derived in Result 1 and obtain $\tilde{\lambda}^{(k)} = (\tilde{\lambda}_1, \tilde{\lambda}_2, \ldots, \tilde{\lambda}_k)$;
4: For i:=1 to k do
5: Add Laplace noise to u_i with ϵ_i based on $GS_{u_i}(H)$ derived in Result 2 to obtain $\tilde{x}_i$;
6: End for
7: Normalize and orthogonalize $\tilde{x}_1, \tilde{x}_2, \ldots, \tilde{x}_k$ to obtain $\tilde{u}_1, \tilde{u}_2, \ldots, \tilde{u}_k$;
8: Output $\tilde{\lambda}_1, \tilde{\lambda}_2, \ldots, \tilde{\lambda}_k$ and $\tilde{u}_1, \tilde{u}_2, \ldots, \tilde{u}_k$

4 Proposed Private Hypergraph Spectrum Publication

In this section, we describe our proposed mechanism EPHS for private hypergraph spectral clustering. We set hypergraph Laplacian matrix as the similarity matrix for hypergraph spectral clustering and we select the eigenvalues and eigenvectors of the original similarity matrix by exponential mechanism and then publish the perturbed ones. Our key idea is to approximate eigenvectors and eigenvalues of similarity matrix in a differential private manner. If we are interested in the top-k eigenvectors of the similarity matrix, we only need to publish a low dimensional matrix consists of perturbed top-k eigenvectors, thus reduces computation cost and storage space, as well as provide differential privacy for hypergraph data publishing.

4.1 Exponential Perturbed Hypergraph Spectrum Algorithm

In algorithm EPHS, we adopt eigenvalue as the score function of exponential mechanism, which means that any two eigenvectors more closely, the more probability they have to swap. Because the output eigenvectors are still orthogonalized to each other, we needn't orthogonalize the eigenvectors.

Algorithm 2. Exponential perturbed hypergraph spectrum algorithm EPHS

Input: Hypergraph Laplacian matrix L, dimension parameter k, privacy parameter ϵ;
Output: The noisy top-k eigenvalues $\lambda^{(k)} = (\lambda_1, \lambda_2, ..., \lambda_k)$ and corresponding noisy eigenvectors $u^{(k)} = (u_1, u_2, ..., u_k)$
1: Decomposition A to obtain the top-k eigenvalues $\lambda^{(k)} = (\lambda_1, \lambda_2, ..., \lambda_k)$ and the corresponding eigenvectors $u_1, u_2, ..., u_k$;
2: For each eigenvalue λ_i in original eigenvectors set λ_k do
3: For each eigenvalue λ_j in candidate set λ' do
4: swap probability is: $P_r = \frac{\exp\left(\frac{\epsilon \cdot |\lambda_i - \lambda_j|}{2GS_\lambda}\right)}{\sum_{i=1}^{|\lambda'|} \exp\left(\frac{\epsilon \cdot |\lambda_i - \lambda_j|}{2GS_\lambda}\right)}$
5: End for
6: Select a eigenvalue λ_j with largest P_r from λ' to swap with λ_i a
7: Delete λ_j from λ'
8: End for
9: Output $\tilde{\lambda}_1, \tilde{\lambda}_2, ..., \tilde{\lambda}_k$ and $\tilde{u}_1, \tilde{u}_2, ..., \tilde{u}_k$

4.2 Privacy Analysis

Definition 1 (Adjacent Hypergraph). If two hypergraphs differ in one hyperlink L_x, they are adjacent hypergraphs. In this paper, we denote a **hyperlink** as a connection between node and hyperedge.

Definition 2 (Hyperlink-privacy). A private query satisfies hyperlink-privacy for all pairs of **adjacent hypergraph** $H_1 = (V_1, E_1, L_1)$, $H_2 = (V_2, E_2, L_2)$ where $V_1 = V_2$, $E_1 = E_2$ and $L_1 = L_2 - L_x$ where $|L_x| = R$. R is the maximum rank of hyperedge E_x where $L_x = (V_x, E_x)$.

Our algorithm achieves hyperlink-privacy by a two-step private operation: eigenvalues perturbation and eigenvectors perturbation, which can provide the guarantee that the query results is almost same on two adjacent hypergraphs. The privacy budget is consequently divided into $k+1$ pieces. Based on the sequential composition [17] and the privacy budget allocation, we measure the privacy level of our algorithm as follows:

- The eigenvalues perturbation is performed on the whole eigenvalues sequence with the privacy budget ϵ_0. This operation preserves ϵ_0-differential privacy.
- The eigenvectors perturbation is performed on each eigenvectors sequence with the privacy budget ϵ_i. According to sequential composition, the eigenvectors perturbation preserves $(\epsilon_i + \cdots + \epsilon_m)$-differential privacy as a whole.

Consequently, we can conclude that our algorithm preserves ϵ-differential privacy.

4.3 Utility Analysis

For eigenvalues, we measure the output accuracy with the absolute error by E_Λ,

$$E_\Lambda = \left|\overline{\lambda}(k) - \lambda(k)\right|_1 = \sum_{i=1}^{k} \left|\lambda_i - \overline{\lambda}_i\right|. \tag{9}$$

For eigenvectors, we define the absolute error as $E_U = \left\|\widetilde{U}_k - U_k\right\|_1$. We also define the cosine similarity to measure the accuracy of each private eigenvector as

$$\cos\langle\widetilde{u}_i, u_i\rangle = \widetilde{u}_i \cdot u_i, i \in [i, k]. \tag{10}$$

5 Experimental Evaluation

We use four type of datasets, two real-life datasets and two synthetic datasets. The real-life datasets is collected from two location-based social network websites, namely, Brightkite and Gowalla, where users shared their locations by checking-in. We extract two sub-hypergraph with rank R = 9 form the original real-life datasets. The one synthetic dataset is a random simple graph with rank R = 2, another is a 3-uniform hypergraph with rank R = 3. In all tables, the result reported are the average of 20 runs.

All the experiments are performed on a computer with 3.30 GHz Intel Corei3-2120 CPU and 4 GB memory. The program is written in Java, and run on JRE 8.

5.1 Performance Comparison with Varying R

In this section, we evaluate the performance of the two output perturb-based privacy preserving method LPHS and EPHS on the accuracy of the eigenvectors and eigenvalues with varying hypergraph rank R. We set $R = 2, 3, 9$ and $k = 5$. In this section, we take a simple ϵ-distribution strategy, distributing ϵ as $\epsilon_0 = 10$ to the eigenvalues and $\epsilon_i = 90$, $(\mathrm{i} \in [1, k])$ equally to each eigenvector. Therefore DPHS satisfies $\epsilon = 460$ differential privacy.

From Table 3, we can see that, with the decrement of R, both eigenvalues and eigenvectors tend to close to the original ones, this is because the rank R of hypergraph has a more significantly effect on both the global sensitivity of eigenvalues and eigenvectors queries. A larger R results in a larger global sensitivity in hypergraph, which lead to introducing more noise to eigenvalues and eigenvectors perturbation.

From Tables 4 and 5, we can see that, EPHS is generally outperforms LPHS. The reason is that EPHS is designed to provide a privacy matrix for original hypergraph Laplacian matrix, while LPNP mask the original data completely. In LPNP, for eigenvectors, we can observe that it tends to achieve a larger cosine similarity when the eigen-gap is large. Thus a better strategy will be distributing privacy parameter ϵ according to the magnitudes of eigen-gaps, instead of equal distribution.

Table 3. Eigen comparison with varying R

Method	Rank	$\cos\langle\widetilde{u}_i, u_i\rangle$	E_u	E_λ
LPHS	R = 2	0.9455	4.6760	3.2295
	R = 3	0.8754	8.1076	3.8619
	R = 9	0.7402	16.5461	4.7379
EPHS	R = 2	0.9923	0.1063	0.1914
	R = 3	0.9243	2.1277	1.0979
	R = 9	0.8126	2.2754	3.1032

Table 4. Eigenvalues comparison

Method	Top-k Eigenvalues					E_λ
	λ_1	λ_2	λ_3	λ_4	λ_5	
λ(k)	10.4003	7.9465	3.9762	3.4539	1.4088	–
eigen-gap	2.4538	2.4538	0.5223	0.5223	2.0451	–
LPHS	11.8325	8.1747	5.3810	3.2326	−0.6411	4.7379
EPHS	10.4003	7.9465	3.9762	3.4539	1.4088	3.1032

Table 5. Eigenvectors comparison

Method	$\cos\langle\widetilde{u}_i, u_i\rangle = \widetilde{u}_i \cdot u_i$					E_u
	u_1	u_2	u_3	u_4	u_5	
LPHS	0.8968	0.8413	0.5567	0.5360	0.8704	16.5461
EPHS	0.8243	0.7833	0.8238	0.7976	0.8340	2.2754

5.2 Spectral Clustering Result

In this section, we evaluate the performance of the proposal and other approaches on the accuracy of spectral clustering with varying privacy budget ϵ, and we set hypergraph rank R to 9. We select a general metric NMI to measure the clustering results. NMI (Normalized Mutual Information) is commonly used to measure the similarity of two clustering results in clustering. The range of NMI is 0 to 1, and the higher it is, the more accurate the division. From Fig. 1, we can observe that, EPHS is significantly outperforms other traditional methods in clustering result.

5.3 Efficiency Evaluation

In addition to utility concerns, scalability is also an important factor for selecting a proper sanitization solution. Figure 2 shows the run-time of both UE, OUE, LPHS and EPHS on two real-life datasets with different datasize. The graph derived from BK has more than 50 thousand nodes, and the graph derived from GW has more than 190 thousand nodes. In general, EPHS is significantly faster than other traditional methods. In particular, it takes only less than 2 s for EPHS to handle with BK, whereas it takes

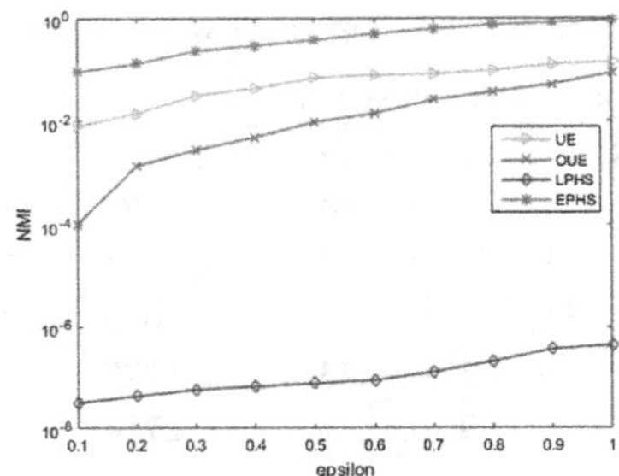

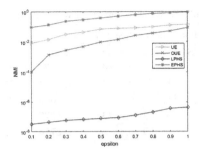

Fig. 1. Clustering result (measured in NMI)

26 s for LPHS and even more than 120 s for UE and OUE to process the smaller dataset BK. When dealing with the larger dataset GW, the proposal runs less than 1.8 s. On the contrary, the UE and OUE methods take even more than 7 min. Therefore, our algorithm is more adaptable to large scale graph data, and the algorithm is more extensible. The reason is that UE and OUE are input perturb-based method which are related to the size of the original dataset. While EPHS and LPHS is based on output perturbation, which is only related to the top-k eigenvalues and eigenvectors. As a whole, our method is more efficient in efficiency.

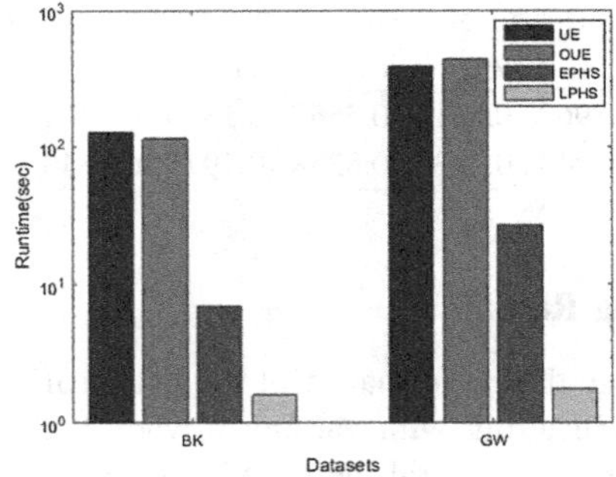

Fig. 2. Running time

6 Conclusion and Future Works

In this work, we propose a spectrum-centric differential privacy scheme for hypergraph spectrum publishing. We evaluate our approach on the differential privacy preserving eigen decomposition of hypergraph Laplacian matrix. In the future work, we will investigate how to enforce differential privacy for some hypergraph spectral analysis tasks, such as spectral clustering based on Laplace mechanism. We will also study the use of smooth sensitivity and how to better distribute privacy budget in the proposed DPHS method. In order to maintain more data utility.

Acknowledgment. This work is supported by National Science Foundation of China Grant #61672088, Fundamental Research Funds for the Central Universities #2016JBM022 and #2015ZBJ007.

APPENDIX

Proof of Result 1

Definition 3 (Hypergraph Rank): In hypergraph, rank R indicates the amount of vertices a hyperedge associated. In this paper, we denote R as the upper rank of hypergraph, which indicates the largest amount of vertices a hyperedge associated.

We denote adding/deleting an link between node v_i and hyperedge e_r on the original hypergraph H as a perturbation matrix P_H added to the original related matrix H. P_H is a matrix where only $P_{(r,i)}$ have value 1/−1 and all other entries are zeros. In other words, it can be seem as adding a perturbation matrix P_L to the original Laplacian matrix, P_L is a symmetric matrix where only $P_{(i,j)}$ and $P_{(j,i)}$ have value 1/−1 and all other entries are zeros. j is the index of every node associated to hyperedge e_r, $v_j \in e_r$. Based on the Definition 3, we have $\|P_L\|_F \leq (2R-1)$.

We denote λ_i as the eigenvalue of the matrix L and $\widetilde{\lambda}_i$ as that of matrix $L+P_L$. Based on the Hoffman-Wielandt theorem [7], we have $GS_{\lambda_i}(H) \leq max\left|\lambda_i - \widetilde{\lambda}_i\right| \leq max\|P_L\|_2$ and $GS_{\lambda^{(k)}}(H) = \sum_{i=1}^{k}\left|\lambda_i - \widetilde{\lambda}_i\right| \leq \sqrt{k}\sqrt{\sum_{i=1}^{k}\left(\lambda_i - \widetilde{\lambda}_i\right)^2} \leq \sqrt{k}\|P_L\|_F \leq \sqrt{k(2R-1)}$.

Proof of Result 2

We denote the eigenvectors of matrix H, $H+P$ respectively as column vectors u_i and $\widetilde{u}_i, i \in [1,k]$. Based on the stochastic perturbation theory [19] (Chap. 4.3), for each eigenvector u_i, it can be viewed as a specific situation for invariant space perturbation when k adopt 1, so we have $\|\widetilde{u}_i - u_i\|_s = \frac{\sigma}{\delta_F}$. According to stochastic perturbation theory [19] (Theorem 4.4), we have $\delta_F > min\{|\lambda_i - \lambda_{i-1}|, |\lambda_i - \lambda_{i+1}|\}$. According to stochastic perturbation theory [19] (Chap. 4.2), we have $\sigma = \|P_L\|_s$, $s = 1, 2, F$, thus $GS_{u_i}(H) = \|\widetilde{u}_i - u_i\|_1 \leq \sqrt{n}\,\|\widetilde{u}_i - u_i\|_1 \leq \frac{\sqrt{n}\|P_L\|_2}{\min(\lambda_i - \lambda_{i-1}, \lambda_i - \lambda_{i+1})}$. Specifically for $i=1$ (similarly for $i=n$), $GS_{u_1}(H) = \frac{\sqrt{n}\|P_L\|_2}{\lambda_1 - \lambda_2}$.

$\|P_L\|_2$: For different link perturbation, P_L has different Euclidean norm. We calculate the Euclidean norm of P_L as $\|P_L\|_2$: in the case of adding/deleting a link between node v_i and hyperedge e_r with largest hyperedge rank.

References

1. Narayanan, A., Shmatikov, V.: De-anonymizing social networks. In: 2009 30th IEEE Symposium on Security and Privacy, pp. 173–187. IEEE (2009)
2. Dwork, C.: Differential privacy. In: Bugliesi, M., Preneel, B., Sassone, V., Wegener, I. (eds.) ICALP 2006, Part II. LNCS, vol. 4052, pp. 1–12. Springer, Heidelberg (2006). https://doi.org/10.1007/11787006_1
3. Pihur, V., Korolova, A.: RAPPOR: randomized aggregatable privacy-preserving ordinal response. In: ACM SIGSAC Conference on Computer and Communications Security, pp. 1054–1067. ACM (2014)
4. Li, Y., Wu, X., Lu, A.: Analysis of spectral space properties of directed graphs using matrix perturbation theory with application in graph partition, pp. 847–852 (2015)
5. Kasiviswanathan, S.P., Nissim, K., Raskhodnikova, S.: Analyzing graphs with node differential privacy **7785**, 457–476 (2013)
6. Day, W.-Y., Li, N., Lyu, M.: Publishing graph degree distribution with node differential privacy. In: Proceedings of the 2016 International Conference on Management of Data, SIGMOD 2016, pp. 123–138. ACM, New York (2016)
7. Wang, Y., Xintao, W.: Preserving differential privacy in degree correlation based graph generation. Trans. Data Privacy **6**(2), 127 (2013)
8. Michoel, T., Nachtergaele, B.: Alignment and integration of complex networks by hypergraph-based spectral clustering. Phys. Rev. E **86**(5), 056111 (2012)
9. Seary, A.J., Richards, W.D.: Spectral methods for analyzing and visualizing networks: an introduction. na (2003)
10. Chen, R., Fung, B.C., Yu, P.S., Desai, B.C.: Correlated network data publication via differential privacy. VLDB J. **23**(4), 653–676 (2014)
11. Dwork, C., Mcsherry, F., Nissim, K.: Calibrating noise to sensitivity in private data analysis. Proc. VLDB Endowment **7**(8), 637–648 (2006)
12. Chaudhuri, K., Sarwate, A., Sinha, K.: Near-optimal differentially private principal components. In: Advances in Neural Information Processing Systems, pp. 989–997 (2012)
13. Kapralov, M., Talwar, K.: On differentially private low rank approximation. In: Proceedings of the Twenty-Fourth Annual ACM-SIAM Symposium on Discrete Algorithms, pp. 1395–1414. SIAM (2013)
14. Ahmed, F., Jin, R., Liu, A.X.: A random matrix approach to differential privacy and structure preserved social network graph publishing. arXiv preprint arXiv:1307.0475 (2013)
15. Wang, T., Blocki, J., Li, N., et al.: Locally differentially private protocols for frequency estimation. In: Proceedings of the 26th USENIX Security Symposium, pp. 729–745 (2017)
16. Mcsherry, F., Talwar, K.: Mechanism design via differential privacy. In: 2007 IEEE Symposium on Foundations of Computer Science, FOCS 2007, pp. 94–103 (2007)
17. Mcsherry, F.D.: Privacy integrated queries: an extensible platform for privacy-preserving data analysis. Commun. ACM **53**(9), 89–97 (2010)
18. Hoffman, A.J., Wielandt, H.W.: The variation of the spectrum of a normal matrix. In: Selected Papers of Alan J Hoffman: With Commentary, pp. 118–120. World Scientific (2003)
19. Stewart, G.W.: Stochastic perturbation theory. SIAM Rev. **32**(4), 579–610 (1990)

Efficient Scheduling Strategy for Data Collection in Delay-Tolerant Wireless Sensor Networks with a Mobile Sink

Zhansheng Chen[1,2(✉)], Hong Shen[3,4], Xiaofan Zhao[5], and Tingmei Wang[2]

[1] School of Computer and Information Technology, Beijing Jiaotong University, Beijing, China
11112085@bjtu.edu.cn

[2] School of Applied Science and Technology, Beijing Union University, Beijing, China

[3] School of Data and Computer Science, Sun Yat-sen University, Guangzhou, China

[4] School of Computer Science, University of Adelaide, Adelaide, Australia

[5] People's Public Security University of China, Beijing, China

Abstract. Wireless sensor networks based on mobile sink (MS) can significantly alleviate the problem of network congestion and energy hole, but it results in large delay because of restriction of moving speed and lead to the loss of data due to the limited communication time. In this paper, a grid-based efficient scheduling and data gathering scheme (GES-DGS) is proposed for maximizing the amount of data collected and reducing energy consumption simultaneously within the delay of network tolerance. The main challenges of our scheme are how to optimize the trajectory and the sojourn times of MS and how to deliver the sensed data to MS in an energy-efficient way. To deal with the above problems, we first divide the monitoring field into multiple grids and construct the hop gradient of each grid. Second, we design a heuristic rendezvous point selection strategy to determine the trajectory of MS and devise a routing protocol based on hops and energy. With extensive simulation, we demonstrate that GES-DGS scheme not only significantly extends network lifespan compared with MS-based data gathering schemes, but also pro-actively adapts to the changes in delay in specific applications.

Keywords: Wireless sensor networks · Trajectory · Mobile sink
Data gathering · Delay

1 Introduction

The use of energy-efficient routing protocols in wireless sensor networks (WSNs) can reduce energy consumption to a certain extent, but it can not avoid the problem of network congestion and energy hole caused by the fixed position of sink. The introduction of mobile sink (MS) can solve the above problems, and greatly improve network lifespan. However, it result in large delay because of restriction of moving

J. H. Park et al. (Eds.): PDCAT 2018, CCIS 931, pp. 15–27, 2019.
https://doi.org/10.1007/978-981-13-5907-1_2

speed and lead to the loss of data due to the limited communication time. Taking the monitoring field of $L \times L$ ($L = 500\,\text{m}$) as an example, MS will takes more than 4 h, which is intolerable for many applications, to access 200 sensor nodes randomly deployed in the region even if it moves at a higher speed ($2\,\text{m/s}$). In addition, the length of MS's patrolling path is restricted by battery endurance, the size of node's buffer and the tolerable delay of application, etc., makes it impossible for MS to access every sensor node. To deal with the above problems, we propose a grid-based energy-efficient scheduling and data gathering scheme (GES-DGS), which joint consider the selection of rendezvous point (RP) and the routing from ordinary node to MS, for maximizing the amount of data collected and reducing energy consumption simultaneously within the delay of network tolerance.

The main contributions of our scheme can be summarized as follows:

- We first analysis the minimum virtual unit (MVU) of data acquisition, ensuring that all nodes in MVUs are within the communication radius of MS. Then, we propose a MVU-based partition method for division of WSNs and construct the hop gradient of each MVU.
- MVU is divided into multiple girds based on node's sensing radius for ensuring full coverage of network monitoring, and only one sensor is kept in active state for reducing network traffic load.
- We design a virtual point-based adaptive patrol route algorithm (VP-APR), using proactive adjustment strategy based on node's energy and location to adapts to the change of delay, to optimize the trajectory of MS.
- In order to reduce and balance energy consumption among nodes, we present a efficient routing protocol based on residual energy level of active node.

The performance of our scheme is evaluated through MATLAB simulations. Simulation results show that our scheme significantly extends network lifespan compared with existing MS-based data gathering schemes, and also achieve other better performance such as less energy consumption, less packet loss create, better throughput and better adaptability.

The rest of the article is organized as follows. In Sect. 2, we summarize the related work of MS-based data gathering schemes briefly. The problem statement is given in Sect. 3. In Sect. 4, we present and describe the details of our data gathering scheme GES-DGS. The simulation results are shown in Sect. 5. Finally, we give our peroration in Sect. 6.

2 Related Work

The issue of data gathering in WSNs has been solved by many researches from different methods. In this section, we give an overview of MS-based data gathering.

Chen et al. [1] proposed a grid-based reliable multi-hop routing protocol, optimizes the cluster head election process by combining individual ability and local cognition via a consultative mechanism based on cluster head's lifespan expectancy, while considering data forwarding delay and reliable transmission of data. Xing et al. [2] presented a rendezvous-based approach in which a subset of nodes serves as the rendezvous points (RPs) that buffer data originated from source nodes and transfer to mobile elements (MEs) capable of carrying data automatically when they arrived. RPs enable MEs to collect a large volume of data every time without traveling long distance, which can achieve a desirable balance between network energy saving and data collection delay. Wen et al. [3] introduced an an energy-aware path construction (EAPC) algorithm, which selects an appropriate set of data collection points, constructs a data collection path, and collects data from the points burdened with data. EAPC accounts for the path cost from its current data collection point to the next point and the forwarding load of each sensor node. Yun et al. [4] have discussed about how to maximize the lifespan of the WSNs by using a mobile sink. Within a prescribed delay tolerance level, each node can store the data temporarily and transmit it when MS is at the most favorable location for achieving the longest WSNs lifespan. The framework proposed integrates energy-saving techniques, multipath routing, a mobile sink, delayed data delivery, and active region control, into a single optimization problem. Whether the proposal should be adopted in practice depends on the trade-off between the lifespan gain and the actual cost. Cayirpunar et al. [5] have considered the optimal mobility patterns of multiple mobile sinks for achieving the maximum WSNs lifespan. In this paper, three representative, i.e., grid, random, and spiral, are used for investigating the characteristics of the optimal mobility patterns and a novel mixed integer programming framework is proposed for characterizing network lifespan under different mobility patterns for multiple mobile BSs. Cheng et al. [6] proposed a timeliness traveling path planning (TTPP) algorithm which focus on how to plan a traveling path that meets the delay requirement of time-sensitive applications for data collection and reduces the amount of relay packets in the WSNs. Based on the least squares curve approach, the proposed TTPP algorithm can find the bestfitting curve for any given set of sensors by reducing the amount of relay packets in the WSNs.

3 Problem Statement

In this section, we introduce some terminologies, network model, energy model and then give our objectives.

Data gathering cycle (*DGC*): It is defined as the process that MS departs from O_s, moves inside $\mathcal{A}$ for collecting sensed data from nodes and back to O_s, forming a closed loop, which is also referred as a round.

Network lifespan T^{α}_{sys}: It is defined as the period until the network fails to provide an acceptable monitoring effect, and parameter α represents acceptable level which depends on the specific application requirements. In this paper, the basic unit of T^{α}_{sys} is *DGC*.

3.1 Network Model

Assumed N sensors are uniformly deployed in a monitoring field $\mathcal{A}$ of $L \times Lm^2$ to collect data periodically and send sensed data to the mobile sink (MS). To simplify the network model, a few reasonable assumptions are adopted as follows:

- All sensor nodes are assumed to be static and have unique ID number, have the same data acquisition rate k_1 bit/s, have the same data transmission rate k_2 bit/s. The heterogeneity of nodes is shown as different limited initial energy. Besides, all sensor nodes have the same buffer size B_1 in packets.
- The data sensed is assumed to be not time-sensitive and they are stored in buffer of sensor nodes. Sensor nodes do not deliver their data to MS until it arrive at the given sojourns within a limited delay T_{delay}, where T_{delay} is a variable that represents application delay.
- MS is located at starting point O_s far away from $\mathcal{A}$ and has constant moving velocity v. MS has unlimited energy and large storage space.
- Intensive network is considered in this paper, all nodes can be connected by single hop or multi-hop. MS and sensors have the same restricted communication radius R_c, and all sensors have the same sensing radius R_s.

3.2 Delay Analysis

In a realistic application, it may not be able to deliver information to MS within direct radio range, thus local multihop routing is adopted in this paper. Compared with the multihop forwarding path in sink fixed network, the path length of local multihop routing is much shorter and the latency overhead of local multihop routing is much smaller. According to the definition of end-to-end delay which is composed of queuing delay d_Q, transmission delay d_T and propagation delay d_P [7], we just focus on d_T but ignore d_Q for its small value and d_P due to the fact that electromagnetic waves propagate at the speed of light in the air. Since d_T of one hop is related to the amount of data transmitted. Therefore, the more data forwarding, the greater the delay of data transmission.

3.3 Energy Model

Simple energy consumption model [8], which means all nodes use the same fixed transmit power and receive power, is adopted in this paper. We assume that e_t and e_r are the energy consumption of one hop required for sending and receiving per packet respectively. Besides, let e_{idle} be the energy consumption required for storing data when node is kept in idle state, and the energy consumption in dormant state is negligible. Therefore, the energy consumption of data collection per round can be expressed as:

$$E_{sys} = \sum\nolimits_{j=1}^{N_g} \sum\nolimits_{i=1}^{N_j} \left(e_t k_t^i + e_r k_r^i\right) \tag{1}$$

where N_g is the amount of girds, N_j is the amount of nodes in grid j, k_t^i and k_r^i denote the amount of packet sent and received by each node, respectively. The amount of data generated by node per round is calculated as follows:

$$k_{self}^i = \begin{cases} B_1 & \textit{if } k_1 \times T_{delay} > B_1 \\ k_1 \times T_{delay} & \textit{if } k_1 \times T_{delay} \leq B_1 \end{cases} \tag{2}$$

Where the relationship between k_t^i and k_r^i is $k_t^i = k_r^i + k_{self}^i$.

Due to the limitation of communication radius R_t, the energy consumption of data transmission is independent of transmission distance per hop but only related to hop count, and the total amount of data received in the network per *DGC* is calculated as follows:

$$\sum\nolimits_{j=1}^{N_g} \sum\nolimits_{i=1}^{N_j} k_r^i = \sum\nolimits_{j=1}^{N_g} \sum\nolimits_{i=1}^{N_j} h_{RP_i}^i k_{self}^i \tag{3}$$

where $h_{RP_i}^i$ denotes the minimum hops from node i to rendezvous point RP_i. Thus, the total energy consumption of data collection per *DGC* can be expressed as the sum of minimum hops,

$$\begin{aligned} E_{sys} = \sum\nolimits_{j=1}^{N_g} \sum\nolimits_{i=1}^{N_j} (e_t k_t^i + e_r k_r^i) &= e_t \sum\nolimits_{j=1}^{N_g} \sum\nolimits_{i=1}^{N_j} (k_r^i (1+\alpha) + k_{self}^i) \\ &= e_t \sum\nolimits_{j=1}^{N_g} \sum\nolimits_{i=1}^{N_j} (h_{RP_i}^i k_{self}^i (1+\alpha) + k_{self}^i) \\ &= e_t k_{self}^i \sum\nolimits_{j=1}^{N_g} \sum\nolimits_{i=1}^{N_j} (h_{RP_i}^i (1+\alpha) + 1) \end{aligned} \tag{4}$$

where $\alpha' = 1 + \alpha = 1 + \frac{e_r}{e_t}$. According to the above analysis, we can see that $h_{RP_i}^i$ depends on the distance between node i and rendezvous point RP_i that MS stays, thus it is crucial to optimize the trajectory of MS for reducing system energy consumption.

In particular, data fusion mechanism is not considered in this paper.

4 Grid-Based Energy-Efficient Scheduling and Data Gathering Scheme

In order to reduce and balance energy consumption, optimizing the patrolling route of MS as far as possible is worth considering within T_{delay}. In this section, we propose an efficient scheme GES-DGS, which can determine the location and different levels of virtual points, select corresponding level of virtual point and dynamically adjust trajectory of MS according to T_{delay}, and forward data sensed by nodes to MS using a simple and efficient routing algorithm.

4.1 Determination of Virtual Points

After randomly deployment of sensors in the monitoring field, they are divided into multiple virtual units. Assuming the coverage area of each node is a disk with a radius of R_c, then we can obtain the minimum virtual unit (MVU) of data acquisition, and the details are illustrated in Fig. 1. From Fig. 1, all nodes in MVU which consists of four sub-units (S_1, S_2, S_3, S_4) can directly communicate with MS in the centre of disk and the size of MVU l_v is $\sqrt{2}R_c$. In order to ensure monitoring performance and save network energy using redundant data elimination method, each sub-unit S_i is divided into m^2 minimum cover units (MCU) using Algorithm 2 (PISA) in [1]. Assuming the value of m is given $(m = 2)$, then the size of each MCU l_g is $\frac{\sqrt{2}}{4}R_c$. Next, monitoring field $\mathcal{A}$ is divided into multiple MVUs from centre to edge, and different levels of virtual points (VPs) will be marked to help optimize the trajectory of MS within a given delay T_{delay}. The number of levels l_e depends on L and R_c. Figure 2 shows the visualization of different levels of VPs $(l_e = 4)$.

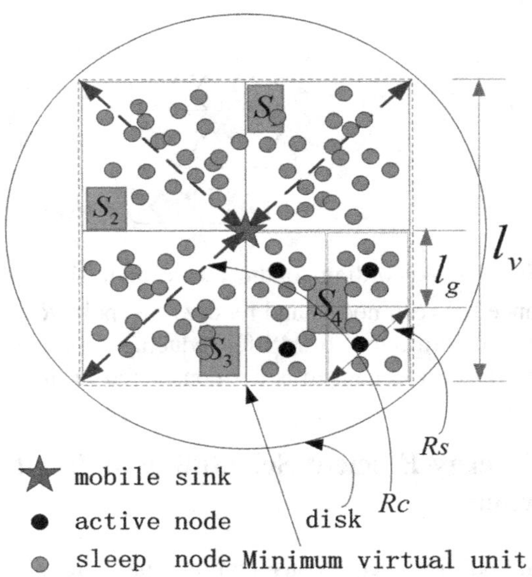

Fig. 1. Minimum virtual unit

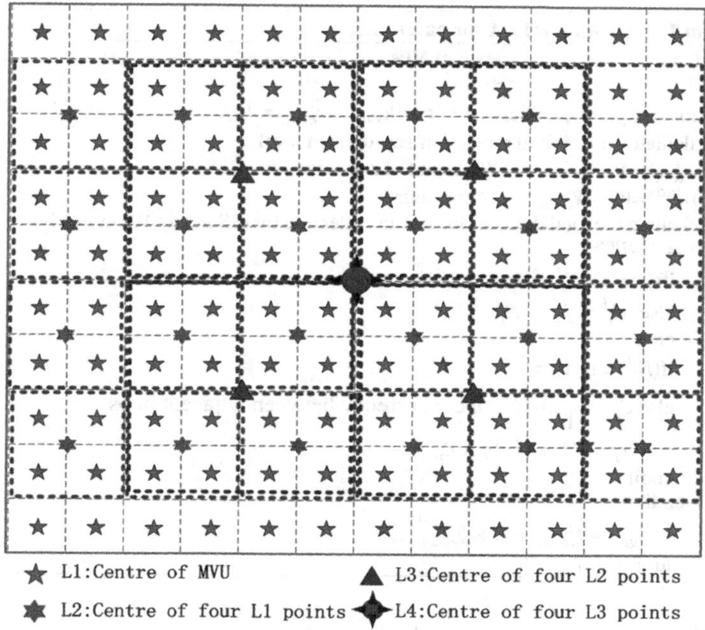

Fig. 2. VPs of different levels

4.2 Theoretical Analysis of Virtual Points

We can determine the amount of MVU according to the values of L and R_c, and obtain the minimum amount of MCU that meets network monitoring performance, which means only one node in active in each MCU, according to the relationship between R_c and R_s. Since the amount of active nodes is constant and same in each sub-unit, we can theoretically estimate the residence time at each VP and calculate the total moving distance traversed by MS.

Suppose that MS has enough receivers to accept all data delivered by all active nodes within one hop range simultaneously, all active nodes in the same sub-unit send data in time slot sequentially and communication interference between different sub-units and MS is not considered, thus we propose a residence time estimation algorithm (RTEA) to approximate calculate the data acquisition time t^i_{level} and a moving distance estimation algorithm (MDEA) to approximate calculate the total moving distance d^i_{level}. The details of RTEA and MDEA are presented in Algorithms 1 and 2.

Algorithm 1. Algorithm RTEA for calculating t_{level}^{i}

```
Input: i (>1)                    //level of VPs
Output: t_level^i
Initialization: t_level^1 = φ where φ = tm^2 k_self; t_level^i = Inf
Step 1: Calculate the 1/4 total amount of sub-units in level i
    1.1 N_level^i = 4^(i-1);            // amount of sub-units
    1.2 H_i^max = log2 N_level^i;       // hop count
Step 2: Calculate the total delay of local data collection by MS at any VP in level i
  2.1 for j = 1: H_i^max
    2.2     if(j == 1) N_i^j = 1;
    2.3     else N_i^j = j^2 - N_i^(j-1);
    2.4     endif
    2.5     if(N_i^j == 1) t_i^j = 1;
    2.6     else C_p = ⌈N_i^j / N_i^(j-1)⌉;    // concurrency between adjacent hops
    2.7         t_i^j = (1 + t_i^(j-1)) × C_p ;
    2.8     endif
    2.9   endfor
    2.10  t_level^i = t_level^i + φ × Σ_{k=1}^{H_i^max} t_i^k;
    2.11 return t_level^i;
```

In algorithm RTEA, ideal concurrent transmission mechanism for corresponding MCUs in adjacent sub-units is adopted in this paper.

Algorithm 2. Algorithm MDEA for calculating d_{level}^{i}

```
Input: i                                  // level of VPs
Output: d_level^i                         //total distance of level i
Initialization: d_level^i = 0, S_MVU = R_c^2;
   1. N_level^i = 4^(i-1);                //amount of MVU at level i
   2. S_level^i = N_level^i × S_MVU;      //local area of data collection at any VP_i
   3. if (mod(L^2, S_level^i) == 0)
   4.     N_level^i = L^2 / S_level^i;
   5. else
   6.     N_level^i = ⌊L^2 / S_level^i⌋;  //minimum amount of VP_i
   7. d_adj^{VP_i} = √(4^(i-1)) l_v;      //distance between two adjacent VP_i
   8. d_inner^i = d_adj^{VP_i} × (N_level^i - 1);
   9. d_outer^i = dist(O_s, VP_i^in) + dist(O_s, VP_i^out);
   10. d_level^i = d_inner^i + d_outer^i;
Return d_level^i;
```

In algorithm MDEA, VP_i^{in} and VP_i^{out} are the two closest VP from point O_s. From algorithm MDEA, the total patrol distance of MS and walking time can be estimated approximately.

4.3 VP-APR: An VP-Based Adaptive Patrol Route for Mobile Sink

Considering the time-varying property of application delay T_{delay}, an VP-based adaptive patrol route algorithm (VP-APR) for MS is proposed for maximizing amount of data collected and reducing energy consumption per *DGC* simultaneously within the given period T_{delay}. Due to the high consistency of the Hilbert curve with our VP construction method proposed in this paper. In VP-APR, we first determine the level of VPs based on Algorithm 1 and determine which level of VP is selected as the reference point, which guides the movement of MS, according to Algorithm 2 and T_{delay}, then obtain the trajectory of MS efficiently. The details of algorithm VP-APR are presented in Algorithm 3.

Algorithm 3. Algorithm VP-APR for determin prtrol route of MS

Input: T_{delay}, L, v, l_v;
Output: ψ_{vp}; //the trajectory of MS
Initialization: $\psi_{vp} = \emptyset$; $l_{vp}^{max} = Inf$; $\psi_{mobile} = \emptyset$;
Step 1. Determine the level of VPs
  1.1 $N_{level}^{max} = \left\lceil \log_4 (\frac{L}{l_v})^2 \right\rceil + 1$; //the maximum level of VP
  1.2 $l_{vp}^{max} = N_{level}^{max}$;
  1.2 for i=1:N_{level}^{max}
  1.3 $N_{vp}^{i} = 4^{N_{level}^{max} - i}$; //the amount of VP_i
  1.4 $t_{vp}^{i} = t_{level}^{i} \times N_{vp}^{i}$; //data acquisition time at all VP_i
  1.5 $t_{move}^{i} = d_{level}^{i} / v$; //motion waste time
  1.6 if ($T_{delay} > t_{vp}^{i} + t_{move}^{i}$)
  1.7 $l_{vp}^{max} = i$;
  1.8 end //end if
  1.9 end //end for
Step 2. Determine trajectory of MS
  2.1 $\psi_{mobile} = \{VP_1^{l_{vp}^{max}}, VP_2^{l_{vp}^{max}}, \ldots, VP_{N_{vp}^{i}}^{l_{vp}^{max}}\}$;
  2.2 $t_{alloc}^{vp} = T_{delay} / N_{vp}^{l_{vp}^{max}}$; //actual allocation time at each VP_i
  2.3 merge peripheral MVUs so that the value of N_{level}^{i} is a power of 2
  2.4 select the moving strategy based on hilbert curve
  2.5 select mobile strategy from ψ_{mobile} and construct ψ_{vp} by (5)
Return ψ_{vp};

In algorithm **VP-APR,** the level of VP l_{vp}^{max} is first determined according to input parameters (T_{delay}, L, v, l_v), then MS traverses the regions in which l_{vp}^{max}*th* VPs are located one by one in the form of hilbert curve. In essence, we convert the energy optimization problem in fixed-sink WSNs into the optimal-route minimum energy consumption problem in mobile sink-based WSNs. In algorithm **VP-APR, VPs are just moving reference and MS** actively adjust the actual rendezvous point (ARP) according to the priority of sub-unit. The priority of sub-unit is calculated as follows:

$$P_i^j = \gamma \left(1 - \frac{E_{max}^i - E_{res}^j}{E_{max}^i} \right) + (1 - \gamma) \left(1 - \frac{d_{vp}^j}{2^{\delta - 1} R_c} \right), \; if \; E_{res}^j > \overline{E_{res}} \tag{5}$$

Where $\delta = l_{vp}^{max}$, i indicates serial number of region, j is the serial number of sub-unit in region i, E_{max}^{i} is the maximum initial energy in region i, E_{res}^{j} is the residual energy of sub-unit j, d_{vp}^{j} is the distance between centre of sub-unit j and VP of region i, $\overline{E_{res}}$ is the minimum threshold value of sub-unit for computing priority. In (5), local energy factor and global hop factor inspired by (4) are considered. In algorithm VP-APR, the first *DGC* is completed by MS under the guidance of VPs, and then MS proactively changes ARPs for finding the optimal trajectory via self-adaptive behavior based on the priority of sub-unit.

4.4 Simple and Efficient Routing Method

In the early stage of network operation, MS starts data gathering tour and moves along ψ_{vp} which is obtained by algorithm VP-APR. When MS arrives at any VP, it loiters and broadcasts 'Hello' *message* to active nodes (*Atns*) within the communication radius R_c. When *Atns* receive 'Hello' message, they will continue to spread 'Hello' message until all *Atns* in the same region are received. Due to the restricted communication radius and limited neighbor *Atns* for data forwarding, a simple and efficient routing method based on the ability of *Atn* is proposed for saving and balancing energy consumption. The ability of *Atn-i* refers to residual representative energy (Rre) of sub-unit in which *Atn* is located and is calculated as follows:

$$f_{Atn}^{Rre}(i) = \sum\nolimits_{j=1}^{N_{Atn-i}^{sur}} E_{res}^{j} \tag{6}$$

where $f_{Atn}^{Rre}(i)$ denotes residual energy of sub-unit in which active node i is located, N_{Atn-i}^{sur} is the amount of surviving nodes in sub-unit that *Atn-i* belongs to. From (6), MS proactively moves to neighbor *Atn* with the largest energy $f_{Atn}^{Rre}(i)$ for achieving dynamic balance of energy among surviving nodes in the same region, i.e., sub-unit with high density of nodes will be preferred as a relay region.

5 Performance Evaluation and Result Analysis

To evaluate performance of GES-DGS, extensive simulations with different evaluation criterias are conducted and the results of GES-DGS scheme with randow waypoint model (RW), shortest path method (SPM) in static WSNs are compared via simulations using MATLAB (version 2014b).

5.1 Experimental Setup

In the simulation, 4000 sensor nodes are scattered randomly in the square monitoring area, 500 m $\times$ 500 m. Sensors continuously generate data at the speed of one packet per second and the size of packet is 500 bits, all nodes and MS have the same maximum communication radius $R_c = 50$ m. The isomorphic network with initial energy

$E_{init} = 1\,\text{J}$ of all nodes, data transmission rate $k_1 = 1\,\text{kb/s}$, the buffer size of node $B_1 = 20\,\text{kbit}$, the value of e_t and e_r one hop per packet are $0.5\,\mu\text{J/bit}$ and $0.4\,\mu\text{J/bit}$ respectively.

In order to reduce the error, all simulation data are the mean of 30 random tests.

5.2 Network Lifespan

Efficient use of node energy is critical for extending network lifespan because of the constrained energy resource on the sensor nodes. We evaluated the *NL* performance of the three scheme as shown in Fig. 3. The value of α is defined as the period until first node fails to work according to the definition of *NL*. Figure 1 reveals that GES-DGS scheme extends about 3 times as long as that of SPM scheme and extends about 2 times as long as that of RW scheme. In general, the changes in network lifespan T_{sys}^{α} of GES-DGS and SPM are also stable than that of RWM scheme with the change in application delay T_{sys}.

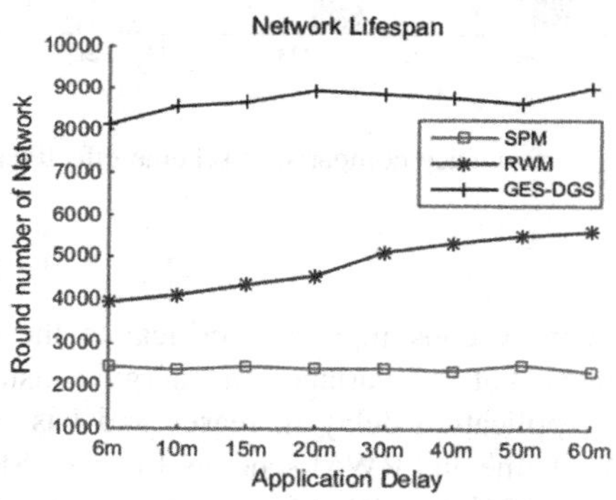

Fig. 3. Performance comparison: network lifespan

5.3 Effectiveness of Scheme

In Fig. 4, we present the measurement of effectiveness performance among SPM, RWM and GES-DGS. It can be found that SPM scheme and GES-DGS scheme obtain better monitoring performance than RWM scheme. Due to the fast speed of wireless communication, SPM scheme achieves good monitoring performance but drastically shortened *NL* as shown in Fig. 4. For RWM scheme, it can not guarantee the full coverage of the monitoring area due to the limited moving speed and T_{sys}, resulting in poor monitoring effectiveness than SPM scheme. However, RWM scheme achieves longer *NL* than SPM scheme for its larger moving distance and random walk feature, which effectively balances energy consumption among sensors. By contrast, we can observed that SPM scheme requires 5 times as much the amount of data acquisition as that of GES-DGS scheme to achieve the same monitoring performance. This is because

GES-DGS scheme uses redundant data elimination method avoiding unnecessary consumption of node energy and significantly extended *NL* as shown in Fig. 3.

Combined with Figs. 3 and 4, we can conclude that GES-DGS scheme significantly extends network lifespan, drastically reduces energy consumption per *DGC*, and ensures network monitoring performance.

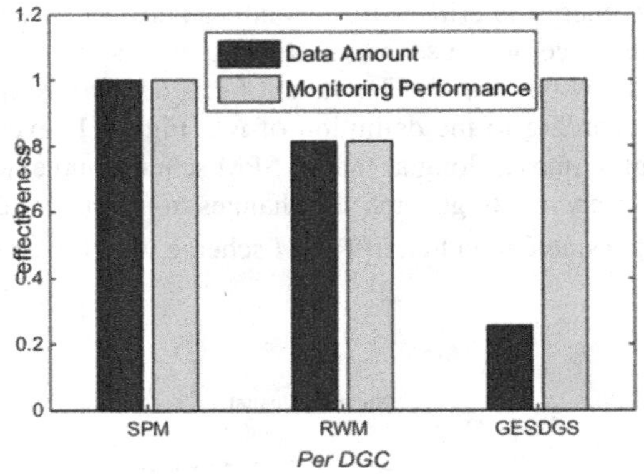

Fig. 4. Performance comparison: scheme effectiveness

5.4 Energy Balance

The imbalance of node energy consumption is critical to the extension of network lifetime. Figure 5 illustrates that the variance in energy consumption of GES-DGS scheme decreases as the application delay increases and has better effect in energy balance that that of SPM scheme and RWM scheme. Besides, SPM has better stability of energy consumption per DGC than RWM scheme and GES-DGS scheme.

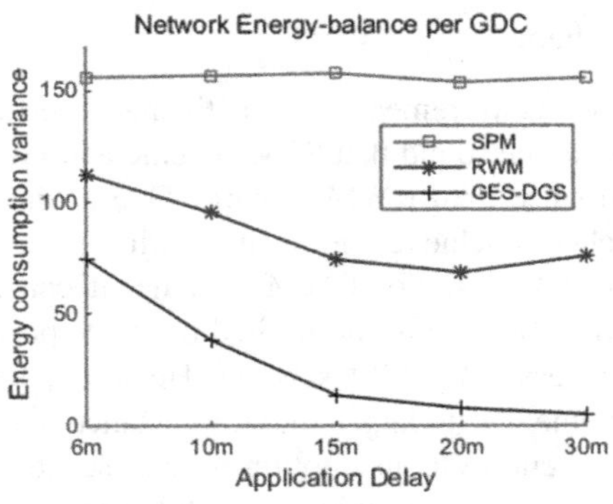

Fig. 5. Performance comparison: energy balancing

6 Conclusion

To maximize the amount of data collected and reduce energy consumption within T_{sys}, we propose a GES-DGS scheme to guide MS patrolling in the monitoring field and collect data efficiently. Next, we intend to study the data acquisition problem of MS with limited energy.

Acknowledgments. This work was supported in part by the National Natural Science Foundation of China under Grant 61170232 and 81160183, Australian Research Council Discovery Project under Grant DP150104871 and the Fundamental Research Funds for the New Teachers' Scientific Research Fund of People's Public Security University of China under its projects number 2018JKF609.

References

1. Chen, Z., Shen, H.: A grid-based reliable multi-hop routing protocol for energy-efficient wireless sensor networks. Int. J. Distrib. Sens. Netw. **14** (2018). https://doi.org/10.1177/1550147718765962
2. Xing, G., Wang, T., Xie, Z., et al.: Rendezvous planning in wireless sensor networks with mobile elements. IEEE Trans. Mob. Comput. **7**(12), 1430–1443 (2008)
3. Wen, W., Zhao, S., Shang, C., et al.: EAPC: Energy-aware path construction for data collection using mobile sink in wireless sensor networks. IEEE Sens. J. **PP**(99), 1 (2017)
4. Yun, Y.S., Xia, Y.: Maximizing the lifespan of wireless sensor networks with mobile sink in delay-tolerant applications. IEEE Trans. Mob. Comput. **9**(9), 1308–1318 (2010)
5. Cayirpunar, O., Tavli, B., Kadioglu-Urtis, E., et al.: Optimal mobility patterns of multiple base stations for wireless sensor network lifespan maximization. IEEE Sens. J. **17**(21), 7177–7188 (2017)
6. Cheng, C.F., Li, L.H., Wang, C.C.: Data gathering with minimum number of relay packets in wireless sensor networks. IEEE Sens. J. **PP**(99), 1 (2017)
7. Huynh, T.T., Dinh-Duc, A.V., Tran, C.H.: Delay-constrained energy-efficient cluster-based multi-hop routing in wireless sensor networks. J. Commun. Netw. **18**(4), 580–588 (2016)
8. Somasundara, A.A., Kansal, A., Jea, D.D., et al.: Controllably mobile infrastructure for low energy embedded networks. IEEE Trans. Mob. Comput. **5**(8), 958–973 (2006)

Green vs Revenue: Data Center Profit Maximization Under Green Degree Constraints

Huaiwen He[1,3] and Hong Shen[1,2](✉)

[1] School of Data and Computer Science, Sun Yat-Sen University, Guangzhou, China
he_huai_wen@aliyun.com, hongsh01@gmail.com
[2] School of Computer Science, University of Adelaide, Adelaide, Australia
[3] School of Computer, University of Electronic Science and Technology of China, Zhongshan Institute, Zhongshan, China

Abstract. With the soaring personal and enterprise computation demands, the scale of cloud centers has been rapidly increasing, which leads to massive amounts of greenhouse gas emission. From the cloud service provider's (CSP) perspective, profit is the key factor to maintain the development of cloud centers, which potentially conflicts with the goal of achieving green data centers for environment protection because the expensive renewable energy will add more cost. This paper addresses the problem of maximizing profit while meeting the green degree constraints for a large scale data center on renewable energy source. Taking into account of the bursty randomness of workload, time-varying electricity price and intermittent green energy, we first formulate the problem in an optimization framework with stochastic constraints for delay-tolerant workload. Then, we show how the deployment of Lyapunov optimization technique can leverage to obtain a low-complexity online solution for profit maximization by combining request admission control, workload scheduling and power management. Moreover, we adopt a non-linear submodular revenue function to optimize the throughput of the system. By decoupling the optimization function of a time average problem into three sub-problems, we solve them to obtain the optimal control strategies. Our proposed algorithm achieves a desirable profit-green tradeoff. At the end, we provide the performance bound of our algorithm, and evaluate its performance through extensive trace-driven simulations.

Keywords: Profit maximization · Stochastic constraints · Renewable energy Lyapunov optimization

1 Introduction

Cloud computing has become a matured popular computing service for individuals and companies in the recent years [1]. However, with the soaring increase of scale of cloud centers, many challenges appear that hinder the sustainability of large scale cloud centers. One of the key challenges is how to earn higher profits while keeping the desired green degree of a cloud center [2]. According to the latest report [3],

J. H. Park et al. (Eds.): PDCAT 2018, CCIS 931, pp. 28–40, 2019.
https://doi.org/10.1007/978-981-13-5907-1_3

Google has emitted 2.99×10^6 of carbon in 2015, which leads to serious damages to the fragile global climate such as global warming. Hence, it is a critical issue to optimize the tradeoff between profit (revenue) and green degree for cloud centers, these two potentially conflicting objectives.

More and more CSPs are aggressively seeking off-site renewable energy from the professional green energy suppliers to reduce carbon emission for cloud centers [4, 5]. The renewable energy is influenced by the local climate condition, so it is more expensive than the "brown energy" (e.g. traditional grid power). Thus, CSPs need to optimize the scheduling strategy to achieve an optimal revenue-green tradeoff. Moreover, the renewable energy is usually intermittently available and supplied for a random length of duration, which imposes great challenge in power management and resource scheduling. In this paper, we focus on the delay tolerant workload which can cooperate with the dynamic renewable energy by opportunistic scheduling, i.e., its execution can be postponed until the renewable energy is available or electricity price is low.

To balance a profit-green tradeoff, we need to answer three important questions: (1) How to optimize the throughput of a system to obtain more revenue while avoiding system congestion? (2) How to schedule the delay-tolerant workload under unknown electricity price and intermittent renewable energy? (3) How to evaluate the green degree of a data center and satisfy the green degree constraint?

To address the above challenges, our paper proposes an online algorithm with jointly request admission control, workload scheduling and power management to maximize the profit of data center while satisfying the long-term green degree constraint. We first formulate the long term profit optimization problem with stochastic constraints. Then, based on Lyapunov optimization technique, we develop a novel online control framework GAMP which makes the following control decisions: (1) request admission control; (2) workload scheduling control; (3) power management control. Moreover, we also provide the performance bound of our algorithm and evaluate the performance by extensive numerical simulations based on the real-world trace data.

In summary, our key contributions are listed as follows:

(1) We propose a profit-green tradeoff online framework for delay-tolerant workload to maximize net profit while satisfying the long term green degree constraint. Different from adopting a linear function as in some relate studies [2, 6], we use a non-linear submodular utility function for calculating revenue. To the best of our knowledge, this is the first study to address the problem.
(2) We propose a new metric of green degree index (GDI) to evaluate the green degree of a cloud center and formulate a novel mathematical model for profit maximization of the cloud center. Moreover, we take into account the randomness of the time-varying electricity price, workload arrival and green energy supply.
(3) We design a multiple-layer optimal online control algorithm that decouples the optimization problem into three sub-problems, makes online request admission control, workload scheduling, power management decisions on them respectively without need of any prior knowledge on future system information.

(4) We test our algorithm on various practical data e.g. workload trace, real-time electricity price, renewable energy generation, to evaluate the effectiveness of our proposed algorithm in practice.

The rest of this paper is organized as follows. We present the related work in Sect. 2. The system model is defined and the optimization problem is formulated in Sect. 3. We propose an online solution for the problem and analyze its performance in Sect. 4. The simulation results are provided to show the good performance of our algorithm in practice in Sect. 5. Section 6 summaries our work and discusses the future direction.

2 Related Work

There are many existing works on the profit maximization for cloud center in the recent years. In [7], Wang et al. have proposed a randomized online algorithm named Random-Fit to maximize the net profit of green data center and analyze its competitive ratio. In [8], Zhou et al. have designed a three layer control framework for optimizing profit for SaaS cloud, they considered the profit-congestion trade-off via workload admission control and VM scheduling, but they did not take the renewable energy into account. Ghamkhari et al. [6] have proposed a profit maximization strategy for optimizing the trade-off between minimizing energy costs and maximizing its revenue for green data center. In [9], the authors have developed an optimization framework to maximize profit for geo-distributed green data center by using the G/D/1 queue to capture the workload distribution.

Different from the above works, our work focuses on profit maximization under coordinating the delay-tolerant workload with the intermittent renewable energy. Moreover, we consider the profit-green tradeoff rather than the tradeoff between energy costs and revenue. We maximize the profit while satisfying the long term green degree constraint of data center.

3 Model and Problem Formulation

3.1 System Model

We consider a time-discrete system model for green cloud center in time horizon with time slot $t = (1, 2, \cdots, \tau, \cdots)$. Herein we only focus on delay-tolerant jobs (e.g., map-reduce jobs, large-scale simulations and video processing jobs [4]). The workload arrival of cloud center at slot t is denoted as $W_i(t)$, where i presents the type of workload, $i \in \mathcal{I}$, and $|\mathcal{I}| = N$. There is an upper bound for ith type of workload, e.g. $0 \leq W_i(t) \leq W_{i,\max}$, and the average rate of $W_i(t)$ is denoted as $\overline{W_i} = \mathbb{E}\{W_i(t)\}$. At each time slot t, the system makes admission control strategy to decide the amount of workload allowed to enter the system. The system should improve its throughput to obtain more revenue while avoiding the system congestion. We denote the admitted requests at slot t as $\lambda_i(t)$, e.g. $W_i(t) - \lambda_i(t)$ are dropped.

The ith type of admitted request first enters a buffer queue and waits for scheduling in the first come first server (FCFS) manner [10]. We use $U_i(t)$ to represent the buffer queue of $\lambda_i(t)$, whose backlog dynamic evolves as follows:

$$U_i(t+1) = max[U_i(t) - E_i(t), 0] + \lambda_i(t) \tag{1}$$

where $E_i(t)$ is the service rate for $\lambda_i(t)$ with an upper bound $E_i(t) \le E_{i,\max}$.

To avoid the congestion and keep the stability of system, $U_i(t)$ shall be mean rate stable, so we have:

$$\lim_{T\to\infty} sup \frac{1}{T}\sum_{t=0}^{T-1} \mathbb{E}\{U_i(t)\} \le \infty \tag{2}$$

Note that our model can also easily extend to handle delay-sensitive workload by maintaining another queue.

Like many leading IT companies, we assume that cloud center purchases off-site green energy to reduce carbon footprint. The amount of wind power, solar power purchased at slot t are denoted as $E_w(t)$ and $E_s(t)$, and the brown power drawn form traditional grid at slot t is denoted as $E_b(t)$ with an upper bound $E_w(t) \le E_{w,\max}$. Thus, we have

$$E_b(t) + E_s(t) + E_w(t) = \sum_{i\in\mathcal{I}} E_i(t) \tag{3}$$

There is an power capacity of cloud center, thus we have $E_b(t) + E_s(t) + E_w(t) \le E_{\max}$, which $E_{\max}$ is the maximum power capacity of data center.

Since renewable energy is unpredictable and seriously depends on the regional climate condition (e.g. wind speed and solar irradiance), so we have the following stochastic constraints at each slot t:

$$0 \le E_s(t) \le E^s_{supply}(t) \tag{4}$$

$$0 \le E_w(t) \le E^w_{supply}(t) \tag{5}$$

where $E^s_{supply}(t)$, $E^w_{supply}(t)$ denote the maximum energy generation of solar energy and wind energy at slot t.

To reduce carbon emission and make cloud center green, the CSP need to improve the usage ratio of renewable energy [4]. Herein we introduce a new metric named green degree index (GDI) to evaluate the green of cloud center, which is defined as the ratio of brown energy to renewable energy. The GDI in slot t is defined as $\frac{1}{\delta(t)}$, which $\delta(t)$ is calculated as $\delta(t) = \frac{E_b(t)}{E_s(t)+E_w(t)}$. The decreasing of $\delta(t)$ means that cloud center becomes green. Note that GDI is similar to carbon neutrality which aims at reducing the net carbon footprint to zero via purchasing renewable energy credits (RECs) to offset the brown energy usage. But GDI is more flexible which can quantitative evaluation the green of data center.

In reality, the CSP should guarantee the time-average of $\delta(t)$ without exceeding a given budget to keep data center green, so we have the following constraint:

$$\lim_{T\to\infty} sup \frac{1}{T}\sum_{t=0}^{T-1} \mathbb{E}\{\delta(t)\} \le \varphi \tag{6}$$

where φ is a long term average GDI budget for cloud center.

3.2 Problem Formulation

The revenue of cloud center is normally a non-decreasing utility function of the throughput of system, which is defined as $\mathcal{R}(\overline{\lambda_i}) = \log(1 + \omega_i\overline{\lambda_i})$, where ω_i is the weight of ith type workload, and $\overline{\lambda_i}$ the time-average throughput defined as $\overline{\lambda_i} = \lim_{T\to\infty} \sup \frac{1}{T} \sum_{t=0}^{T-1} \mathbb{E}\{\lambda_i(t)\}$. Note that the function $\mathcal{R}(\overline{\lambda_i})$ is a submodular function and accords with the law of diminishing marginal law in the real-world [8, 11]. We assume that electricity price is dynamic fluctuating in time horizon and keeps stable in a slot. Thus, the total power cost at each slot t is expressed as

$$\mathcal{F}(t) = p_w(t)E_w(t) + p_s(t)E_s(t) + p_b(t)E_b(t) \tag{7}$$

where $p_w(t)$, $p_s(t)$, $p_b(t)$ are the price of wind energy, solar energy and brow energy at slot t, and $p_b(t) = \min[p_b(t), p_w(t), p_s(t)]$.

The time average power cost is defined as

$$\overline{\mathcal{F}} = \lim_{T\to\infty} \sup \frac{1}{T}\sum_{t=0}^{T-1} \mathbb{E}\{\mathcal{F}(t)\} \tag{8}$$

Thus, we have the long-term net profit of cloud center as follows:

$$\phi = \sum_{i\in\mathcal{I}} \mathcal{R}(\overline{\lambda_i}) - \overline{\mathcal{F}} \tag{9}$$

The optimization problem in our paper is described as follows: make admission control, workload scheduling decision, power management strategy at each time slot t to maximize the time average net profit of cloud center while satisfying the long term

GDI constraint. Note that green and profit is two potentially compete objectives, make data center greener also means less profit. We formulate the profit maximization problem as a constraint stochastic optimization problem **P1** as follows:

$$(\mathbf{P1}) \qquad \max_{\overrightarrow{\boldsymbol{\lambda}}(t),\overrightarrow{\boldsymbol{E}}(t),E_b(t),E_s(t),E_w(t)} \phi \tag{10}$$

s.t.

$$\lim_{T\to\infty}\sup\frac{1}{T}\sum_{t=0}^{T-1}\mathbb{E}\{U_i(t)\}\leq\infty \tag{11}$$

$$\lim_{T\to\infty}\sup\frac{1}{T}\sum_{t=0}^{T-1}\mathbb{E}\{\delta(t)\}\leq\varphi \tag{12}$$

$$0\leq E_s(t)\leq E^s_{supply}(t) \tag{13}$$

$$0\leq E_w(t)\leq E^w_{supply}(t) \tag{14}$$

$$\overline{\lambda}_i\leq\overline{W}_i, \sum_{i\in\mathcal{I}}\overline{\lambda}_i\leq E_{\max} \tag{15}$$

The control decisions in problem **P1** includes three parts: (1) admission control of user request: $\overrightarrow{\boldsymbol{\lambda}}(t)=\{\lambda_i(t),\forall i\in\mathcal{I}\}$; (2) the amount of energy usage of each type of workload $\overrightarrow{\boldsymbol{E}}(t)=\{E_i(t),\forall i\in\mathcal{I}\}$, $\forall i\in\mathcal{I}$; (3) the amount of each type of energy to purchase, which includes $E_b(t)$, $E_w(t)$, $E_s(t)$.

The challenge of P1 lies in three aspects: (1) the future system information (e.g. workload, renewable energy supply and electricity price) is unknown and unpredictable; (2) the GDI constraint is time-couple which means current decision will affect the following future decisions; (3) the profit objective function is non-linear of a time average. The above challenges make the problem P1 hard to solve by the existing algorithm [2, 6–8] or only simply extending them. Thus, we need to design a new online solution for P1 which is illustrated in the following section.

4 Algorithm Design

4.1 Problem Relaxation

Note that the objective function of P1 is a non-linear function of a time average which does not conform to the structure of Lyapunov drift-plus-penalty framework, so we can not directly apply Lyapunov optimization theory to solve it. According to [8, 12], we introduce an auxiliary variable γ_i and rectangle constraint to transform **P1** to the following problem **P2**:

$$(\mathbf{P2})\ \max \quad \sum_{i\in\mathcal{I}} \overline{\mathcal{R}(\gamma_i)} - \overline{\mathcal{F}} \tag{16}$$

$$\text{s.t.} \quad \overline{\gamma}_i \le \overline{\lambda}_i,\ \forall i \in \mathcal{I} \tag{17}$$

and constraints (11), (12), (13), (14), (15).

Since $\mathcal{R}(*) = \log(1+\omega_i^*)$ is a convex and non-decreasing function, so the solution of **P2** is also the optimal solution of **P1** [8].

To solve problem P2, we need to transfer constraint (17) into a queue stability problem [12]. For each type of workload $\lambda_i(t)$, we construct a virtual queue $G_i(t)$ with initializing $G_i(0) = 0$, and the backlog of $G_i(t)$ updates as following:

$$G_i(t+1) = max[G_i(t) + \gamma_i(t) - \lambda_i(t), 0] \tag{18}$$

Similar to previous analysis, we rewrite inequality (12) as $\lim\sup_{T\to\infty} \frac{1}{T}\sum_{t=0}^{T-1} \mathbb{E}\{E_b(t) - \varphi(E_s(t)+E_w(t))\} \le 0$, then we introduce another virtual queue $Z(t)$ to transform the time-average constraint (12) into a queue stability problem [12]. Actually, $Z(t)$ represents how far the current energy usage deviates form the GDI constraint. With initializing $Z(0) = 0$, the queue $Z(t)$ dynamic evolves as

$$Z(t+1) = max[Z(t) + E_b(t) - \varphi(E_s(t) + E_w(t)), 0] \tag{19}$$

We use two vectors $\boldsymbol{U}(t) = (U_i(t))$ and $\mathbf{G}(t) = (G_i(t))$ to link all queues of $U_i(t)$ and $G_i(t)$, $\forall i \in \mathcal{I}$. By defining a combined queue vector $\boldsymbol{\Theta}(t) = [\boldsymbol{U}(t), \boldsymbol{G}(t), Z(t)]$, we define the quadratic Lyapunov function [12] as:

$$L(\boldsymbol{\Theta}(t)) \triangleq \frac{1}{2}\left[\sum_{i\in\mathcal{I}} U_i^2(t) + \sum_{i\in\mathcal{I}} G_i^2(t) + Z^2(t)\right] \tag{20}$$

Then, we obtain the *one-period conditional Lyapunov drift* [12] as following:

$$\Delta(\boldsymbol{\Theta}(t)) \triangleq \mathbb{E}\{L(\boldsymbol{\Theta}(t+1)) - L(\boldsymbol{\Theta}(t))|\boldsymbol{\Theta}(t)\}$$

Note that $L(\boldsymbol{\Theta}(t))$ represents the congestion of all virtual queues $G_i(t)$, $Z(t)$ and actual queues $U_i(t)$, $\forall i \in \mathcal{I}$. Following the Lyapunov optimization theory [12], we obtain the one slot *drift-plus-penalty* expression as:

$$\Delta_V(t) \triangleq \Delta(\boldsymbol{\Theta}(t)) - V\mathbb{E}\left\{\sum_{i\in\mathcal{I}} \overline{\mathcal{R}(\gamma_i)} - \overline{\mathcal{F}}|\boldsymbol{\Theta}(t)\right\} \tag{21}$$

where V is the parameter to control green-profit tradeoff.

From the following Lemma 1, we get the supremum bound of one time-slot *drift* $\Delta(\boldsymbol{\Theta}(t))$.

Lemma 1: For any control strategy $(\vec{\boldsymbol{\lambda}}(t), \vec{\boldsymbol{E}}(t), E_b(t), E_w(t), E_s(t))$ at each time slot t, the one slot Lyapunov drift $\Delta(\boldsymbol{\Theta}(t))$ is bounded by the following:

$$\begin{aligned}\Delta(\boldsymbol{\Theta}(t)) \leq B &+ \sum_{i\in\mathcal{I}} G_i(t)\mathbb{E}\{\gamma_i(t) - \lambda_i(t)|\boldsymbol{\Theta}(t)\} \\ &+ \sum_{i\in\mathcal{I}} U_i(t)\mathbb{E}\{\lambda_i(t) - E_i(t)|\boldsymbol{\Theta}(t)\} \\ &+ Z(t)\mathbb{E}\{E_b(t) - \varphi(E_s(t) + E_w(t))|\boldsymbol{\Theta}(t)\}\end{aligned}$$

where $B = \frac{\sum_{i\in\mathcal{I}}(3W_{i,\max}^2 + E_{i,\max}^2) + E_{b,\max}^2 + \varphi^2(E_{s,\max} + E_{w,\max})^2}{2}$.

Proof: Squaring the queue backlog update Eqs. (1), (18) and (19), the Lemma 1 holds, here we omit the details for brevity.

According to Lemma 1 and combining with Eq. (21), yields the supremum bound of $\Delta_V(t)$ as follows:

$$\begin{aligned}\Delta(\boldsymbol{\Theta}(t)) - V\mathbb{E}\left\{\sum_{i\in\mathcal{I}} \mathcal{R}(\overline{\gamma_i}) - \overline{\mathcal{F}}|\boldsymbol{\Theta}(t)\right\} &\leq B \\ - \sum_{i\in\mathcal{I}} \mathbb{E}\{V\mathcal{R}(\overline{\gamma_i}) - G_i(t)\gamma_i(t)|\boldsymbol{\Theta}(t)\} & \\ - \sum_{i\in\mathcal{I}} \mathbb{E}\{G_i(t)\lambda_i(t) - U_i(t)\lambda_i(t)|\boldsymbol{\Theta}(t)\} & \\ + \mathbb{E}\{(Z(t) - Vp_b(t))E_b(t) + (Z(t) - Vp_s(t))E_s(t) + (Z(t) - Vp_w(t))E_w(t) & \\ + \sum_{i\in\mathcal{I}} U_i(t)E_i(t)|\boldsymbol{\Theta}(t)\} &\end{aligned}$$

4.2 Online Control Algorithm Design

According to the Lyapunov optimization theory [12], we seek to greedily minimize the upper bound of $\Delta_V(t)$ at each time slot t, then we can obtain a close-to-optimal objective value of problem P1. According to the expression of upper bound of $\Delta_V(t)$, the minimization problem of $\Delta_V(t)$ can be decoupled into the following three sub-problems, which can be dependently calculated.

(1) Sub-problem 1: Choose auxiliary variable $\gamma_i(t)$ to

$$\max_{\gamma_i(t)} \quad V\log(1 + \omega_i\gamma_i(t)) - G_i(t)\gamma_i(t)$$

$$\text{s.t. } 0 \leq \gamma_i(t) \leq W_{i,\max}, \forall i \in \mathcal{I}$$

The above problem is a univariate convex optimization problem, we can obtain the optimal solution by calculating the first derivatives of objective function.

(2) Sub-problem 2: Choose request admission control decision $\lambda_i(t)$ to

$$\max_{\lambda_i(t)} \quad G_i(t)\lambda_i(t) - U_i(t)\lambda_i(t)$$

$$\text{s.t.} \quad 0 \leq \lambda_i(t) \leq W_i(t), \forall i \in \mathcal{I}$$

Sub-problem 2 is a simple linear programming problem, in which $\lambda_i(t)$ can be deduced by the following threshold rule:

$$\lambda_i(t) = \begin{cases} W_i(t), & G_i(t) > U_i(t) \\ 0, & \text{else} \end{cases} \tag{22}$$

(3) Sub-problem 3: Choose workload scheduling decision $\vec{\boldsymbol{E}}(t)$ and energy purchasing decision ($E_b(t)$, $E_w(t)$, $E_s(t)$) as

$$\begin{aligned} \min_{E_b(t),E_s(t),E_w(t),\vec{\boldsymbol{E}}(t)} \quad & (\mathrm{Z}(t) - Vp_b(t))E_b(t) + (\mathrm{Z}(t) - Vp_s(t))E_s(t) \\ & + (\mathrm{Z}(t) - Vp_w(t))E_w(t) + \sum_{i \in \mathcal{I}} U_i(t)E_i(t) \end{aligned}$$

s.t. constraints (3), (13), (14).

Sub-problem 3 is a linear programming problem, which can be solved by some mature algorithms such as the simplex method.

Now, we design an online algorithm, named GAMP (Green-Aware online control Algorithm for Maximization Profit) to jointly determine request admission control, workload scheduling and power management decision for green data center. In each time slot t, our algorithm observes the current system information (backlog of queues, electricity price, workload arrival and renewable energy generation), then make the following decisions: (1) admission control strategy $\vec{\boldsymbol{\lambda}}(t)$; (2) workload scheduling decision $\vec{\boldsymbol{E}}(t)$; (3) energy purchase decision $E_b(t)$, $E_w(t)$, $E_s(t)$. The detail of GAMP is summarized as follows.

Algorithm 1: GAMP

Input: $W_i(t)$, $E^s_{supply}(t)$, $E^w_{supply}(t)$, $p_w(t)$, $p_s(t)$, $p_b(t)$, $\forall t$

Output: $\vec{\lambda}(t)$, $\vec{E}(t)$, $E_b(t)$, $E_w(t)$ and $E_s(t)$

1: Initialization: Initialize the backlog of virtual queues $G_i(0)=0$, $Z(t)=0$ and actual queue $U_i(0)=0$ control parameters V.

At each time slot t:

2. Observe the current system state: $U_i(t)$, $G_i(t)$, $Z(t)$, $W_i(t)$, $E^s_{supply}(t)$, $E^w_{supply}(t)$, $p_w(t)$, $p_s(t)$, $p_b(t)$.
3. Solve sub-problem P1 to obtain decision $\gamma_i(t)$
4. Calculate decisions $\lambda_i(t)$ according to the thresh-hold rule (22).
5. Solve sub-problem 3 to obtain decisions $\vec{E}(t)$, $E_b(t)$, $E_w(t)$ and $E_s(t)$.
6. Update the queue length of $U_i(t)$, $G_i(t)$ and $Z(t)$ according to Eq.(1), Eq. (18) and Eq. (19).
7. $t \leftarrow t+1$

4.3 Performance Analysis of GAMP

Theorem 1: If the process ($E_b(t)$, $E_w(t)$, $E_s(t)$, $E^s_{supply}(t)$, $E^w_{supply}(t)$, $W_i(t)$, $\forall i$) is i.i.d. over slots and GAMP is run at each slot t, for any control parameter $V > 0$, then the gap between its achieved by GAMP and the optimal profit ϕ^{opt} satisfy the follows:

$$\liminf_{t\to\infty} \phi \geq \phi^{\text{opt}} - B/V \tag{23}$$

where ϕ^{opt} is the maximum profit of problem P1.

Proof: The above result can be deduced form Theorem 5.1 in [12], we omit the details for brevity.

5 Experiment Result

5.1 Simulation Parameters Setup

We simulate a large scale cloud center with using off-site green energy. The time horizon in simulation is 4464 slots with 10-min in one slot. We simulate 6 heterogeneous workload whose weight vector $\vec{\omega} = \{2,3,4,5,4,3\}$ [8]. We choose the locational marginal price (LMPs) from NYISO [13] to simulate the fluctuating electricity price $p_b(t)$, and the price of wind energy and solar energy are higher 1.5 cents and 18.0 cents than $p_b(t)$ [14]. We retrieve the climate data of Loyola Marymount University form National Renewable Energy Laboratory [15] to model the wind energy and solar energy generation. The electricity price of power grid is depicted in Fig. 1(a), the wind energy generation and solar energy generation are depicted in Fig. 1(b) and (c).

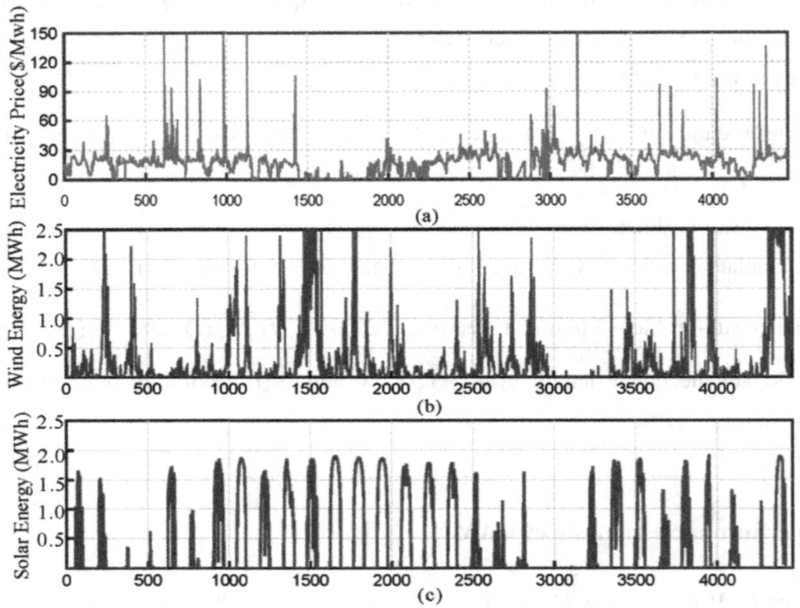

Fig. 1. Dataset used in simulations

5.2 Numerical Result

We analyze the impact of different values of control parameter V under our algorithm GAMP with setting the GDI budget $\varphi = 150\%$, as shown in Fig. 2(a), we observer that: (1) The varying curves fluctuate violently at the beginning of simulation, but they become stable with the increasing of time t, which demonstrate that GAMP works efficiently (2) With the increasing of value of V, the time average profit improves which conforms to Theorem 1 because that GAMP can achieve an optimal profit with a diminishing gap ($1/V$).

With setting control parameter $V = 70000$, we show the profit-tradeoff of achieved by GAMP as plotted in Fig. 2(b). Note that φ is the ratio of brown energy to the total green energy, so cloud center is greener with the smaller values of φ. In Fig. 2(b), we see that larger GDI budget φ incurs higher profits; this is because that more cheap

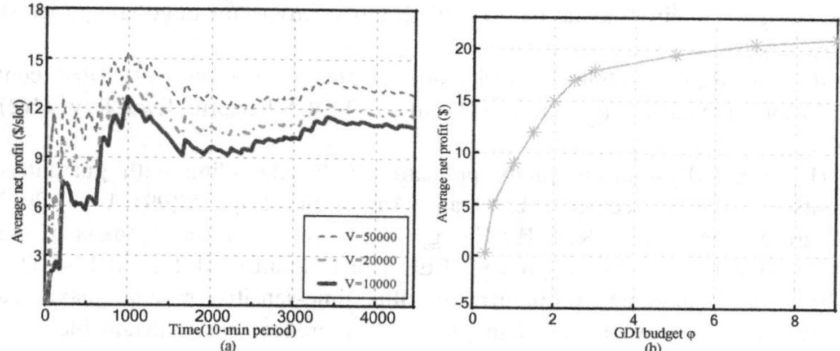

Fig. 2. (a) Time average profit under different values of control parameter V; (b) Time average profit under different values of GDI budget φ (Color figure online)

brown energy is used which also means the increasing of carbon emission. So the CSP need to balance the tradeoff between profit and green. Figure 2(b) also illustrates that our propose algorithm can help the data center to achieve a desirable profit-green tradeoff.

6 Conclusion

In this paper, we propose a multiple layer online control framework for maximizing the net profit of cloud center while satisfying the long term green constraints. We first formulate a novel profit-green tradeoff mathematical model with stochastic constraints on use of renewable energy. Then we design an online control algorithm named GAMP which can deploy easily and works efficiently without need of future system information knowledge. Moreover, we also provide the performance bound of GAMP and evaluate its effectiveness by extensive simulations based on the real-world trace. We plan to study this problem with the consideration of the variable workload length and delay constraints of user requests in the future.

Acknowledgment. This work was supported by the National Key R&D Program of China Project under Grant 2017YFB0203201, Australian Research Council Discovery Projects funding DP150104871, Training Program for Outstanding Young Teachers in Higher Education Institutions of Guangdong Province (No. YQ2015241), Guangdong Natural Science Foundation (2016A030313018), Science and Technology Project of Zhongshan City (No. 2017B1130, No. 2015B2307).

References

1. Kong, F., Liu, X.: A survey on green-energy-aware power management for datacenters. ACM Comput. Surv. **47**(2), 1–38 (2014)
2. Zhou, A., et al.: Maximizing the profits of cloud service providers via dynamic virtual resource renting approach. EURASIP J. Wirel. Commun. Netw. **2015**(1), 71 (2015)

3. Google, Google's environmental report (2016). https://environment.google/projects/environmental-report-2016/
4. Mahmud, A.H., Ren, S.: Online capacity provisioning for carbon-neutral data center with demand-responsive electricity prices. ACM SIGMETRICS Perform. Eval. Rev. **41**(2), 26–37 (2013)
5. Yuan, H., et al.: Time-aware multi-application task scheduling with guaranteed delay constraints in green data center. IEEE Trans. Autom. Sci. Eng. **PP**(99), 1–14 (2017)
6. Ghamkhari, M., Mohsenian-Rad, H.: Energy and performance management of green data centers: a profit maximization approach. IEEE Trans. Smart Grid **4**(2), 1017–1025 (2013)
7. Wang, H., et al.: Randomization improving online time-sensitive revenue maximization for green data centers. In: Green Computing Conference and Sustainable Computing Conference (2015)
8. Liu, F., et al.: On arbitrating the power-performance tradeoff in SaaS clouds. IEEE Trans. Parallel Distrib. Syst. **25**(10), 2648–2658 (2014)
9. Kiani, A., Ansari, N.: Profit maximization for geographical dispersed green data centers. IEEE Trans. Smart Grid **PP**(99), 1 (2015)
10. He, H., Shen, H.: Green-Aware online resource allocation for geo-distributed cloud data centers on multi-source energy. In: International Conference on Parallel and Distributed Computing, Applications and Technologies (2017)
11. Campaign Monitor. http://www.campaignmonitor
12. Neely, M.J.: Stochastic network optimization with application to communication and queueing systems. Synth. Lect. Commun. Netw. **3**(1), 211 (2010)
13. Nyiso (2018). http://www.nyiso.com/public/
14. Zhang, Y., Wang, Y., Wang, X.: GreenWare: greening cloud-scale data centers to maximize the use of renewable energy. In: ACM/IFIP/USENIX International Conference on Distributed Systems Platforms and Open Distributed Processing (2011)
15. Measurement and instrumentation data center (2018). http://www.nrel.gov/midc/

Design and Implementation of a Novel SDN-Based Architecture for Wi-Fi Networks

Lei Liu, Mingzheng Li, Lei Mei, and Ye Tian(✉)

School of Computer Science and Technology,
University of Science and Technology of China, Hefei 230026, Anhui, China
{lliu0610,salmz,meixsh}@mail.ustc.edu.cn,
yetian@ustc.edu.cn

Abstract. Wi-Fi technologies have attracted great attention due to the low cost of construction and maintenance. However, the explosive growth of mobile applications bring new challenges in terms of difficult service quality requirement, such as ultra bandwidth, seamless mobility and high reliability. In this paper, we propose an efficient SDN-based architecture for Wi-Fi networks, which can exploit MPTCP to enable simultaneously transmission for a specific application, and achieve fine-grade path assignment for a single subflow. To evaluate our design, we build a real-world testbed that consists of commodity AP devices, the experimental results demonstrate that our system can significantly improve the wireless throughput and reduce handover delay at the same time.

Keywords: Software-Defined Networking · Wi-Fi · Multipath TCP Subflows

1 Introduction

Wi-Fi Access Points (APs) have been widely deployed to offer a better wireless coverage, providing users with ubiquitous Internet access in public places, such as campus, hospital and enterprise. However, current wireless access points, usually referred to the business wireless routers, are produced by different vendors with specialized operating system, which have created an insurmountable barrier to cooperate among multi-vendor access points. When users move through the wireless coverage area, they are unable to enjoy seamless handover experience, and they also failed to use multiple access points to transmit data simultaneously for improving the throughput performance of Wi-Fi networks.

Software-Defined Networking (SDN [1, 2]) is an emerging paradigm of future network that simplify the management of entire network by decoupling control plane and data plane. By extending the concept of SDN to Wi-Fi networks, several architectures, CloudMac [3] and Odin [4] realize the central control of network traffic in a programmatic way and provide some features that are not included in existing solutions, such as seamless handover. However, the research thrust of these approaches is to introduce programmability into WLANs without considering performance enhancement on throughput. MP-SDWN [5] accelerates packet delivery for a single

J. H. Park et al. (Eds.): PDCAT 2018, CCIS 931, pp. 41–49, 2019.
https://doi.org/10.1007/978-981-13-5907-1_4

client with different applications over multiple access points. However, MP-SDWN is only applicable to different application levels and can not to cope with the situations where all traffic belongs to a given application.

In this paper, we design a SDN-based Wi-Fi architecture that allows client to transmit simultaneously on multiple access points, which improves the throughput of Wi-Fi network in wireless overlapping area. By leveraging Multipath TCP, we deploy strategies which divide traffic from a specific application into different subflows, and dynamically assigns subflows on APs according to the receiver signal strength indicator (RSSI). Furthermore, to exhibit our proposed SDN-based Wi-Fi architecture in real world, we build a testbed extended from Odin. The experimental results show that the system can effectively improve the throughput while supporting seamless handover in coverage Wi-Fi areas.

The remainder of this paper is organized as follows: In Sect. 2, we present the design of our SDN-based Wi-Fi architecture and evaluate performance of system in Sect. 3. Finally, we conclude the paper in Sect. 4.

2 System Architecture

2.1 Overview

As shown in Fig. 1, our SDN-based Wi-Fi architecture is composed of a centralized controller in the control plane and multiple widely dispersed physical access points in the data plane. Each physical access point runs an AP manager for managing user access and a flow manager to apply traffic policies. In the following, we describe the main components of SDN-based Wi-Fi architecture.

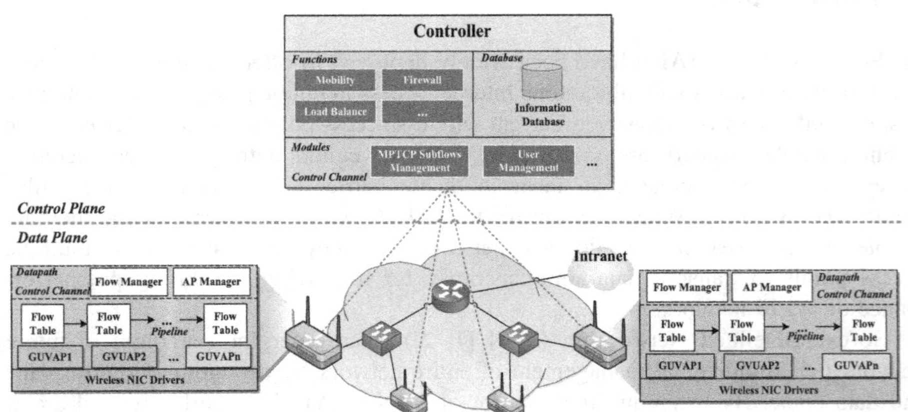

Fig. 1. Overview of SDN-based Wi-Fi architecture

2.1.1 Controller

The controller is the core part of SDN and it serves as the operating system for the network. In our architecture, we define two new modules named user management and MPTCP subflows management, to provide the basic functions of the system. The user management module is responsible for user access management, mainly including identity authentication, spawning and arranging GUVAP (*global unique virtual access point*) for each user. In addition, it also provides interfaces for the network operator to access real-time information, such as the number of users currently carried on the specific physical access point. The MPTCP subflows management module dynamically controls the number of subflows for each MPTCP session, and add or remove subflows according to the values of RSSI reported by access points.

2.1.2 AP Manager

The AP manager running on physical access point takes charge of managing GUVAPs of the users, and exchanges messages with the user management module of the controller through a custom protocol. The AP manager performs corresponding operations on the physical access point after receiving the messages that spawn or remove user's GUVAP from the controller, which determines whether the user can transmit through current physical access point. Moreover, since the wireless interfaces of physical access point run in monitor mode, the AP manager can receive all 802.11 frames, including management frames and data frames, and then extracts valuable information such as the RSSI by sniffing the *radiotap* header of native IEEE 802.11 frames.

2.1.3 Flow Manager

The flow manager also runs on the physical access point and acts like a firewall. The flow manager receives forwarding rules of MPTCP subflows from the MPTCP subflows management module in the controller through the OpenFlow protocol, and then adds, updates or deletes matching flow entries of flow table, thus implementing fine-grained control of each flow identified by the five-tuple.

2.2 Multipath Transmission Mechanism at Application-Level

2.2.1 Interaction Procedure of User Association

Based on the concept of Wi-Fi split-MAC proposed by Cisco [6], we implement a unified processing of IEEE 802.11 management frames on the control plane. We reply the probe response frames passively instead of sending beacon frames periodically in the scan phase, when the AP manager receives probe request frame specified with the broadcast SSID from the user device, and it informs the controller with probe request notification, the controller then generates a globally unique *basic service set identifier* (BSSID) related to the MAC address of user device, and instructs the AP manager to reply probe response frame with corresponding BSSID. In this way, the user will know available Wi-Fi network and the association phase will be triggered sequentially when user device attach to the network. As shown in Fig. 2, the controller is responsible for handling probe request, authentication request and association request and instructing the AP Manager to return probe, authentication or association response to corresponding user device. Besides, the controller also generates a new GUVAP for the user device and instructs the AP manager to spawn the GUVAP on the physical access point after receiving the association request. In authentication phase, an extra 4 MAC-Layer packets will be exchanged if any security mechanism is enabled by the controller. Finally, the user device obtains the IP address by interacting with DHCP server, which can be a standalone server deployed in the distance.

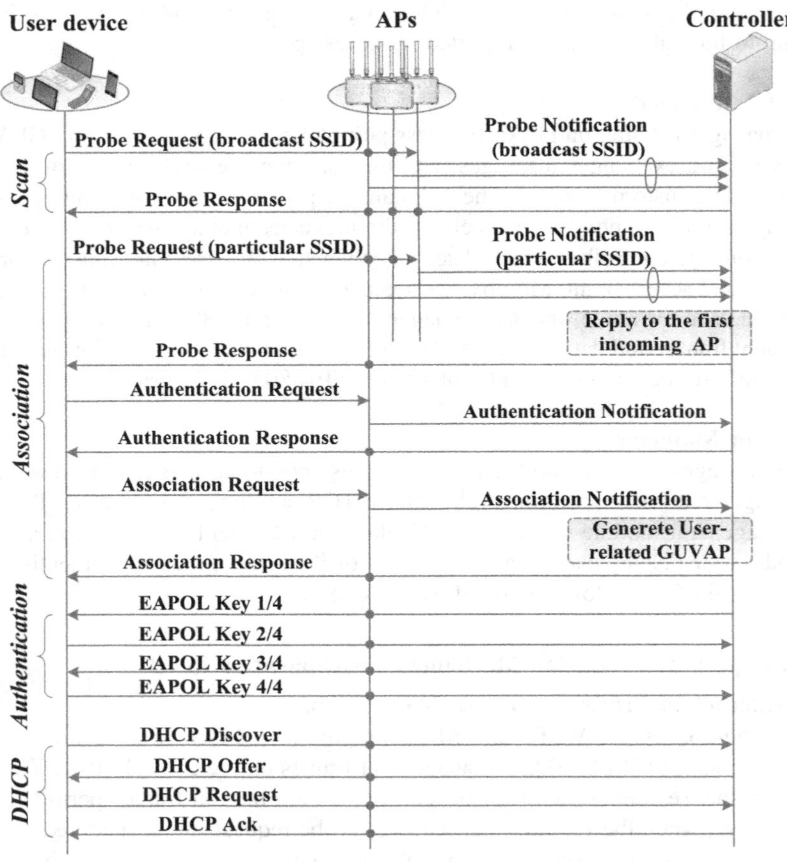

Fig. 2. Wi-Fi connection set-up procedure

2.2.2 Mechanism of Multipath Transmission

On the basis of the Multiple TCP, we separate a single TCP connection into multiple derivative subflows. As illustrated in the Fig. 3, when the user initially creates an MPTCP session named *mainflow* with specific five-tuple, it sends a SYN packet with new MP_CAPABLE option, after receiving the new MPTCP SYN on the physical access point, the Flow Manager sent it out to destination host directly. If the destination host supports MPTCP, it will add the option to SYNACK and reply with the information such as the receiver key and flags. Upon receiving the final ACK which also carries the MP_CAPABLE option, the controller extracts the five-tuple information from the header and instructs the flow manager to generate matching flow entries in flow tables.

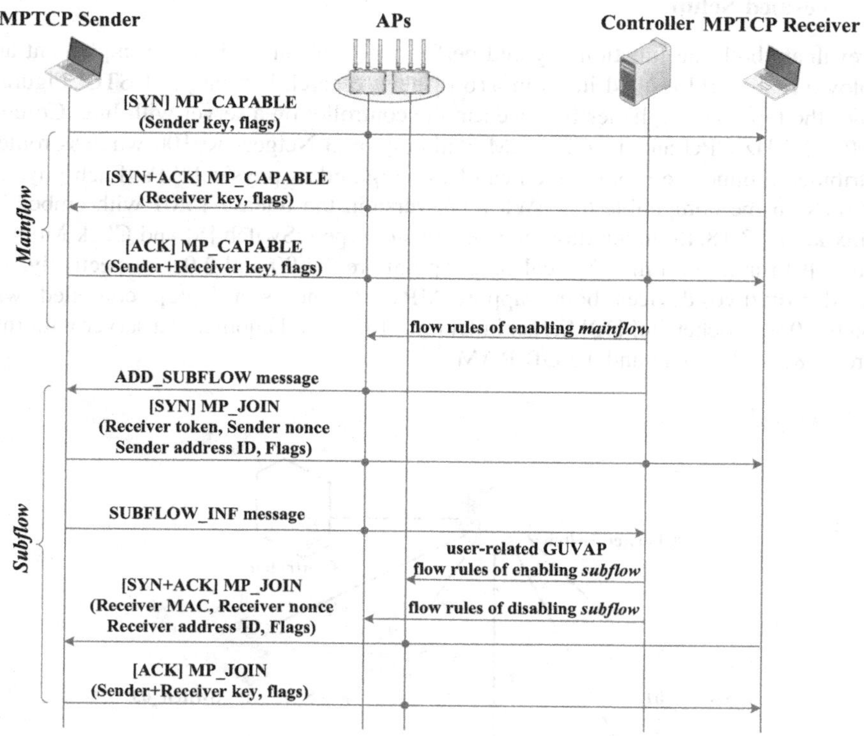

Fig. 3. MPTCP session establishment procedure

After the *mainflow* has been established, the controller periodically inspects whether there are other physical access points that are suitable to transmit data for each user. If exists, the controller then sends an "ADD_SUBFLOW" message to inform the application running on the user device to add new subflows. The application adds *subflow* to the MPTCP session by sending a SYN packet with MP_JOIN option after receiving the message, at the same time, it sends the "SUBFLOW_INF" message which carries the information of the new subflows to the controller. When receiving the SUBFLOW_INF message, the controller instructs the AP manager to spawn the user-related GUVAP on the selected physical access point and guides the flow manager to enable this subflow while disabling it on other physical access points involved in the transmission for the user. It is notable that all messages exchanged between the controller and the user device are implemented with Packet-In and Packet-Out, which greatly reduces the overhead of controller compared with establishing a TCP or UDP connection between the controller and the user device.

3 Performance Evaluation

3.1 Testbed Setup

To evaluate both the functionality and performance of our design, we implement and deploy a real-world testbed in room 416 of new research building at USTC. Figure 4 shows the topology of the testbed, we run the controller on a server with Intel Core i7-8700 3.2 GHz CPU and 16 GB RAM, and employ 3 Netgear R6100 wireless routers distributed around the room, which can be configured as physical APs. Each physical AP runs on the compatible OpenWRT [7] version, the release 15.01 with embedded Linux kernel 3.18.45, in addition, the versions of OpenvSwitch [8] and Click Modular Router [9] running on the physical access point are 2.3.90 and 2.0.1 respectively. We also deploy two devices both support MPTCP, one is a laptop compiled with mptcp_v0.92_socket_api [10] kernel in Ubuntu 14.04, and another is a server with Intel Core i7-8700 3.2 GHz and 16 GB RAM.

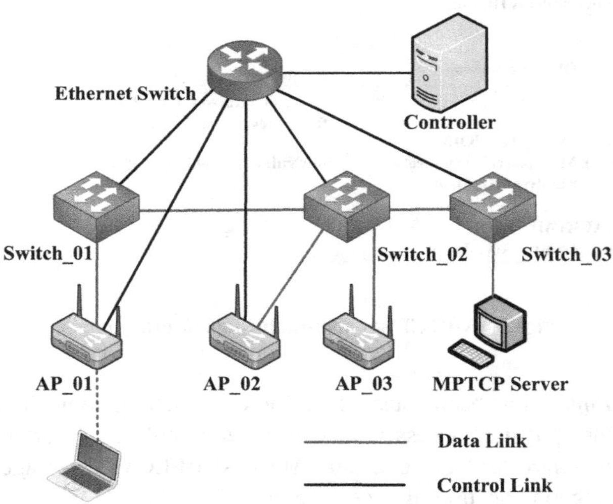

Fig. 4. Physical topology of Testbed

3.2 Throughput

We evaluate the performance on throughput of our proposed SDN-based architecture for Wi-Fi network. In our experiment, the laptop initially accesses the network through the AP_01 and establishes a *mainflow* with the server to transfer file. Next, the controller sends an "ADD_SUBFLOW" message to inform the laptop to create new subflow and enables this subflow by spawning GUVAP and sending flow rules to the AP_02. We take the same approach to create a new subflow transmitted through AP_03 in the end.

Figure 5 shows the throughput changes throughout the process, when there is only one subflow transmitted through a single access point, the throughput is about 7 Mbps, and the throughput increases to 12 Mbps after adding subflow 2 transmitted through AP_02. Finally, when adding subflow 3 transmitted through AP_03, the total throughput of this MPTCP session increase to 16 Mbps. The experimental results prove that our proposed SDN-based Wi-Fi architecture can effectively increase the throughput for a single MPTCP session.

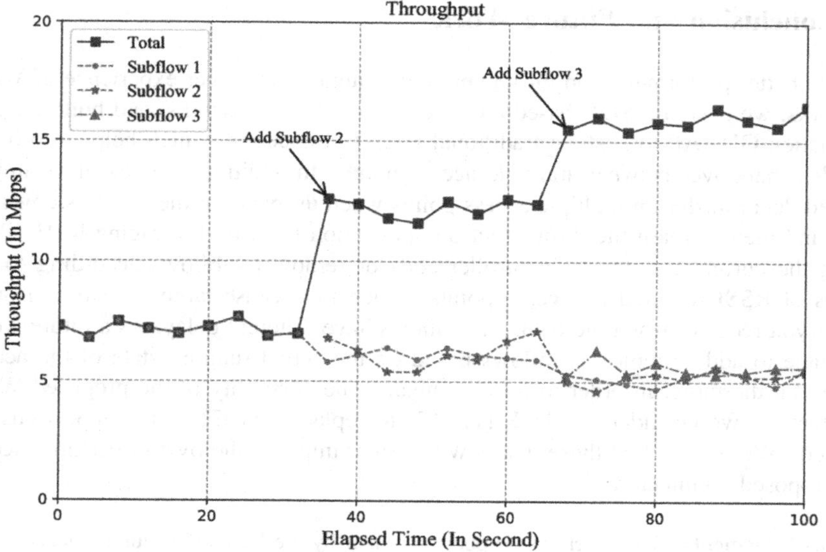

Fig. 5. Throughput improvement with three Subflows

3.3 Handover

In order to prove that our proposed SDN-based Wi-Fi architecture can support mobility better than regular Wi-Fi, we compare the throughput changes during the handover process. In Fig. 6, the client takes about 4 s to re-establish the association in regular Wi-Fi, which will result in throughput down to 0. For our proposed Wi-Fi architecture, the throughput remains stable during the handover process.

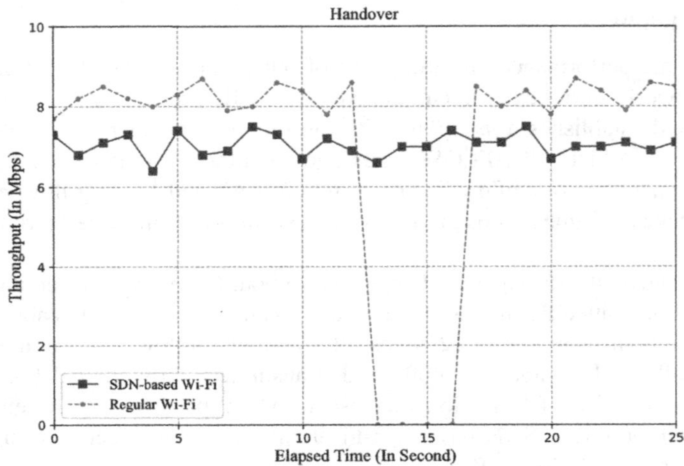

Fig. 6. Throughput changes in handover process

4 Conclusion and Future Work

For the better performance on throughput and seamless handover experience of Wi-Fi networks, we design a SDN-based architecture for Wi-Fi networks, and build a novel abstraction GUVAP to abstract traditional 802.11 connection, which helps to provide seamless handover between multiple access points. In addition, we also have implemented data transfer on multiple access points when users are in the wireless coverage area, and then increase the throughput on application level by leveraging MPTCP.

In the current design, the controller adds or removes subflows according to the values of RSSI reported by access points, which may cause large-scale out-of-order delivery at receiver when the paths of subflows have different delay. As a future work, we plane to add or remove subflows based on the actual running state of the access points. Furthermore, in order to further enhance the versatility of the proposed Wi-Fi architecture, we consider use POF [11, 12] to replace OpenFlow to support custom protocol. We believe that these works will further improve the overall performance of our proposed architecture.

Acknowledgements. This work was supported in part by the National Natural Science Foundation of China (No. 61672486), subtask of the Chinese National Key Project on New Generation Broadband Wireless Mobile Communication Networks (No. 2017ZX03001019-004), and Anhui Provincial Natural Science Foundation (No. 1608085MF126).

References

1. Nunes, B.A.A., Mendonca, M., Nguyen, X.-N., Obraczka, K., Turletti, T.: A survey of software-defined networking: past, present, and future of programmable networks. IEEE Commun. Surv. Tutor. **16**(3), 1617–1634 (2014)
2. Kreutz, D., Ramos, F.M., Verissimo, P.E., Rothenberg, C.E., Azodol-molky, S., Uhlig, S.: Software-defined networking: a comprehensive survey. Proc. IEEE **103**(1), 14–76 (2015)
3. Vestin, J., Dely, P., Kassler, A., Bayer, N., Einsiedler, H., Peylo, C.: CloudMAC: towards software defined WLANs. In: Proceedings of the 18th Annual International Conference on Mobile Computing and Networking, Istanbul, Turkey, 22–26 August 2012
4. Suresh, L., Schulz-Zander, J., Merz, R., Feldmann, A., Vazao, T.: Towards programmable enterprise WLANs with Odin. In: Proceedings of the First Workshop on Hot Topics in Software Defined Networks, Helsinki, Finland, 13 August 2012
5. Xu, C., Jin, W., Zhao, G., Tianfield, H., Yu, S., Qu, Y.: A novel multipath-transmission supported software defined wireless network architecture. IEEE Access **5**, 2111–2125 (2017)
6. Stanley, D., Calhoun, P., Montemurro, M.: Control and provisioning of wireless access points (CAPWAP) protocol specification (2009)
7. O.D. Team, Openwrt wireless freedom (2014)
8. O. VSWITCH, Open vswitch (2013)
9. Kohler, E., Morris, R., Chen, B., Jannotti, J., Kaashoek, M.F.: The click modular router. ACM Trans. Comput. Syst. (TOCS) **18**(3), 263–297 (2000)
10. MPTCP v0.92 socket api. https://github.com/hoang-tranviet/mptcp/tree/mptcp_v0.92_socket_api. Accessed 9 May 2018
11. Song, H.: Protocol-oblivious forwarding: unleash the power of SDN through a future-proof forwarding plane. In: Proceedings of the Second ACM SIGCOMM Workshop on Hot Topics in Software Defined Networking, Hong Kong, China, 16 August 2013
12. Tan, X., Zou, S., Guo, H., Tian, Y.: POFOX: towards controlling the protocol oblivious forwarding network. In: Park, J., Yi, G., Jeong, Y.S., Shen, H. (eds.) Advances in Parallel and Distributed Computing and Ubiquitous Services. LNEE, vol. 368, pp. 21–30. Springer, Singapore (2016). https://doi.org/10.1007/978-981-10-0068-3_3

Location Based One-Time Conference Protocol Without Personal Information

Jiyoung Lim(✉) and SangHyun Lim(✉)

Korean Bible University, Seoul, Korea
{jylim,youngmoon92}@bible.ac.kr

Abstract. Nowadays, we produce massive digital data using smartphones and need to share them with others. However, we feel tired of sharing our personal information repeatedly since we live in the society where the excessive personal information is disclosed. We propose a protocol for the location-based one-time communication service using a beacon. In the service, people at the same place could meet at online to share some digital data without revealing their personal information. After meeting, any personal information does not be traceless to each other. People can communicate without disclosure of personal information.

Keywords: Beacon · BLE (Bluetooth Low Energy) signal
One-time communication · Location-based communication

1 Introduction

As smartphones have propagated rapidly, we frequently use digital data as forms of communication. People produce a lot of digital data such as text, picture, video, location information and voice and try to exchange and share them. To do this people should reveal their phone number or ID of SNS application with each other. Sometimes we do not want to share our personal information such as private and public phone numbers and IDs of SNS app. We propose a service protocol that people at the same place could attempt the one-time digital data exchange with each other not to share their personal information.

We choose a beacon among various ways to know people's location information. Our proposal is the offline to online service based on the location information when people working for the same purpose at the same place want to communicate. We use only beacon signal information for the service. Since a beacon sends the signal including location information periodically, people with same beacon signal request the channel for the signal, enter the channel and communicate securely at the same channel. After working, people terminated the service not to store their personal information.

The remainder of this work is organized as follows. Section 2 describes the related work. System architecture and protocol for our proposal are described in Sect. 3. Implementation is described in Sect. 4. In Sect. 5, we conclude our work.

J. H. Park et al. (Eds.): PDCAT 2018, CCIS 931, pp. 50–55, 2019.
https://doi.org/10.1007/978-981-13-5907-1_5

2 Related Works

2.1 Beacon

Beacon uses the Bluetooth 4.0, Bluetooth Low Energy (BLE) technology [1]. While former Bluetooth technology provides in pairing between two devices, a beacon broadcasts BLE signals periodically and continuously. Devices such as smart phones receive signals, send the request with the signal information to the server through wireless networks such as LTE and WIFI. and get a response from a server.

A beacon sends a fixed BLE signals with a small amount of data such as the location information. iBeacon provides the location information through UUID, Major and Minor which can be assigned as values determined randomly by the service provider [2]. A service provider can divide a place into several blocks and determine the value of UUID, major, and Minor for each block. Table 1 shows an example of UUID, Major and Minor. A beacon product provides the functions of the iBeacon and EddyStone, since iBeacon of Apple and EddyStone of Google [3] are considered as the Beacon standard.

Table 1. An example of UUID, major and minor

Korean Bible Univ.		Bldg #1	Bldg #2	Bldg #3
UUID		123A4B3-FC18-1234-9F12-C90DB1F207F1		
Major		1000	2000	3000
Minor	Room #1	101	101	101
	Room #2	102	102	102
	Room #3	103	103	103

3 Our Proposed Protocol

Our protocol aims to communicate with each other through the secure channel not to exchange their own personal information. We describe the initial phase, the channel connection phase and the beacon and session management phase.

3.1 Initial Phase

First of all, if a device runs the client app, it sends the initial message to the server using HTTPS rest to request the port for SSL socket and the token for the one-time session maintenance. The server responses the client's request, allocates the resource and wait the client's SSL TCP socket connect request. At the initial phase, completes that the client tries to connect the SSL socket and the server accepts it. Figure 1 shows the initial procedure.

The initial phase restarts when the client that has used the channel sends another new beacon signal asking to connect a new channel corresponding the signal. In the protocol, we do not keep any client's personal information and need only the token to maintain the current session. Token means that the client belongs to a special group at the same place and is one of means that determining the session life time.

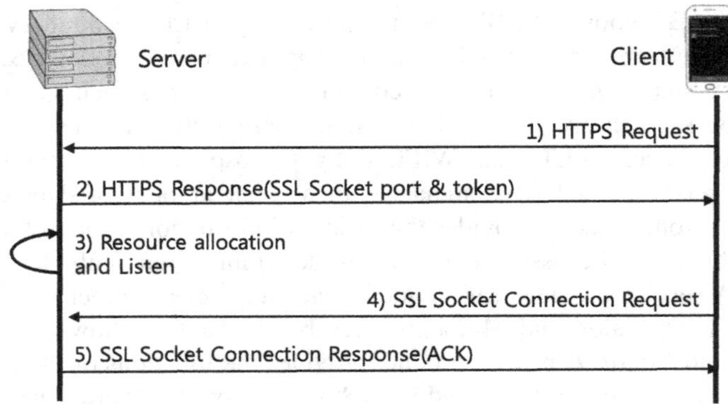

Fig. 1. Initial procedure between server and client

3.2 Channel Connection Phase

Because the beacon signal includes their position information, people in the same room or block have the same beacon signal. If they send the same signal information to the server, they communicate with the same channel through the SSL socket.

If a device receives the beacon signal, it sends the beacon signal information to the server through the SSL socket connected in the initial phase. The server checks if a channel for the beacon signal information is allocated or not. If it does not exist, the server creates a channel for the beacon signal information. After verifying the existence of the valid channel, the device is added to the channel user list. Hence forward, people with same beacon signal can transmit and receive the data among themselves through the SSL socket. Figure 2 shows the connection procedure.

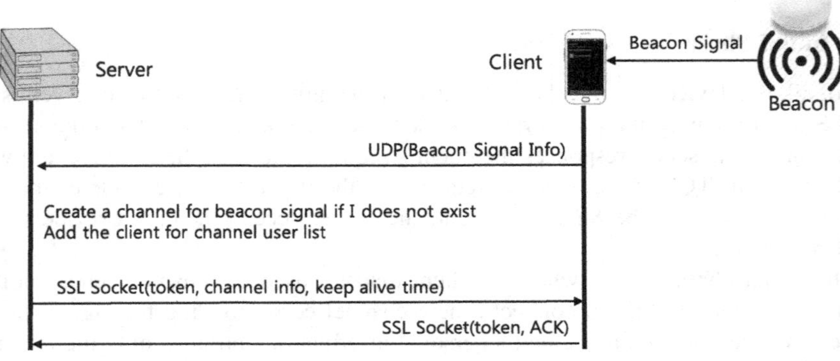

Fig. 2. Channel connection procedure

3.3 Beacon Signal and Session Management

Periodic beacon signal reception is important to maintain the channel connection. Devices receiving beacon signals should send the signal information to the server through UDP whenever the beacon signal arrives. Because the beacon signal information could not arrive at the server according to UDP property, the server checks the channel status, extend the channel session life time, inform the client of keep alive time and response ACK every 30 s through SSL sockets. When the server does not receive it, the server request the client tolerates during 120 s. After timer expires, the server decides that the client leaves the place, disconnects the channel from the device and nullify the token of the client. Figure 3 shows the beacon signal and session management procedure.

The client wants to disconnect the channel by sending a terminating message to the server. The server releases the channel and nullify the token. Hence, if the client wants to connect the server again, the server does not store the client's information at the repository such as database and should issue the token newly.

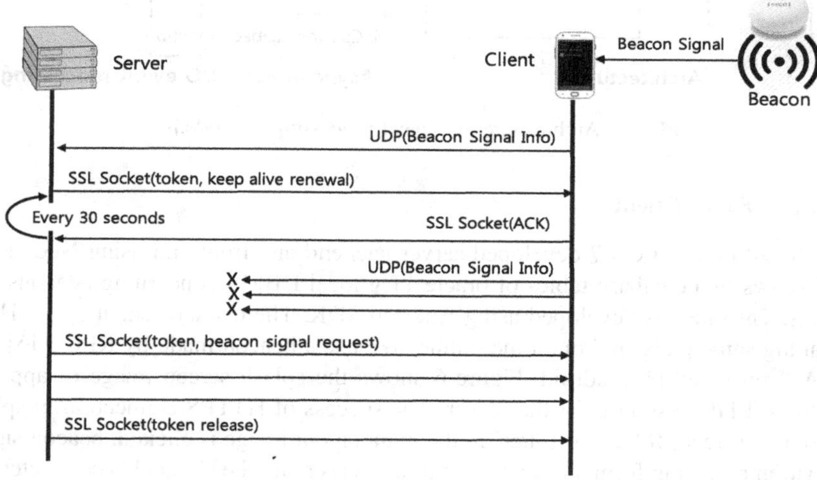

Fig. 3. Beacon signal and session management procedure

4 Implementation

4.1 Node.js

Node.js is a server middleware based on the google chrome V8 java script engine providing the asynchronous I/O. It is a high performance server development technology based on a single thread. Global service provider such as PayPal and Groupon changed their service platform based on Node.js.

To process I/O such as file and network I/O, a program sends I/O requests to Node.js and processes other tasks until arriving the event of I/O completion in Fig. 4. Because CPU time is used only for processing without waiting, Node.js increases the utilization and performance. Since only a single thread processes several requests, Node.js is considered as one of a solution for C10k problem [4].

Node.js always does not shows the higher performance than that of java and others. In case of low CPU processing and a lot of concurrent connection processing, Node.js is superior to others.

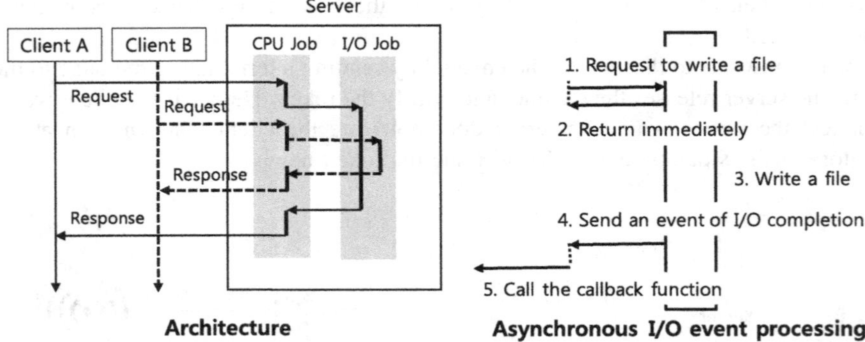

Fig. 4. Architecture and event processing of Node.js

4.2 Server and Client

Server based on Cent OS 7 developed server backend and front end using Node.js. In Fig. 5 shows two database tables of oracle 11 g local DBMS concerning beacons and channels. The client is developed using Android SDK. The name of client app is Here. O meaning same place and one-time online. We use a beacon made by RECO [5] and BLE API provided by Android. Figure 6 shows the splash screen image of app that waits the HTTPS response for the server. The success of HTTPS connection in splash goes to next screen. If button 'Enter' in the main layout image is clicked, beacon signal information receiving from a beacon is sent to a server and the channel is connected for the beacon. Then people with same beacon signal can communicate with each other.

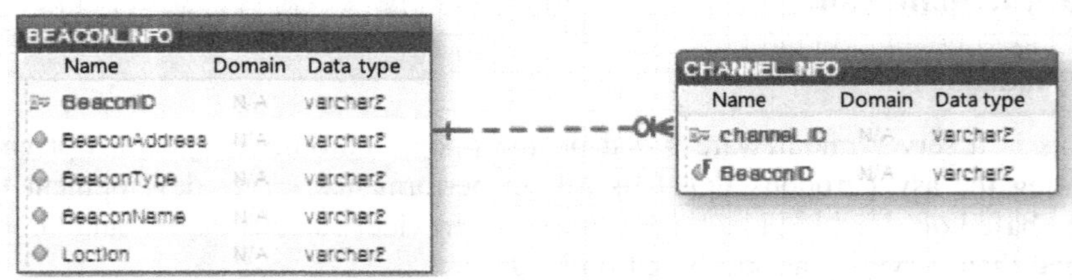

Fig. 5. Beacon and channel database table

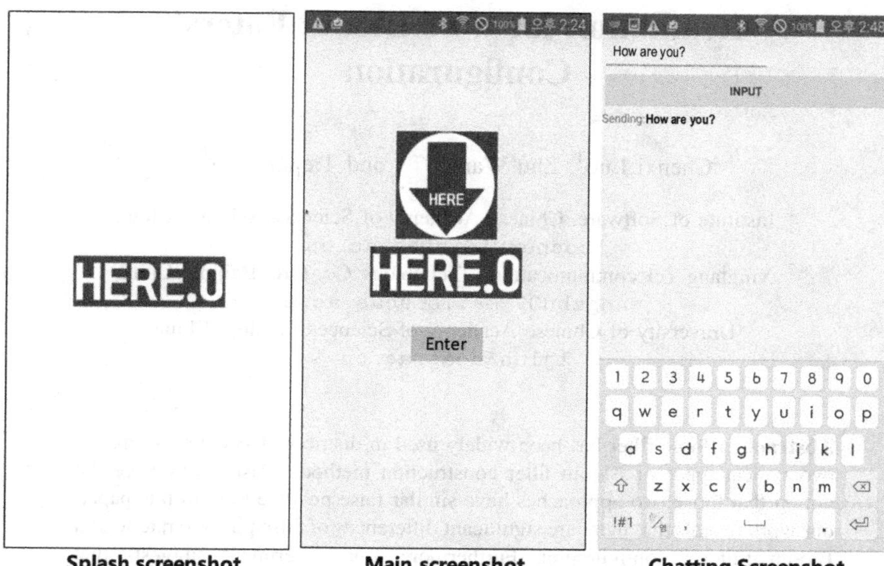

Fig. 6. Client screenshots

5 Conclusion

We proposed a secure protocol for the location-based one-time communication service using a beacon without fear of disclosure for personal information. The session was maintained by the secure SSL socket and one-time token based on beacon signal and smartphone mac address and another beacon signal information through UDP. Digital data did not be encrypted however additional encryption is not difficult. We thought our protocol is useful to people that want the conference without exchange of personal information.

References

1. Kalliola, K., High accuracy indoor positioning based on BLE. Nokia research center presentation (2011)
2. Gast, M.S.: Applications with IBeacon: Proximity and Location Services with Bluetooth Low Energy. O'Reilly Media Inc., Sebastopol (2014)
3. Dasgupta, A., Nagaraj, R., Nagamani, K.: An internet of things platform with Google eddystone beacons. J. Softw. Eng. Appl. **9**(06), 291 (2016)
4. The C10K problem. http://www.kegel.com/c10k.html
5. RECO Beacon. http://reco2.me/reco-beacon/

Evaluation for Two Bloom Filters' Configuration

Chenxi Luo[1], Zhu Wang[2(✉)], and Tiejian Luo[3]

[1] Institute of Software, Chinese Academy of Sciences, Beijing, China
luochenxi@iscas.ac.cn
[2] Xingtang Telecommunications Technology Co. Ltd., Beijing, China
wangzhu09@mails.ucas.ac.cn
[3] University of Chinese Academy of Sciences, Beijing, China
tjluo@ucas.ac.cn

Abstract. Bloom filter has been widely used in distributed systems. There are two typical types of Bloom filter construction methods. Past works have the notion that those two approaches have similar false positive rate. In this paper, our work reveals that there are significant differences of false positive rate within Bloom Filter's configuration. Furthermore, our experiment suggests that parameters' setting up demonstrates an impact on Bloom Filter's behaviour. According to the results, it is vital for adjusting the Bloom filter parameters while the index space is limited.

Keywords: Bloom filter · False positive rate

1 Introduction

Bloom filter, which is a compact indexing algorithm, has been widely used in distributed and pervasive networks. Due to its simple structure and smooth integration characteristic, the mathematical format allows considerable potential improvement for system designers to develop new variations for their identical application requirements. Counting Bloom filters [1–3] can be used to improve network router performance [4]. Other variations are adopted in state machines [5], IP trace back [6], Internet video [7], distributed storage system index [8], flow monitoring [9], software defined networks [10] and publish/subscribe networks [11].

In pervasive computing area, there are also Bloom filter applications. Research [12] and [13] use Bloom filter to represent items or policies during communication process. The algorithm can also serve as an efficient content management component like [14–17]. The Bloom filters have very compact size, which fit well in embedded systems with limited memory space. Also, the lookup time complexity is O(C), and thus saves energy consumption, especially suitable for pervasive embedded nodes with constraint energy supply. Our previous work [17] focuses on how to design a compact index service for distributed systems using Bloom filters. In this paper we analyse the Bloom filter itself: the influence of Bloom filter parameters.

J. H. Park et al. (Eds.): PDCAT 2018, CCIS 931, pp. 56–62, 2019.
https://doi.org/10.1007/978-981-13-5907-1_6

2 Related Work

There are several indexing and locating techniques that deal with the object lookup task.

2.1 Table-Based Approach

Traditional database stores the object as an entry in a table and performs item lookup by searching the item key in the table. The database offers a straightforward way to store the item itself and its metadata and receives a widespread usage on the Internet. Until now, database is still an irreplaceable storage method in most of the online applications. However, owing to the linearity nature of tables, database suffers from low lookup speed and large storage cost. The space consumption is O(N), where N is the total number of items. The optimal time complexity of a table is O(logN).

2.2 Hash-Based Approach

Hash table transforms object lookup into calculation by mapping the item onto a hash value and replaces the traditional metadata table with a hash table. In this way, the storage space is reduced and item lookup is accelerated. However, in order to maintain the consistency of hash functions, the data structure restricts the freedom of content placement and requires the object migration when storage nodes are inserted or deleted. Moreover, the hash-based method assigns the storage pattern for each item and lacks storage flexibility. That causes trouble in systems like content delivery network where objects may be transferred from node to node.

2.3 Perfect Hashing

Probabilistic index is widely used in today's distributed systems. Perfect hashing maps objects of a set onto a set of sequential buckets and stores the buckets. Then it puts the objects' fingerprint into the buckets. When a query arrives, it first finds the corresponding bucket and checks if the fingerprint matches the query. The data structure has a false rate of $0.5^{m/n}$, where m is the total index space and n is the object number. However, perfect hashing can only be used in static sets.

3 Bloom Filter Algorithm and Performance

Bloom filter [18] works as an index which records all elements of a set. We may assume that the set $S = \{x_1, x_2, \ldots, x_n\}$, which consists of n elements. A Bloom Filter vector (BFV), which consists of m bits, is used to represent elements of set S. All bits of the vector are set to zero initially. For each element, the algorithm uses k hash functions $\{h_i\}_{i=1 \ldots k}$ to map the element onto k positions of the vector and sets the bit on the position to 1. The k functions, ranging from 1 to m, are independent from each other and can map elements of the set S to a random place on the vector. During the insertion period, the algorithm maps all elements of the set to load the BFV with all the information of the elements. Figure 1 shows the element insertion.

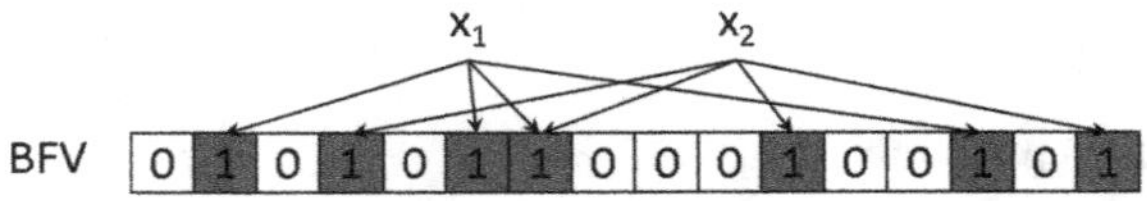

Fig. 1. Insert elements of set S into Bloom filter vector.

In lookup procedure which we want to check whether an element y belongs to the set S, the algorithm uses the same hash functions to map y onto k locations and check whether all $h_i(y)$ equal to 1. If the answer is no, we conclude that y doesn't belong to S, otherwise, we say y belongs to S. The time complexity of Bloom filter lookup is O(C). Figure 2 shows the look up procedure.

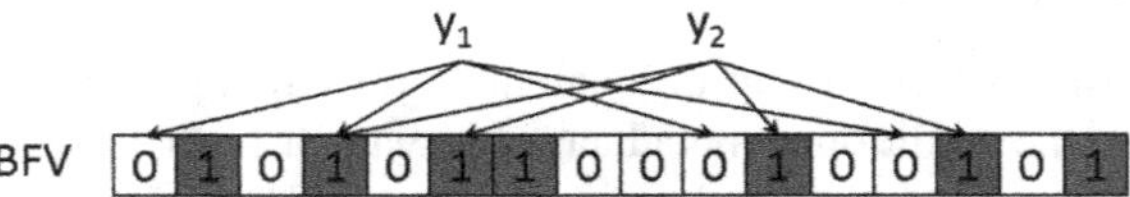

Fig. 2. Lookup procedure. $y_2 \in S$, y_1 doesn't belong to S.

It needs to be mentioned that there is a probability that elements don't belong to S be judged as inside S by Bloom filter. That is to say, Bloom filter has a false positive rate. Literature [19] shows that the false positive rate of a Bloom filter with m bits, n elements and k hash functions is

$$f_p = \left(1 - \left(1 - \frac{1}{m}\right)^{kn}\right)^k \tag{1}$$

where *m* is the number of bits in the bloom filter, *n* is the number of elements and *k* is the number of hash functions.

Later in this paper, we refer to that kind of Bloom filter as Standard Bloom Filter (SBF).

Literature also says that there is another construction of Bloom filter. Instead of having one array of size m shared among all the hash functions, each hash function has a range of m/k consecutive bit locations disjoint from all the others. The total number of bits is still m, but the bits are divided equally among the k hash functions. Later in this paper, we refer to that kind of Bloom filter as Partial Bloom Filter (PBF). The false positive rate of PBF is

$$f_p = \left(1 - \left(1 - \frac{k}{m}\right)^{n}\right)^k \tag{2}$$

Since $k \geqq 1$,

$$\left(1 - \frac{k}{m}\right)^{n} \leq \left(1 - \frac{1}{m}\right)^{kn} \quad (3)$$

The false positive rate of PBF is at least as large as that of SBF. The PBF approach may be useful for implementation reasons; for example, dividing the bits among the hash functions may make parallelization of array accesses easier. Literature [19] also says that the false positive rate of SBF and PBF is very close to each other. In this paper we want to compare the false positive rate of the two constructions of Bloom filter.

4 Performance Comparison of Two Bloom Filter Types

In this section we compare the false positive rate of SBF and PBF. We want to find under what circumstances the two types are similar and under what circumstances the two types differ a lot. We will check the influence of m, k and n separately.

4.1 Influence of Hash Function Number

Here we set m = 1000 and n = 100. We change hash number k from 1 to 20 and calculate the false positive rate of SBF and PBF. Then we compare the two results and find the difference of them. The result is shown in Fig. 3.

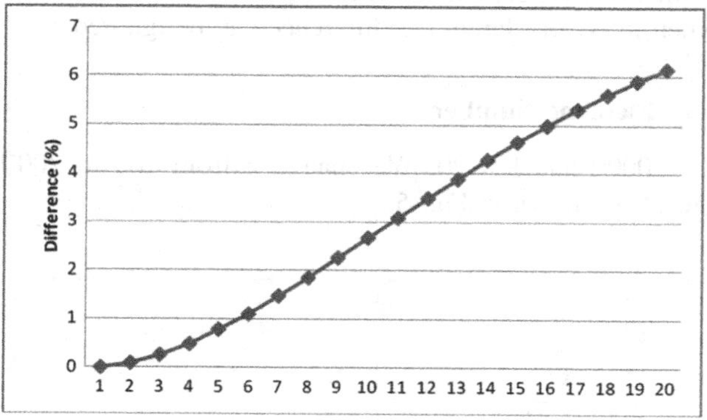

Fig. 3. Influence of hash function number.

In the figure above the horizontal axis is the hash number, from 1 to 20. The vertical axis is the difference of the two algorithms in percentage. We can see that the difference grows with the hash number. It reaches 7% when hash number is 20.

4.2 Influence of Bloom Filter Size

Here we set m values 100, 1000, 10000 and 100000 respectively. m/n values 10 and k ranges from 1 to 20. The result is shown in Fig. 4.

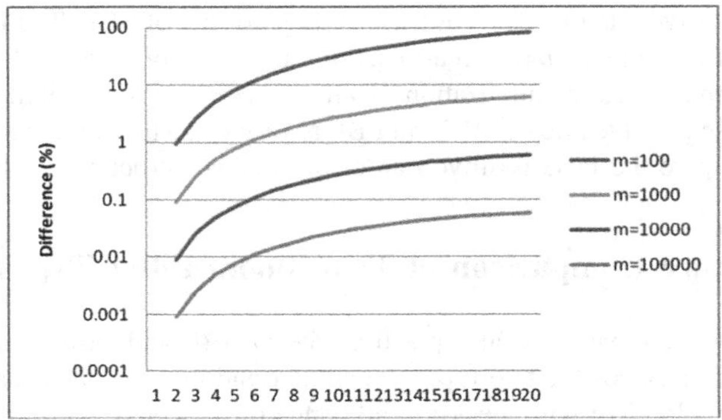

Fig. 4. Influence of Bloom filter size.

In the figure, the horizontal axis is the hash number and the vertical axis is the difference of the two algorithms in percentage. We can see that the size of Bloom filter has very large impact on the performance difference. The difference decreases with the increase of m. When m is 100, the difference even reaches 83%. That indicates that when the Bloom filter size is very small, the false positive rate of SBF and PBF differs a lot. However, when m is very large, the difference can be ignored.

4.3 Influence of Element Number

Here we set m = 10000 and k = 20. We change n from 100 to 2000 and find the difference. The result is shown in Fig. 5.

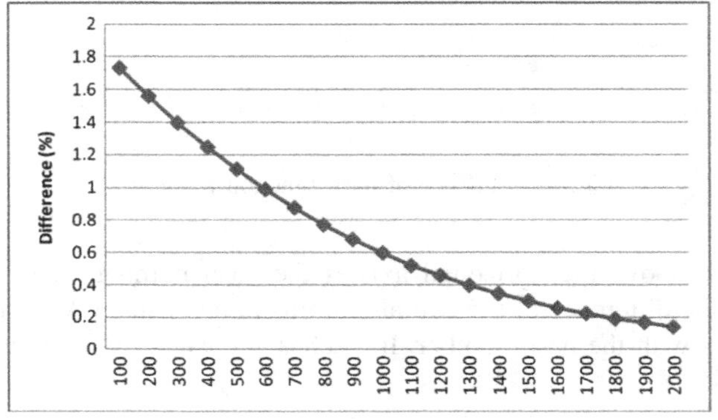

Fig. 5. Influence of element number.

The horizontal axis is n and the vertical axis is the difference. We can see that the difference decreases with the increase of element number.

5 Conclusion and Future Work

In this paper we have shown the two construction methods of Bloom filter. We have analysed the performance difference of the two. We have found that the number of bits in Bloom Filter (m) has very large impact of the difference. When m is very small, the difference can be very explicit and when m is large, the difference can be ignored. Hash function number and element number also have impact on performance difference. The increase of hash function number and the decrease of element number cause the increase of false positive difference. We have given suggestions on how to choose the Bloom filter parameters when using the data structure: to avoid large error when using Partial Bloom filters, the most important thing is to choose a large Bloom filter size, with element number fixed. Hash number can reach its optimal value.

Our future work includes using the suggestions to design highly-efficient Bloom filters and applying the data structure to different scenarios.

References

1. Bonomi, F., Mitzenmacher, M., Panigrahy, R., Singh, S., Varghese, G.: An improved construction for counting bloom filters. In: Azar, Y., Erlebach, T. (eds.) ESA 2006. LNCS, vol. 4168, pp. 684–695. Springer, Heidelberg (2006). https://doi.org/10.1007/11841036_61
2. Fan, L., Cao, P., Almeida, J., Broder, A.Z.: Summary cache: a scalable wide-area web cache sharing protocol. IEEE/ACM Trans. Netw. **8**(3), 281–293 (2000)
3. Ficara, D., Giordano, S., Procissi, G., Vitucci, F.: Multilayer compressed counting bloom filters. In: Proceedings of the 27th Conference on Computer Communications (INFOCOM), pp. 311–315. IEEE (2008)
4. Song, H., Dharmapurikar, S., Turner, J., Lockwood, J.: Fast hash table lookup using extended bloom filter: an aid to network processing. In: Proceedings of the 2005 Conference on Applications, Technologies, Architectures and Protocols for Computer Communications (SIGCOMM), pp. 181–192. ACM (2005)
5. Bonomi, F., Mitzenmacher, M., Panigrah, R., Singh, S., Varghese, G.: Beyond bloom filters: from approximate membership checks to approximate state machines. In: Proceedings of the 2006 Conference on Applications, Technologies, Architectures and Protocols for Computer Communications (SIGCOMM), pp. 315–326. ACM (2006)
6. Sung, M., Xu, J., Li, J., Li, L.: Large-scale ip traceback in high-speed internet: Practical techniques and information-theoretic foundation. IEEE/ACM Trans. Netw. **16**(6), 1253–1266 (2008)
7. Wang, Z., Luo, T.: Intelligent video content routing in a direct access network. In: Proceedings of the 3rd Symposium on Web Society, pp. 147–152. IEEE Computer Society (2011)
8. Wang, Z., Luo, T., Xu, Y., Cheng, F., Zhang, X., Wang, X.: A fast indexing algorithm optimization with user behavior pattern. In: Zu, Q., Hu, B., Elçi, A. (eds.) ICPCA/SWS 2012. LNCS, vol. 7719, pp. 592–605. Springer, Heidelberg (2013). https://doi.org/10.1007/978-3-642-37015-1_52

9. Du, Y., Wang, S.: Two-stage adaptive bloom filters for per-flow monitoring in software defined networks. In: Proceedings of the 2018 IEEE International Conference on Communications (ICC), pp. 1–7. IEEE (2018)
10. Lei, K., Li, K., Huang, J., Li, W., Xing, J., Wang, Y.: Measuring the control-data plane consistency in software defined networking. In: Proceedings of the 2018 IEEE International Conference on Communications (ICC), pp. 1–7. IEEE (2018)
11. Jokela, P., Zahemszky, A., Rothenberg, C.E., Arianfar, S., Nikander, P.: Lipsin: line speed publish/subscribe inter-networking. In: Proceedings of the 2009 Conference on Applications, Technologies, Architectures and Protocols for Computer Communications (SIGCOMM), pp. 195–206. ACM (2009)
12. Chen, T., Guo, D., He, Y., Chen, H., Liu, X., Luo, X.: A bloom filters based dissemination protocol in wireless sensor networks. Ad Hoc Netw. **11**(4), 1359–1371 (2013)
13. Takiguchi, T., Saruwatari, S., Morito, T., Ishida, S., Minami, M., Morikawa, H.: A novel wireless wake-up mechanism for energy-efficient ubiquitous networks. In: Proceedings of the 2009 IEEE International Conference on Communications Workshops, pp. 1–5. IEEE (2009)
14. Ghosh, M., Özer, E., Biles, S., Lee, Hsien-Hsin S.: Efficient system-on-chip energy management with a segmented bloom filter. In: Grass, W., Sick, B., Waldschmidt, K. (eds.) ARCS 2006. LNCS, vol. 3894, pp. 283–297. Springer, Heidelberg (2006). https://doi.org/10.1007/11682127_20
15. Jimeno, M.: Saving Energy in Network Hosts with an Application Layer Proxy: Design and Evaluation of New Methods that Utilize Improved Bloom Filters. PhD thesis, University of South Florida (2010)
16. Qwasmi, N., Liscano, R.: Bloom filter supporting distributed policy-based management in wireless sensor networks. In: Proceedings of the 4th International Conference on Ambient Systems, Networks and Technologies, pp. 248–255. Elsevier (2013)
17. Wang, Z., Luo, C., Luo, T.: Selection optimization of bloom filter-based index services in ubiquitous embedded systems. In: Jin, H., Wang, Q., Zhang, L.-J. (eds.) ICWS 2018. LNCS, vol. 10966, pp. 231–245. Springer, Cham (2018). https://doi.org/10.1007/978-3-319-94289-6_15
18. Bloom, B.H.: Space/time trade-offs in hash coding with allowable errors. Commun. ACM **13**(7), 422–426 (1970)
19. Broder, A., Mitzenmacher, M.: Network applications of bloom filters: a survey. Internet Math. **1**(4), 485–509 (2004)

A Fast Global AVF Calculation Methodology for Multi-core Reliability Assessment

Jiajia Jiao(✉) and Dezhi Han

Shanghai Maritime University, Shanghai 201306, China
jiaojiajia@shmtu.edu.cn

Abstract. Soft error induced bits upset has received increasing attention in reliable processor design. To measure the processor reliability, Architectural Vulnerability Factor (AVF) is often calculated by fast Architectural Correct Execution (ACE) analysis or accurate fault injection in a CPU core (e.g., alpha, x86, ARM) processor or GPU. However, the AVF calculation in the entire multicore system composed of several CPU cores, GPU, caches and memory banks, mostly depends on time consuming realistic laser tests or complex fault injection (days or years). To shorten the evaluation time, this paper presents a partition-based AVF calculation methodology from local to global. This approach combines local AVF for each component and Input-Output Masking (IOM) calculation between components to calculate the global AVF fast using probabilistic theory in a cost-effective way. The comprehensive simulation results of a 7-way parallelized motion detection application demonstrate the error location and error propagation path affects global AVF values. The probabilistic theory driven global AVF estimation time is only the order of magnitude in seconds.

Keywords: Multicore · Soft error · Architectural Vulnerability Factor Input-Output Masking · Reliability assessment

1 Introduction

Decades of rapidly scaling technology have brought the challenge of soft errors to modern multicore computing systems. Evaluating and analyzing multicore resilience to transient soft errors caused by high-energy particle strikes grows increasingly important. Multicore designers should develop tools or methodologies to understand the soft errors impacts on different architectures or applications.

To measure the processor reliability, Architectural Vulnerability Factor (AVF) [1] is often calculated by fast Architectural Correct Execution (ACE) analysis [2] or accurate fault injection in a CPU core (e.g., alpha, x86, ARM) processor or GPU [3]. However, the AVF calculation in the entire multicore system composed of several CPU cores, GPU, caches and memory banks, mostly depends on realistic laser tests or complex fault injection. To shorten the evaluation time, sampling technique or parallel simulation can be used. However, the evaluation accuracy and simulator development time increase. How to design a cost-effective method for the multicore reliability assessment has become a critical problem.

J. H. Park et al. (Eds.): PDCAT 2018, CCIS 931, pp. 63–72, 2019.
https://doi.org/10.1007/978-981-13-5907-1_7

From a 'reuse' perspective, we try to combine some outstanding works on soft error analysis together seamlessly to satisfy the complex multicore system. This paper presents a partition-based AVF calculation methodology from local to global. This approach combines local AVF for each component and Input-Output Masking (IOM) calculation between components to calculate the global AVF fast in a cost-effective way.

2 Related Work

How to solve the complex multicore reliability assessment issue has been attractive for academic and industry researchers. At present, the existing AVF estimation works in processors are usually classified into two categories as follows.

(1) ACE analysis method proposed by Intel research group, requires only once simulation for assessing the lower bound of AVF [2]. It has been used for fast AVF calculation in alpha-based single core reliability analysis platform sim-SODA [3] and GPU-based platform gpgpu-SODA [4]. To characterize more masking effects, probabilistic graphical models (PGM) is used to improve AVF estimation accuracy [5]. To further evaluate the multi-cell upsets impacts, a ACE analysis based two-level model driven framework is proposed. These fast ACE analysis approaches are recently used to support the online reliability management in some reconfigurable architecture [6], task mapping algorithms [7] and task migration strategies [8].

(2) Fault injection, is a pervasive accurate estimation method in all processor architectures. Few cases use the laser as a fault injector [9] or processors located in the satellite are sent to aerospace for realistic faults injection experiments [10]. Considering the long evaluation time and high cost of realistic experiments, processor simulators are mostly integrated with a software fault injection tool. An accelerated fault injection methodology is proposed for complex Multi-bit Upsets in processor communication component [11]. Some architecture-level fault injection tools for GPU (e.g., GUFI by France TIMA library [12], SASSIFI by NVIDA group [13]) have been also developed. To better understand GPU soft-error resilience, the hybrid approach is designed for combining architectural fault-injection and neutron beam testing [14]. However, the CPU-GPU collaborative computing is increasingly important in the modern chips [15, 16]. Fault injection surely can estimate the AVF value in heterogeneous multicore architectures [17] as well as single CPU core and GPU.

In all, fault injection is accurate but time consuming while the fast ACE analysis is hard to capture the complex error masking effects in a large scale multicore system.

As a complementary work to the existing works, this paper presents a fast reliability assessment methodology from local AVF calculation to global AVF calculation. Some efficient works on local AVF calculation can be reused very well. Similarly, the some works on capturing the masking effects between different components are also considered. This approach takes advantages of both local AVF for each component and IOM calculation between components together to calculate the global AVF fast using the basic probabilistic theory.

3 Proposed Global AVF Calculation Methodology

The proposed approach for calculating global AVF in a multicore architecture includes three parts: local AVF estimation for each component, IOM estimation between two components, and global AVF estimation based on data flow graph using probabilistic theory.

3.1 Computing Local AVF

AVF is defined as the probability of wrong output observation when a soft error induced bit upsets occurs. If the input error location and output observation point are in the same component, e.g., alpha core or data cache, we consider the corresponding AVF is a local value. Otherwise, e.g., the input error occurs in alpha core observation, while output observation point is set to GPU, this requires a global AVF.

Unlike time consuming fault injection, here local AVF can be estimated by ACE analysis fast. That means we can use counters to record the component utilization fast as local AVF. There are computing structure and storage structure in all components. For example, a router in Network on Chips has storage structure like input buffers and computing structure like routing computing unit to determine the next hop direction. Therefore, we provide the computing rules for computing structure and storage structure in the following Table 1.

Table 1. Handling rule for local AVF in multicore

Type	Handling rule for local AVF	Example
Computing structure	(1) Cut the entire timeline into busy pieces or free pieces	e.g., local AVF for ALU is equal to 0.4, which is calculated by 40 M cycles/100 M cycles
	(2) Sum up all busy pieces for total busy cycles	
	(3) The ratio of busy cycles in total execution cycles is local AVF	
Storage structure	(1) Cut the entire timeline into read-to-read, read-to-write, write-to-write or write-to-read pieces	e.g., local AVF for instruction cache is equal to 1 because all of storage accesses are read operations
	(2) Sum up all read to read and write to read pieces for the total X-to-read cycles	
	(3) The ratio of accumulative X(read or write)-to-read cycles in total execution is local AVF	

3.2 Computing IOM

The IOM term was proposed in to calculate data cache AVF with masking effects in core [18]. IOM value is obtained by fault injection, the ratio of the number with correct output in total number of errors. E.g., injection 1000 errors in all load operations, but

800 errors do not cause wrong outputs in stored results. The corresponding IOM is 0.8. An analytical model based IOM calculation is proposed for fast evaluation [19].

Here, we redefine the IOM value calculation for any components. In a multicore system, there are three types of components: communication component, memory component and core. The core masking uses our previous work [5, 19] while communication and memory are defined to zero for no error masking during error propagation. Therefore, we provide the computing rules for different components in the following Table 2.

Table 2. Handling rule for IOM in multicore

Type	Handling rule for IOM	Example
Communication component	If the destination component is for communication, the IOM is 0	e.g., the data flow from Core to Network on Chip, the IOM is 0
Memory component	If the destination component is for memory, the IOM is 0	e.g., the data flow from Core to data cache, the IOM is 0
Core	(1) Define instruction masking rules for specific Instruction Set Architecture (ISA)	e.g., the data flow from data cache to CPU core, the IOM is 0.8
	(2) Calculate average IOM based on instruction masking for specific application during once simulation	

3.3 Computing Global AVF

Based on the data flow between components, we try to use the probabilistic theory for global AVF calculation. For example, Fig. 1 shows a simple application data flow. The local AVF for register file in CPU0 and IOM can be obtained from given handling rules. Here, Com represents the communication component like bus or other interconnections; Mem depicts the memory component like cache or main memory. CPU or GPU is the computing core.

From this data flow graph in Fig. 1, there are two error propagation paths: CPU0->Com->Mem->GPU and CPU0->Com->CPU1. If an error in two paths is both masked, the error is disappeared in the global AVF calculation and does not perform a failure. Otherwise, it keeps its own impacts from the local AVF to the global AVF and cause the wrong output. Therefore, we can calculate the factual global AVF for register file of CPU0 in Eq. (1).

$$\begin{aligned} AVF_{rf\_cpu0\_global} = AVF_{rf\_cpu0\_local} \times (1- \\ (1 - IOM_{cpu0\_com}) \times (1 - IOM_{com\_mem}) \times IOM_{mem\_gpu} \\ \times (1 - IOM_{cpu0\_com}) \times IOM_{com\_cpu1}) \\ = 0.136 \end{aligned} \tag{1}$$

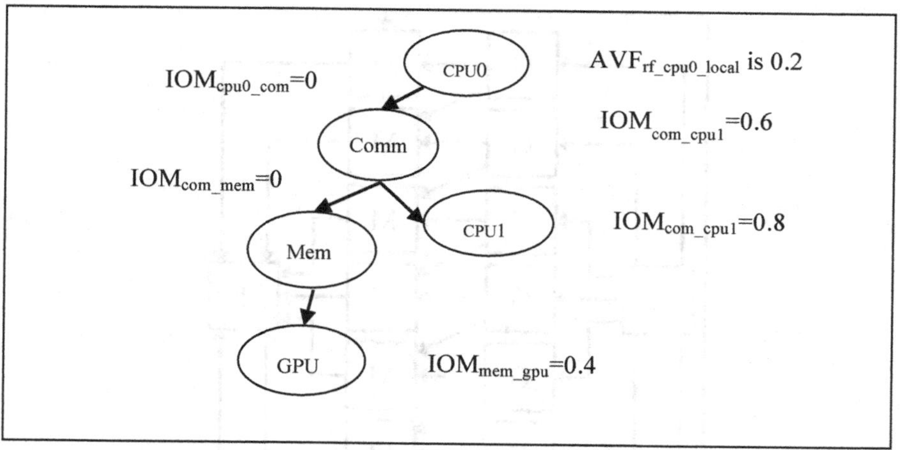

Fig. 1. One example of global AVF calculation in register files of CPU0 in a CPU-GPU collaborative multicore system

The comparative results of 0.2 and 0.136 demonstrate the different soft error impacts from local region to global region due to error masking in error propagation.

4 Simulation and Results Analysis

This section describes the simulation configurations and comprehensive results analysis respectively.

4.1 Simulation Configuration

To evaluate the proposed reliability assessment methodology, we do a case study of the video processing application in Fig. 2 [20]. The motion detection application consisted of three phases, of which in the first one the video stream is processed with a noise-mitigating 5×5 pixel Gaussian filter, followed by subtraction of consecutive frames and 5-pixel median filtering.

In multicore experiments Gaussian filtering was performed in a dedicated actor, whereas the median filtering frame subtractions were performed in the same actor to achieve the load balance. The 7-way parallelized application has 1, 2, 3, …7 separate processing paths such that each processing path was dedicated to processing of one frame depicted in Fig. 2, where S is the source actor, T is the sink, G stands for Gaussian filtering and M for combined median filtering and frame subtraction. The links between actors integrates the buffers with wires for memory and communication together. The actors executed on the GPU were *G*, *M* while *S* and *T* executed on CPU cores.

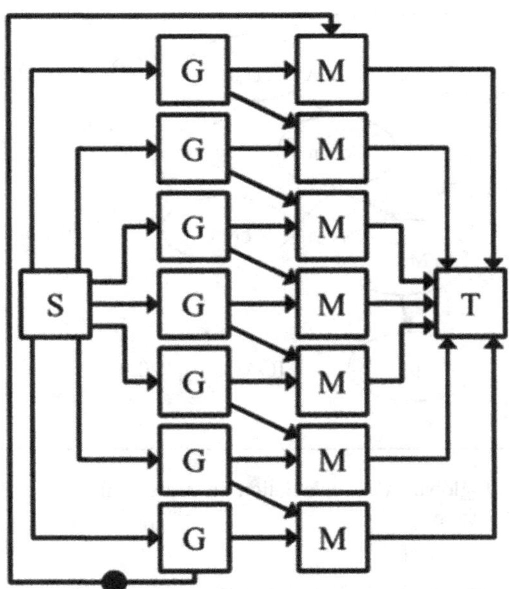

Fig. 2. Data flow graph of the 7-way parallelized motion detection application.

The CPU core is instanced as the alpha while GPU is instanced as the NVIDA Fermi structure. Lack of the unified platform to evaluate the GPU-CPU multicore reliability, we use the synthesized trace for the approximate processing.

4.2 Results Analysis

Local AVF for alpha CPU and Fermi GPU can be available respectively for typical SPEC2000-int in sim-SODA and workloads for GPU from Parboil benchmark, NVIDIA CUDA SDK package, Rodinia benchmark suite and the third party applications in GPGPU-SODA as respectively in Fig. 3.

Due to the limited space the AVF similarity, the AVF values for other structure like reorder buffer, load-store queue are not listed. The average local AVF of multi-benchmark is approximate to the 7-way parallelized motion detection application trace. In other words, the average local AVF value based on 12 benchmarks of SPEC2000-int is approximate the local AVF value is assumed as local AVF value in CPU core for 7-way parallelized motion detection application.

Similarly, IOM for alpha CPU can be available for typical SPEC2000-int in Fig. 4. The available IOM value from calculation rules are also compared with fault injection results. It is demonstrated that the accuracy loss is limited to about 10% [21]. IOM for NVIDIA GPU parallel thread execution (PTX) ISA has no existing work no support directly. Average IOM value in GPU architecture is assumed equal to IOM in alpha core. We assume the average is approximate to the 7-way parallelized motion detection application trace.

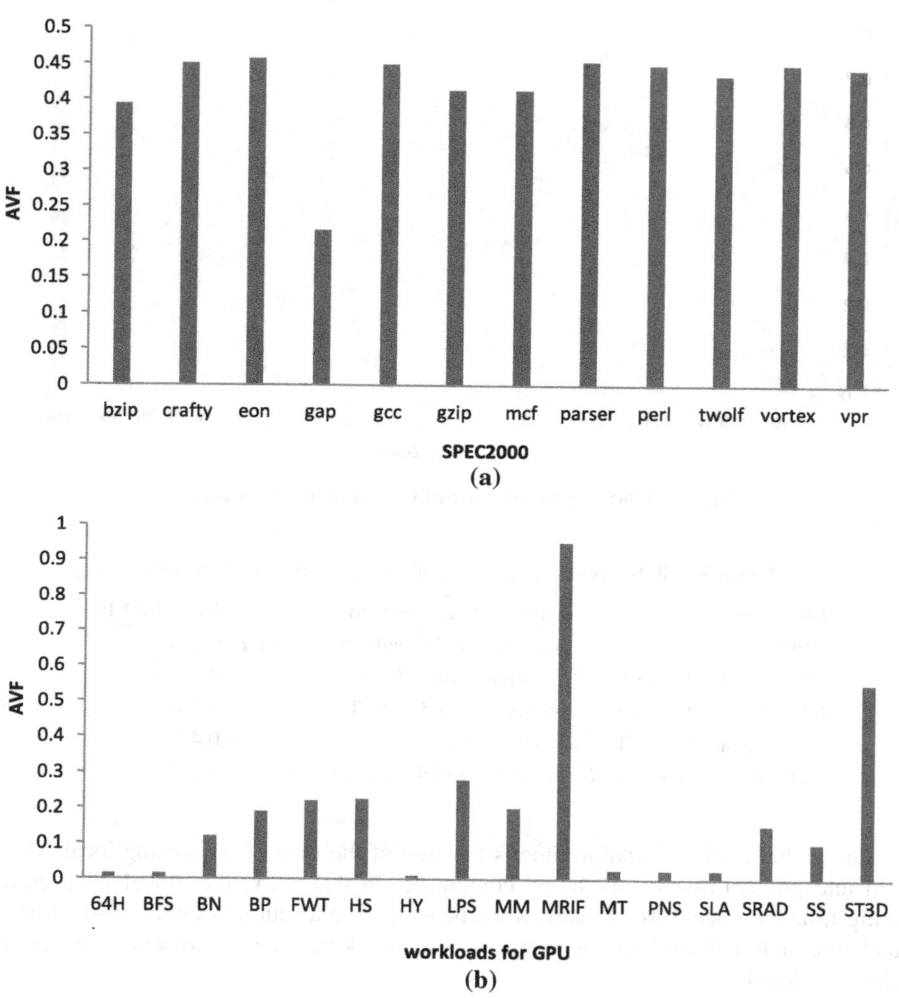

Fig. 3. Local AVF estimation for register files in alpha core and GPU. (a) Local AVF values for SPEC2000 in alpha core. (b) Local AVF values for GPU

Finally, the global AVF values can be calculated from local AVF, IOM and data flow using the probabilistic theory. Table 3 lists several global AVF values for different error propagation paths. It is observed that both error propagation path and error location determine the global AVF. For example, the global AVF equal to 0.277 if CPU register file in S. Unlike the complex propagation paths, the global AVF equal to 0.42 if CPU register file in T. Changing the error location from register file to recorder buffer also result in global AVF from 0.277 to 0.138. Similarly, if error locates in GPU structure, the global AVF is also affected by error location and error propagation paths.

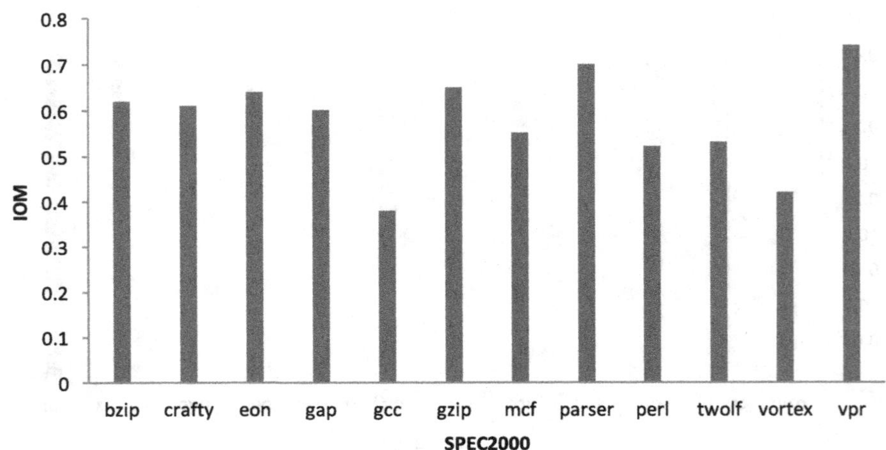

Fig. 4. IOM estimation for SPEC2000 in an alpha core.

Table 3. Global AVF values for different error propagation paths

Error location	Error propagation path	Global AVF
CPU register file in S	Several parallel paths from S to T	0.277
GPU register file in G	Two paths from G to T	0.0634
GPU register file in M	one paths from G to T	0.082
CPU register file in T	Inner path in T	0.42
CPU recorder buffer in S	Several parallel paths from S to T	0.138

The global AVF calculation time is the sum of the time of computing local AVF, IOM and probabilistic theory based computing. In this work, the global AVF computing time for each 100 M instructions benchmark execution is about 10–15 s in a quad-core laptop. Therefore, the proposed framework provides a cost-effective way to calculate global AVF.

5 Conclusion

This paper presents a global AVF calculation methodology for fast reliability assessment in multicore architecture. Unlike the system vertical analysis layer by layer, the proposed approach makes good use of local AVF and IOM seamlessly for characterize the error masking effects via probabilistic theory. The case study of 7-way parallelized motion detection application demonstrates the global AVF measures the multicore reliability in dozens of seconds.

To further verify the effectiveness of proposed method, we will try to setup a full system and computing the corresponding IOM values of GPU for more accurate estimation in the future work. More cases study of applications running in CPU-GPU cooperative architecture will be also done using the proposed framework.

Acknowledgement. This work is supported by National Natural Science Foundation, numbered 61502298 and Innovation Program of Shanghai Maritime University.

References

1. Mukherjee, S.S., Weaver, C., Emer, J., et al.: A systematic methodology to compute the architectural vulnerability factors for a high-performance microprocessor. In: 36th Annual IEEE/ACM International Symposium on Microarchitecture, pp. 29–40. IEEE (2003)
2. Mukherjee, S.S., Emer, J., Reinhardt, S.K.: The soft error problem: an architectural perspective. In: 11 International Symposium on High-Performance Computer Architecture, 2005. HPCA-11, pp. 243–247. IEEE (2005)
3. Fu, X., Li, T., Fortes, J.: Sim-SODA: a unified framework for architectural level software reliability analysis. In: Workshop on Modeling, Benchmarking and Simulation (2006)
4. Tan, J., Yi, Y., Shen, F., et al.: Modeling and characterizing GPGPU reliability in the presence of soft errors. Parallel Comput. **39**(9), 520–532 (2013)
5. Jiao, J., Juan, D.C., Marculescu, D., et al.: Exploiting component dependency for accurate and efficient soft error analysis via probabilistic graphical models. Microelectron. Reliab. **55**(1), 251–263 (2015)
6. Srinivasan, S., Koren, I., Kundu, S.: Online mechanism for reliability and power-efficiency management of a dynamically reconfigurable core. In: 33rd IEEE International Conference on Computer Design (ICCD), pp. 327–334. IEEE (2015)
7. Namazi, A., Abdollahi, M., Safari, S., et al.: LORAP: low-overhead power and reliability-aware task mapping based on instruction footprint for real-time applications. In: Euromicro Conference on Digital System Design (DSD), pp. 364–367. IEEE (2017)
8. Pouyan, F., Azarpeyvand, A., Safari, S., et al.: Reliability aware throughput management of chip multi-processor architecture via thread migration. J. Supercomput. **72**(4), 1363–1380 (2016)
9. Vargas, P.F.R.: Evaluation of the SEE sensitivity and methodology for error rate prediction of applications implemented in Multi-core and Many-core processors. Université Grenoble Alpes (2017)
10. Jiao, J., Fu, Y., Wen, S.: Accelerated assessment of fine-grain AVF in NoC using a multi-cell upsets considered fault injection. Microelectron. Reliab. **54**(11), 2629–2640 (2014)
11. Tselonis, S., Gizopoulos, D.: GUFI: a framework for GPUs reliability assessment. In: IEEE International Symposium on Performance Analysis of Systems and Software (ISPASS), pp. 90–100. IEEE (2016)
12. Hari, S.K.S., Tsai, T., Stephenson, M., et al.: SASSIFI: an architecture-level fault injection tool for GPU application resilience evaluation. In: IEEE International Symposium on Performance Analysis of Systems and Software (ISPASS), pp. 249–258. IEEE (2017)
13. Previlon, F.G., Egbantan, B., Tiwari, D., et al.: Combining architectural fault-injection and neutron beam testing approaches toward better understanding of GPU soft-error resilience. In: IEEE 60th International Midwest Symposium on Circuits and Systems (MWSCAS), pp. 898–901. IEEE (2017)
14. Vega, A., Buyuktosunoglu, A., Bose, P.: Transparent CPU-GPU collaboration for data-parallel kernels on heterogeneous systems. In: 22nd International Conference on Parallel Architectures and Compilation Techniques (PACT), pp. 245–256 (2013)
15. Sun, Y., Gong, X., Ziabari, A.K., et al.: Hetero-mark, a Benchmark suite for CPU-GPU collaborative computing. In: IEEE International Symposium on Workload Characterization (IISWC), pp. 1–10. IEEE (2016)

16. Rosa, F., Ost, L., Reis, R., et al.: Fast fault injection to evaluate multicore systems soft error reliability (2017)
17. Vargas, V., Ramos, P., Velazco, R., et al.: Evaluating SEU fault-injection on parallel applications implemented on multicore processors. In: IEEE 6th Latin American Symposium on Circuits and Systems (LASCAS), pp. 1–4. IEEE (2015)
18. Haghdoost, A., Asadi, H., Baniasadi, A.: Using input-to-output masking for system-level vulnerability estimation in high-performance processors. In: 15th CSI International Symposium on Computer Architecture and Digital Systems (CADS), pp. 91–98. IEEE (2010)
19. Jiao, J., Han, X., Fu, Y.: A PGM based multi-level reliability analysis method for data cache. IEICE Electron. Express **12**(16), 20150453 (2015)
20. Boutellier, J., Nyländen, T.: Design flow for GPU and multicore execution of dynamic dataflow programs. J. Signal Process. Syst. **89**(3), 469–478 (2017)
21. Jiao, J., Marculescu, D., Juan, D.C., Fu, Y.: A two-level approximate model driven framework for characterizing multi-cell upsets impacts on processors. Microelectron. J. **48**, 7–17 (2016)

Adaptive Control and Optimization on Dynamic Load Balancing Model

Tinglei Zhao(✉), Jianzhong Qiao, Shukuan Lin, and Yanhua Wang

College of Computer Science and Engineering, Northeastern University, Shenyang, China
zhaotinglei_0928@163.com

Abstract. Load distribution of distributed system is not balanced. The imbalance could cause low efficiency. There are many studies on how to balance loads in a distributed system, among which are multiple studies achieve the goal by using linear time delay system. In the paper, we propose an impulsive and switching load balancing model with time delay based on control theory. In order to describe various current states of a node, we construct corresponding sub-system according to the dynamics of node resources. The model reallocates tasks in light of task's real-time running status so as to improve the efficiency of dynamic load balancing. We propose an adaptive optimal control strategy, which makes the feedback control method of our model more practical. Experimental results demonstrate that the proposed model can make the system balanced, and verify the feasibility of the model.

Keywords: Time delay · Impulsive and switching system
Dynamic load balancing · Adaptive control · Distributed system

1 Introduction

Load balance is an important research topic in distributed system, for load unbalance among nodes would affect system efficiency seriously. A distributed system makes use of discrete computing resources, so the bottom communication, transmission model and available resource show a high degree of heterogeneous status, which leads to extreme unbalance of nodes and performance degradation of system. A distributed system consists of several nodes. Each node can enter and exit dynamically, and submit different kinds of new tasks many times. For new arrival tasks, the available resource information need to be collected and managed. Next, load balance are carried out on the basis of the condition. However, the concrete running status of a task can only be obtained after start-up, which makes the dynamic load balancing method must reallocate the parallel tasks in line with its real-time running status [1]. When dealing with a task on a large scale, the system divides it into several sub-tasks and maps them into each node. Then each node returns its computing result, which will be mixed together. That means the whole task complete. During the process, if some nodes are too busy to return its result in time, then the whole system will fall into waiting obviously. So system performance would be affected seriously. Therefore, load balance is necessary for distributed system. Load balance belongs to NP problem [2], but can reach a better level with various optimization strategies.

J. H. Park et al. (Eds.): PDCAT 2018, CCIS 931, pp. 73–83, 2019.
https://doi.org/10.1007/978-981-13-5907-1_8

In order to solve the problem of load unbalance among nodes in distributed system, we propose an impulsive and switching load balancing model with time delay. When the system is starting up, each node preserves the state and load information which are broadcasted by others. Each node computes its state, switches to the corresponding sub-system according to the state, and migrates load by load migration rule. Then system reaches balance. Switching to a sub-system is triggered by the own state of a node. The model can make load balancing among nodes efficiently and improve system performance with low overhead.

We propose an adaptive optimal control strategy aimed at the model constructed in the preceding part. Because the specific running status of a task can only be obtained after starting-up. For a new arrival task, some information are dynamical and unpredictable. Such as the number of nodes, length of load queue, CPU utilization, utilization of memory and other available resources of the system. So these resource information need to be collected and managed dynamically. The model should study information dynamically and accumulate knowledge as time goes on, then make predicting and adjusting actions, which leads the model to be more practical in real environment.

In the paper, the contributions of our work can be concluded as follows:

(1) We propose an impulsive and switching load balancing model with time delay based on control theory. The model consists of 4 sub-systems which correspond with the states of a node on various moment.
(2) A node makes its own state as impulsive signal to switch to a corresponding sub-system. A node needs not to preserve the state information of other nodes.
(3) Migration proportion of a node is calculated according to its own and others processing ability, which makes the model local and global.
(4) An adaptive optimal control strategy is proposed. The strategy studies information and accumulates knowledge as time goes on, then makes predicting and adjusting actions, which combines prediction and optimization, so leads to better performance of system load balancing.
(5) We make simulation based on multi-thread which is under various initial conditions and load migration proportion. By choosing appropriate time delay, system scale and other parameters flexibly, the practicability and efficiency of our model are proved.

2 Related Works and Problem Formulation

In recent years, a number of works have been done on load balance. Reference [3] proposed a time delay model for distributed dynamic load balance system based on diffusion model. The model supposed that processing rate and producing rate obeyed exponential distribution and Poisson distribution respectively, and made theoretical analysis for a Delay Dynamic Load Balancing (DDLB) system owned 2 nodes with random regeneration theory in queuing theory. Then they obtained a relative optimal load balance gain according to minimizing the parallel processing time, and confirmed the theory result by Monte Carlo stochastic simulation experiment. It is obvious that the method belongs to the domain of stochastic load balance. Although the specific theoretical load balance value when the system balancing was not given in [4], they verified the stability of the model in theory. Reference [5] adopted time domain analysis method, got the load balance gain of DDLB system on the condition of random communication delay and transmission delay, improved the scalability and flexibility of dynamic load balancing method. However, the shortcoming of the model is that they can get the suitable theory load balance gain by a certain initial condition under a serial complex calculation. Reference [1] proposed a frequency domain analysis method for DDLB system and its sufficient and necessary condition, found the approximation relation among time delay, system scale and load balance gain, revealed the effect on stability of dynamic load balance by various time delay and system scale as well. They assumed that the system was linear continuous in the reference, considering that the processing rate λ_i equals to the producing rate μ_i, and let $d_i(t) = 0$. So the model could not describe dynamic load balancing system accurately for that is constructed on so much assumptions.

The contributions for DDLB system above have made deep analyses in theory and accurate practical verification, but there are some insufficient:

(1) A node needs to communicate with other nodes to guarantee the correctness when carrying on load migration. It is hard to acquire the current load queue length and other information of each node for the dynamic of resources and the uncertainty of task running state. And there lacks research on extendibility of heterogeneous DDLB system with different time delay.
(2) The constructed models ignore several factors that will lead to unbalance in the process of load balancing, so they could not describe the process of dynamic load balancing accurately.
(3) In the process of theoretical analysis, [4] and [5] assume that the communication delay and transmission delay are constants. So the analysis result is conservative for DDLB system with large time delay obviously.
(4) During the verification, in virtue of the concrete time delay cannot be determined in actual distributed environment, the experiment result can only verify the practicability of the proposed model. In many cases it could not reflect the effect of load balance stability by a variety of time delay in system.

In view of the above shortcomings, we synthesize the characteristics of impulsive switching system model [6, 7] and time delay system model [8–10], and propose an impulsive and switching load balancing model with time delay which aims at distributed dynamic load balancing process. The model constructs 4 sub-systems which are corresponding to the states of node in the system. When the node state changes, an impulsive signal is triggered. The impulsive signal makes a node switch to a corresponding sub-system, which avoids message transmission frequently. So massive communication delay is eliminated and system overhead is reduced. We take consideration of processing ability of a node and other nodes comprehensively when making load migration. So the system will not generate more over-load nodes after migration and the efficiency of load balancing will improve.

3 Impulsive and Switching Load Balancing Model with Time Delay

In distributed system, each node owns a private memory to access directly. But each node can only access the private memory of other nodes by the method of explicit message passing. So it is necessary to make dynamic load balancing for available resources and the running status of tasks.

In the load balancing model, the compute node is either a bearer of computing tasks or a collector of available resources information of the system. The system consists of n nodes $P_0, P_1,\ldots, P_{n-1}$. The processing ability a_i $(0 \le i \le n-1)$ of node P_i can be calculated as follows:

a_i is an aggregative indicator of the number of cores Nc_i, frequency F_i, memory M_i and current utilization $U_i(t)$ of node P_i. We deduce the approximate relation among a_i, Nc_i, F_i, M_i and $U_i(t)$ according to much data in tests. $U_i(t)$ has little effect on a_i when it is low to α. And a_i is descending slowly with $U_i(t)$ ascending from α to β. But it falls rapidly when $U_i(t)$ reaches to β. Node P_i cannot run regularly when $U_i(t)$ reaches to 1, so we define $a_i = 0$ now. So the relation between a_i and $U_i(t)$ is nonlinear. Besides, a_i is proportionate to Nc_i, F_i and M_i. Equation (1) shows the approximate expression for processing ability of node P_i. The transition points α and β are 0.2 and 0.55 in experiment.

$$a_i = \begin{cases} Nc_i*(F_i + \varepsilon*M_i) & 0 \le U_i(t) \le \alpha \\ Nc_i*(F_i + \varepsilon*M_i) * \frac{1-(\beta-\alpha)^2}{\ln(2-\beta)} * \ln(2 - U_i(t)) & \beta < U_i(t) \le 1 \\ Nc_i*(F_i + \varepsilon*M_i) * (1 - (U_i(t) - \alpha)^2) & \alpha < U_i(t) \le \beta \end{cases} \tag{1}$$

The work load of node P_i is recorded as w_i $(0 \le i \le n-1)$. The work load which can be processed in the kth time interval $[t_k, t_{k+1})$ is recorded as ideal load w_i' and

$$w_i' = (t_{k+1} - t_k) * \frac{\int_{t_k}^{t_{k+1}} a_i(t)dt}{t_{k+1} - t_k} \tag{2}$$

$[t_k, t_{k+1})$ is the lingering time of a sub-system.

The state of node P_i can be judged by Definition 1.

Definition 1. (Judging criterion for node state) if the load of node P_i subjects to $w_i < w^{inf}$, $w^{inf} < w_i < w^{sup}$, $w_i > w^{sup}$ or $w_i = 0$, then P_i belongs to a light-load node, a moderate-load node, an over-load node or an empty-load node respectively. The load upper bound w^{sup} and lower bound w^{inf} are $w^{sup} = w_i' * (1 + \xi)$ and $w^{inf} = w_i' * (1 - \xi)$ respectively. ξ ($\xi = 0.05$) is a value set in advance according to actual distributed environment.

When the system is starting-up, node P_i broadcasts its own state information and load information to other nodes, and collects information from other nodes. Node P_i calculates and maintains the current state of system. The state of system *state* can be calculated in Eq. (3). If *state* = 1, we deal with the ideal load of P_i with $w_i' = \omega * w_i'$ and convert *state* to 0, then carry on by the impulsive and switching load balancing model with time delay in Eq. (4). Else when *state* = 0, we carry on the impulsive and switching load balancing model with time delay in Eq. (4) directly.

$$state = \begin{cases} 1 & \omega > t \\ 0 & \omega \leq t \end{cases}, \quad \omega = \frac{\sum_1^n w_i}{\sum_1^n a_i} \tag{3}$$

$$\begin{cases} \dot{w}_i(t) = f_{\sigma_i(k)}(t, w_i(t), w_i(t-\tau_i)) & t \in [t_k, t_{k+1}), \sigma_i(k) \in \{\Gamma_{OLN}, \Gamma_{MLN}, \Gamma_{LLN}, \Gamma_{ELN}\} \\ \Delta w_i(t_{k+1}) = w_i(t_{k+1}) - w_i(t_{k+1}^-) & k = 0, 1, 2, \ldots \\ w_i(\theta) = w_i(t_0) & \theta \in [t_0 - \tau_i, t_0] \end{cases} \tag{4}$$

In Eq. (4), $w_i(t)$ is the state vector of system and $w_i(t) \in R^n$. $\sigma_i(k)$ is the identifier of sub-system corresponding with the state of P_i on different time. The activated sub-system is expressed as $\sigma_i(k)$ in the kth time interval $[t_k, t_{k+1})$. The switching time sequence subjects to $0 \leq t_0 < t_1 < \ldots t_k < t_{k+1} < \ldots$. $t_{k+1} - t_k$ is the lingering time of a sub-system. The time delay τ_i of system subjects to $\tau_i > 0$. $\Delta w_i(t_{k+1})$ stands for state jumping. The initial condition subjects to $w_i(\theta) = w_i(t_0)$.

Each node chooses a suitable sub-system in Eqs. (7)–(10) according to its own load state judged by itself. After judging its own load state, an over-load node migrates its over load to adjacent nodes in light of Eq. (7). A light-load node or an empty-load node transforms its state in light of Eqs. (8) or (10) if it has received some migration load. However, if a light-load node has not received any migration load, then it will transform its state in light of Eq. (9). An empty-load node will not make state transition which has received no load. A moderate-load node neither migrates load to other nodes nor receives load from other nodes. It makes state transition according to Eq. (9). All the load migration process carried on must obey load migration rule, and the proportion of load migration is calculated in Eq. (5).

Definition 2. (load migration rule) for two arbitrary nodes P_i and P_j which are in the state of $\sigma_i(k)$ and $\sigma_j(k)$, the load can be migrated by the method of $\{P_i \rightarrow P_j | \sigma_i(k) = \Gamma_{OLN}, \sigma_j(k) \in \{\Gamma_{LLN}, \Gamma_{ELN}\}\}$.

Because each node P_i owns its special processing ability a_i, so the size of load in migration of each time is different. So does the effection for system performance. If a over load is divided into a fixed size without taking consideration of processing ability of a node, then the node state would transform frequently. The load received every time of the node with weak processing ability should be fewer than the node with strong processing ability for the former owns a smaller ideal load w_i'. The system state changes when its load exceeds its ideal load. In order to guarantee the system stability, the load migrated from a node with strong processing ability should not be excessive. Therefore, we use Eq. (5) for an adaptive load migration proportion.

$$p_{ij} = \frac{1}{n} * \left(1 - \frac{a_i(t)}{\sum_1^n a_i(t)}\right) \tag{5}$$

Then the size of load migration one time which we definite as load migration unit can be shown in Eq. (6).

$$unit_{ij}(\mathrm{t}) = p_{ij}(\mathrm{t}) * (w_i - w_i') \tag{6}$$

Judge the state of node by Definition 1. If the node P_i is in the state of over-load, then the node should only migrate its over load to others but not receive load from others, except for processing its own tasks and producing new tasks. We construct sub-system model based on control theory for the process. The state space is described as follows concretely.

$$\Gamma_{OLN}: \begin{cases} \frac{dw_i(t)}{dt} = \lambda_i(t) - \mu_i(t) - \sum\limits_{i \neq j} p_{ij} u_i(t - \eta_{ij}), \\ y_i(t) = w_i(t) - w_i'(t), \\ u_i(t) = k_i y_i(t). \end{cases} \tag{7}$$

When the node P_j is in the state of light-load, then the node should only receive load from other over-load node P_i but not migrate its load to others, except for processing its own tasks and producing new tasks. We construct sub-system model based on control theory for the process. The state space is described as follows concretely.

$$\Gamma_{LLN}: \begin{cases} \frac{dw_j(t)}{dt} = \lambda_j(t) - \mu_j(t) + \sum\limits_{j \neq i} p_{ji} u_i(t - \eta_{ji}), \\ y_j(t) = w_j(t) - w_j'(t), \\ u_j(t) = 0. \end{cases} \tag{8}$$

When the node P_i is in the state of moderate-load, then the node should neither receive load from others nor migrate its load to others, except for processing its own tasks and producing new tasks. We construct sub-system model based on control theory for the process. The state space is described as follows concretely.

$$\Gamma_{MLN}: \begin{cases} \frac{dw_i(t)}{dt} = \lambda_i(t) - \mu_i(t), \\ y_i(t) = w_i(t) - w_{\mathrm{i}}'(t), \\ u_i(t) = 0. \end{cases} \tag{9}$$

When the node P_j is in the state of empty-load, then the node should only receive load from other over-load node P_i but not migrate its load to others because it has not any own tasks and would not produce new tasks. We construct sub-system model based on control theory for the process. The state space is described as follows concretely.

$$\Gamma_{ELN}: \begin{cases} \frac{dw_j(t)}{dt} = \sum\limits_{j \neq i} p_{ji} u_i(t - \eta_{ji}), \\ y_j(t) = w_j(t) - w_j'(t), \\ u_j(t) = 0. \end{cases} \tag{10}$$

State variable w_i stands for the length of load queue of node P_i.

Input variable u_i stands for the over load that is migrated from node P_i to other nodes in an unit time.

Output variable y_i stands for the over load of node P_i.

State variable $w_i^{'}$ stands for the length of ideal load queue of node P_i.

k_i stands for the closed loop gain of load balancing system in an unit time.

p_{ij} is the load migration proportion calculated by Eq. (5).

η_{ij} stands for the transmission delay from node P_i to node P_j.

λ_i and μ_i stand for the task producing rate and task processing rate of node P_i respectively. We calculate them as follows:

$$\lambda_i = r_i * w_i(t) \tag{11}$$

$$\mu_i = \frac{a_i(t)}{\int_{t_k}^{t_{k+1}} a_i(t)dt} * w_i(t) \tag{12}$$

Among of which $r_i \in R^+$. Its value is fixed by the intrinsic attribute of a task. $[t_k, t_{k+1})$ is the lingering time of a sub-system.

4 Adaptive Control and Optimization of the Model

Adaptive dynamic programing (ADP) aims at studies on optimization and prediction. It studies information and accumulates knowledge as time goes. Then makes predicting and adjusting actions, maximizes the utility function and achieves goals eventually [11].

In the system, we choose performance index function J as follows:

$$J = (J_1\ J_2\ \ldots\ J_n)^T$$

Among of which, $J_i = \frac{1}{2}\int_{t_k}^{t_{k+1}} \varepsilon_i^2(t)dt,\ i \in n,\ k = 0,1,2,\ldots.$ $t_{k+1} - t_k$ stands for the time needed by each dynamic load balancing in system. $\varepsilon_i(t) = y_i^{reality}(t) - y_i^{model}(t)$, $y_i^{reality}(t)$ stands for the real output of system, $y_i^{model}(t)$ stands for output of the system model.

When the over load of the system is 0, $y_i^{model}(t) = 0$, the system achieves the goal status of load balancing. The nodes status is $\sigma_i(k) \in \{\Gamma_{MLN}, \Gamma_{LLN}, \Gamma_{ELN}\}$.

As Fig. 1 shows, in the proposed ADP designing, the goal of the referential network is providing predicted ability for study and optimization procedures. Two kinds of reinforcement signal (primary reinforcement signal of external environment, secondary reinforcement signal of referential network) are used to improve the generalization ability and study ability. Primary reinforcement signal $r(t)$ is usually a binary signal. 0 and −1 are used to stand for 'overload' and 'not overload' respectively. To provide abundant reinforcement signal, the system uses quaternary values reinforcement signals (0, −0.4, −0.7, −1) stand for the load status of system. Reference [12] proposed a kind of secondary reinforcement signal about ADP designing. It provides abundant information for each sampling time: $r(t) = -\sum_{i=1}^{n} \left(\frac{x_i - x_{i,d}}{x_{i,\max}}\right)^2$. Where x_i stands for the ith status of state vector x, $x_{i,d}$ is the expected referential status, $x_{i,\max}$ is the formal maximized status value. By this method, ADP architecture performs good in generalization and study ability. Internal reinforcement signal *s*(*t*) is constructed by referential network. On the other hand, it cooperates with evaluating network to do optimization and study.

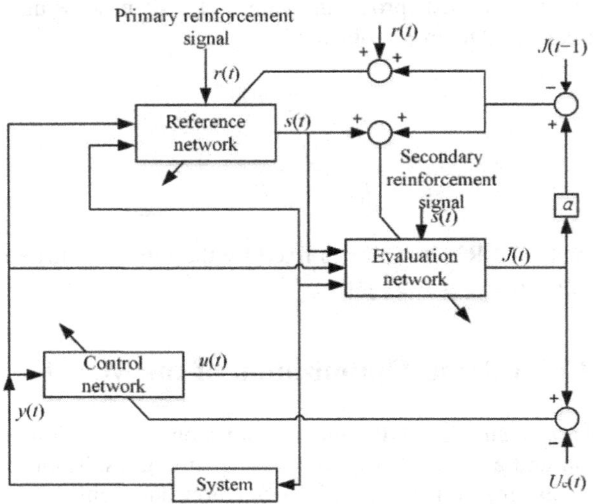

Fig. 1. Structure of adaptive control system.

Figure 1 uses anti-propagation rule. 3 parameters converted paths are used to adjust 3 kinds of network parameters [13]. Control network can make anti-propagation for the deviation between expecting goal U_c and *J* function that is from evaluation network indirectly. The deviation can be used to updates the parameters of control network, $e_a(t) = J(t) - U_c(t)$. The primary reinforcement signal *r*(*t*) is on a higher level, which uses binary signal to stand for 'overload' or 'not overload'. The secondary reinforcement signal *s*(*t*) can be a consecutive value with abundant information, which can improve the study and generalization performance. Therefore, it can be used to update the parameters of referential network: $e_f(t) = \alpha J(t) - [J(t-1) - r(t)]$. Once the referential network output signal *s*(*t*), it will be the input of evaluation network, which is used to adjust the parameters of evaluation network, $e_c(t) = \alpha J(t) - [J(t-1) - s(t)]$.

In the paper, we choose collaborative learning strategy. In each phase of the strategy, firstly, we use anti-propagation to adjust the referential parameters based on primary reinforcement signal *r*(*t*). Then, the referential network output secondary reinforcement signal *s*(*t*), which adjusts the parameters of evaluation network by anti-propagation. Once the evaluation network is adjusted in the phase, it will provide a new *J*(*t*) evaluation. On the contrary, it will be used to adjust referential network in the next phase. Therefore, the referential network and evaluation network can train as a method of cooperation.

5 Experiment and Result Analysis

With the design above, we make simulation in multi-thread environment for the proposed model under specific initial condition. The software for simulation is win64 multi-thread on VC++, hardware is Intel Core i7 with 2.30 GHz frequency and 4G memory. The conservative synchronization strategy is realized by *mutex*. And we use *sleep* to simulate the transmission time delay η on the same time scale. The system scale is controlled by setting global variable. Figure 2 describes the specific simulation result of a heterogeneous system. In order to analyze the node state and the change of queue length in the system conveniently, we take 6 node for example to explain. The specific initial condition is like: $n = 6$. Initial queue length are: $w_1 = 410$, $w_2 = 1300$, $w_3 = 0$, $w_4 = 960$, $w_5 = 620$, $w_6 = 2160$. Load balancing gain $K = 0.8$.

The state and queue length of each node change on the time of T1 = 92.511 ms, T2 = 143.247 ms, T3 = 170.43 ms, T4 = 184.607 ms, T5 = 191.584 ms. With the migration of load, time needed by each phase decreases progressively. The over load are all migrated to light-load node and empty-load node exactly. The overhead of each phase consist of the executing time of load balancing algorithm and transferring time of load migration chiefly. From the experiment, we know the time of load balancing algorithm in each phase is less than 300 μs except for the first phase with 411 μs. It can be seen that the overhead comes from load transferring chiefly. In the figure, the time spent on overload migration in each phase are 92.1 ms, 50.4 ms, 26.9 ms, 13.9 ms and 6.7 ms respectively. As time goes on, the over load decrease quickly. On T5, the overload in system becomes to 0. So the goal of load balancing is achieved.

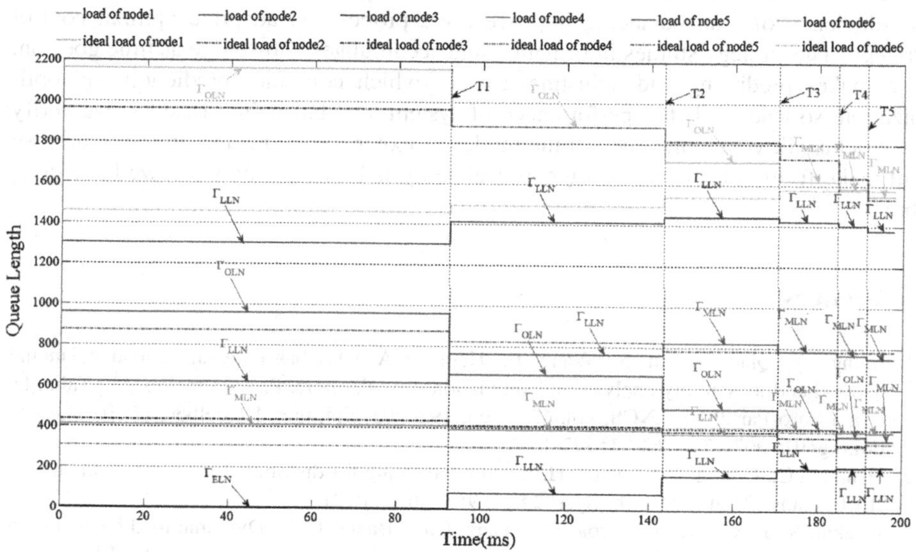

Fig. 2. State and queue length change of nodes in process of load balancing.

Each node judges whether need to load balancing periodically. At the initial time T0, the main thread activates other threads. The statistical thread makes statistics for instant load information and state of each node, and judges the system state then deals with ideal load. Each node thread calculates its own processing ability a_i and ideal load $w_i^{'}$. The nodes are on the state of moderate-load, over-load, light-load, empty-load light-load and over-load. Node 4 and node 6 migrate one unit workload u_4(T1) and u_6(T1) to node 2, node 3 and node 5 respectively. After that on T1, the state of node 2 has not changed for it owns a strong a_2. Node 3 receives the load from node 4 and node 6, which makes its state change to light-load from empty-load. Node 5 receives some load. So its queue increases. But this does not change its state. On T2, over-load nodes continue migrating their over loads to other light-load. The state of node 5 changed to moderate-load from light-load. Because node 1 is moderate-load on T1, it does not receive load. It also consumes load by itself. To T3, node 4 and node 6 migrate over load to light-load nodes 1, 2 and 3. To T4, only node 4 has over load. It migrates over load to light-load nodes 2 and 3. After a new period of migration, there is no over-load node on T5, the system is balanced.

6 Conclusion and Future Work

In the paper, we propose an impulsive and switching load balancing model with time delay based on description of state space by control theory. We construct corresponding sub-systems according to the various states of each node in system. When making load migration, we take account of the processing ability of a node and other nodes. So there are not more nodes which will change into the state of over load, and the efficiency of load balancing improves. We propose an adaptive optimal control strategy. The strategy studies information and accumulates knowledge as time goes on, then makes predicting and adjusting actions, which combines prediction and optimization, so leads to better performance of system load balancing. Besides, we verify the practicability and efficiency of the model by experimental results. In the future, we will make further deep study on theoretical analysis based on the work we have done above.

References

1. Meng, Q., Qiao, J., Lin, S., Wang, E., Han, P.: A delay-based dynamic load balancing method and its stability analysis and simulation. In: D'Ambra, P., Guarracino, M., Talia, D. (eds.) Euro-Par 2010. LNCS, vol. 6271, pp. 192–203. Springer, Heidelberg (2010). https://doi.org/10.1007/978-3-642-15277-1_19
2. Hsiao, H.C., Chung, H.Y., Shen, H.: Load rebalancing for distributed file systems in clouds. IEEE Trans. Parallel Distrib. Syst. **24**(5), 951–962 (2013)
3. Dhakal, S., Hayat, M.M., Pezoa, J.E., Yang, C.D., Bader, D.A.: Dynamic load balancing in distributed systems in the presence of delays: a regeneration-theory approach. IEEE Trans. Parallel Distrib. Syst. **18**(4), 485–497 (2007)

4. Tang, Z., Birdwell, J.D., Chiasson, J., Abdallah, C.T., Hayat, M.: Resource-constrained load balancing controller for a parallel database. IEEE Trans. Control Syst. Technol. **16**(4), 834–840 (2008)
5. Tang, Z., White, J., Chiasson, J.: Closed-loop load balancing: comparison of a discrete event simulation with experiment. In: Proceedings of the 2005 American Control Conference, Portland, pp. 2721–2726 (2005)
6. Liu, X., Zhong, S.M., Ding, X.Y.: Robust exponential stability of impulsive switched systems with switching delays: a Razumikhin approach. Commun. Nonlinear Sci. Numer. Simul. **17**(4), 1805–1812 (2012)
7. Shim, H., Tanwani, A.: Hybrid-type observer design based on a sufficient condition for observability in switched nonlinear systems. Int. J. Robust Nonlinear Control **24**(6), 1064–1089 (2014)
8. Liu, X.W.: Robust H∞ filtering for switched discrete-time systems with time-delays. In: Proceedings of the 26th Chinese Control Conference, Zhangjiajie, pp. 660–664 (2007)
9. Liu, J., Liu, X.Z., Xie, W.C.: Delay-dependent robust control for uncertain switched systems with time-delay. Nonlinear Anal. Hybrid Syst. **2**(1), 81–95 (2008)
10. Egorov, V., Mondi, S.: Necessary conditions for the exponential stability of time-delay systems via the Lyapunov delay matrix. Nonlinear Anal. Hybrid Syst. **24**(16), 1760–1771 (2014)
11. Werbos, P.J.: Intelligence in the brain: a theory of how it works and how to build it. Neural Netw. **22**(3), 200–212 (2009)
12. Si, J., Barto, A.G., Powell, W.B., Wunsch, D.C.: Handbook of learning and approximate dynamic programming. IEEE Trans. Autom. Control **51**(10), 1730–1731 (2006)
13. He, H.-B., Ni, Z., Jian, F.: A three-network architecture for on-line learning and optimization based on adaptive dynamic programming. Neurocomputing **78**(1), 3–13 (2012)

Design Considerations for an Intelligent Video Surveillance System Using Cloud Computing

Kyung-Soo Lim(✉), Seoung-Hyeon Lee, Jong Wook Han, and Geon-Woo Kim

Information Security Research Division, Electronics and Telecommunications Research Institute, Daejeon 34129, Korea
{lukelim, duribun2, hanjw, kimgw}@etri.re.kr

Abstract. Recently, deep neural network and cloud computing based intelligent video surveillance technology are growing interests in the industrial and academia. The synergy with both technologies emerges as a key role of the public safety and video surveillance in the field. Reflecting these trends, we have been studying a cloud-based intelligent video analytic service using deep learning technology. INCUVAS (cloud-based INCUbating platform for Video Analytic Service) is a platform that continuously enhances the video analysis performance by updating real-time dataset with the deep neural network on a cloud environment. The goal of this cloud service can provide continuous performance enhancement and management using image dataset from the real environment. In this paper, we consider the design requirements for online deep learning based intelligent video analytics service.

Keywords: Video surveillance · Intelligent video analysis · Deep learning

1 Introduction

The interests in recent ICT services based on deep neural networks are currently higher than ever. Especially, the research on intelligent image analysis technology, which is one of the most used fields in Artificial Intelligent (AI), is underway actively. On the other hand, the cloud-based CCTV solutions such as VSaaS (Video Surveillance as a Service) are also being popularized by mitigating increases for effective video management of large-scale control systems. The VSaaS is a web-based video surveillance service that includes video recording, video storage, remote viewer, intelligent video analysis, and so forth.

VSaaS is a business model where those services are offered under the software-as-a-service model. For service providers, these various services of VSaaS are a model of continuous profit making, so they are making efforts to publicize and popularize them. Meanwhile, the global market for VSaaS is estimated to reach the US $1.7 billion by 2020 [1]. Currently, it has based on the cloud platform to manage resources and enable elasticity and flexibility nowadays. It means that the existing video management system (VMS) has moved to a cloud solution.

Cloud computing has become one of the most discussed IT paradigms of recent years. With cloud computing, organizations can consume shared computing and

J. H. Park et al. (Eds.): PDCAT 2018, CCIS 931, pp. 84–89, 2019.
https://doi.org/10.1007/978-981-13-5907-1_9

storage resources rather than building, operating and improving infrastructure on their own. Cloud computing enables organizations to obtain a flexible, secure, and cost-effective IT infrastructure, in much the same way that national electric grids enable homes and organizations to plug into a centrally managed, efficient, and cost-effective energy source [2]. The most famous cloud service platform is Amazon Web Service (AWS) that provides on-demand cloud computing platforms to individuals, companies, and governments with computing power, database, storage, content media transfer, and so forth [3].

On the other hand, NVIDIA has launched NVIDIA GPU CLOUD (NGC) that is concentrated on the cloud service with GPU-accelerated high-performance computing for deep learning. NGC provides a comprehensive GPU -accelerated cloud service including deep learning software, third-party HPC applications, and other partner applications [4].

Likewise, cloud computing and AI-based intelligent video surveillance solution have emerged as a core technology in the field of city safety, and the interest in industry and academia is increasing. Recently, the video surveillance system is evolving into a large-scale intelligent CCTV system that integrates with core ICT technology trends such as cloud, mobile, and big data analysis. For example, DAS (Domain Awareness System) developed by New York City and Microsoft in the United States has been extended to San Francisco, Los Angeles and Brazil with the proven use of intelligent video surveillance technology [5].

Despite emerging the all-mighty CCTV technology, the intelligent video analytics analytic (IVA) solution is not quite acceptable or competent in real-world situation. In other words, the lab-level algorithms for computer vision, even if it has developed with the proving the performance on the test set, had resulted in drastic performance degradation when it applied to real environments. Although the image recognition technology is developed for various environments, it cannot be identical with the camera scene, which is installed in the field. Therefore, there is a need for a technique that can directly apply the image analysis engine developed by performing deep learning (DL) with the scene of a camera installed in the field, on the spot.

On the other hand, domestic video surveillance companies are hard to invest and develop the deep learning based IVA technology because of low-level of R&D capability and the high price of hardware especially GP/GPU(General Purpose Graphic Processing Unit). Thus, they are simply transforming or utilizing the open source library of IVA. It means that the performance and reliability of their IVA will might be quite low to apply for the real environment. Therefore, it is necessary to support the platform or service by the government or public organization to nurture related industries to follow recent global trends.

Reflecting these trends, we have been researching a cloud-based intelligent video analytic service using deep learning technology. INCUVAS (cloud-based INCUbating platform for Video Analytic Service) is a platform that continuously enhances the video analysis performance by updating the real dataset with the deep neural network on a cloud environment, as shown in Fig. 1. A client uploads the images of the site to the cloud server on the INCUVAS; it creates the image analysis engine based on the uploaded image by deep learning technology and applies it to the field. The goal of this cloud service can provide continuous performance enhancement using image dataset

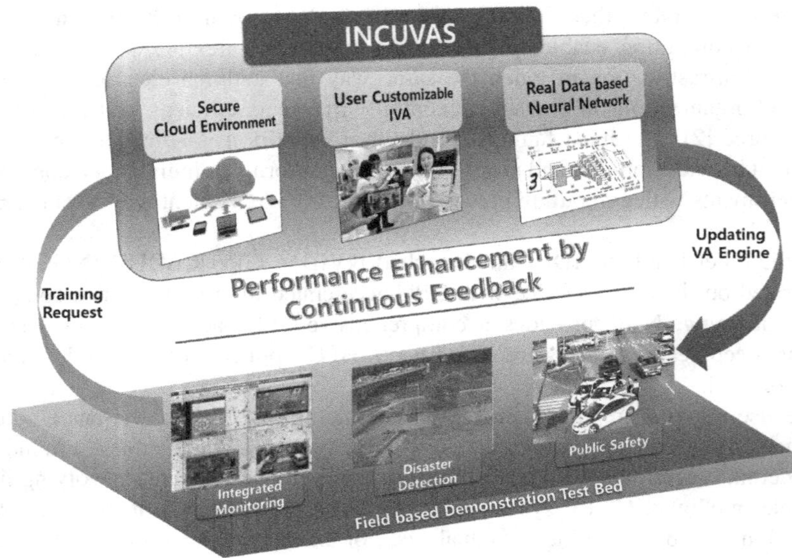

Fig. 1. The Goal of INCUVAS

from the real environment. In this paper, we consider the design consideration for online deep learning based intelligent video analytic service.

The biggest difference between INCUVAS and exiting cloud service are that INCUVAS trains and develops real datasets and it supports user-customized video analytic service, for example, face recognition, behavior analysis, and so forth [6, 7]. The organization for city safety, such as a police department, an integrated video surveillance center, who want an IVA with the proof guarantee for city safety. They can upload their video clips in the field and deploy the neural network engine after the learning and verification are completed.

2 Design Considerations for an Intelligent Video Surveillance System Using Cloud Computing

The definition of the incubator in this paper is one of the various forms of the virtualized machine (VM). It will be a VM instance based on the cloud platform or a container such as Docker. It will depend on the capability of the cloud cluster based on hardware specification. The INCUVAS assigns an incubator to the client who wants those services.

The purpose of ICUVAS provides cloud service, which is online deep learning from the uploaded dataset by the user to provide the most reliable and powerful intelligent video surveillance what he/she wants. For example, the client wants to launch the finding a missing child service in a national park. The client logged into INCUVAS server and chose the re-identification service for the missing child. Then, the, he/she uploads video or images in the scenes to the incubating cloud server. After the uploading completed, the allocated incubator will learn the object recognition and detection techniques based on the scene of the field. When the neural network engine is created, the client will download and apply it to the IVA system. If its performance of downloaded IVA does not yield as expected, the client will request and update more appropriate IVA engine on the cloud service of INCUVAS.

There are four core requirements for providing an intelligent video surveillance platform using cloud computing. First, a cloud-based incubating platform supports a technology that continuously enhances and optimizes the performance of deep learning based IVA by real-time updating on a cloud environment. It can apply to all AI-inspired services using neural network technology. However, the INCUVAS aims to provide the IVA service on video surveillance for city safety. Second, neural network based IVA should provide the DL-based face/vehicle recognition, traffic accident detection, and license plate detection by deep-resolution technology, which are to support automatically recognize the real-time situation. Third, INCUVAS should be able to construct the image database for the online deep learning based on the real surveillance dataset in the field. It includes the semiautomatic feature for generating the ground truth (GT) dataset and online learning technology on a cloud platform with interoperability. Finally, it should support the field-based demonstration for intelligence public safety to establish the infrastructure for the provision of intelligence physical security through the cooperation of the National Police Agency and the local government. Thus, the completion stage of INCUVAS can contribute the real-time traffic accident recognition, identifying crime suspects (people or vehicles), and deep resolution for the plate number for the vehicle.

- Provide a neural network engine for IVA
- Support cloud platform for handling various clients

Fig. 2. The Structure of Openstack based INCUVAS

- Provide online deep learning service distributing various IVA
- Create a Field-based image database and semiautomatic GT dataset
- Provide Field-based test bed demonstration service for the law enforcement and city safety

Currently, we have built a prototype of Openstack based cloud platform. The instance created on this prototype supports the neural network for IVA, which used public video surveillance dataset for training as shown Fig. 2. Among the many features of the Openstack project, Keystone-based user access control meets the INCUVAS requirements for managing various clients with security [8, 9].

3 Conclusion and Future Works

As we described above, we have been researching a cloud-based intelligent video analytic service using deep learning technology. We called INCUVAS (cloud based INCUbating platform for Video Analytic Service) is a platform that continuously enhances the video analysis performance by updating the real dataset with the deep neural network on a cloud environment. The biggest difference between INCUVAS and exiting cloud service are that it trains and develops real surveillance datasets and it supports user-customized video analytic service. Those dataset provided e organization for city safety, such as a police department, an integrated video surveillance center, who want an IVA with the proof guarantee for city safety. They can upload their video clips in the field and deploy the neural network engine after the learning and verification are completed. It collects images through cooperation with those government organizations and uses them as a training dataset for the IVA development. Furthermore, those image dataset maintains security and access control to prevent the personal privacy infringement.

On the other hand, it is currently being tested on a prototype of INCUVAS. However, our research is underway for the developing the infrastructure and techniques to manage the thousands of incubators in the future.

Acknowledgments. This work was supported by the Institute for Information & Communications Technology Promotion (IITP) grant funded by the Korea government (MSIP). (2017-0-00207, Development of Cloud-based Intelligent Video Security Incubator Platform).

References

1. The Global Video Surveillance As A Service (VSaaS)_ market, Global Industry Analysts, Inc. September 2015. http://www.strategyr.com/MarketResearch/Video_Surveillance_As_A_Service_VSaaS_Market_Trends.asp
2. Jackson, K. R., et al.: Performance analysis of high performance computing applications on the Amazon web services cloud. In: 2010 IEEE Second International Conference on Cloud Computing Technology and Science, Indianapolis, IN, pp. 159–168 (2010)
3. Jinesh, V., Sajee, M.: Overview of Amazon Web Service, Amazon Web Service. January 2014. https://aws.amazon.com

4. NVIDIA GPU Cloud. https://www.nvidia.com/en-us/gpu-cloud/
5. Domain Awareness System. https://en.wikipedia.org/wiki/Domain_Awareness_System
6. Oh, S.H., Han, S.W., Choi, B.S., et al.: Deep feature learning for person re-identification in a large-scale crowdsourced environment. J. Supercomput. **74**, 6753 (2018). https://doi.org/10.1007/s11227-017-2221-5
7. Oh, S.H., Kim, G.W., Lim, K.S.: Compact deep learned feature-based face recognition for visual internet of things. J. Supercomput. **74**, 6729 (2018). https://doi.org/10.1007/s11227-017-2198-0
8. Cui, B., Xi, T.: Security analysis of Openstack keystone. In: 2015 9th International Conference on Innovative Mobile and Internet Services in Ubiquitous Computing, Blumenau, pp. 283–288 (2015)
9. Keystone Security GAP and Threat Identification (Quick Study), OpenStack Folsom Release. https://wiki.openstack.org/w/images/c/c9/OpenStack_Keystone_Analysis.pdf

Investment Intention Towards Online Peer-to-Peer Platform: A Data Mining Approach Based on Perceived Value Theory

Xizi Wang(✉) and Li Wang

Shanghai University, Shanghai, China
katewang91@163.com, 1015811070@qq.com

Abstract. The financial industry has experienced a wide range of changes and innovations that bring from technology and Internet. The online Peer-to-Peer (P2P) lending platform is a relatively new phenomenon that thoroughly change the finance and e-commerce industries. In fact, in China the number of P2P online lending platforms has grown rapidly and will probably continue to increase over the next decade. However, researches in this field is still in its infant stage. Deeply understanding this business pattern has significant implications both theoretically and practically. We studied the investors' investment intention towards P2P platform based on perceived value theory. Unlike most empirical researches that employ questionnaire to collect data, this paper crawled data from WDZJ (www.wdzj.com) which is a third-party online loan information platform. We collected 517 platforms data to investigate the relationship between platform characteristics and investment intention. Results derived from data suggest that simply offering higher return rate is less attractive for investors. Those platforms which possess high registered capital and have successful financing history along with low risk are more favored by investors.

Keywords: Online lending platforms · Perceived value theory
Investment intention

1 Introduction

P2P lending platforms enable individual lenders make unsecured loans to other individual borrowers over the Internet without traditional middlemen [1, 2]. As a novel financing model, P2P has attracted increasingly attention since its emerging. Compared with traditional finance industry, P2P online lending platforms, which eliminate the involvement of financial intermediary, can raise funds quickly, increasing the efficiency of finance landscape. With a high return on investment and a low investment threshold, P2P lending platforms satisfy individuals' investment needs under the current economic situation [3].

Due to these benefits, a wide range of P2P platforms have come to appear over the last decade in China, such as RenrenDai, LuP2P, Jinyingmao and so forth. According to data from WDZJ (www.wdzj.com), by June 2018 there have been approximately 1800 P2P lending platforms in China. The latest monthly report shows that the overall volume of online lending turnover has exceeded 100 billion yuan. P2P lending market

J. H. Park et al. (Eds.): PDCAT 2018, CCIS 931, pp. 90–99, 2019.
https://doi.org/10.1007/978-981-13-5907-1_10

plays an important role in improving the efficiency of financial system and encouraging innovation and entrepreneurship.

Although P2P lending platforms has developed rapidly in terms of financing scale and diversity, it has not been deeply investigated. In the perspective of literature, previous studies mainly discuss this topic from two aspects, one is the determinants of P2P platform success [4–6], another relates to participant behaviors [7, 8]. Besides, most of previous studies examined lending intention towards different projects on an identical platform. Conversely, we considered several dimensions of P2P lending platform and focused on the question: which platform features are related to users' investment intention towards the platform. We wrote a program via Python to collect data from the Internet Lending Home (WDZJ). Multiple regression analysis was employed in this paper to test the research model which based on perceived value theory. To the best of our knowledge, this study is the first effort to apply it in the context of Chinese P2P lending market. Our will be of interest to both academia and industries. For researchers, we proved that platforms with good operating capability are more attractive to investors. For practitioners, the findings can serve as an instruction for platforms' continuous improvement and contribute to strengthening lenders' investment intention.

The remainder of the paper is organized as follows: Sect. 2 provides the theoretical foundation of P2P lending and perceived value. In Sect. 3, we propose our research model and hypotheses, followed by a summary of the results. In the last section, we conclude this paper by discussing the study's implications for research and practice.

2 Literature Review

2.1 Chinese P2P Lending Background

P2P platform is of great importance for Chinese small and micro-enterprise (SME) because of the lack of mature finance mechanism in most developing countries [9]. A considerable number of enterprises especially SMEs and individuals suffer from the repression of traditional finance mechanism in China. The limited private capital investment channels make this problem even severe. Therefore, the contemporary financial environment has resulted in a rapid advancement of Chinese P2P lending platforms.

There are some general challenges that need to be solved in P2P lending market. Risks such as information asymmetry risk, default risk and legal risk threat the healthy development of P2P lending market. Lin discussed the causes of default risk in the context of Chinese lending platform [10]. They also presented some characteristics of borrowers with low default risk. These features include female gender, young adults, long working time, stable marital status, high educational level, working in large company, low monthly payment, low loan amount, low debt to income and no default history. To mitigate the information asymmetry problem, suggestions like introducing "soft" information into P2P lending marketplace were proposed [11]. Healy and Palepu [3] suggested that the participation of third-party information mediators can also reduce the asymmetry of information.

In addition to the common problems mentioned above, Chinese lending platforms also face some other challenges. It has been found that some criminals use the P2P platform to deliberately defraud and launder money. According to the latest data from WDZJ (https://shuju.wdzj.com/problem-1.html), over 2000 P2P lending platforms were categorized as "problematic platforms" by June 2018. A majority of them were investigated for business termination or runaway. The fact is that Chinese investors should not only consider how to choose a right investment project but also a trustworthy P2P platform.

2.2 Related Studies on P2P Lending

2.2.1 Behavior Patterns

Behavior science, as the mainstream research of IS field, including topics like analyzing motivations of a particular behavior or behavior patterns in a particular context. Therefore, many IS scholars have worked on the behavior analysis on P2P lending platform. As P2P lending market belongs to crowdfunding community [12], in this section we reviewed the related behavior research either on the P2P lending platforms or on crowdfunding platforms.

Generally, there are four types of investors: active backers, trend followers, altruists, and the crowd [13]. Trend followers paid relatively less attention to the project creators' social capital and were more likely to invest in projects that were in later stage of the funding process. Similarly, [14] found that donor's support of the project is in line with the U-shaped model, which means that project fundraisers receive more support at the beginning and the end of the project. Mohammadi analyzed the data on Fundedbyme, an equity crowdfunding platform in Sweden [15]. Their study provided evidence of gender difference with investor behavior. Female investors are unlikely to invest in start-ups (such as high-tech companies). This may be due to female investors have a higher degree of risk aversion. Moreover, they found that projects with a higher proportion of male investors are more likely to attract female investors.

Behaviors like herding effects are relatively common in both P2P lending environment and crowdfunding context. Burtch [16] studied the dynamic of herding effects with the project process of financing. They found that herding effects decreased along with the project financing rate increased. Investors feel that their investment become less important, which is called crowding-out effect. Similarly, Brekovich confirmed the impact of herding effects by analyzing data from Prosper. They also emphasized the significance of hard facts and soft facts in explaining investors' decisions [17]. Lee also found the strong evidence of herding behavior on Popfunding, one of the biggest P2P lending platforms in South Korea [8].

2.2.2 Determinants of Lending Intention

About what factors affecting lending intention, prior scholars mainly consider individual's lending intention towards a particular project. Thus, individual information and trust relationship between borrowers and lenders were extensively employed in prior literature. For example, Wang's research suggested that the lender's trust of the borrower had impact on lending willingness [18]. Larrimore investigated how to establish and increase trust in P2P lending interactions [19]. According to their work, extending the length of loan description as well as more narrative and concrete description were effective to increase trust and the likelihood of funding success. Herzenstein found that the borrower's debt-to-income ratio has a negative effect on bidding [20]. Chen investigated the factors affecting online lending intention in Chinese P2P environment [9]. They argued that trust is the most significant factor influencing individuals' lending intention. While trust is affected by borrowers' perceived information quality, perceived social capital, and perceived risk. Their conclusions are consistent with prior studies which focused on lending behavior.

2.2.3 Perceived Value Theory

It has been acknowledged that online users' value perceptions steer their purchase decisions [21]. Perceived value, which has been extensively applied to behavior studies, are strong correlated to customer loyalty, satisfaction and continuance intention. Parasuraman argued that perceived value is one of the most important factor for service providers to gain competitive advantage [22]. Given the fact that many Chinese P2P platforms compete with each other, offering similar rewards. It's imperative to help platforms survive in the fierce competition. In the e-commerce context, [23] verified that product price, perceived product quality, and valence of experience have impact on perceived customer value. While perceived value is an effective predictor of customer purchase intention. Motivated by their work, we assume that users' lending intentions towards platform are influenced by the features of the platform. Despite of the valuable findings in prior work, rare studies have highlighted the relationship between investors' lending intention with platform attributes. Our paper fills this knowledge gap.

3 Research Method

3.1 Research Model and Hypotheses

Young Kim studied customers purchase intention by identifying the dimensions of online shopping attributes [24]. They used eight attributes, such as credit card security, fast delivery time etc., to represent the transaction factor. Other factors included incentive programs, site design and interactivity. While the impact of site design on purchase intention were not supported, other factors were proved as important predictors of purchase intentions concerning clothes. Besides, results of [23] provided evidence of a highly significant positive relationship between perceived customer value and online purchase intention. They summarized some key factors with a particular e-store, such as customer service, ease of use of the website, e-retailer's reputation. As investor's lending intention concerning a platform resemble to purchase intention towards a particular e-store. We thus examined lending intention towards platform by four factors: perceived risk, interest return, experience of service, and operational performance. The research model is shown in Fig. 1.

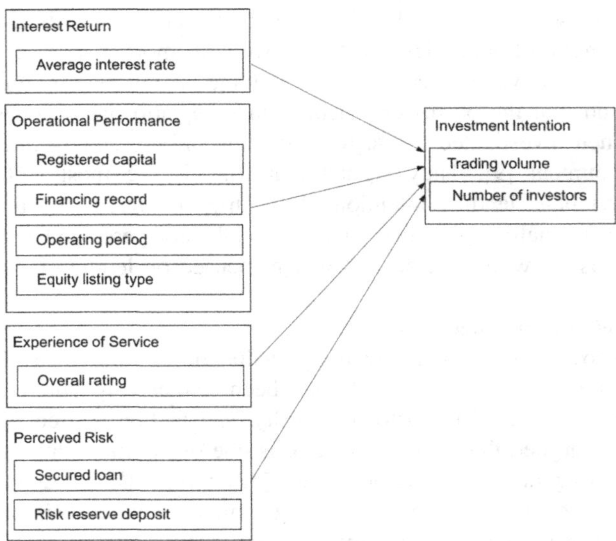

Fig. 1. Research model

The dependent variable (investment intention) were composed of two attributes: trading volume and the number of lenders. The first factor (perceived risk) included secured loan and risk reserve deposit. The experience of service was measured by the overall rating, which derived from individuals' rating towards each platform. We used average interest rate of the platform to represent platform interest return. The platform operational performance was associated with four attributes, including registered capital, financing record, operating period, equity listing type. The description of each attribute can be found in Table 1.

Some P2P lending platforms serve as comprehensive financial platforms. They not only offer P2P lending project, but also providing users with investment projects. The interest rate of the former project type is determined either through an auction or by the platform, while interest rate of investment projects is usually decided by the platform. Previous literature argued that investors were easily lured by P2P platform's high interest rate [25]. This implies the following hypothesis:

H1: Interest return is positively associated with investment intention towards platform.

[26] suggested that customer's perception of brand have positively impact on purchase intention. For the same product, customers prefer to purchase them from brand that has high reputation. [27] proved that perceived reputation had significant impact on users' initial trust which was a positive determinant of users' initial investing intention. The operational indicators such as listing company and remarkable financial record are conductive to long-term reputation establishment. So, our hypothesis is proposed as follows:

Table 1. Descriptions of measurements

Factors	Attributes	Interpretation
Investment intention	Trading volume (TV)	Total transaction amount on a particular platform
	Number of lenders (NL)	The accumulated number of people who participated in transactions within the latest month
Interest return	Average interest rate (AIR)	The average number of interest rate reference (weighted by transaction amount)
Operational performance	Registered capital (RC)	Legal registered capital
	Financing record (FR)	Discrete variable The platform has financing record was marked as 1. Otherwise, it was 0
	Operating period (OP)	Operating period since official launched online
	Equity listing type (ELT)	Discrete variable There are six listing types: unlisted, new third board, growth enterprise market, small and medium enterprise board, main board, and international listing, marking as 1–6 accordingly
Experience of service	Overall rating (OR)	Users (including the third-party users and users from a particular platform) rating towards the platform
Perceived risk	Secured loan (SL)	Discrete variable Platforms that guarantee the secured loan was marked as 1. Otherwise, it was 0
	Risk reserve deposit (RRD)	Discrete variable Platform that has deposit system was marked as 1. Otherwise, it was 0

H2: Operational performance is positively associated with investment intention towards platform.

When investor faces the same benefit and risk, an important consideration for them to choose a platform is the service provided by the platform. Customer's behavior research reckoned that shopping service affects purchases intention. We thus posit the following hypothesis:

H3: Experience of service is positively associated with investment intention towards platform.

Literature in economics suggest that individuals' risk appetite can be divided into risk preference, risk neutral and risk aversion [27]. According to Featherman's definition, perceived risk refers to the potential for loss in the pursuit of a desired outcome of using an e-service [28]. It was considered as an inhibitor to purchase behavior [29]. Platforms with risk reserve systems can secure investors' capital, which probably improve individuals' investment intention. Therefore, our hypothesis is proposed as follows:

H4: Perceived risk is negatively associated with investment intention towards platform.

3.2 Data Analysis

The Internet Lending Home (http://www.wdzj.com) is one of the largest and most authoritative third-party platforms that provides online P2P platform information and loan information. We crawled 517 P2P platforms data from Internet Lending Home. Multiple regression approach is employed for data analysis. Corresponding equations areas follows:

$$Y_1 = \beta_0 + \beta_1 X_1 + \beta_2 X_2 + \cdots + \beta_8 X_8 + \varepsilon$$
$$Y_2 = \beta_0 + \beta_1 X_1 + \beta_2 X_2 + \ldots + \beta_8 X_8 + \varepsilon$$

Where Y_1 represents the trading volume (TV), and Y2 refers to the number of lenders (NL).

The total 5,170 platform features were collected. Table 2 gives the result of descriptive statistics.

Table 2. Descriptive statistics

	Min	Max	Mean	SD
TV	5.20	1092935.11	31098.48	95963.39
NL	.000	301075.0	7662.82	26015.35
AIR	4.84%	21.04%	10.885%	2.783%
RC	76	315804	7719.54	25197.073
FR	0	1.0	.211	.408
OP	4.42	8.20	6.846	.412
ELT	1	6	1.46	1.264
OR	0	4.8	3.233	1.311
SL	0	1	.61	.488
RRD	0	1	.11	.318

Table 3 shows the correlation coefficient and significance level of each variable. The correlation coefficient between the independent variables is less than 0.7, so there is no multicollinearity among variables.

Table 3. Correlation analysis results

	AIR	RC	FR	OP	ELT	OR	SL	RRD
AIR	1							
RC	−.172**	1						
FR	−.112**	.024	1					
OP	.088*	.176**	.139**	1				
ELT	−.035	.138**	.112*	.067	1			
OR	−.041	.038	.081	.105*	.052	1		
SL	−.174**	.056	.072	.079	.074	.156**	1	
RRD	−.080	.118**	.068	.104*	.071	.086	.124**	1

Significance level *P < 0.1, **P < 0.05, ***P < 0.01

3.3 Result Discussion

We applied multiple regression analysis on trading volume and the number of lenders respectively. The R square of trading volume is 39.2%, the adjusted R square is 38.2%. While the R square of the regression analysis on the number of lenders is 34.1%, and the adjusted R square is 33.0%. Table 4 summarized regression results.

Table 4. Regression results

	Interest return	Operational performance				Experience of service	Perceived risk	
Attributes	AIR	RC	FR	OP	ELT	OR	SL	RRD
TV	−.265**	.238**	.213**	.223**	.118**	.110**	0.032	.125**
NL	−.162**	.235**	.216**	.165**	.093*	.164**	0.048	.172**

Significance level *P < 0.1, **P < 0.05, ***P < 0.01

The results show that average interest rate is negatively correlated with the trading volume and the number of lenders, rejecting H1. A possible explanation for this result may be that high interest rate are usually accompanied with high risks. The higher interest rate the platform offers, the higher level of risk the customers perceive, decreasing their lending intention. In fact, the correlation results in Table 3 reveals that the platform's interest rate is negatively correlated with the secured loan and risk reserve deposit, proving evidence for the explanation.

In the dimension of platform operational performance, all attributes show great impact on both trading volume and number of lenders, supporting H2. This is consistent with prior study that claim the significant influence of platform reputation on users' initial investing intention [27]. Overall rating was also found to have significantly influence on both trading volume and number of lenders. As such, H3 is supported.

Regarding the two attributes in perceived risk, risk reserve deposit shows strong relationship with trading volume as well as number of lenders. However, the direct influence from secured loan to lending intention was not significant. As such, H4 is partially supported.

4 Conclusion

In this study, we analyzed the data from WDZJ through multiple regression approach to examine the factors affecting lenders' lending intention towards P2P lending platform. Results of our study show that the significant factors are interest rate, overall rating, risk reserve deposit and operational performance. Contrary to our hypotheses, the impact of secured loan on investment intention towards platform is not significant.

Since the overall interest rate of P2P marketplace is higher than the traditional investment channels, investors pay more attention to the risk of investment. They are likely to choose projects with relatively low risks, revealing that most investors are in risk neutral.

This work also provides several insights for P2P industry. First of all, improving platform operational performance. Investors believe that platform with strong operating capability and high reputation offers better lending and investment projects, reducing the probability of investment failure. P2P Platforms can improve their operating capability by increasing the amount of registered capital, conducting successful financing activities, etc. Secondly, providing better service. P2P online lending is a web-based service. An excellent experience of service, coupled with platform's high reputation in the P2P community, can enhance investors' investment intention towards the platform. Efforts, such as offering a wide variety of investment projects and facilitating the lending process, are highly encouraged. Moreover, it's necessary to reduce users' perceptions of investment risks. For customers, there is a trade-off between risk and return rate. Our findings suggest that most of them prefer relatively low risks regardless of the interest rate. Building risk reserve deposit system along with cooperation with third-party evaluation institutions can contribute to improving lending intention.

References

1. Lin, M., Prabhala, N.R., Viswanathan, S.: Judging borrowers by the company they keep: friendship networks and information asymmetry in online peer-to-peer lending. Manage. Sci. **59**(1), 17–35 (2013)
2. Chen, D., Han, C.: A comparative study of online P2P lending in the USA and China. J. Internet Bank. Commer. **17**(2), 12–15 (1970)
3. Healy, P.M., Palepu, K.G.: Information asymmetry, corporate disclosure, and the capital markets: a review of the empirical disclosure literature. J. Account. Econ. **31**(1–3), 405–440 (2001)
4. Ceyhan, S., Shi, X., Leskovec, J.: Dynamics of bidding in a P2P lending service: effects of herding and predicting loan success. In: Proceedings of the 20th International Conference on World Wide Web, pp. 547–556. ACM (2011)
5. Wen, X., Zhang, Z., Wu, X.: A research on the influence factors of P2P lending market (2017)
6. Wang, M., Wang, T., Kang, M., Sun, S.: Understanding perceived platform trust and institutional risk in peer-to-peer lending platforms from cognition-based and affect-based perspectives. In: PACIS, p. 208 (2014)
7. Gao, X., Liu, Y., Gao, Y., Zhao, J., Yang, F., Wang, T.: Understanding the role of commitments in explaining P2P lending investing willingness: antecedents and consequences. In: PACIS, p. 170 (2015)
8. Lee, E., Lee, B.: Herding behavior in online P2P lending: an empirical investigation. Electron. Commer. Res. Appl. **11**(5), 495–503 (2012)
9. Chen, D., Hao, L., Van Slyke, C.: Toward an understanding of online lending intentions: evidence from a survey in China. In: CAIS 36, p. 17 (2015)
10. Lin, X., Li, X., Zheng, Z.: Evaluating borrower's default risk in peer-to-peer lending: evidence from a lending platform in China. Appl. Econ. **49**(35), 3538–3545 (2017)
11. Xu, Y., Luo, C., Chen, D., Zheng, H.: What influences the market outcome of online P2P lending marketplace?: A cross-country analysis. J. Glob. Inf. Manag. (JGIM) **23**(3), 23–40 (2015)

12. Borello, G., De Crescenzo, V., Pichler, F.: The funding gap and the role of financial return crowdfunding: some evidence from European platforms. J. Internet Bank. Commer. **20**(1), 1–20 (1970)
13. Lin, Y., Boh, W.F., Goh, K.H., How different are crowdfunders? Examining archetypes of crowdfunders and their choice of projects. In: Academy of Management Proceedings, vol. 2014, no. 1, p. 13309. Academy of Management, Briarcliff Manor (2014)
14. Kuppuswamy, V., Bayus, B.L.: Crowdfunding creative ideas: the dynamics of project backers. In: Cumming, D., Hornuf, L. (eds.) The Economics of Crowdfunding, pp. 151–182. Springer, Cham (2018). https://doi.org/10.1007/978-3-319-66119-3_8
15. Mohammadi, A., Shafi, K.: Gender differences in the contribution patterns of equity-crowdfunding investors. Small Bus. Econ. **50**(2), 275–287 (2018)
16. Burtch, G., Ghose, A., Wattal, S.: An empirical examination of the antecedents and consequences of contribution patterns in crowd-funded markets. Inf. Syst. Res. **24**(3), 499–519 (2013)
17. Berkovich, E.: Search and herding effects in peer-to-peer lending: evidence from prosper. com. Ann. Finance **7**(3), 389–405 (2011)
18. Wang, P., Zheng, H., Chen, D., Ding, L.: Exploring the critical factors influencing online lending intentions. Financ. Innov. **1**(1), 8 (2015)
19. Larrimore, L., Jiang, L., Larrimore, J., Markowitz, D., Gorski, S.: Peer to peer lending: the relationship between language features, trustworthiness, and persuasion success. J. Appl. Commun. Res. **39**(1), 19–37 (2011)
20. Herzenstein, M., Dholakia, U.M., Andrews, R.L.: Strategic herding behavior in peer-to-peer loan auctions. J. Interact. Mark. **25**(1), 27–36 (2011)
21. Pura, M.: Linking perceived value and loyalty in location-based mobile services. Manag. Ser. Qual. Int. J. **15**(6), 509–538 (2005)
22. Parasuraman, A.: Reflections on gaining competitive advantage through customer value. J. Acad. Mark. Sci. **25**(2), 154 (1997)
23. Chen, Z., Dubinsky, A.J.: A conceptual model of perceived customer value in e-commerce: a preliminary investigation. Psychol. Mark. **20**(4), 323–347 (2003)
24. Young Kim, E., Kim, Y.K.: Predicting online purchase intentions for clothing products. Eur. J. Mark. **38**(7), 883–897 (2004)
25. Mild, A., Waitz, M., Wöckl, J.: How low can you go?—Overcoming the inability of lenders to set proper interest rates on unsecured peer-to-peer lending markets. J. Bus. Res. **68**(6), 1291–1305 (2015)
26. Aaker, J.L., Garbinsky, E.N., Vohs, K.D.: Cultivating admiration in brands: Warmth, competence, and landing in the "golden quadrant". J. Consum. Psychol. **22**(2), 191–194 (2012)
27. Li, J., Zheng, H., Kang, M., Wang, T., Chen, S.: Understanding investment intention towards P2P lending: an empirical study. In: PACIS, p. 82 (2016)
28. Featherman, M.S., Pavlou, P.A.: Predicting e-services adoption: a perceived risk facets perspective. Int. J. Hum. Comput. Stud. **59**(4), 451–474 (2003)
29. Peter, J.P., Ryan, M.J.: An investigation of perceived risk at the brand level. J. Mark. Res., 184–188 (1976)

Pipeline Patterns on Top of Task-Based Runtimes

Enes Bajrovic(✉), Siegfried Benkner, and Jiri Dokulil

Faculty of Computer Science, University of Vienna, Vienna, Austria
{enes.bajrovic, siegfried.benkner, jiri.dokulil}@par.univie.ac.at

Abstract. Task-based runtime systems have gained a lot of interest in recent years since they support separating the specification of parallel computations from the concrete mapping onto a parallel architecture. This separation of concerns is considered key to coping with the increased complexity, performance variability, and heterogeneity of future parallel systems and to facilitating portability of applications across different architectures. In this paper we present our work on a programming framework that enables the expression of pipeline patterns at a high-level of abstraction by adding pragma directives to sequential C++ codes. Such high-level abstractions are then transformed to a runtime coordination layer, which utilizes different task-based runtime systems including StarPU and OCR to realize efficient parallel execution on single-node multi-core architectures. We describe the major aspects of our approach for mapping pipeline patterns to task-based runtimes and present experimental results for a real-world face-detection application indicating that a performance competitive with low-level programming approaches can be achieved.

Keywords: Parallel programming · Runtime systems · Multicore architectures

1 Introduction

Many recent research efforts have focused on the use of dynamic task-based runtime systems (e.g., StarPU [1], HPX [2], OCR [3]) to realize efficient applications on increasingly complex parallel architectures. Commonly, these approaches represent an application as a directed acyclic graph (DAG) where nodes represent computational tasks, and edges represent dependences between tasks. From such a representation, a runtime system can dynamically determine tasks that can be executed in parallel and use sophisticated scheduling algorithms to map these tasks to available execution units of the target system. With such an approach, programmers usually only specify what can be potentially executed in parallel, while deferring to the runtime how parallel execution is organized on a specific parallel architecture. Separating the specification of parallelism from its implementation enables the runtime system to better adapt an application to a specific parallel architecture taking into account changing performance characteristics of execution units or dynamically changing workloads.

While task-based runtime systems offer the required flexibility to deal with emerging and future architectures, they are often at a very low level of abstraction and

J. H. Park et al. (Eds.): PDCAT 2018, CCIS 931, pp. 100–110, 2019.
https://doi.org/10.1007/978-981-13-5907-1_11

consequently very complex and difficult to use directly for application development. Usually programmers are required to rewrite their applications in a task-parallel way, identifying the individual tasks, describing dependences between them and selecting a suitable scheduling strategy to achieve efficient execution.

In our work we aim at raising the level of abstraction when dealing with task-parallel programming of modern hardware. We have developed a programming framework that supports expression of high-level pipeline patterns within sequential C/C++ codes while automatically generating a low-level task-parallel program for efficient execution on different parallel architectures using either the StarPU or OCR runtime system. Initially we have developed the support for high-level pipeline patterns within the European PEPPHER project [4, 15], which relied on a direct transformation of annotated C++ code to the StarPU runtime system. In our recent work we have redesigned our system to support different low-level task-based runtime systems through the introduction of an intermediate runtime coordination layer (RCL) which decouples our programming system from the concrete underlying low-level runtime.

Specifically this paper makes the following contributions: (1) a programming system that maps high-level pipeline patterns to a task-based runtime system hiding the details of parallel execution from the user while providing some means for influencing the degree of parallelism; (2) an intermediate runtime coordination layer (RCL) that utilizes application-specific knowledge to control how tasks are submitted to a lower-level runtime system; (3) a discussion of the major issues in targeting different runtime systems like StarPU and OCR from the RCL and the major differences between those runtimes; and (4) an experimental evaluation of our framework using a real-world face-detection application showing that performance comparable to much more complex, low-level programming approaches can be achieved.

The remainder of the paper is organized as follows: Sect. 2 discusses related work. Section 3 provides an overview of our programming framework and outlines pipeline patterns. Section 4 describes the runtime coordination layer and presents the main issues in targeting StarPU and OCR, respectively. Section 5 presents experimental results followed by a conclusion and discussion of future directions in Sect. 6.

2 Related Work

Addressing the challenges of programming increasingly complex parallel architectures, a variety of dynamic task-based runtime systems and programming approaches have been developed [1–3, 5–10]. We discuss some of these systems below.

StarPU [1] is an asynchronous task-based runtime system and programming library for heterogeneous systems. StarPU introduces *codelets* that abstract *tasks* and allow programmers to provide different implementation variants of a computational kernel (e.g. for CPUs or GPUs). Moreover, StarPU provides data management facilities that automate the data transfer between processing units. StarPU automatically determines data dependences by analyzing task arguments and supports various scheduling strategies for deciding when and where to execute tasks taking into account historic performance data, resource readiness, and data availability.

The Open Community Runtime (OCR) [3] is an open specification for a low-level, event-driven, asynchronous, task-based runtime system designed for extreme scale parallel computing. A parallel application in OCR is expressed as a DAG where nodes represent tasks (or events) and edges dependences. Event-driven tasks (EDTs) are decoupled from their actual loci of execution, which are determined dynamically by the runtime system. Events are used to control the execution and synchronization of tasks. The data in OCR is organized in *data* blocks, which describe data independently from their actual physical memory location. This allows OCR to dynamically relocate tasks and data in order to improve locality, optimize performance and support fault tolerance. As opposed to StarPU, which automatically infers dependences between tasks, in OCR all dependences between tasks need to be explicitly specified.

OmpSs [5] extends OpenMP with support for asynchronous task-parallel execution. The OmpSs memory model assumes a shared address space to enable automatic data movement across the system. It provides support for multiple implementation variants of tasks when targeting heterogeneous systems. Many features introduced in OmpSs have been adopted by the OpenMP standard [6].

Legion [7] is a data-centric, task-based programming system that supports dynamic hierarchical data partitioning based on the concept of logical regions. Tasks are bound to regions and may access regions with different privileges and subject to different data coherence modes. Legion provides a mapping interface that enables programmers to control the mapping of tasks and data regions in order to optimize an application for a specific architecture.

The HPX runtime system [2] is a C++-based implementation of a task-based programming model within an active global address space, which supports migration of objects between the nodes of clusters.

Most of the mentioned runtime systems are being used directly by application programmers, leading to a low-level programming methodology similar or even more complex than the dominating MPI model. While for some runtime systems also high-level approaches exist (e.g., the Regent language targeting Legion, or OpenMP tasking extensions), none of the discussed runtimes currently supports high-level patterns comparable to our pipeline patterns.

Thread Building Blocks (TBB) [8] is a task-based C++ template library. It provides support for parallel patterns (including pipelines) and concurrent data structures that hide some of the complexity of parallel programming. TBB's task-based approach utilizes work-stealing task schedulers to map tasks and perform load balancing among processing resources. While TBB is a template library and provides an integrated scheduler, our approach relies on annotating sequential C++ codes and allows targeting different low-level runtime systems transparently.

3 High-Level Parallel Pipeline Patterns

We provide a set of preprocessing directives that enable expression of pipeline patterns at a high level of abstraction in sequential C/C++ codes and a transformation system and runtime layer for executing pipelines on top of task-parallel runtime systems.

3.1 Overview of the Programming Framework

Figure 1 provides an overview of our programming framework. We support applications written in C/C++ with extensions for high-level pipeline patterns. A source-to-source compiler translates sequential C++ code with pipeline patterns to an explicitly parallel representation that utilizes the *Runtime Coordination Layer* (RCL) for managing execution on top of a task-based runtime system (OCR or StarPU).

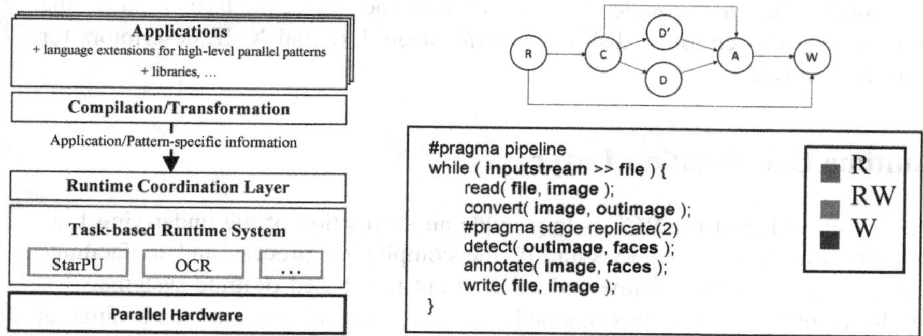

Fig. 1. Programming framework (left), face detection pipeline (right)

Currently our framework targets single-node multicore systems. Support for heterogeneous architectures (e.g., CPU/GPU systems) is currently restricted to the StarPU runtime. The framework also provides hooks for interfacing it with an external autotuner [16] to support tuning of application-specific parameters like stage replication factors and runtime-specific parameters like the number of worker threads or the scheduling strategy [11].

3.2 High-Level Pipeline Patterns

A pipeline consists of several interconnected stages, where data items flowing through the pipeline are processed at every stage. Usually, stages are connected via buffers from which stages access the data. While buffered pipelines require additional memory, they allow to decouple stages, to mitigate relative performance differences, and to better control parallel execution and pipeline throughput. In our approach buffers between stages are hidden from the user and generated automatically by the compiler.

A pipeline is constructed with the *pipeline* pragma directive from *while-loops*. The *stage* directive is used to indicate that one or more function calls correspond to a pipeline stage and may be omitted if a stage contains only a single function call. For each stage function multiple implementation variants may be provided (e.g., a CPU or a GPU variant). An external XML-based descriptor has to be provided for each stage function comprising its interface specification, the intent of function arguments (*read, write, readwrite*), and information about available implementation variants.

Within the stage directive a *replication factor R* may be specified, indicating that *R* instances of a stage should be generated, operating in parallel on different data items. The specification of replication factors enables users to control the degree of parallelism. The stage directive may also be used to merge multiple function calls into a single stage in order to increase the computational granularity of stages.

Figure 1 shows a face detection pipeline. Images are read from disk (*read* stage), converted to a grayscale (*convert* stage) and analyzed for human faces (*detect* stage). The detect stage, which is the computationally most intensive one, is replicated 2 times. In the *annotate* stage a rectangle is drawn around each detected face. Finally, the resulting image is written to the disk in the *write* stage. External XML descriptors for stages are not shown.

4 Runtime Coordination Layer

The primary objective of the RCL is to provide an abstraction of the underlying low-level runtime system in order to simplify the compilation process and to facilitate retargeting the programming framework to different task-based runtime systems.

The RCL utilizes an object-oriented representation of patterns and aims at exploiting pattern-specific knowledge available in the high-level code in order to optimize task-parallel execution. The main objectives of RCL are: (1) to manage the details of submitting tasks to the low-level runtimes, (2) to manage data transfers between pipeline stages, and (3) to control the degree of parallelism by restricting the number of tasks concurrently submitted to the low-level runtime.

4.1 Representation of Pipeline Patterns in RCL

During the transformation process, the compiler analyzes each pipeline and generates code that constructs a corresponding object-oriented representation of the pipeline. The compiler automatically derives the stage interconnections by analyzing the data flow between stages and the intent (*read*, *write*, or *readwrite*) of the arguments of each stage function. Stage interconnections provide the basis for automatically generating buffers between stages and for setting up explicit task dependences as required by OCR.

For each type of data item that is being processed in a pipeline, a *data descriptor* is generated, comprising the details of how these items are represented (e.g., size and location in memory) and managed. From these data descriptors corresponding data handling code for the different low-level runtimes is generated.

Similarly, since for each invocation of a stage function a task needs to be submitted to the low-level runtime system, the RCL generates corresponding *task descriptors*. A *task descriptor* comprises all information needed to create a concrete task for a low-level runtime system. Besides information about data consumed and produced by a pipeline stage, the task descriptor captures the stage dependences, and contains pointers to one or more implementation variants of the stage function. Task descriptors are then used by RCL classes to dynamically generate tasks for the concrete underlying runtime system.

4.2 Execution of Pipelines with RCL

The compiler generates for each pipeline RCL code that executes the pipeline in an asynchronous fashion. For each pipeline stage the following *stage mechanism* is executed by a so-called *runner*: (1) if all required input data of the stage are available, the data are acquired from the corresponding buffers; otherwise the stage mechanism waits (sleeps) until data becomes available; (2) using the information in the associated task descriptor, a task is generated and submitted asynchronously to the underlying runtime system (StarPU or OCR); (3) when the task finishes execution, the runner is notified, resumes execution, and the generated output is pushed to the corresponding output buffer(s). If the output buffer of a stage is full, the runner will wait (sleep) until there is a free slot in the target output buffer.

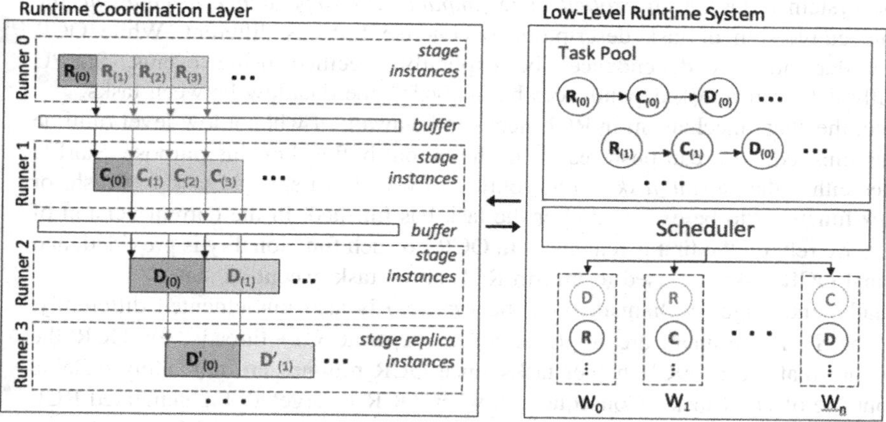

Fig. 2. RCL and low-level runtime system

Figure 2 sketches how the first three stages of the face detection pipeline are managed by the RCL. For each stage, the stage mechanism is executed by a *runner*, submitting for each instance of a stage a task to the low-level runtime system. In case of a replicated stage (e.g., the *detect* stage), a runner is generated for each stage replica. The low-level runtime system schedules the execution of tasks to worker threads.

The number of tasks submitted to the low-level runtime at the same time (and potentially running in parallel) is controlled by stage replication factors and buffer sizes. If a stage is replicated *R* times, then *R* stage objects are generated, executing the above-mentioned stage mechanism in parallel. Consequently, for replicated stages tasks are submitted in parallel to the underlying runtime system. As all replicas of a stage share input and output buffers, buffers have been implemented such that they can be accessed efficiently in parallel while maintaining ordering of data items to preserve the original semantics of the application. RCL buffers have been implemented on top of concurrent queues supporting *fifo*, *filo*, and *priority* ordering.

Another way to control the degree of parallelism is via the sizes of buffers. If the output buffer of a stage is full, the stage will wait (sleep) until there is a free slot in the target output buffer. Consequently, the size of buffers not only determines the memory footprint, but also the number of tasks that are being submitted to the runtime.

For each type of data item processed corresponding data representations for the low-level runtime systems are derived based on the associated data descriptor. Both OCR and StarPU require that data consumed or produced by a task is contiguous in memory and that the access mode (intent) is explicitly specified (*read, write, readwrite*). We avoid explicit copying of data and handle all data transfers between stages only by means of manipulating pointers (i.e., data items in buffers are represented by pointers to corresponding data descriptors).

The RCL provides special classes for dynamically generating tasks from task descriptors associated with the stages. Depending on whether the targeted low-level runtime system relies on an *explicit* or an *implicit task dependence mechanism*, the dynamic conversion of task descriptors to concrete tasks is different. While OCR requires that all task dependences be explicitly specified using events, StarPU implicitly determines task dependences by analyzing the dataflow between tasks.

Also, the stage mechanism in RCL needs to be aware of when a low-level runtime task has finished, so it can push results to the output buffer. For this purpose, StarPU provides either the *starpu_task_wait()* routine to wait for a specific task to finish, or *callback* functions to be executed after the task has finished. In the current version of the RCL we rely on the first mechanism. In OCR no such functionality is provided, and additional OCR tasks are used to inform RCL about task execution state.

Finally, the stage mechanism (i.e., the *runners)* is also implemented differently. While for StarPU runners are implemented as separate C++ threads, for OCR the runners are realized as OCR helper tasks since OCR routines are not safely callable from outside of OCR tasks. Consequently, when OCR is targeted, the generated RCL program will start with the OCR *mainEdt()* and not with the standard C++ *main()*.

4.3 Task Generation for StarPU

For each instance of a stage, RCL generates a task for StarPU. A StarPU task is described by a *codelet* which usually comprises a pointer to the stage function, a set of handles for all data items consumed and produced, and information about performance models used for scheduling decisions.

In order to enable StarPU to analyze data dependences between tasks, all the input and output data of a kernel has to be registered with StarPU using data handles generated from RCL data descriptors. If the access mode of a stage argument is *read* or *readwrite,* RCL passes a pointer to the argument to StarPU, otherwise it acquires a pointer from StarPU. Registering data also allows StarPU to automatically manage data transfers across different execution units, e.g., from CPU to GPU memory.

For each stage the RCL uses a separate C++ thread (runner) to execute the stage mechanism. Once all input data are available, data handles are registered with StarPU and the generated task is submitted asynchronously to StarPU by calling *starpu_task_submit()* followed by *starpu_task_wait()* to wait until the task has finished. Upon task completion, input data of the task are deregistered and output data pushed to the output buffer(s) allowing subsequent stages to continue executing.

The StarPU runtime system provides different scheduling strategies and for the experiments reported in this paper we have used the HEFT scheduling strategy [1], which relies on performance models built from historic execution data of tasks.

4.4 Task Generation for OCR

The OCR task-parallel model relies on *event driven tasks* (*EDTs*), which are connected via *events* with other tasks, to represent a parallel application as a task graph (DAG) that is dynamically scheduled for parallel execution by the runtime system. EDTs operate on contiguous, relocatable blocks of data, called *data blocks*, which are managed by the runtime system. An EDT may have multiple pre-slots for input data and one post-slot for the output. Instances of EDTs are created from EDT templates, which comprise information about the function an EDT executes, its parameters, and its dependences.

As opposed to StarPU where dependences between tasks are automatically determined, OCR requires all dependences to be explicitly specified. Once all data blocks an EDT depends on are available, a task becomes runnable and will eventually be executed on some execution unit selected by the runtime and run to completion regardless of the behavior of other tasks. The RCL provides the *OCRRuntime* class that comprises a set of EDT templates used to orchestrate the execution of pipelines, as well as for calling *ocrShutodown()* when an application terminates.

Due to restrictions of OCR (OCR owns the *main()*) the generated RCL program does not start with the standard main function but with the OCR *mainEdt()* function. Also, because calling OCR routines from outside of OCR tasks is not possible, we utilize for each stage instance an OCR *helper task* to run the stage mechanism and to submit tasks for executing the stage function. The helper task creates OCR data blocks for all data items (by calling *ocrDbCreate()*), an *executor* EDT (*ocrEdtCreate()*) to execute the stage function, and it sets up dependences (*ocrAddDependence()*) between the data blocks and the *executor EDT*. It then calls corresponding OCR routines to satisfy all the dependences of the *executor EDT*, which then becomes runnable and is eventually run by the OCR runtime executing the stage function. Upon completion, another *helper* task will be generated managing the execution of the next stage instance.

5 Evaluation

For evaluation we use the face detection application outlined previously in Fig. 2. The application utilizes routines from the OpenCV library [12], which have been slightly reengineered to conform to the tasking model. For the measurements, the application processes a set of 500 images of 360p resolution, each containing an arbitrary number of faces.

We present speedup measurements on a machine with 2 octa-core Intel Xeon E5-2650 CPUs (2.0 GHz, 128 GB RAM). We compare our high-level pipeline code using either StarPU or OCR runtime to a sequential C++ version, hand coded StarPU and OCR versions, and a TBB version that uses the pipeline algorithm of TBB. Additionally, we evaluated the impact of stage replication factors on the overall pipeline performance. We have used GCC 5.3.0 compiler, OpenCV version 3.3.1, StarPU 1.2.4, TBB 4.2, and our own OCR implementation [13, 14].

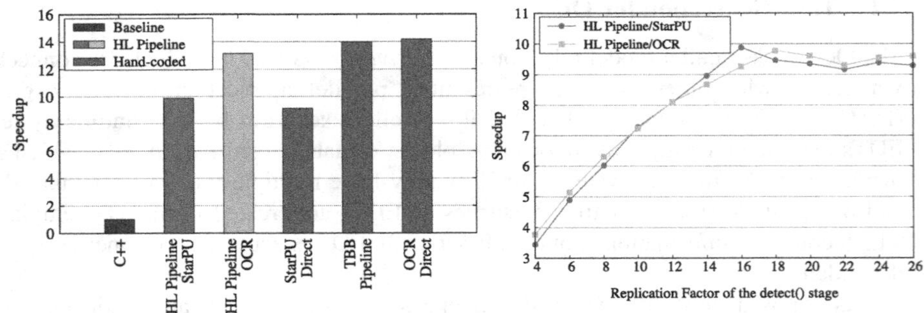

Fig. 3. Speedup of face detection pipeline over sequential C++ on 16 cores (left) and impact of replication factors (right) (Color figure online)

Figure 3 (left) shows the speedup on 16 cores for different versions over the sequential C++ version (red bar), which is the same as the high-level pipeline code without pragma directives. Our high-level pipeline code on top of StarPU (blue bar) achieved a speedup of 9.9 and on top of OCR (green bar) a speedup of 13.1. Interestingly, the highest performance was achieved for OCR when replicating the *convert* and *annotate* stage 5 times each and *detect* stage 16 times, while for StarPU when replicating only the *detect* stage 16 times.

The hand-coded StarPU and OCR versions (StarPU/OCR Direct), which have been parallelized manually, achieved speedups of 9.1 and 14.2, respectively. The TBB version, which uses the TBB pipeline algorithm achieved a speedup of 14. These comparisons show that the performance achievable with our high-level patterns is very close (in case of StarPU even better) to hand-coded versions that utilize the low-level runtimes directly or to TBB, which is a highly optimized template.

Our high-level pipeline code has 50 lines of code (4 lines/pragmas more than the sequential C++ version), the TBB version has 85 lines, the hand coded StarPU version 294 lines, and the hand coded OCR version 209 lines. These line counts do not include the used OpenCV functions, which were the same for all these versions.

Figure 3 (right) shows how replicating only the *detect* stage affects the overall performance. As can be seen, in this configuration our high-level pipeline code performs best with a replication factor of 16 using StarPU and of 18 using OCR.

6 Conclusions and Future Work

In this paper we have presented our work on a programming framework that supports high-level pipeline patterns within sequential C++ codes, which are transformed by our source-to-source compiler to an intermediate runtime coordination layer (RCL) that realizes task-parallel execution on top of the StarPU or OCR runtime system. Our experiments with a face detection application show that performance competitive with much more complex low-level programming approaches can be achieved.

Our future work will focus on improving the tuning support for our framework (e.g. tuning of stage replication factors) and on support for heterogeneous systems.

Acknowledgement. The work was supported in part by the Austrian Science Fund (FWF) project P 29783 Dynamic Runtime System for Future Parallel Architectures.

References

1. Augonnet, C., Thibault, S., Namyst, R., Wacrenier, P.-A.: StarPU: a unified platform for task scheduling on heterogeneous multicore architectures. Concurr. Comput.: Pract. Exp. - Euro-Par **2009**, 187–198 (2011)
2. Kaiser, H., Heller, T., Adelstein-Lelbach, B., Serio, A., Fey, D.: HPX - a task based programming model in a global address space. In: PGAS 2014: the 8th International Conference on Partitioned Global Address Space Programming Models (2014)
3. Cledat, R., Mattson, T.: OCR, the open community runtime interface. OCR specification 1.2.0 (2016)
4. Benkner, S., et al.: PEPPHER: efficient and productive usage of hybrid computing systems. IEEE Micro **31**(5), 28–41 (2011)
5. Bueno, J., et al.: Productive programming of GPU clusters with OmpSs. In: 2012 IEEE 26th International Parallel Distributed Processing Symposium (IPDPS) (2012)
6. OpenMP Architecture Review Board. OpenMP Application Programming Interface v4.5 (2015)
7. Bauer, M., Treichler, S., Slaughter, E., Aiken, A.: Legion: expressing locality and independence with logical regions. In: Proceedings of the International Conference on High Performance Computing, Networking, Storage and Analysis, Salt Lake City, Utah (2012)
8. Pheatt, C.: Intel® threading building blocks. J. Comput. Sci. Coll. **23**(4), 298 (2008)
9. Robson, M.P., Buch, R., Kale, L., Runtime coordinated heterogeneous tasks in charm++. In: ESPM2 Workshop, in Conjunction with SC16, Salt Lake City (2016)
10. Majeti, D., Sarkar, V.: Heterogeneous Habanero-C (H2C): a portable programming model for heterogeneous processors. In: 2015 IEEE International Parallel and Distributed Processing Symposium Workshop (2015)
11. Bajrovic, E., Benkner, S.: Automatic performance tuning of pipeline patterns for heterogeneous parallel architectures. In: The 2014 International Conference on Parallel and Distributed Processing, Techniques and Applications (2014)
12. Bradski, G., Kaehler, A.: Learning OpenCV 3: computer vision in C++ with the OpenCV Library. O'Reilly Media, Sebastopol (2016)
13. Dokulil, J., Sandrieser, M., Benkner, S.: OCR-Vx - an alternative implementation of the open community runtime. In: International Workshop on Runtime Systems for Extreme Scale Programming Models and Architectures, in conjunction with SC15, Austin, Texas, November 2015

14. Dokulil, J., Sandrieser, M., Benkner, S.: Implementing the open community runtime for shared-memory and distributed-memory systems. In: 24th Euromicro International Conference on Parallel, Distributed and Network-Based Processing (PDP), Heraklion, Greece. IEEE Computer Society, February 2016
15. Benkner, S., Bajrovic, E., Marth, E., Sandrieser, M., Namyst, R., Thibault, S.: High-level support for pipeline parallelism on many-core architectures. In: Kaklamanis, C., Papatheodorou, T., Spirakis, Paul G. (eds.) Euro-Par 2012. LNCS, vol. 7484, pp. 614–625. Springer, Heidelberg (2012). https://doi.org/10.1007/978-3-642-32820-6_61
16. Gerndt, M., Cesar, E., Benkner, S. (eds.): Automatic tuning of HPC applications - the periscope tuning framework (PTF). Shakar Verlag, Herzogenrath (2015)

A Real-Time Routing Protocol in Wireless Sensor-Actuator Network

Yue Lu[1], Hui Tian[1,2(✉)], and Jiajia Yin[1]

[1] School of Electronics and Information Engineering, Beijing Jiaotong University, Beijing, China
Hui.Tian@griffith.edu.au

[2] School of Information and Communication Technology, Griffith University, Southport, Australia

Abstract. Wireless Sensor-Actuator Network (WSAN) is a network embedded with a few powerful actuators on the basis of the original WSNs. Based on perception of the WSAN, those actuators can aggregate and process the obtained information in real time, and thus reduce the transmission of redundant information in the network. This helps to save energy consumption. However, due to time delay in data analysis and fusion in a WSAN, the service sensitive to time is provided with poor performance. Therefore, how to balance energy saving and time delay is the main problem in a WSAN. A time-sensitive WSAN network model is established in this paper. To realize the hierarchical structure of the network, the clustering algorithm is used to cluster the sensor nodes. The design of the network routing algorithm based on clustering is formed to the classical Traveling Salesman Problem (TSP). In our WSAN model, we propose the Shuffled Frog Leaping and Ant Colony Algorithm (SFL-ACA) node clustering algorithm. In some practical application scenarios, we proposed an improved scheme to further reduce the delay of network transmission. Based on the WSAN network simulation, we study the influence of different number of actuator nodes on the performance of WSANs, which therefore produces an effective number of actuator nodes deployment scheme in this paper.

Keywords: WSAN · Routing protocol · TSP · SFL-ACA

1 Introduction

Recent technological advances have led to the development of wireless sensor-actuator networks (WSANs) which are capable of observing the physical word, making decisions based on the observations and performing appropriate actions. WSAN is an extension of wireless sensor networks (WSN) [1]. The components of WSAN include sensor nodes, actuator nodes and sink nodes [2]. The sensor nodes sense the events in the physical world and forward the data to actuator nodes or sink nodes. The actuator nodes are responsible to take decisions and react in the event area according to the received data. The function of the sink nodes is to instruct the actuator nodes to perform related operations. WSAN has many merits in applications [3]. For example, wireless sensor nodes are not restricted by cable, which are portable and have low requirements

J. H. Park et al. (Eds.): PDCAT 2018, CCIS 931, pp. 111–120, 2019.
https://doi.org/10.1007/978-981-13-5907-1_12

for external environment and facilities [4]. What's more, low costs, good fault tolerance, good flexibility and scalability are also the reasons it can be widely used.

However, WSAN is also facing many new challenges [5], such as real-time requirement, limited energy, redundant nodes and security problems. One of the main challenges of WSAN is to improve the real-time performance of the network. The real-time performance of the network directly determines the performance of the system. This paper focuses on real-time routing protocol of WSAN.

In this paper, we propose an improved WSAN routing protocol based on SFL-ACA. Our protocol considers the mobile actuator node which may reduce communication delay. We study the influence of different number of actuator nodes to the performance of WSANs, which produces an effective number of actuator nodes configuration scheme.

The remainder of the paper is organized as follows: Sect. 2 introduces the model assumption and clustering algorithm; Sect. 3 further presents the improved WSAN routing protocol based on SFL-ACA; Sect. 4 gives the experimental results and analysis and Sect. 5 finally concludes this paper.

2 Network Model and Problem Definition

2.1 Network Model

Assume that WSAN is a $Y \times Y$ square event area and the number of sensor nodes and actuator nodes is N and M respectively. The properties of the network are as follows:

(1) Each sensor node has a probability W_i (range from 0 to 1), determined by the frequency of the occurrence of the event area.
(2) If a sensor node senses an event, it would report the incident to the actuator node in one hop, otherwise it would not report.
(3) The actuator nodes are removable. The actuator nodes perform corresponding tasks according to the information uploaded by the sensor nodes.

Each actuator node has its own mobile path and it does not interfere with each other. When the actuator node is close to the sensor node, the message is sent to ask if the event is detected. And if the sensor node detects an event, the data would be uploaded to the actuator node.

2.2 Network Nodes Clustering

In a WSAN, actuator nodes start from a random point, then move to the corresponding node according to the node sequence on its own mobile path, and finally return to the starting point. We assume that mobile path of actuator nodes is a cluster. The actuator nodes are the cluster head node and the sensor nodes passing through the path are cluster members.s_i represents the i th sensor node and the set of sensor nodes is $Sensor = \{s_1, s_2, \ldots, s_N\}$ ma_j stands for the j th actuator node and the set of actuator nodes is $Actuator = \{ma_1, ma_2, \ldots, ma_M\}$ M_P_j represents the movement path of ma_j. The steps of the clustering algorithm are as follows:

Step1: Calculate the node weight K of the sensor nodes.

$$K_i \approx W_i \times M$$

Step2: Assign cluster members.

(1) Select the sensor nodes with the same node weight K to form the set S_K, $K = 1, 2, \ldots, M$;
(2) The sensor nodes in S_M are allocated to all paths M_P_j, $j = 1, 2, \ldots, M$
(3) The sensor nodes in S_{M-1} are allocated so that they belong to any $M - 1$ path M_P_j, $j = 1, 2, \ldots, M$. The current path is denoted by $|M\_P_j|$ in the allocation. We select the minimum number of $|M\_P_j|$ to ensure that the number of nodes in each path is approximately the same.
(4) Repeat (3) until the allocation is completed (Fig. 1).

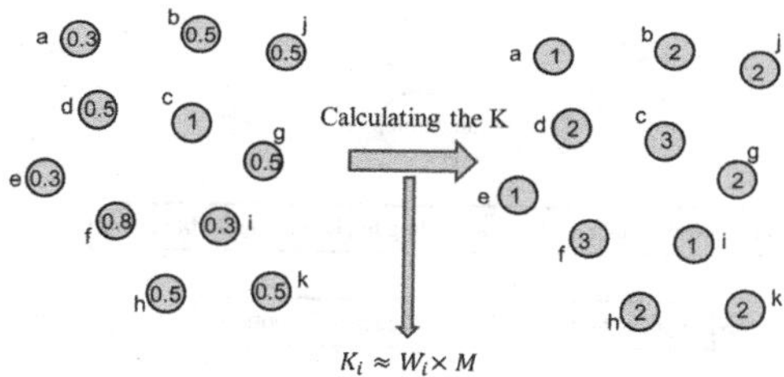

Fig. 1. Calculate the value of K by clustering algorithm

We get the following paths when $M = 3$ and the results of the distribution are shown in Table 1.

$$s_1 = \{a, e, i\} \quad s_2 = \{b, d, g, h, j, k\} \quad s_3 = \{c, f\}$$

Table 1. Node assignment by clustering algorithm

M_P_1	=	{	c	f	b	d		h	j		a			}
M_P_2	=	{	c	f	b		g	h		k		e		}
M_P_3	=	{	c	f		d	g		j	k			i	}

2.3 Path Selection

The basic framework of the WSAN model has been completed when the sensor nodes are clustered. Path selection is the most important step of the model, which is related to the real-time and energy consumption of the network.

Assume that the mobile path of the actuator node ma_M is M_P_M and contains $t(t \leq N)$ sensor nodes. Since the path M_P_M can only pass through each sensor node once and must return to the starting point, so there are $t!$ paths in the path scheme.

Suppose the starting point are randomly selected. According to the previous analysis, it is a closed loop which passes through cluster member only once. Therefore, the shortest path problem can be seen as a Traveling Salesman Problem (TSP) [6].

2.4 Shuffled Frog Leaping and Ant Colony Algorithm

There are many algorithms for solving TSP. In general, TSP algorithm has two parts: approximate algorithms and exact algorithms. Most of the existing methods use approximate algorithms to solve TSP, such as Shuffled Frog Leaping Algorithm (SFLA) [7], Ant Colony Algorithm (ACA) [8], Genetic Algorithm (GA) [9] and so on.

The SFL-ACA node clustering algorithm suitable for WSAN is proposed by combining the local update search ability of the SFLA [10] and the positive feedback mechanism and distributed computing method of ACA, which effectively realizes the real-time information collection and transmission of WSAN.

The SFL-ACA algorithm flow chart is shown in Fig. 2.

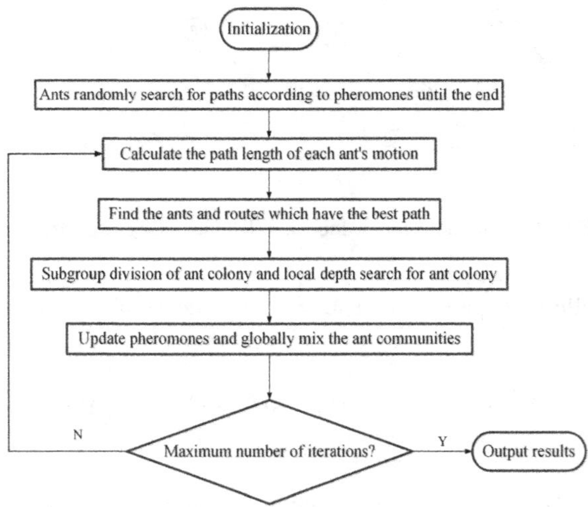

Fig. 2. The flow chart of SFL-ACA

3 An Improved WSAN Routing Protocol Based on SEL-ACA

The improved WSAN routing protocol based on SFL-ACA is a real-time routing protocol. We set probability of sensor nodes and cluster the nodes by SEL-ACA based on the improved WSAN network model.

3.1 Probability Setting of Sensor Nodes

The sensor nodes are randomly distributed in the improved WSAN network model. Different probability values are set for each sensor node and the probability represents how often the abnormal event may happen in monitored area of each sensor node. In some area, the probability of anomaly detected by sensor nodes are high and the frequencies of sensor nodes uploading information are also high. So an improved WSAN routing protocol based on SFL-ACA is proposed for various situations with different even probabilities.

3.2 Communication Between Sensor Nodes and Actuator Nodes

The network model used in our paper assumes the similar real-time model of WSAN routing protocol with multiple actuators (MA) [11] node models.

In order to minimize the delay of sensor nodes and actuator nodes transmission, we select the mobile actuator nodes as the cluster head and the sensor nodes on the path as cluster members. When the actuator nodes pass through the sensor nodes and the actuator nodes are in propagation range of the sensor nodes, the sensor nodes delivers all the detected information to the actuator nodes in one-hop manner so that the cluster heads aggregate all the information in the cluster.

3.3 Communication Among Actuator Nodes

Ad-hoc communication is adopted under the node assistance of the actuator nodes and the actuator nodes, considering actuator nodes have strong mobility and computation power.

Firstly, the simplified mode of work is to broadcast the information of the node, such as IP address, location coordinate, and so on, to all nodes by Hello messages periodically.

When one of the actuator nodes receives the Hello message from the adjacent actuator nodes, or the broadcast message passing through the other actuator nodes, it can build the topology of the surrounding local actuator distribution. Thus, the routing table of nodes is established, and the routing table is maintained and updated to ensure valid communications among actuator nodes.

4 Experiment and Analysis

4.1 Experiment of Communication Delay Performance

We apply four different routing protocols to simulate the communication delay performance. The Improved WSAN routing protocol based on SFL-ACA (I-SFL-ACA), General WSAN routing protocol based on SFL-ACA (G-SFL-ACA), Improved WSAN routing protocol based on Energy Efficient and QoS Aware Routing protocol (EEQAR) [12] (I-EEQAR) and General WSAN routing protocol based on EEQAR (G- EEQAR).

The parameters setting for the communication delay experiment is shown in Table 2.

Table 2. Parameters for delay experiment

Parameters	Values
Network coverage	1000 m*1000 m
Number of mobile actuator nodes	5
Number of sensor nodes	16, 30, 50, 70, 100
Probability of sensor nodes	0.0–1.0
MAC protocol	IEEE 802.15.4
Initial energy of sensor nodes	10 J
Transmission distance of sensor nodes	50 m
Transmission distance of actuator nodes	158 m

The simulation results of four routing protocols are shown in Fig. 3. With the increased number of sensor nodes, the communication delay of the network is increased. The network communication delay of the general WSAN routing protocol is about two times of the improved WSAN routing protocol under the same number of sensor nodes. Comparing G-EEQAR and I-EEQAR, similar results are obtained. The improved WSAN routing protocol reduces the frequency of the sensor nodes reporting information to the mobile actuator nodes with small probability of event occurrence, which thus reduces the average communication delay.

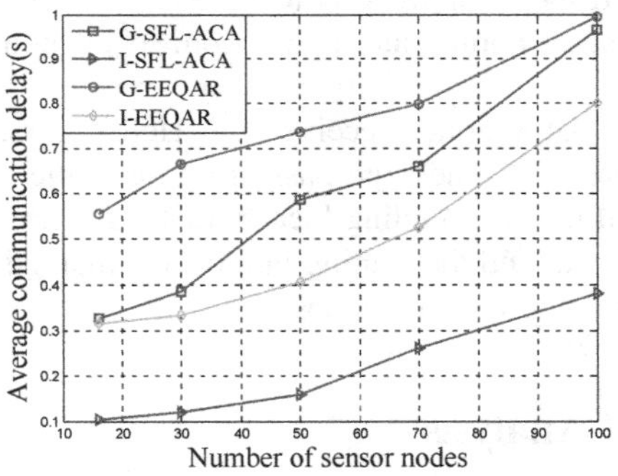

Fig. 3. Simulation diagram of communication delay

With the increased number of sensor nodes, the communication delay of the network is increased. With the same number of sensor nodes in the same network, the delay of network communication in G- EEQAR is about 2 times of G-SFL-ACA. A comparative analysis of two simulation models I-SFL-ACA and I-EEQAR produces similar results. Because I-SFL-ACA cluster head and the mobile actuator nodes are set directly. I-EEQAR considers the distance between the hop distance from the actuator nodes to the sensor nodes, and the sensor nodes in the I-SFL-ACA directly upload information to the actuator node in single hop, which greatly reduces the communication delay between sensor nodes and actuator nodes.

Therefore, it is concluded that with the same protocol and the same number of sensor nodes, the improved WSAN routing protocol has less network communication delay than the general WSAN routing protocol. With the same number of sensor nodes in the same network, the delay of the WSAN routing protocol based on SFL-ACA is smaller than that of the WSAN routing protocol based on EEQAR.

4.2 Deployment of Actuator Nodes

We study the performance of the improved WSAN routing protocol based on SFL-ACA by deploying different numbers of actuator nodes. The average time delay in propagation varies as the number of actuator nodes changes. The parameters are shown in Table 3.

Table 3. Parameter setting table for number of actuator experiment

Parameters	Values
Network coverage	1000 m*1000 m
Number of sensor nodes	100, 200, 300, 400, 500 and 600
Probability of sensor nodes	0.0–1.0
MAC protocol	IEEE 802.15.4
Initial energy of sensor nodes	10 J
Transmission distance of sensor nodes	50 m
Transmission distance of actuator nodes	158 m

The frequency of events in this experiment is divided into low frequency, medium frequency and high frequency. The frequency is set at 0.5 times per second, 2 times per second and 10 times per second respectively. The frequencies of events are different and the probability of event collision between nodes and the network congestion would be different. Therefore, we may study the appropriate deployment scheme of the number of actuator nodes in different network scenarios.

The experiment provides that the acceptable range of average communication delay is where the average communication delay of the network changes very little with the increase of actuator nodes.

From Fig. 4, it can be seen that with the increase of the number of actuator nodes, the trend of the network average communication delay decline is the fastest, faster and slower. Finally, the communication delay of the network will be stable at a lower time. Therefore, under the premise of reducing communication delay and not increasing too much network cost, the number of actuator nodes is proportional to the number of sensor nodes.

We can achieve an effective number configuration of actuator nodes from the above simulation experiment. The table for the number configuration of actuator nodes under the condition of low frequency, medium frequency and high frequency are shown from Tables 4, 5 and 6.

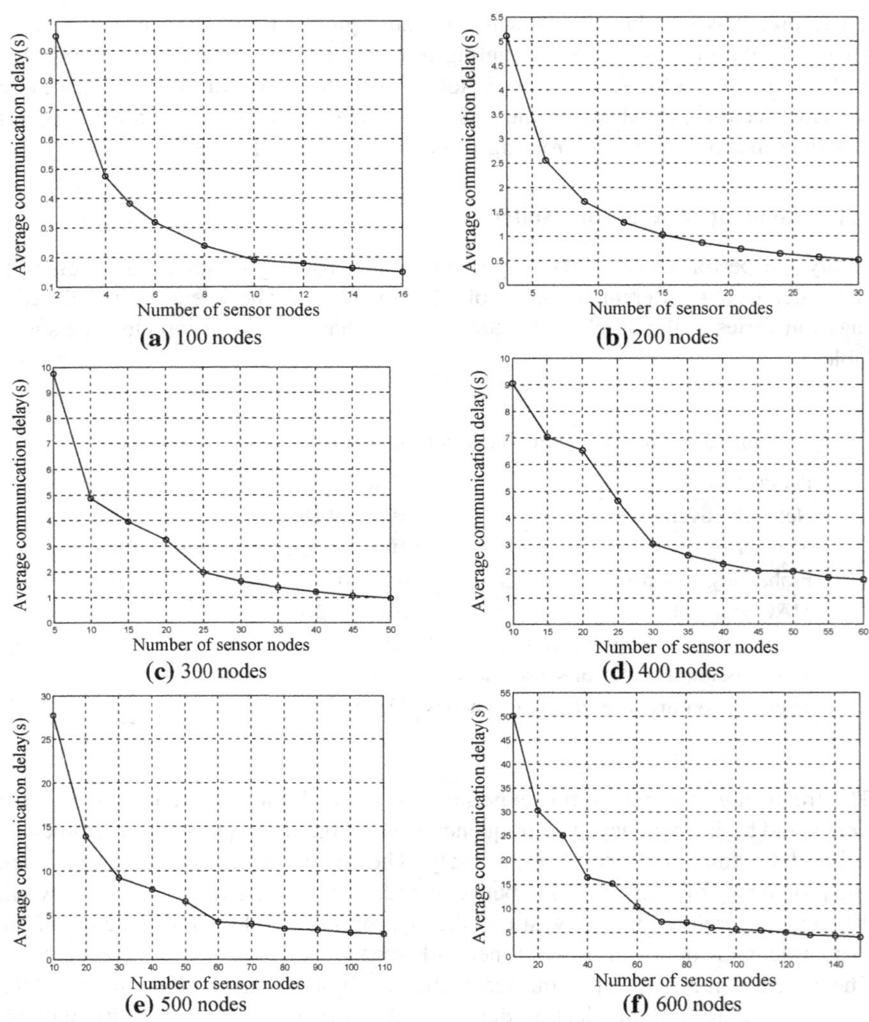

Fig. 4. Time delay simulation of different number nodes under low frequency

Table 4. Configuration of the number of actuator nodes under low frequency

Number of sensor nodes	Number of actuator nodes	Ratio
100	10	10%
200	21	10.5%
300	25	8.33%
400	45	11.25%
500	70	14%
600	100	16.67%

Table 5. Configuration of the number of actuator nodes under medium frequency

Number of sensor nodes	Number of actuator nodes	Ratio
100	10	10%
200	21	10.5%
300	30	10%
400	40	10%
500	80	16%
600	100	16.67%

Table 6. Configuration of the number of actuator nodes under high frequency

Number of sensor nodes	Number of actuator nodes	Ratio
100	14	14%
200	24	12%
300	35	11.67%
400	45	11.25%
500	90	18%
600	110	18.33%

Under low frequency and medium frequency condition, when the number of sensor nodes is less than 500 and the number of actuator nodes is about 10% of the number of sensor nodes, the transmission reliability of the network can be guaranteed and the average transmission delay can be limited to an acceptable range. What's more, under the condition of high frequency, the number of actuator nodes should be about 12% of the number of sensor nodes. The reliability and real-time of the transmission can be guaranteed under low frequency when the number of actuator nodes is greater than 500 and less than 600, the number of actuator nodes is about 15% of the number of sensor nodes. And under medium frequency, the number of actuator nodes should be about 16% of the number of sensor nodes. The number of actuator nodes should be about 18% of the number of sensor nodes when under high frequency.

The simulation experiment of the number configuration of the actuator nodes under three different frequency conditions can draw the following conclusions: for the different number of the sensor nodes in the network, the number of actuator nodes will be different. Too few actuator nodes will affect the communication delay of the whole network while too many actuator nodes will not improve the real-time performance, but it will increase the network construction cost.

5 Conclusion

We proposed an improved WSAN routing protocol based on SFL-ACA in this paper. We assumed the occurrence probability of event in area monitored by each sensor nodes and clustered the nodes according to solution to TSP problem. Different from the general WSAN routing protocol, our protocol considered the mobile actuator node which may reduce communication delay. Through experiments on communication delay performance and number configuration of the actuator nodes, the performance of our WSAN routing protocol and its advantages over other algorithms are verified. It is shown that our protocol can guarantee the transmission reliability of the network while the average transmission delay of the network is limited to a satisfactory threshold. The proposed actuator deployment schemes is able to deploy an appropriate number of actuators in the actual network so as to avoid unnecessary overhead in network configuration.

Acknowledgement. This work is supported by Beijing Natural Science Foundation Grant No.4172045, Research Initiative Grant of Australian Research Council Discovery Projects funding DP150104871, and National Science Foundation of China Grant No. 61501025.

References

1. Akyildiz, I., Kasimoglu, I.: Wireless sensor and actor networks: research challenges. Ad Hoc Netw. J. **2**, 351–367 (2004)
2. Vassis, D., Kormentzas, G., Skianis, C.: Performance evaluation of single and multi-channel actuator to actuator communication for wireless sensor actuator network. Ad Hoc Netw. **4**, 487–498 (2006)
3. Melodia, T., Popili, D.: A distributed coordination framework for wireless sensor-actuator networks. In: Proceedings of the 6th ACM International Symposium on Mobile Ad Hoc Networking and Computing, pp. 99–110 (2005)
4. Pandian, P., Safeer, K.: Wireless sensor network for wearable physiological monitoring. J. Netw. **3**, 20–30 (2008)
5. Pallavi, R., Prakash, G.C.B.: A review on network partitioning in wireless sensor and actor networks. In: International Conference on Applied and Theoretical Computing and Communication Technology, pp. 771–782. IEEE (2016)
6. Lin, S., Kernighan, B.W.: An effective heuristic algorithm for the TSP. Oper. Res. **21**, 498–516 (1973)
7. Eusuff, M., Lansey, K., Pasha, F.: Shuffled frog-leaping algorithm: a memetic meta-heuristic for discrete optimization. Eng. Optim. **38**, 129–154 (2006)
8. Duan, H., Wang, D., Zhu, J., et al.: Development on ant colony algorithm theory and its application. Control Decis. **19**, 1321–1326 (2004)
9. Maulik, U., Bandyopadhyay, S.: Genetic algorithm-based clustering technique. Pattern Recogn. **33**, 1455–1465 (2004)
10. Nisha, D.M.: Actor node based on-demand rekeying scheme for wireless sensor and actor networks. In: International Conference on Parallel, pp. 357–362. IEEE (2017)
11. Khan, M.A., Shah, G.A., Ahsan, M., et al.: An efficient and reliable clustering algorithm for wireless sensor actor networks (WSANs). In: IEEE International Midwest Symposium on Circuits and Systems, pp. 332–338. IEEE (2010)
12. Yahiaoui, S., Omar, M., Bouabdallah, A., et al.: An energy efficient and QoS aware routing protocol for wireless sensor and actuator networks. AEU – Int. J. Electron. Commun. **22**, 345–368 (2017)

Privacy Preserving Classification Based on Perturbation for Network Traffic

Yue Lu[1], Hui Tian[1,2(✉)], Hong Shen[3,4], and Dongdong Xu[1]

[1] School of Electronics and Information Engineering, Beijing Jiaotong University, Beijing, China
hui.tian@griffith.edu.au

[2] School of Information and Communication Technology, Griffith University, Southport, Australia

[3] School of Computer Science, University of Adelaide, Adelaide, Australia

[4] School of Data and Computer Science, Sun Yat-Sen University, Guangzhou, China

Abstract. Network traffic classification is important to many network applications. Machine learning is regarded as one of the most effective technique to classify network traffic. In this paper, we adopt the fast correlation-based filter algorithm to filter redundant attributes contained in network traffic. The attributes selected by this algorithm help to reduce the classification complexity and achieve high classification accuracy. Since the traffic attributes contain a large amount of users' behavior information, the privacy of user may be revealed and illegally used by malicious users. So it's demanding to classify traffic with certain segment of frames which encloses privacy-related information being protected. After classification, the results do not disclose privacy information, while may still be used for data analysis. Therefore, we propose a random perturbation algorithm based on relationship among different data attributes' orders, which protects their privacy, thus ensures data security during classification. The experiment results demonstrate that data perturbed by our algorithm is classified with high accuracy rate and data utility.

Keywords: Network traffic classification · Privacy preserving
Machine learning

1 Introduction

Network traffic classification has attracted research interest recently, because it largely improved efficiency of the management of networks and Quality of Services (QoS) [1]. With the development of Internet technology, new application types are emerging rapidly, especially the application based on non-standard ports and protocol encryption, which reduces the efficiency of port-based method [2] and deep payload inspection (DPI) [3]. Using machine learning (ML) methods to classify network traffic has been widely used. ML-based methods use some statistical features such as message length and duration to identify the type of flow as different application types such as WWW, MAIL, etc. ML techniques may classify new-coming network traffic efficiently and greatly improve the accuracy.

J. H. Park et al. (Eds.): PDCAT 2018, CCIS 931, pp. 121–132, 2019.
https://doi.org/10.1007/978-981-13-5907-1_13

The statistical features of network traffic carry a lot of sensitive information which may represent behaviors and other critical information of users. For example, the size of the data packets sent by a user when downloading and browsing the web are different. If an attacker determines the statistical information in a stream, he may determine the user's behavior and recognize the user at a high probability. Therefore, there are risks of privacy disclosure by applying ML-based classification. How to achieve a high classification accuracy while not disclosing users' privacy is the key issue that we study in this paper. This problem is also known as how to tradeoff the data's utility and security.

2 Related Work

ML-based methods for classification mainly include unsupervised and supervised learning approaches. Unsupervised approaches classify unlabeled data based on the similarities between sample data, such as K-means clustering [4], EM [5], Auto Class [6], DBSCAN [7] and so on. They are suitable for identifying new application types in dynamic environment. The classification accuracy is hard to define and classification results may not include much useful information. Supervised learning algorithms are more popular in this field. Naive Bayes proposed by Moore et al. [8].

C4.5 Decision Tree [9] and K-Nearest Neighbor (KNN) [10] are all popular supervised algorithm. SVM [11] may overcome the problem of localized optimization. Supervised classification algorithms can greatly improve the accuracy of the classification results because there have been useful prior-knowledge before classification. In this paper we focus on using supervised classification algorithms for network traffic classification.

Privacy preserving techniques are applied to protect the privacy of users when publishing data. There are mainly three methods to preserve privacy. The first method is using *k*-anonymity [12] to reduce the granularity of data representation, such that any given record maps onto at least *k*-1 other records in the data. This method is vulnerable to malicious attackers and only provide limited protection on sensitive data. Data encryption [13] is the second method for privacy protection which applies encryption methods in the data mining process. This method is more secure than the first one, but hard to be applied in many scenarios because it involves complex computation. Randomized perturbation is the third method in which noise is added for masking the values, such as differential privacy [14]. This method is easy to implement, which is less complex than data encryption. On the other hand, it provides higher security level than *k*-anonymity.

Therefore, in order for data classification to guarantee a high accuracy rate and preserve privacy, we propose a random perturbation algorithm based on the data attributes' order relationship. The experimental results indicate that the perturbed data are classified with a satisfactory accuracy while privacy is preserved.

The remainder of the paper is organized as follows: Sect. 3 introduces the process of supervised classification system and describes the algorithm of attributes order adjustment in detail; Sect. 4 presents the perturbation-based network traffic privacy preserving method; Sect. 5 gives the experimental results and analysis; Sect. 6 concludes this paper.

3 The Process of Supervised Classification System

Supervised classification system is consisted of four parts as shown in Fig. 1: training or test sets, attribute selection, classifier and classification results evaluation. The supervised classification system uses a partial set of attributes to identify the type of flow application, and then use the ML-based methods to construct the classifier and classify the network traffic.

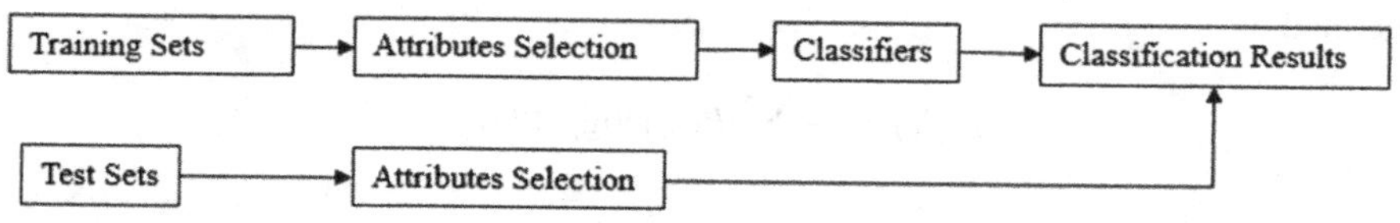

Fig. 1. Supervised classification system

3.1 Dataset Statistics

The experimental data sets proposed by Moore et al. [15] is adopted in our paper. The data sets includes 377526 network flow samples and is divided into 10 types. The statistical information of the data sets is shown in the Table 1.

Table 1. Dataset statistics

Type of flow	Number of flows	Percent
WWW	328091	86.91%
MAIL	28567	7.56%
BULK	11539	3.05%
DATABASE	2648	0.70%
SERVICE	2099	0.56%
P2P	2094	0.55%
ATTACK	1793	0.48%
MULTIMEDIA	1152	0.31%
INTERACTIVE	110	0.3%
GAME	8	0.002%
Total	377526	100%

3.2 Attributes Selection

Each sample in the data sets is captured from a complete TCP flow and includes 248 attributes, among which a large number of features do not contribute to flow classification and thus regarded as redundant. These features do not improve the accuracy of classification, but only increase the search complexity for classification algorithms. So we adopt a Fast Correlation-Based Filter algorithm (FCBF) [16] to filter redundant attributes contained in network traffic.

In FCBF algorithm, each attribute in the network traffic is regarded as a set of random events. We measure the correlation in order to filter the attributes with high redundancy. The main parameter adopt in FCBF is Symmetry Uncertainty (S), which is defined as

$$SU(X,Y) = 2\left[\frac{IG(X|Y)}{H(X)+H(Y)}\right]$$

In above formula, $H(X)$ stands for the information entropy of the variable X, which is defined as

$$H(X) = -\sum_{i} P(x_i)\log_2(P(x_i))$$

Where $P(x_i)$ represents the probability that variable X has value x_i.

$IG(X|Y)$ is used to measure the closeness of the relationship between variables X and Y. It is expressed as

$$IG(X|Y) = H(X) - H(X|Y)$$

Where $H(X|Y) = -\sum_{y_j} P(y_j)\sum_{x_i} P(x_i|y_j)\log_2(P(x_i|y_j))$ represents the mutual information between variables X and Y.

The value of $SU(X,Y)$ is between [0, 1]. 0 means that two variables are independent and 1 means that two variables determine each other. SU can not only measure the correlation between two attributes, but also measure the correlation between attributes and categories.

The steps of FCBF algorithm are as follows. First, calculate SU between attributes as well as between attributes and categories. Second, select a feature subset S' by setting a threshold δ of SU. We thereby select S' by following steps:

(1) Select the attributes according to the threshold δ, that is, select $A_i \in S'$, $1 \le i \le N$ satisfies$SU_{i,c} \ge \delta$. $SU_{i,c}$ denotes the value of SU that measures the correlation between the attribute i and the class C.
(2) Remove the redundant attributes. For the attribute i and attribute j that selected by (1), if $SU_{i,j} \ge SU_{i,c}$, the attribute j is regarded as a redundant attribute where the $SU_{i,j}$ denotes the value of SU that measures the correlation between the attribute i and the attribute j.

Finally, the attributes in feature subset S' are output in descending order.

3.3 Classifier

Supervised learning methods mainly classify the network traffic according to the marker sample data, and establish the relationship between the network flow features and the training set sample categories. Naïve Bayes, C4.5 Decision Tree, KNN, SVM are supervised learning methods. In this paper, we would use the above supervised classification algorithms to classify network traffic and compare the performance of different classification algorithms.

3.4 Classification Results Evaluation

The purpose of network traffic classification is to correctly identify the types of network applications. We use accuracy as an indicator to evaluate the quality of traffic classification methods. It is computed as below:

$$accuracy = \frac{\textit{the number of correctly labeled flows}}{\textit{the total number of flows}}$$

4 Perturbation-Based Network Traffic Privacy Preserving Method

Deep mining of network traffic will lead to leakage of users' privacy. So we propose a perturbation-based network traffic privacy preserving method, which uses the data after perturbation of the raw data for classification. The flow chart is shown as Fig. 2.

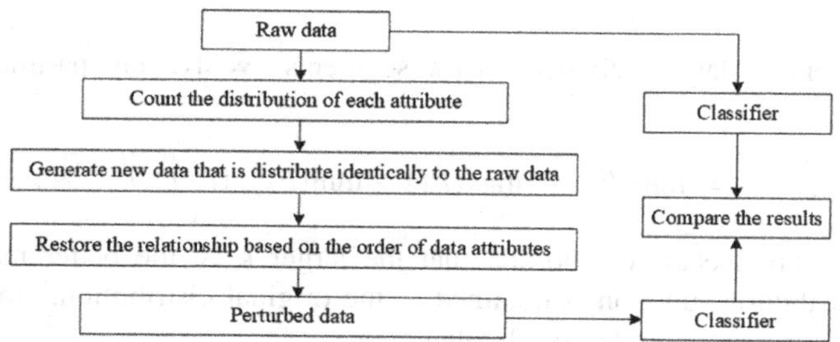

Fig. 2. Perturbation-based privacy preserving method

Assume that A is the set of raw attribute data, the number of samples contained in A is n (the number of rows in the data table), that is,

$$A = (X_1, X_2, \ldots, X_n)$$

Let the newly generated attribute data set be B after perturbation, the number of samples contained in B is also n, that is,

$$B = (Y_1, Y_2, \ldots, Y_n)$$

Assume that each sample contains m attributes, that is, each the n-th sample can be represented as $X_n = (x_n^1, x_n^2, \ldots, x_n^m)$. The value corresponding to the m-th attribute in the data table A is expressed by $P_m = (x_1^m, x_2^m, \ldots, x_i^m, \ldots, x_n^m)$. Similarly, the perturbed data table B should also contain m attributes in each sample. The n-th sample of the perturbed data is expressed as $Y_n = (y_n^1, y_n^2, \ldots, y_n^m)$ and the corresponding value of m-th attribute in the data table B is described by $Q_m = (y_1^m, y_2^m, \ldots, y_i^m, \ldots, y_n^m)$.

4.1 Count Distribution of Each Attribute

We take the first attribute $P_1 = (x_1^1, x_2^1, \ldots, x_i^1, \ldots, x_n^1)$ in A as an example to count the distribution. Since each attribute may include many different values, if we calculate probability for each value, the complexity of the algorithm will increase. What's more, this method cannot be applied to each dimension data so the generality is limited. Therefore, we propose a method by dividing the values into several intervals and the expressions of the corresponding distribution functions in each interval are obtained respectively. The specific process is as follows.

(1) Find the maximum value as $\max(P_1)$ and the minimum value as $\min(P_1)$ in P_1.
(2) Divide the data in $[\min(P_1), \max(P_1)]$ into k segments uniformly and get every partition value respectively. Intern partition values satisfy that:

$$\min(P_1) = s_1 \leq s_2 \leq \ldots s_i \leq \ldots \leq s_{k+1} = \max(P_1)$$

Since the entire data is divided into k segments evenly, any partition value is calculated by

$$s_i = \min(P_1) + (\max(P_1) - \min(P_1))\,(i-1)/k$$

It is seen from the above formula that the larger k is, the better match of the calculated distribution function is obtained to the original distribution. However, this results in the increased complexity of calculation.

(3) Count the number of data contained in each partition. The total number of data is denoted as num_{total}. The number within $(-\infty, s_1]$ is denoted as num_1 and the number in the interval $(s_k, s_{k+1}]$ is denoted as num_k.

Then we can get the probability at partition value s_1 is

$$P(x = s_1) = \frac{num_1}{num_{total}}$$

The probability at any partition value s_j can be calculated by

$$P(x = s_i) = \frac{\sum_1^i num_i}{num_{total}}$$

4) Use the linear interpolation method to get the expressions of the corresponding distribution functions in each partition respectively after calculating the probability at any partition value. Then we can get the distribution function of the original data. The distribution function $F(x)$ of P_1 is defined as follows.

$$F(x) = \begin{cases} 0 & x < s_1 \\ \frac{\sum_1^i num_i}{num_{total}} & x = s_i \\ F(s_i) + \frac{F(s_{i+1}) - F(s_i)}{s_{i+1} - s_i}(x - s_{i+1}) & s_i < x < s_{i+1} \\ 1 & x > s_{k+1} \end{cases}$$

We use the above distribution function to approximate the original distribution of data. Then the inverse function is used to generate random data that is independent and identically distributed (*iid*) to the original data. The inverse function is to exchange the range and domain of the original distribution function. In addition, we guarantee the number of generated data in each partition is the same as the original data and conforms to a uniform distribution. Then we can get the random data of each attribute by repeating the above steps.

4.2 Restore the Relationship Based on the Order

Although the data generated by the above steps retain approximate identical distribution as the original data, the randomly generated data cause damage to utility of the data. In order to restore the correlation among the data as much as possible, we adjust the perturbed data based on their attributes' order relationship. The following example describes the adjustment process.

Assume that two attributes of the raw data are $P_1 = (x_1^1, x_2^1, \ldots, x_i^1, \ldots, x_n^1)$ and $P_2 = (x_1^2, x_2^2, \ldots, x_i^2, \ldots, x_n^2)$. $P_1' = (x_1^{1'}, x_2^{1'}, \ldots, x_i^{1'}, \ldots, x_n^{1'})$ is the generated sequence by above steps that has identical distribution to P_1, $P_2' = (x_1^{2'}, x_2^{2'}, \ldots, x_i^{2'}, \ldots, x_n^{2'})$ is the perturbed data from P_2. We use the order relationship between P_1 and P_2 to adjust the perturbed data P_1' and P_2'. Firstly, we obtain the orders in each data sequence. The goal is to keep the perturbed data as close as possible to raw data. Then we use the order relationship between P_1 and P_2 as a reference, the data in P_2' is adjusted based on P_1'. If the order in P_1 corresponds to the order in P_2, the order in P_1' also corresponds to the same order in P_2', that is, the value in P_2' should be adjusted to the corresponding position. For instance, if the smallest value in P_1 corresponds to the largest value in P_2, the largest value in P_2' should adjust to the proper position to correspond the smallest value in P_1'. Repeat the above steps, until all the data adjustments are complete. Q_1 and Q_2 are the adjusted sequences of P_1' and P_2' respectively. So Q_1 and Q_2 are published as perturbed data for research on traffic classification.

The published data is a perturbation to the original data according to the above process. What's more, the random perturbation algorithm based on the order relation can not only prevents data reconstruction but also maintains the correspondence of the original data thus guaranteeing the utility of data.

5 Experiment and Analysis

5.1 Attribute Selection Experiment

We apply the FCBF algorithm to filter redundant attributes in Weka environment and use the default threshold δ to select 10-dimensional attributes. Table 2 shows the results of selecting attributes. We do not take the information about ports in the dataset into consideration so as to avoid the impact of ports.

Table 2. Attributes selected by FCBF algorithm

No.	Attributes
1	initinal_window_bytes (b_a)
2	SYN_pkts_sent (a_b)
3	pushed_data_pkts (b_a)
4	initinal_window_bytes (a_b) (b_a)min_segm_size(b_a)
5	req_sack (a_b)
6	mss_requested (a_b)
7	max_retrans(a_b)
8	min_data_control(b_a)
9	segs_cum_acked(a_b)
10	missed_data (b_a)

In the experiment, the 248-dimensional attributes in the original network traffic data are compared with the 10-dimensional attribute results selected by the FCBF algorithm. The parameters in the experiment are determined by using the enumeration local search method. Naive Bayes adapts default parameters. The C-SVC algorithm is selected in SVM and the kernel function is RBF. The nonlinear mapping and the use of a step size search strategy result in a penalty factor C = 512 and a nuclear parameter $\gamma = 0.05$. An important parameter in the C4.5 decision tree is the significance level α that can be used to set the range of confidence intervals and we set $\alpha = 0.2$. KNN has a parameter K which means the number of neighbors and we set $K = 5$. We use 10-fold cross-validation methods to get classification results.

It can be seen from classification results in Table 3, the classification accuracy and the stability of Naïve Bayes has been greatly improved after the FCBF algorithm. The classification accuracy of the KNN algorithm has also been improved to some extent. Although the accuracy of the C4.5 Decision tree and SVM slightly decreased, the running time of the algorithm was greatly reduced due to the reduction of the dimensions. It can be seen that the 10-dimensional attributes selected by the FCBF algorithm can obtain satisfactory classification results and are superior to the original data classification results in terms of classification efficiency. Therefore, we further analyze the 10-dimensional data in the following experiments.

Table 3. The comparison of classification results

	Raw attributes			Selected attributes		
	Accuracy (%)	Variance (%)	Time (s)	Accuracy (%)	Variance (%)	Time (s)
NB	56.42	6.55	1.5	94.14	2.43	0.3
C4.5	99.21	0.76	3.7	98.99	0.68	1.5
KNN	94.31	8.01	35.2	97.54	0.41	15.3
SVM	99.48	0.1	4651	99.4	0.1	1624

5.2 Data Perturbation Experiment

We first use the first dimension attribute to classify and then increase the dimension of attributes. As the dimension of the attribute increases, the relationship between the dimension of the attribute and the classification accuracy is shown in the Fig. 3.

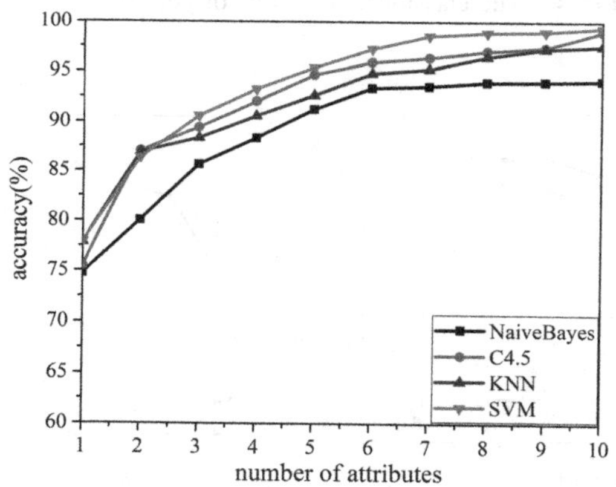

Fig. 3. The classification results of raw data

It is shown in Fig. 3 that the accuracy of each classification algorithm is constantly increasing with the increase of the dimensionality. The accuracy of each classification algorithm would reach more than 90%, which means that data is of high utility. By comparing several classification algorithms, it can be found that the classification accuracy of the SVM algorithm is relatively high.

Next, our proposed perturbation algorithm is applied to the data. First, the perturbation algorithm is applied to the first dimension attribute and generate a sequence that is identically distributed to it. Then the correspondence between the perturbation data and the category is performed according to the correspondence between the original first dimension attribute and the category. After that, the first dimension attribute is used as a criterion to adjust the internal position of the second attribute and then the two-dimensional attribute data are used to classify. Each dimension attribute is perturbed in this way. We repeat the experiment three times and get average in order to further verify the effectiveness of the algorithm. Figures 4 and 5 show the experimental results when we divide the data into 100 segments uniformly, that is, $k = 100$.

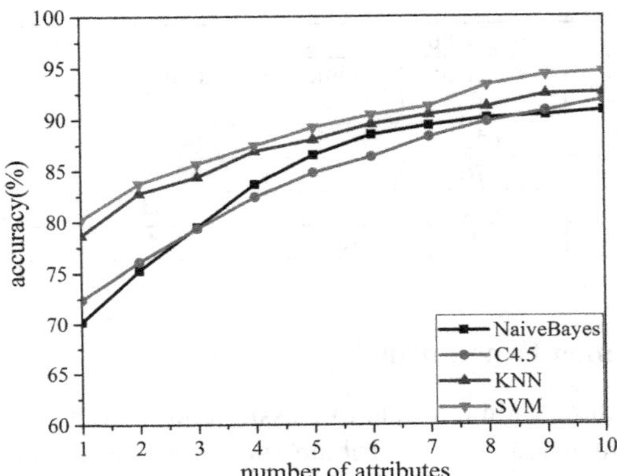

Fig. 4. The classification results of perturbed data

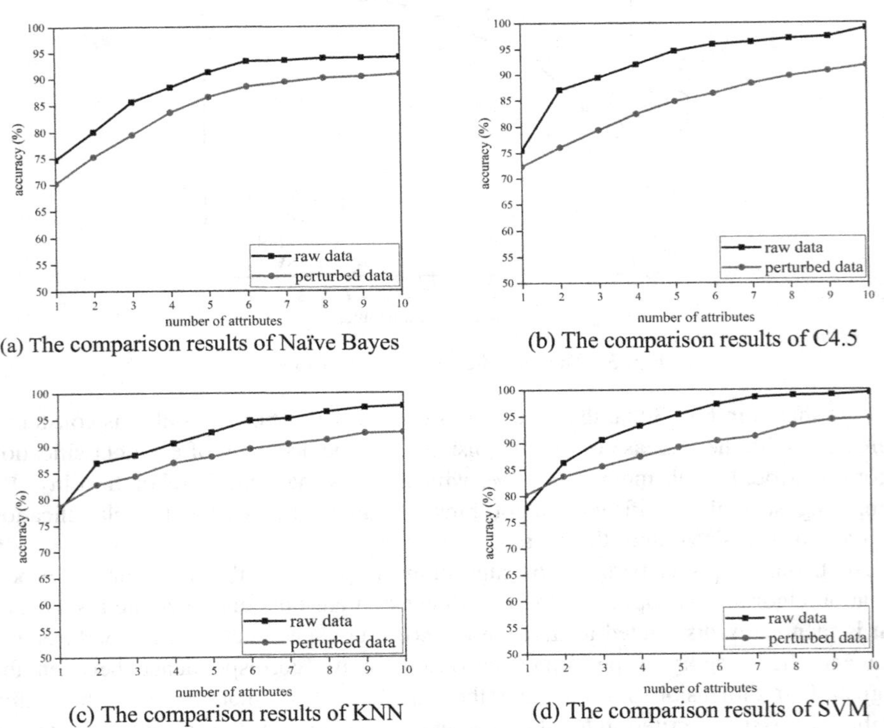

(a) The comparison results of Naïve Bayes

(b) The comparison results of C4.5

(c) The comparison results of KNN

(d) The comparison results of SVM

Fig. 5. The comparison results of different classifiers

It can be seen from the Figs. 4 and 5, the classification accuracy of each algorithm increases as the dimension of the attribute increases. Compared with the results of the classification of the raw 10-dimensional attributes, it can be seen that the classification accuracy rate has slightly decreased. When the 10-dimensional attributes are all disturbed, the classification accuracy rate can still be accepted, indicating that the data still maintains high utility. In addition, the perturbation-based method is suitable because the data after perturbation can be used for various classification algorithms.

6 Conclusion

Network traffic classification is the process of identifying network applications and classifying the corresponding traffic. In this paper, we first adapt the FCBF algorithm to filter the redundant attributes of network traffic and use several supervised learning methods to classify network traffic. The experiment results show that features selected by the FCBF algorithms can maintain satisfactory accuracy and improve the efficiency of classification algorithms. Based on selected attributes, a privacy preserving algorithm using perturbation for network traffic classification is proposed in order to protect users' privacy information in network traffic. The algorithm first generates a random sequence with the identical distribution as the original attributes, and then adjusts the data according to the order relationship of the original attributes. The experimental results indicate that the privacy preserving algorithm can effectively achieve the accuracy of classification, that is, guarantee the utility of data. On the other hand, due to the use of random perturbation method, it is difficult for an attacker to reconstruct the original data that helps to guarantee the security of data. Moreover, the perturbation algorithm is scalable to be used for various classification algorithms.

Acknowledgement. This work was done under the support of Research Initiative Grant of Australian Research Council Discovery Projects funding DP150104871, Beijing Natural Science Foundation Grant No. 4172045 and National Science Foundation of China Grant No. 61501025.

References

1. Guo, L., Shen, H.: Privacy-preserving internet traffic publication. In: IEEE Trustcom/BigDataSE/ISPA, pp. 884–891 (2017)
2. Moore, A.W., Papagiannaki, K.: Toward the accurate identification of network applications. In: Dovrolis, C. (ed.) PAM 2005. LNCS, vol. 3431, pp. 41–54. Springer, Heidelberg (2005). https://doi.org/10.1007/978-3-540-31966-5_4
3. Madhukar, A., Williamson, C.: A longitudinal study of P2P traffic classification. In: 14th IEEE International Symposium on Modeling, Analysis, and Simulation of Computer and Telecommunication Systems, pp. 179–188 (2006)
4. Kanungo, T., Mount, D.M., Netanyahu, N.S.: An efficient *K*-means clustering algorithm: analysis and implementation. IEEE Trans. Pattern Anal. Mach. Intell. **24**, 881–892 (2002)
5. McGregor, A., Hall, M., Lorier, P., Brunskill, J.: Flow clustering using machine learning techniques. In: Barakat, C., Pratt, I. (eds.) PAM 2004. LNCS, vol. 3015, pp. 205–214. Springer, Heidelberg (2004). https://doi.org/10.1007/978-3-540-24668-8_21

6. Zander, S., Nguyen, T., Armitage, G.: Automated traffic classification and application identification using machine learning. In: IEEE Conference on Local Computer Networks, pp. 250–257 (2005)
7. Erman, J., Arlitt, M., Mahanti, A.: Traffic classification using clustering algorithms. In: Proceedings of the 2006 SIGCOMM Workshop on Mining Network Data, pp. 281–286 (2006)
8. Moore, A.W., Zuev, D.: Internet traffic classification using bayesian analysis techniques. ACM SIGMETRICS Perform. Eval. Rev. **33**, 50–60 (2005)
9. Williams, N., Zander, S.: Evaluating machine learning algorithms for automated network application identification, Center for Advanced Internet Architectures Technical report (2006)
10. Li, W., Moore, A.W.: A machine learning approach for efficient traffic classification. In: 15th IEEE International Symposium on Modeling, Analysis, and Simulation of Computer and Telecommunication Systems, pp. 310–317 (2007)
11. Deng, H., Yang, A.M.: P2P traffic classification method based on SVM. In: Computer Engineering and Applications (2006)
12. Aggarwal, C.C.: On *k*-anonymity and the curse of dimensionality. In: Proceedings of the 31st International Conference on Very Large Data Bases, pp. 901–909 (2005)
13. Waters, B.: Efficient identity-based encryption without random oracles. In: Cramer, R. (ed.) EUROCRYPT 2005. LNCS, vol. 3494, pp. 114–127. Springer, Heidelberg (2005). https://doi.org/10.1007/11426639_7
14. Dwork, C.: Differential Privacy. In: Bugliesi, M., Preneel, B., Sassone, V., Wegener, I. (eds.) ICALP 2006, Part II. LNCS, vol. 4052, pp. 1–12. Springer, Heidelberg (2006). https://doi.org/10.1007/11787006_1
15. Moore, A.W., Zuev, D.: Discriminators for use in ow-based classification (2005)
16. Yu, L., Liu, H.: Feature selection for high-dimensional data: a fast correlation-based filter solution. In: 20th International Conference on Machine Learning, pp. 856–863 (2003)

Fault Diagnosis of a Wireless Sensor Network Using a Hybrid Method

Farzin Piltan and Jong-Myon Kim(✉)

School of Electrical Engineering, University of Ulsan,
Ulsan 680-749, South Korea
piltan_f@iranssp.org, jmkim07@ulsan.ac.kr

Abstract. This paper proposes a reliable, intelligent, model-based (hybrid) fault detection and diagnosis (FDD) technique for wireless sensor networks (WSNs) in the presence of noise and uncertainties. A wireless sensor network is a network formed by a large number of sensor nodes in which each node is equipped with a sensor to detect physical phenomena such as light, heat, pressure, and temperature. Increasing the number of sensor nodes can cause an increase in the number of faulty nodes, which adversely affects the quality of service (QoS). Herein, the WSN modeling is based on an adaptive method that combines the fuzzy C-means clustering algorithm with the modified auto-regressive eXternal (ARX) model and is utilized for fault identification in WSNs. The proposed adaptive nonlinear ARX fuzzy C-means (NARXNF) clustering technique obtains both an improved convergence and error reduction relative to that of the traditional fuzzy C-means clustering algorithm. In addition, the proportional integral (PI) distributed observation is used for diagnosing multiple faults, where the convergence, robustness, and stability are validated by a fuzzy linear matrix inequality (FLMI). To test the proposed method, this technique was implemented through simulation using Omnet++ and MATLAB.

Keywords: Wireless sensor network · Fault diagnosis
Fuzzy C-means clustering ARX model · PI distributed observation technique
Reliability

1 Introduction

The purpose of this research is to study fuzzy model-based fault diagnosis and tolerant (FDT) algorithms for WSNs and to explore their use in various situations, particularly focusing on the enhancement of WSN quality of service (QoS) and reliability. The Internet of Things (IoT) is the network of physical devices, vehicles, home appliances and other items embedded with electronics, software, sensors, actuators, and connectivity which enables these objects to connect and exchange data, and Wireless communication technologies, especially WSNs, can play an important role in IoT [1]. A Wireless Sensor Network (WSN) is a network containing various types of intelligent sensors, which monitor different types of physical conditions of the environment such as temperature, sound, flow, and pressure. These sensors transfer the sensed data to another location. The WSN consists of multiple nodes, and the node object structures

J. H. Park et al. (Eds.): PDCAT 2018, CCIS 931, pp. 133–142, 2019.
https://doi.org/10.1007/978-981-13-5907-1_14

are the microcontroller, sensor, battery, communication mode, and storing [2, 3]. The WSN faults are divided into two main categories: software faults and hardware faults. Hardware faults are permanent and divided into four main groups: transmitted circuit faults, receiver circuit faults, battery faults, and sensing faults [4]. To analyze the condition of the WSN, different types of condition monitoring techniques are employed in the literature; one is based on the node energy efficiency, but those based on neighboring node information and on the data of individual nodes are the most widely used [5]. Therefore, fault diagnosis and tolerant (FDT) techniques are necessary to identify faults before complete node failure and a degradation of the network.

The main purpose of FDT is to determine whether the WSN is normal or abnormal, or more specifically to determine the position of faults for quick fault recovery. However, the major challenges in this task are to estimate the magnitude and direction of the faults to find the accurate fault location and specification. Fault detection and diagnosis can be divided into three major categories: data-driven approaches, model-based methods, and hybrid techniques. The diagnosis decision in the data-driven approach is completely dependent on the specific and unique dataset and a proper tuning in its various hyper-parameters and the main drawback of this method is its reliability [6–8]. To apply a model-based fault diagnosis method, it is necessary to characterize the dynamic behavior of the WSN. The main advantages of the model-based FDT are stability, reliability and robustness. To extract the dynamic behavior of the WSN, two methods can be introduced, mathematical modelling and a system identification method based on linear or nonlinear estimation techniques. On the other hand, to design a hybrid technique, intelligent methods are applied to model-based techniques of FDT in WSNs. Different comprehensive model reference algorithms have been reviewed by researchers, such as parity space approaches, observer-based methods, and parameter estimation techniques [4, 9, 10].

System identification-based observer techniques can be divided into linear system identification and nonlinear system identification approaches [11]. Linear system identification is used to identify abnormal behavior in linear and basic nonlinear systems. In an earlier approach, the application of system identification for fault detection in a WSN was limited to linear techniques [5], which was possible because of their simple design structure and well-understood dynamic response, but it has been a challenge to model noisy and nonlinear systems. However, to identify or estimate the behavior of a nonlinear system, such as a WSN, it is necessary to design a nonlinear system model. Artificial intelligent (AI)-based theory is widely applied for identification in nonlinear systems [11, 12]. Fuzzy-based identification methods [13] and neural network-based methods [14], are used to improve the performance of linear identification in WSNs. Fuzzy-based methods have been successfully applied in manifold applications of control engineering and in intelligent decision making. However, the linear fuzzy-based technique faces challenges, such as guaranteeing stability and producing a rule base for design accuracy [15]. To improve the performance of linear fuzzy techniques, a network-based adaptive methodology is suggested.

This study is concerned with intelligent model-based fault diagnosis. The main challenges regarding fault diagnosis in a WSN are to estimate the fault magnitude and the direction of the fault to accurately determine the fault type. For this purpose, an appropriate thresholding technique is applied to estimate fault magnitudes of the residual signal, which is generated via the difference between the actual signal and the system's estimation signal, while the system model performs its update operation [4, 5]. In practice, these residuals are highly sensitive to the possible faults in the system that are used to evaluate the fault diagnosis [5]. These signals are certainly independent of the input and output processing in normal (or healthy) conditions because these signals should be zero, or close to zero under normal conditions. The magnitude of the residual signals increases in faulty conditions, deviating from zero by a considerable amount [16]. Therefore, our proposed observer model for the WSN is based on an improved fuzzy C-means clustering ARX-PI-observer, to ensure improve the convergence, stability, and robustness.

Overall, the key contributions of this research are summarized as follows: (1) detecting faults and modelling the WSN based on the sensitive ARX fuzzy C-means clustering network, (2) estimating and identifying the fault, based on the proposed PI observation theory, (3) further generating an improved residual signal to better detect faults and improve the diagnosis rate, and (4) designing a distributed technique to recover the node and improve the reliability and stability. The rest of this paper is organized as follows: Sect. 2 gives the problem statements. Section 3 presents the proposed methodology for fault detection in detail and outlines the fault diagnosis and fault recovery process. Section 4 shows the dataset used for validation of the methodology. The results of the research are presented in Sect. 5.

2 Problem Statements

As the WSN is a highly nonlinear, and multi-input multi-output (MIMO), The WSN model in this research work can be represented as a graph structure, G(S,E), where S is a set of sensor nodes and E is a set of communication links between sensor nodes S_i and S_j, having a sensing range less than the communication range. The assumption is that the node S_i will perform low power transmission for nodes within R_c and may sometimes, as required, perform high powered communication with S_j. In this research, the faults are divided into the three main groups: transmitted circuit faults, battery faults, and receiver circuit faults. The cluster head is used to monitor the transmitter circuit's status, based on the 200-bit lifeline message. The efficiency of the transmitter circuit is calculated by [4]:

$$T(\phi) = \frac{l}{S_i}, \tag{1}$$

where l and S_i are the lifeline message and total time elapsed since the initiation of the WSN, respectively. This value is used to calculate the threshold value for fault detection and diagnosis based on the proposed method. The efficiency of a battery, which is useful for the detection of a battery fault, is calculated as follows [4]:

$$\begin{aligned} &\Lambda_i = (\varepsilon_t + \varepsilon_{A_1}\lambda^2)M_s \quad \lambda < \Phi \rightarrow \Lambda_i = (\varepsilon_t + \varepsilon_{A_2}\lambda^4)M_s \quad \lambda \geq \Phi \\ &\Phi = \sqrt{\frac{\varepsilon_{A_1}}{\varepsilon_{A_2}}}, \end{aligned} \tag{2}$$

where $\varepsilon_t, \varepsilon_A, M_s$ and λ are the energy to run the transmitter, energy utilized for the amplifier, message size, and the difference between the cluster head separation and the sensor's transmission distance, respectively. The efficiency of the receiver circuit is calculated as [4]:

$$\alpha(\phi) = \frac{\mu}{S_t}, \tag{3}$$

where μ and S_t are the number of the received OK lifeline messages and the total time that the receiver has been transmitting, respectively. The state space modeling of the WSN is represented as follows [17]:

$$\begin{cases} W(k) = [\alpha W(k-1) + \beta_z Z(k-1) + \beta_u u(k-1) + \Psi] + f(k-1) \\ Z(k) = (K)^T W(k) \end{cases}, \tag{4}$$

where $W(k), (\alpha, \beta_z, \beta_u, K), u(k), f(k), Z(k)$ and Ψ are the state vector, coefficients, input, faults, output, and uncertainties, respectively. According to (1), this paper defines its three primary objectives: (1) modeling and detecting the fault of the WSN based on the adaptive nonlinear network Tsk fuzzy ARX method and (2) fault estimation and identification for the WSN based on the PI-observation technique and the integration of defects to model the faults. Figure 1 illustrates the problem statements and methods to solve it.

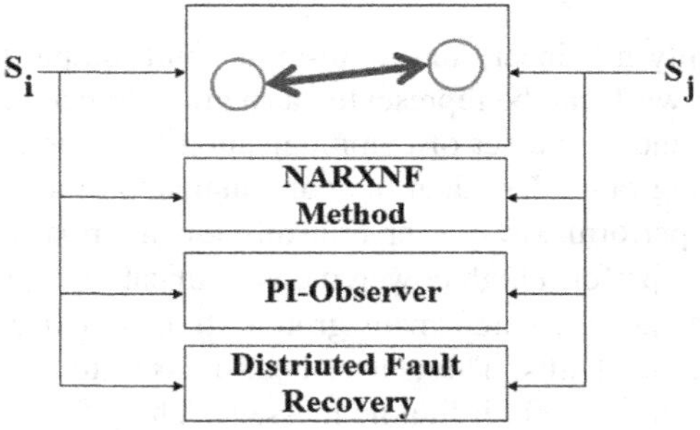

Fig. 1. The block diagram of the proposed fault detection, and diagnosis scheme.

3 Proposed Method

A wireless sensor network is a nonlinear, and multi-input multi-output (MIMO). Thus, to identify the system's behavior, for fault detection, and to estimate the system's output performance, for fault diagnosis, are the main considerations of this research.

3.1 WSN's Fault Detection

Modeling is accomplished by a transformation function to find possible WSN behavior based on the behavior of nodes and neighboring nodes. The adaptive nonlinear autoregressive neuro-fuzzy (NARXNF) algorithm is used for WSN modeling.

This method estimates the output parameters of the WSN. In the first step, a discrete ARX theory is applied to identify the normal behavior of the WSN since this method is efficient for system prediction. To detect the node and communication faults in the WSN, the state-space input-output filter TSK fuzzy ARX function is defined as below:

$$\begin{cases} \hat{W}(k) = [\alpha\hat{W}(k-1) + \beta_z\hat{Z}(k-1) + \beta_u u(k-1) + \hat{\Psi}] + \hat{f}(k-1) \\ \hat{Z}(k) = (K_{n,i})^T\hat{W}(k) \end{cases}, \tag{5}$$

where $\hat{W}(k), \hat{f}(k), \hat{Z}(k)$ and $\hat{\Psi}$ are the estimated system state, estimated fault, estimated measured output, and estimated uncertainty, respectively. The ARX-TSK fuzzy sampling rule is defined by the following definition:

If $P_{x,1}$ is a_{11} and $P_{y,1}$ is b_{11} then $\hat{Y}_{s_i}(1) = P_{x,1}x_1 + P_{y,1}y_1 + r_1$, to update the $\hat{Y}_s(i,j)$, we have

$$\hat{Y}_s(i,j) = \frac{\omega_i\hat{Y}_s(i) + \omega_j\hat{Y}_s(j)}{\omega_i + \omega_j}, \tag{6}$$

where $(P_x, P_y), r$ and Y_s are the respective inputs and states of the linear system identification, the weight of system identification, and the output function of the WSN. Based on the WSN system modelling in (4) and (5), we have

$$\Psi + f - \hat{\psi} - f = 0 \rightarrow \Psi + f = \Delta_N \rightarrow if(f = 0) \rightarrow \Psi = \chi \leq \Delta_N, \tag{7}$$

where Δ_N is the normal threshold value. Thus, the threshold value is calculated as the uncertainty value and this threshold value is used for fault detection. Based on the approach found in (7), in a faulty WSN we have

$$\Psi + f - \hat{\psi} - f = 0 \rightarrow \Psi + f = \Delta_N \rightarrow if(f \neq 0) \rightarrow \Psi > \Delta_N. \tag{8}$$

3.2 WSN's Fault Diagnosis

For fault diagnosis of a WSN, the formulation of proportional-integral (PI) observer is presented, as follows:

$$\begin{cases} \hat{W}(k) = [\alpha\hat{W}(k-1) + \beta_z\hat{Z}(k-1) + \beta_u u(k-1) + \hat{\Psi}] + \hat{f}(k-1) + K_p[Z(k-1) - \hat{Z}(k-1)] \\ \hat{Z}(k) = (K_{n,i})^T\hat{W}(k) \end{cases} \tag{9}$$

As discussed in (9), to improve the performance of the state condition, the proportional part is applied to linear system modeling. To model the faults (e.g., transmitted fault, receiver fault, and battery fault) in PIO, the fault is modelled (estimated) based on the integral term as follows:

$$\hat{f}(k) = f(k-1) + K_{ii}[Z(k-1) - \hat{Z}(k-1)]. \tag{10}$$

Based on (9) and (10), the residual signal in the WSN is calculated as follows:

$$r(k) = [Z]^T - [\hat{Z}]^T = [\, W_T^T(k) \quad W_B^T(k) \quad W_R^T(k)\,]^T - [\, \hat{W}_T^T(k) \quad \hat{W}_B^T(k) \quad \hat{W}_R^T(k)\,]^T \tag{11}$$

where $r(k), W_T^T(k), W_B^T(k), W_R^T(k)$ are the residual signal of WSN, the state signal of the WSN with respect to transmission faults, the state signal of the WSN regarding battery faults, and the state signal of WSN related to receiver faults, respectively. As discussed in [4], if $\Delta_R > \Delta_T > \Delta_B > \Delta_N$, where the variables are the threshold values in receiver condition, transmitter condition, battery condition, and normal state, then the residual signal for each state is calculated as follows:

$$r_n = [W] - [\hat{W}] \rightarrow r_n < \Delta_N, r_n < \Delta_B, r_n < \Delta_T, r_n < \Delta_R \tag{12}$$

$$r_B = [W] - [\hat{W}] \rightarrow r_B > \Delta_N, r_B < \Delta_B, r_B < \Delta_T, r_B < \Delta_R \tag{13}$$

$$r_T = [W] - [\hat{W}] \rightarrow r_T > \Delta_N, r_T > \Delta_B, r_T < \Delta_T, r_T < \Delta_R \tag{14}$$

$$r_R = [W] - [\hat{W}] \rightarrow r_R > \Delta_N, r_R > \Delta_B, r_R > \Delta_T, r_R < \Delta_R, \tag{15}$$

where $\Delta_R > \Delta_T > \Delta_B > \Delta_N, r_R, r_T, r_B$ and r_N are the receiver threshold, transmitter threshold, battery threshold, normal state threshold, receiving residual WSN signal, transmission residual WSN signal, battery residual WSN signal, and normal residual WSN signal, respectively.

4 Dataset, Results and Discussion

The key components of the experimental apparatus are employed in the experiment, and included a limitation in the area, coupled with a limitation in the number of nodes. The communication range and density of nodes should also be identified. Table 1 shows the settings considered for our simulation studies. Figure 2 shows the residual signals for normal and faulty nodes, and the threshold value for fault detection in the WSN. Based on Fig. 3, the proposed method improves the performance of fault detection in the WSN. The faulty battery residual signal is detected and identified in Fig. 2. Based on (13), the battery fault causes a fluctuation in the normal signal, and the threshold signal is used to detect this fault type. Figure 4 shows the fault diagnosis with regard to a transmission fault, based on the proposed method. As apparent in Fig. 2, the proposed method defines a sensitive threshold value to isolate the normal, battery fault, and faulty transmission residual signals. To isolate the faulty receiver signal based on (15), a new threshold value should be optimized. Figure 5 illustrates the fault diagnosis for the receiver signal. According to this figure, the proposed method defines the threshold value to isolate the receiver fault.

Table 1. Values used in the simulations [4]

Parameter	Value
Network area	500 * 500 m^2
Network size	50–300
Base station location	(50, 175)
Sensing range	10 m
Energy (transmitter)	150 nJ
Energy (reciever)	50 nJ
Energy (amplifier)	10 pJ
Energy (multipath)	0.0013 pJ

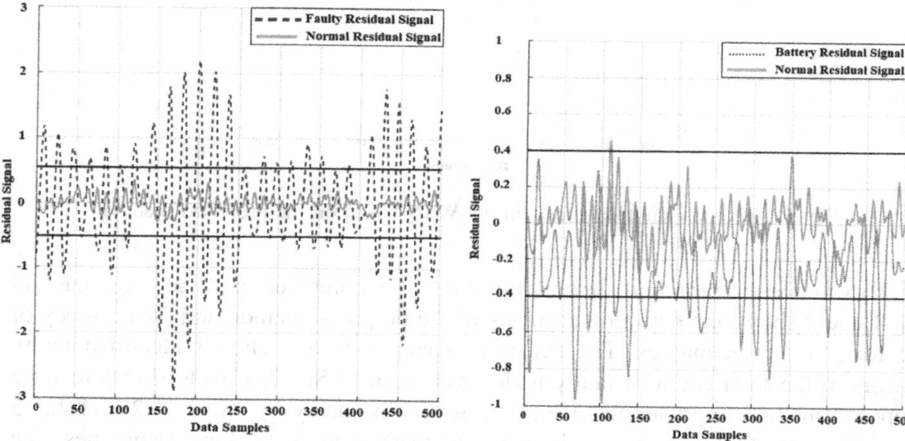

Fig. 2. Fault detection in the WSN based on the proposed method

Fig. 3. Battery Fault diagnosis in the WSN based on the proposed method

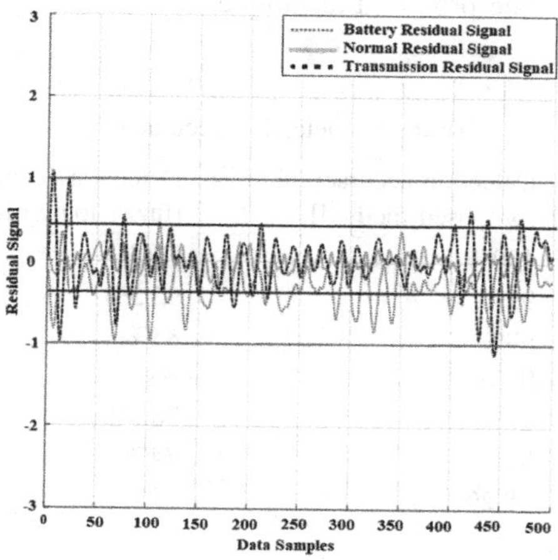

Fig. 4. Transmission fault diagnosis in the WSN based on the proposed method

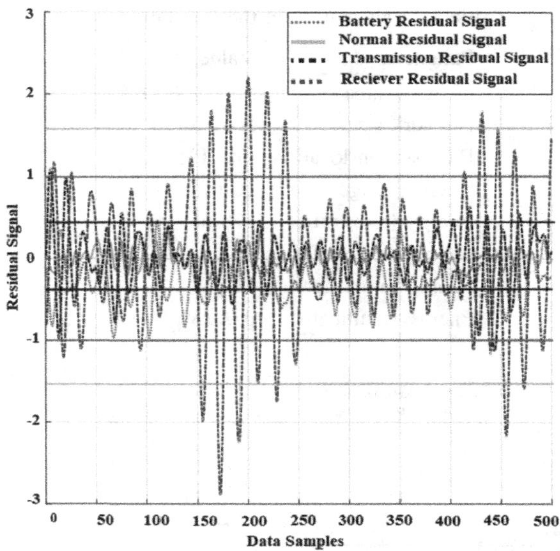

Fig. 5. Receiver fault diagnosis in the WSN based on the proposed method

To validate the proposed method, our method is compared with the fuzzy method [4]. Table 2 shows the detection accuracy of the proposed method and the accuracy of the fuzzy logic techniques. The detection accuracy is the ratio of identified faulty sensors to the total count of faulty nodes within the WSN. The table shows that the proposed method improves the average detection accuracy by about 13.2%. Table 3 shows the rate of false alarms for the proposed method and fuzzy logic techniques. The false alarm rate is defined by the ratio of the number of non-faulty nodes detected as faulty nodes to the number of non-faulty nodes in the WSN. As evidenced by Table 3, the proposed method improves the rate of false alarms from 0.093 using the fuzzy technique to 0.036 using the proposed technique.

Table 2. Detection accuracy

Number of nodes	Detection accuracy % [proposed method]	Detection accuracy % [fuzzy logic method]
50	100%	100%
100	99%	92.5%
150	98%	85%
200	98%	80%
250	97%	78.5%
300	97%	74%
Average	98.2%	85%

Table 3. False alarm

Number of nodes	False alarm [proposed method]	False alarm [fuzzy logic method]
50	0.01	0.04
100	0.02	0.061
150	0.03	0.082
200	0.04	0.11
250	0.06	0.125
300	0.06	0.14
Average	0.036	0.093

5 Conclusions

In this paper, we presented a new fault detection, estimation, and identification scheme with an application in WSNs based on developing a fuzzy ARX for a PI observer. The proposed method was verified with a WSN, proving that it has an improved efficiency in detecting, isolating, and identifying concurrent faults of the battery, of the transmission and receiver. The proposed intelligent nonlinear ARX model was successfully implemented to solve the challenge of system modelling and fault detection in a WSN. To improve the diagnostic accuracy for multi-faults systems, fuzzy PI observation is employed instead of adding extra hardware.

Acknowledgements. This work was supported by the Korea Institute of Energy Technology Evaluation and Planning (KETEP) and the Ministry of Trade, Industry & Energy (MOTIE) of the Republic of Korea (No. 20181510102160, 20162220100050, 20161120100350, and 20172510102130). It was also funded in part by the Leading Human Resource Training Program of Regional Neo Industry through the National Research Foundation of Korea (NRF) funded by the Ministry of Science, ICT and future Planning (NRF-2016H1D5A1910564), and in part by the Basic Science Research Program through the National Research Foundation of Korea (NRF) funded by the Ministry of Education (2016R1D1A3B03931927).

References

1. Akyildiz, I.F., Su, W., Sankarasubramaniam, Y., Cayirci, E.: Wireless sensor networks: a survey. Comput. Netw. **38**, 393–422 (2002)
2. Wu, Y., Stankovic, J., He, T., Lin, S.: Realistic and efficient multi-channel communications in wireless sensor networks. In: Proceedings INFOCOM, pp. 1867–1875 (2008)
3. Chowdhury, K.R., Nandiraju, N., Chanda, P., Agrawal, D.P., Zing, Q.A.: Channel allocation and medium access control for wireless sensor networks. Ad-Hoc Netw. **7**, 307–321 (2009)
4. Jadav, P., Babu, V.K.: Fuzzy logic based faulty node detection in wireless sensor network. In: 2017 International Conference on Communication and Signal Processing (ICCSP). IEEE (2017)
5. Chouikhi, S., et al.: Recovery from simultaneous failures in a large scale wireless sensor network. Ad Hoc Netw. **67**, 68–76 (2017)

6. Gao, Z., Cecati, C., Ding, S.X.: A survey of fault diagnosis and fault-tolerant techniques—Part I: fault diagnosis with model-based and signal-based approaches. IEEE Trans. Ind. Electron. **62**(6), 3757–3767 (2015)
7. Keshtgari, M., Deljoo, A.: A wireless sensor network solution for precision agriculture based on zigbee technology. Wirel. Sens. Netw. **4**, 25–30 (2012)
8. Konstantinos, K., et al.: Topology optimization in wireless sensor networks for precision agriculture applications. In: International Conference on IEEE Sensor Technologies and Applications, Sensor Comm 2007 (2007)
9. Najeh, T., et al.: Input fault detection and estimation using PI observer based on the ARX-Laguerre model. Int. J. Adv. Manuf. Technol. **90**(5–8), 1317–1336 (2017)
10. Agrawal, S., Mohanty, S.R., Agarwal, V.: Bearing fault detection using Hilbert and high frequency resolution techniques. IETE J. Res. **61**(2), 99–108 (2015)
11. Anh, H.P.H., Nam, N.T.: Novel adaptive forward neural MIMO NARX model for the identification of industrial 3-DOF robot arm kinematics. Int. J. Adv. Robot. Syst. **9**(4), 104–112 (2012)
12. Alavandar, S., Nigam, M.J.: Neuro-fuzzy based approach for inverse kinematics solution of industrial robot manipulators. Int. J. Comput. Commun. Control **3**(3), 224–234 (2008)
13. Wu, L., et al.: Fault detection for underactuated manipulators modeled by Markovian jump systems. IEEE Trans. Ind. Electron. **63**(7), 4387–4399 (2016)
14. Al-Dabbagh, R.D., Kinsheel, A., Mekhilef, S., Baba, M.S., Shamshirband, S.: System identification and control of robot manipulator based on fuzzy adaptive differential evolution algorithm. Adv. Eng. Softw. **78**, 60–66 (2014)
15. Aleksovski, D., et al.: A comparison of fuzzy identification methods on benchmark datasets. IFAC-Papers on Line **49**(5), 31–36 (2016)
16. Hartmann, A., et al.: Identification of switched ARX models via convex optimization and expectation maximization. J. Process Control **28**, 9–16 (2015)
17. Lo, C., Lynch, J.P., Liu, M.: Distributed model-based nonlinear sensor fault diagnosis in wireless sensor networks. Mech. Syst. Sig. Process. **66**, 470–484 (2016)

An Optimization Theory of Home Occupants' Access Data for Determining Smart Grid Service

Seung-Mo Je[1] and Jun-Ho Huh[2(✉)]

[1] Department of Computer Science Education, Korea University, Seoul, Republic of Korea
jsm3316@korea.ac.kr

[2] Department of Software, Catholic University of Pusan, Busan, Republic of Korea
72networks@pukyong.ac.kr, 72networks@cup.ac.kr

Abstract. There are a series of nodes in a Smart Grid environment and to let them work efficiently, their tasks should be adequately scheduled. As for the scheduling methods, this study proposes two kinds of scenarios: use of the greedy algorithm or the Floyd-Warshall algorithm both of which have their own merits and demerits. The effectiveness of the scheduling algorithm becomes different depending on the number of nodes. Also, there are two kinds of nodes: mobile nodes and non-mobile nodes. One good example of a node that easily moves is a person. The performing a headcount for the people with their personal information such as their images or whereabouts is not an easy task due to ever strengthening civil rights. It is also difficult to select an effective scheduling algorithm due to the number of dynamic nods. Thus, to determine an efficient scheduling method, some meaningful correlations between the number of AP access, which can be regarded as the number of people, and the number of people in a certain space have been studied by using the AP access record of a Smart Device (Smart Phone, Tablet, etc.) always carried by most of the people these days instead of using personal information. This study then provides a direction of improving network operation by grasping an exact number of nodes in the smart grip service environment based on the correlations revealed.

Keywords: Optimization · Optimization theory · Access data
Smart grid service · Micro grid · OPNET · Python big data protocol ideas

1 Introduction

Smart Grid is a next-generation Power Grid which facilitates efficient use of energy through real-time information exchange between the electricity supplier and users based on the ICT-integrated Power Grid [1, 2]. As per Fig. 1, the power usage largely fluctuates depending on the weather or other conditions so that most of the current power companies usually maintain more than 10% of normal usage as a reserve power but this indeed is a tremendous amount of energy being wasted.

J. H. Park et al. (Eds.): PDCAT 2018, CCIS 931, pp. 143–152, 2019.
https://doi.org/10.1007/978-981-13-5907-1_15

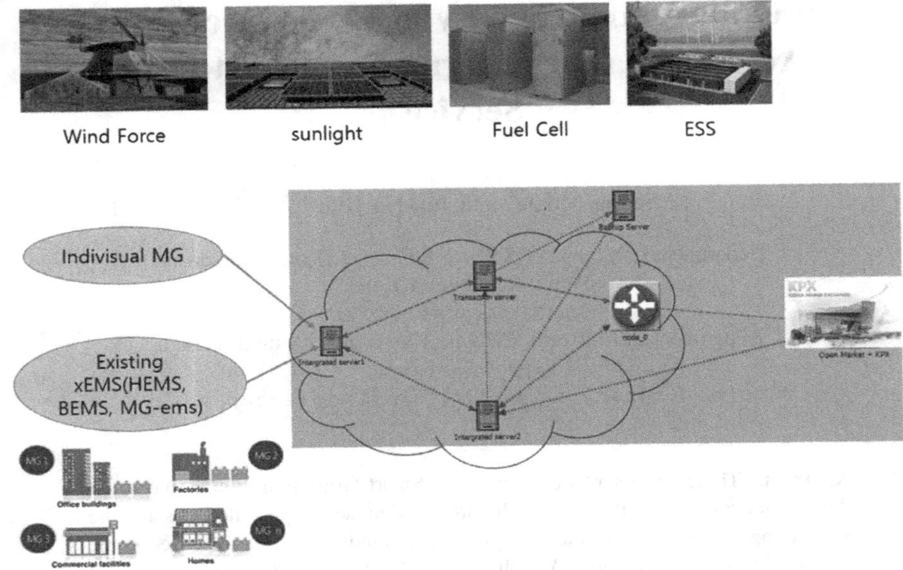

Fig. 1. Future of Smart Grid

The Smart Grid is an excellent alternative to such a system having an advantage of minimizing such an energy waste by flexibly controlling the power supply by allowing the power company to check the real-time power consumption level while the users are able to use the power by automatically checking the time zones to which cheap rates are applied through the exchange of information between them. In short, the power is produced and supplied according to the power usage, minimizing energy waste [3–7].

All the things in a Smart Grid environment communicate with each other, even with humans through the intermediaries such as smartphones, computers, or other communication devices – expanding the domain of communications [8–11].

Normally, a multiple number of nodes are deployed in a smart grid environment and for the communications between things, the number does not change much. That is, in most cases, the number of household appliances or other electronic devices does not change much days, weeks, or months after. However, it is a little different when it comes to humans as they have a will to freely move around with their two feet rather than having the passivity of things. This negatively affects the prediction of the time complexity when configuring a mesh network. The effects can be estimated with the algorithms (e.g., Floyd-Warshall algorithm or greedy algorithm) used for the dynamic programming.

2 Related Works

Figure 2 shows the trend on the Smart Grid systems in Republic of Korea. The Korean government has announced the 'Smart Grid Road Map' and plans to invest 2.75 billion won by 2030 hoping to vitalize relevant industries, starting with construction of a demonstrational complex in Jeju-Do. Their initial Smart Grid model had aimed to advance the element technologies but recently, they are seeking to invigorate the Smart Grid system business internationally by developing several promising business models and more sophisticated convergent systems. The purpose of such Smart Grid system is to intellectualize the Power Grid first and let the supplier of electric power to adjust the level of power supply and its rates and, on the part of consumers, to control their usage by understanding the price of electricity and their own usage patterns through the smart meters and internet. One of the typical methods used in the Smart Metering systems is a Cable TV (CATV) network-based remote metering method [3].

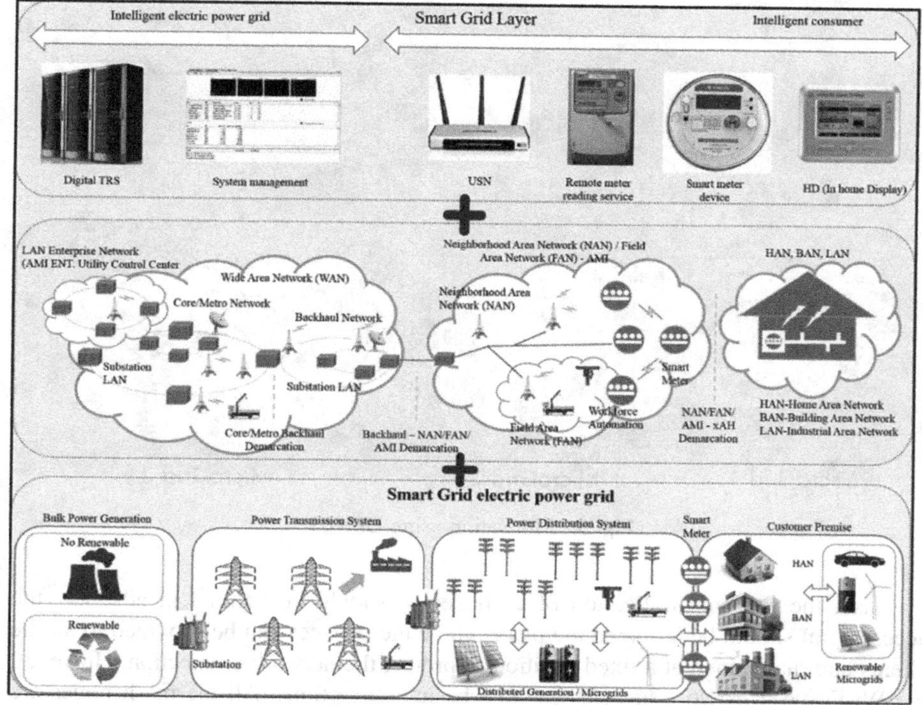

Fig. 2. A trend on the Smart Grid in Republic of Korea [3]

Smart Grid is a Power Grid which pursues a high-level reliability and stability through effective and intelligent power management using its own communication network. The efficient use of power can be achieved with a cutting-edge Smart Grid technology which optimizes the energy efficiency by grafting the ICT technology on to existing Power Grid (Fig. 3).

Recently, intensive research works are being conducted as the smart grid has emerged as a next-generation intelligent Power Grid that can overcome the energy crisis. Although it is now possible to directly connect optical cables to the expensive high-end routers, the Wi-Fi signals are still not quick or strong enough to maintain a stable connection. Even though the communication distance can be increased by using a Wi-Fi extender or a repeater, the desired bandwidth cannot be obtained due to the properties of wireless media [12–16].

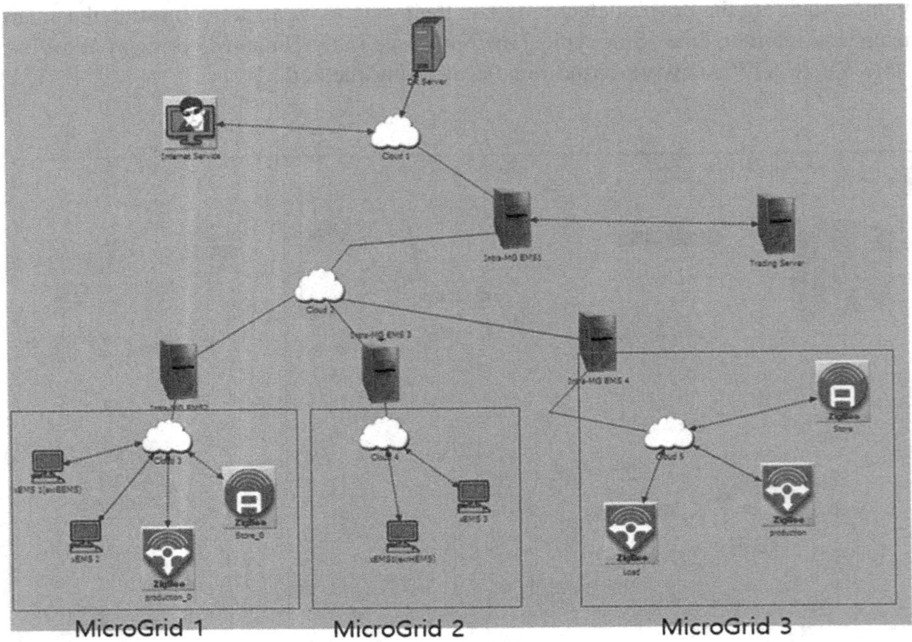

Fig. 3. Motivation using OPNET

Also, the role and obligation of each device should be defined in advance: the access point should be arranged the place where the services can be provided to all the clients and maintained at a fixed position to protect the network connection. However, the Wi-Fi mesh network has shaken such basic assumptions. All the mesh nodes are equal to each other and make the network independent with their self-forming capacity. After completing the initial setup, the device can be positioned in any places necessary to perform the role of an endpoint, distance extender, or even a gateway to the external network. Since the mesh network can reassign itself by using its self-treatment mechanism, it is able to maintain the supporting function when partial nodes have been relocated or the power supply has been powered off [17–21].

Owing to such a function, new paradigms and topologies based on the Wi-Fi technology have become possible. These paradigms include various types of closed circuit systems which support M2M (Machine-to-Machine) topology, the automated household appliances or devices arrangement, and the Smart Grids consisting of a large number of hops and nodes [22–24].

Also, it is possible to construct a mesh network by inserting the basic IEEE 802.11s function only to the device fitted with a meshing function, or other functions such as precise synchronization between devices, simultaneous actuation of meshing and other Wi-Fi-based functions, dynamic conversion between a mesh network and AP-based network can be mounted [25–28]. However, there is one condition which should be considered seriously when it comes to the bandwidth of which the mesh network can process. That is, as the best route for transmitting data from one point to another can change continuously, it is most important to determine the best route as quickly as possible with a dynamic and reliable method. Although there are many routing algorithms, it is not an easy task to achieve the optimal Wi-Fi operation through fine adjustment of these algorithms. In other words, if the anonymous network introduced in 'The world of diverse anonymous networks' is one that has been developed as a technological alternative for the infrastructure of existing networks, some of the new types of networking technologies having a special purpose such as the development of a next-generation mobile communications, home networking, or public safety management are now drawing attention, being called the Mesh Networking.

The existing wired networks adopt a form of all the APs (Access Points) are wired in a way that the signals are continuously linked through the wireless router(s) or other means of relays (cf. AP can be considered as the device similar to the wired/wireless router). Figure 4 shows example problems of scaling up native ethernet and Wi-Fi.

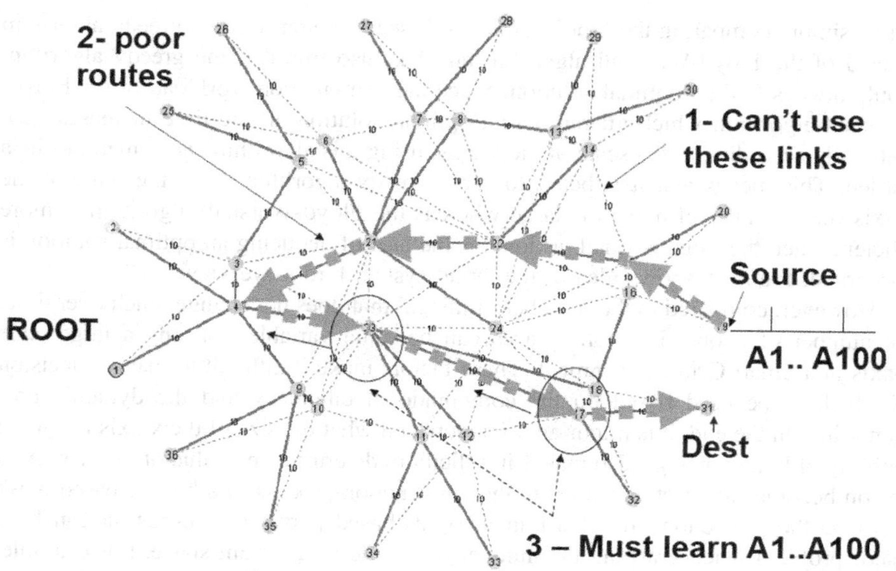

Fig. 4. Example problems of scaling up native ethernet and Wi-Fi [18]

However, the mesh network adopts a system where all the zones will be linked wirelessly by allowing the wireless routers, which assume the role of antennas in the existing radio communication bases, to act as the mesh nodes if the main AP alone is being wired. This is a technology that has made it possible to implement the wired mesh-type network structure in the wireless networks and emerged to surmount the limits of existing wireless LAN. For instance, it is a form which has widened the hotspot zones established by the telecommunication companies. If the hotspot zone is one that allows users to connect with wireless networks within a building, the mesh network adopts a technology which allows them to do the same at the wider spaces such as resorts, large parks or ports. It is indeed possible to apply this technology to much wider spaces depending on the application methods. Meanwhile, the users can connect to a mesh network by taking a similar procedure used for the WiFi connection method. This technology is quite useful for connecting to a communication network in time of crisis in the remote areas where any kinds of cable communication networks are not available. Such a technology has the advantage in home networking as well so that it is regarded as one of the essential technologies for the era of ubiquitous ICT.

3 Utilization of Wi-Fi AP Connection Information for Determination of Headcount for Selection of Scheduling Algorithm

When simply comparing the time complexes, it seems better to use a greedy algorithm instead of the Floyd-Warshall algorithm but it is also true that the greedy algorithm would always find an optimal solution. Since the computation workload of the Floyd-Warshall algorithm which guarantees an optimal solution all the time increases geometrically according to the scale of the x-axis, using this algorithm can sometimes be a burden. This means that it is better to use a greedy algorithm when the scale of the x-axis (i.e., number of nodes) is large whereas the Floyd-Warshall algorithm is more efficient when the same is small, as long as the task of deducing an optimal solution is concerned. Figure 5 shows idea of the target system bird's-eye view.

However, compared to the 'number of things' that does not change much over time, the number of people in public places can be quite variable, forming a large-scale x-axis in a Smart Grid environment. These factors make it difficult to make a decision of which to be used based on the computational efficiency and the dynamic programming. In the end, it is important to understand what the size of the x-axis is but the mobility of human beings also makes it difficult to determine the value of the size. As a person becomes abstract as a node in the construction process of a human-based mesh network, the protection law of a human rights-based personal information can be a major problem when grasping the number of people in a certain space. For example, although it would be possible to do a headcount by using CCTV's, the same law may become an obstacle in some counties.

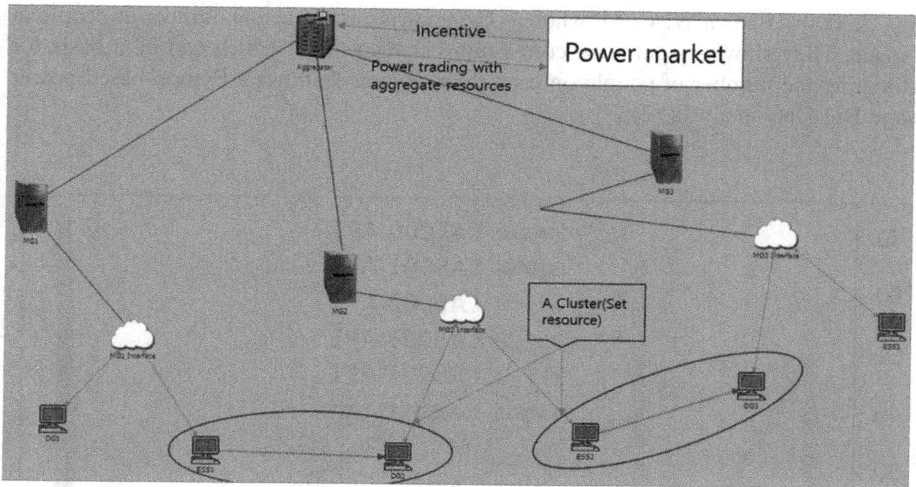

Fig. 5. The idea of the target system and bird's-eye view

4 Performance Evaluation Using Python

To solve such a problem and decide which algorithm should be used to efficiently construct a mesh network, performing a headcount within a Smart Grid environment by utilizing the factors significantly correlated with the number of people is proposed. Figure 6 shows a Python-based power usage Big Data analysis (1).

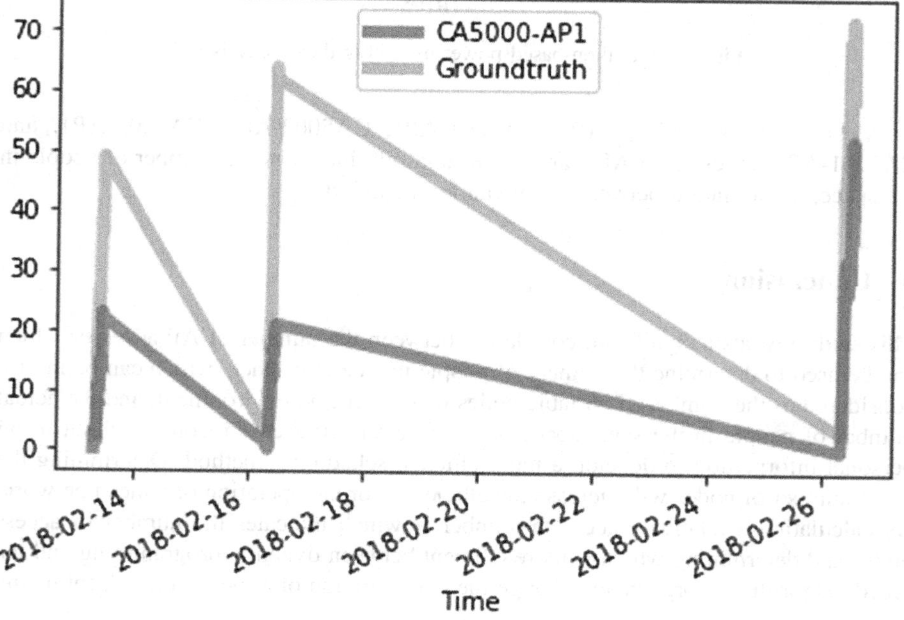

Fig. 6. A Python-based power usage big data analysis (1)

CA5000-AP1 is a Wi-Fi AP whereas Groundtruth is the actual number of people in a space. They show a significant correlation between them, which can be a basis for estimating the number of people on each floor. Figure 7 shows a Python-based power usage Big Data analysis (2).

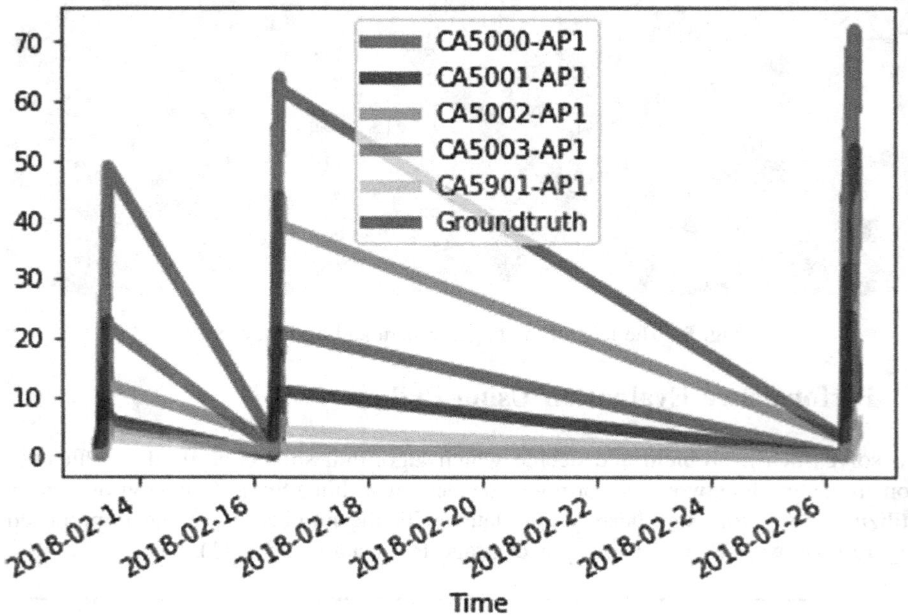

Fig. 7. A Python-based power usage big data analysis (2)

Meanwhile, as CA5000-AP1, CA5001-AP1, CA5002-AP1, CA5003-AP1, and CA5901-AP1 represents 5 AP's and as Groundtruth indicates the number of people in the space, a correlation between them can be identified.

5 Conclusion

This study reveals a significant correlation between the number of AP accesses which can be used to determine the number of people in a certain space, which can be in turn considered as the number of variable nodes in a smart grid environment, and the actual number of people in the same space based on their AP access records without using personal information to develop a more efficient scheduling method. Determining the exact number of nodes will increase the efficiency for the operation of a mesh network. By calculating a relatively accurate number N which indicates the number of access nodes and determining which is more efficient between dynamic programming and the greedy algorithm to organically change the configuration of a mesh network following the changes in the environment, a more efficient smart grid network environment can be created without performing unnecessary computation works wasting the energy.

In the future work, the Authors plans to implement an actual service by using the algorithm proposed in this study.

Acknowledgments. This work was supported by the National Research Foundation of Korea (NRF) grant funded by the Korean government (MSIT) (No. 2017R1C1B5077157).

References

1. Farhangi, H.: The path of the smart grid. IEEE Power Energy Mag. **8**(1), 18–28 (2010)
2. Huh, J.-H., Seo, K.: Blockchain-based mobile fingerprint verification and automatic log-in platform for future computing. J. Supercomput., 1–17 (2018)
3. Huh, J.-H., Otgonchimeg, S., Seo, K.: Advanced metering infrastructure design and test bed experiment using intelligent agents: focusing on the PLC network base technology for Smart Grid system. J. Supercomput. **72**(5), 1862–1877 (2016)
4. Weaver, W.W., Krein, D.P.: Game-theoretic control of small-scale power systems. IEEE Trans. Power Deliv. **24**(3), 1560–1567 (2009)
5. Kasbekar, G.S., Sarkar, S.: Pricing games among interconnected microgrids. In: Proceedings of Power and Energy Society General Meeting, pp. 1–8. IEEE (2012)
6. Mohsenian-Rad, A.H., Wong, V.W.S., Jatskevich, J., Schober, R., Leon-Garcia, A.: Autonomous demand-side management based on game-theoretic energy consumption scheduling for the future smart grid. IEEE Trans. Smart Grid **1**(3), 320–331 (2010)
7. Kshetri, N.: Blockchain's roles in strengthening cybersecurity and protecting privacy. Telecommun. Policy **41**, 1027–1038 (2017)
8. Levin, R.B., Waltz, P., LaCount, H.: Betting blockchain will change everything – SEC and CFTC regulation of blockchain technology. In: Handbook of Blockchain. Digital Finance, and Inclusion, vol. 2, pp. 187–212. Elsevier (2017)
9. Prybila, C., Schulte, S., Hochreiner, C., Webe, I.: Runtime verification for business processes utilizing the Bitcoin Blockchain. Futur. Gener. Comput. Syst., 1–11 (2017)
10. Sikorski, J.J., Haughton, J., Kraft, M.: Blockchain technology in the chemical industry: machine-to-machine electricity market. Appl. Energy **195**, 234–246 (2017)
11. Aguayo, D., et al.: Link-level measurements from an 802.11b mesh network. ACM SIGCOMM Comput. Commun. Rev. **34**(4), 121–132 (2004)
12. Je, S.-M., Huh, J.-H.: Nash equilibrium solution for communication in strategic competition adopted by Zigbee network for micro grid. In: Kim, K.J., Baek, N. (eds.) ICISA 2018. LNEE, vol. 514, pp. 595–606. Springer, Singapore (2019). https://doi.org/10.1007/978-981-13-1056-0_58
13. Lee, S., Huh, J.-H.: An effective security measures for nuclear power plant using big data analysis approach. J. Supercomput., 1–28 (2018)
14. Bychkovsky, V., et al.: A measurement study of vehicular internet access using in situ Wi-Fi networks. In: Proceedings of the 12th Annual International Conference on Mobile Computing and Networking. ACM (2006)
15. Cheng, Y.-C., et al.: Accuracy characterization for metropolitan-scale Wi-Fi localization. In: Proceedings of the 3rd International Conference on Mobile Systems, Applications, and Services. ACM (2005)
16. Bahl, P., et al.: White space networking with Wi-Fi like connectivity. ACM SIGCOMM Comput. Commun. Rev. **39**(4), 27–38 (2009)

17. Kellogg, B., et al.: Wi-Fi backscatter: internet connectivity for RF-powered devices. ACM SIGCOMM Comput. Commun. Rev. **44**(4), 607–618 (2014)
18. Huawei: Shortest Path Bridging IEEE 802.1aq Overview (2011). https://meetings.apnic.net/31/ppt/APRICOT_SPB_Overview.ppt
19. Dorri, A., Kanhere, S.S., Jurdak, R.: Towards an optimized blockchain for IoT. In: Proceedings of the Second International Conference on Internet-of-Things Design and Implementation, pp. 173–178. ACM (2017)
20. Donelli, M., et al.: A planar electronically reconfigurable Wi-Fi band antenna based on a parasitic microstrip structure. IEEE Antennas Wirel. Propag. Lett. **6**, 623–626 (2007)
21. Huh, J.-H.: PLC-based design of monitoring system for ICT-integrated vertical fish farm. Hum. Centric Comput. Inf. Sci. **7**(20), 1–19 (2017)
22. Brachet, N., et al.: Method and system for selecting and providing a relevant subset of Wi-Fi location information to a mobile client device so the client device may estimate its position with efficient utilization of resources, U.S. Patent No. 8,369,264, USA, 5 February 2013
23. Tran, B.: Mesh network personal emergency response appliance, U.S. Patent No. 7,733,224, USA, 8 June 2010
24. Talla, V., et al.: Powering the next billion devices with Wi-Fi. In: Proceedings of the 11th ACM Conference on Emerging Networking Experiments and Technologies. ACM (2015)
25. Pass, R., Shi, E.: Fruitchains: a fair blockchain. In: Proceedings of the ACM Symposium on Principles of Distributed Computing, pp. 315–324. ACM (2017)
26. Huh, J.-H.: Implementation of lightweight intrusion detection model for security of smart green house and vertical farm. Int. J. Distrib. Sens. Netw. **14**(4), 1–11 (2018)
27. Raniwala, A., Chiueh, T. C.: Architecture and algorithms for an IEEE 802.11-based multi-channel wireless mesh network. In: Proceedings of the IEEE 24th Annual Joint Conference of the IEEE Computer and Communications Societies, INFOCOM 2005, vol. 3. IEEE (2005)
28. Sagari, S., et al.: Coordinated dynamic spectrum management of LTE-U and Wi-Fi networks. In: 2015 IEEE International Symposium on Dynamic Spectrum Access Networks (DySPAN). IEEE (2015)

Automatic Classification of Transformed Protocols Using Deep Learning

Changmin Jeong[1(✉)], Mirim Ahn[1], Haengho Lee[1], and Younggiu Jung[2]

[1] Agency for Defense Development, Yuseong, Daejeon, Republic of Korea
{min,mahn,Haengho}@add.re.kr
[2] YM-Naeultech, Inharo, Namgu, Incheon, Republic of Korea
youngq.jung@ym-naeultech.com

Abstract. Protocol reverse-engineering technique can be used to extract the specification of an unknown protocol. However, there is no standardized method and in most cases, the extracting process is done manually or semi-automatically. Since only frequently seen values are extracted as fields from the messages of a protocol, it is difficult to understand complete specification of the protocol. Therefore, if the information about the structure of the unknown protocol could be acquired in advance, it would be easy to conduct reverse engineering. This paper suggests a method of recognizing 8 commercial protocols and transformed protocols of their own using deep learning techniques. When the proposed method is conducted prior to APRE (Automatic Protocol Reverse Engineering) process, it is possible to obtain useful information beforehand when similarities exist between unknown protocols and learned protocols.

Keywords: Automatic protocol reverse engineering · Transformed protocol

1 Introduction

The network-centric battlefield in the future will be a five-dimensional space including ground, sea, air, space and cyberspace. In cyber space, if there is a way to analyze the communication protocol that enemies use in their weapon systems, various types of cyber and electronic warfare attacks can be possible.

Protocol reverse engineering techniques that can extract protocol specifications are used for a structure analysis of unknown protocols [1–7]. However, there is no standardized method and in most cases, the extracting process is done manually or semi-automatically. Since only frequently seen values are extracted as fields from the messages of a protocol, it is difficult to understand specification of the protocol completely. Therefore, if advance information about the structure of an unknown protocol can be acquired before performing protocol reverse-engineering, the analysis of the private protocol structure will be simpler and easier. Manual analysis usually takes a lot of time and require large amount of professional knowledge also.

J. H. Park et al. (Eds.): PDCAT 2018, CCIS 931, pp. 153–158, 2019.
https://doi.org/10.1007/978-981-13-5907-1_16

Analysts generally proceed manually to obtain advance information. However, this procedure requires a lot of time and professional knowledge. Many studies have been done to solve these problems using machine learning or statistic information [8–14].

This paper suggests a method to derive prior information for the analysis of the private protocol using deep learning techniques. We first used a deep learning technique to find out if it is possible to classify the commercial protocols without protocol ID and Port number. Then we studied whether the similar protocol can be recognized based on the deep learned information even if the protocol structure is changed.

The components of the protocol are syntax, semantic, and timing. Therefore, when a new protocol is created using existing protocols, it is done by considering all three factors of the protocol rather than changing only a part of the existing protocol structure. However, the purpose of our work is to acquire advance information about protocol structure before applying APRE process that derives the protocol structure. Therefore, the proposed method will be useful to understand the structure of the new protocol made similar to the structure of the existing protocol.

The composition of the paper is as follows. Section 2 describes the types, characteristics, and transformation of the protocols used in the deep learning. Section 3 describes the structure of deep learning and the recognition rate used in the learning of the deep learning. Finally, the conclusion of the study is described in Sect. 4.

2 Protocol

2.1 Protocols Used in Learning

The 8 protocols used in the learning are widely used commercially and shown in Table 1. After collecting the packets of these protocols, feature vectors are extracted based on them. The feature extractions method is as follows. First, 256 bytes including a header for each protocol packet are extracted. Then the average and frequency values are obtained. The example of obtaining frequency values from the 256 bytes of protocols is shown in Fig. 1.

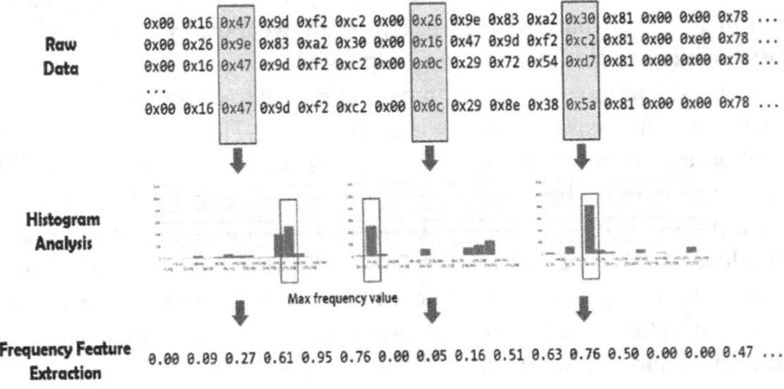

Fig. 1. Process of frequency extraction

Table 1. Protocols used in learning

Criteria	Description
ARP	Address Resolution Protocol
ICMP	Internet Control Message Protocol
HTTP	HyperText Transfer Protocol
LDAP	Lightweight Directory Access Protocol
MYSQL	Open source relational database management system based Structured Query Language
SMTP	Simple Mail Transfer Protocol
TDS	Tabular Data Stream
DNS	Domain Name System

2.2 Transformed Protocol

The definition of a transformed protocol is a new protocol whose structure is similar to that of an existing protocol. The components of the protocol are syntax, semantic, and timing and all of these three elements of a protocol are considered when creating a new protocol from an existing protocol. But in this paper we define a transformed protocol as a protocol of which only syntax is changed.

In order to form transformed protocols, it is assumed that existing protocols are 8 commercial protocols shown in Fig. 1. of which the option field of each message is removed by 0%, 20%, 40% and 60% which we define as the transformation rate. Though this approach does not accurately represent transformed protocols, it is one way to create transformed protocols easily. Figure 2 shows an example to create the transformed protocol. In Fig. 2, the blue shaded area is modified or deleted according to the transformation rate.

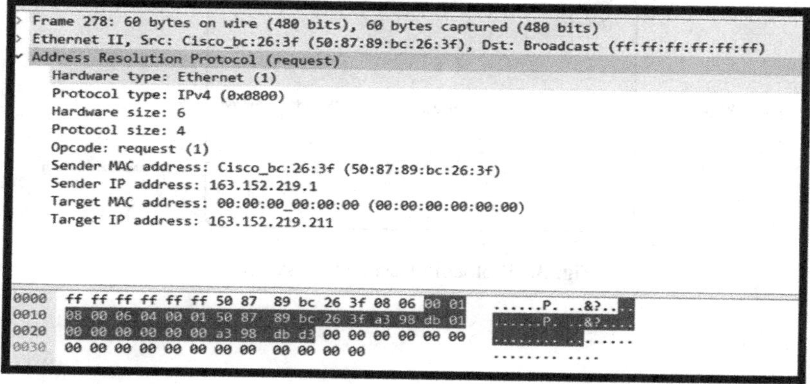

Fig. 2. Example of transformed protocol formation

3 Classification Result

The protocol classification system that learns and recognizes the existing or new protocols is shown in Fig. 3. First, 8 commercial protocols are input in PCAP format, and IDs and port numbers of protocols are deleted, and feature vectors are extracted through statistical information or autoencoder. As the proposed method is to classify protocol based on its structure or header contents, protocol ID and Port Number should be deleted.

The feature vectors of protocols are used as inputs of the neural network. Classification results of transformed protocols are shown in Fig. 4. Protocols are transformed by 0%, 20%, 40% and 60%. In Fig. 4. the "original" means the average classification rate of 8 protocols with protocol IDs and port numbers and "0%", "20%", "40%" and "60%" mean the average classification rate of 8 protocols without protocol IDs and port numbers. The results of Fig. 4. show that the proposed method classifies protocols well without using protocol IDs and port numbers. It can also be seen that this method recommends the most similar protocol even when the existing protocol is transformed. As a result, if the preprocessing is performed using the method proposed in the paper prior to the APRE process on the unknown protocols, meaningful information can be obtained when the unknown protocol has similarity to the learned protocol structure.

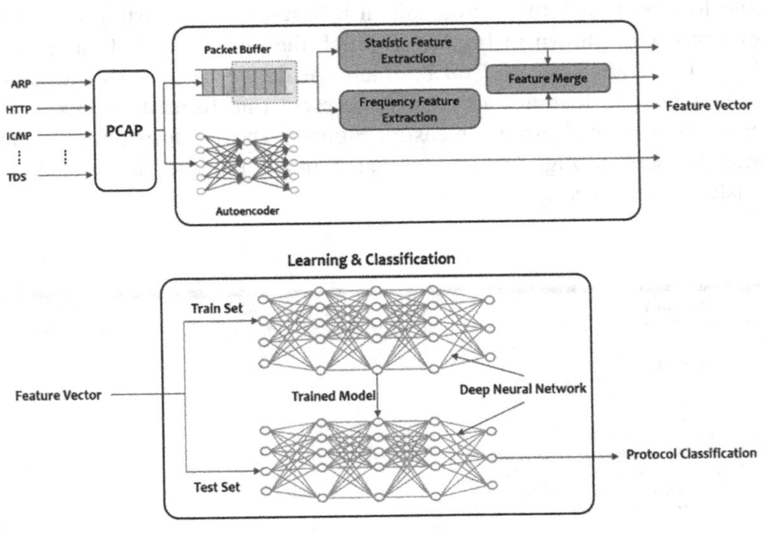

Fig. 3. Protocol classification system

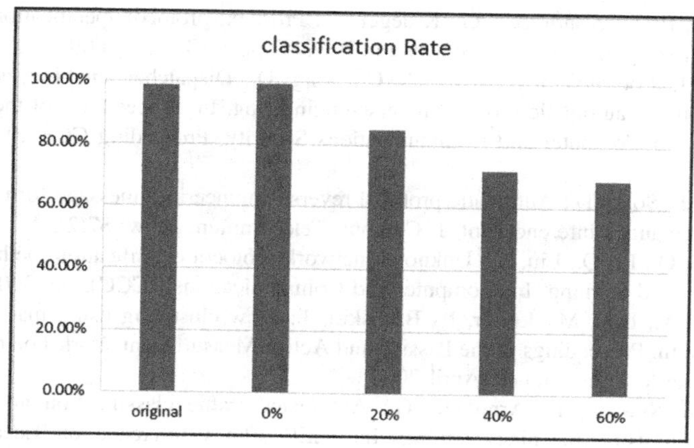

Fig. 4. Classification rate of transformed protocols

4 Conclusion

In this paper, the method to classify 8 commercial protocols and their transformed protocols using the deep learning technology is proposed. By using the method prior to APRE process of the unknown protocols, useful advance information can be obtained when the unknown protocols have the similar structure to the learned protocols. We found out that this method recommends the most similar protocol even when the existing protocol is transformed. As a result, if the preprocessing is performed using the method proposed in the paper prior to the APRE process on the unknown protocols, meaningful information can be obtained when the unknown protocol has similarity to the learned protocol structure.

Currently, only structure transformations are considered in transforming protocols. In the future, semantics, timing, etc. will also be in consideration.

References

1. Weidong, C., Jayanthkumar, K., Wang, H.J.: Discoverer: automatic protocol reverse engineering from network traces. In: USENIX Security Symposium, pp. 199–212 (2007)
2. Caballero, J., Yin, H., Liang, Z., Song, D.: Polyglot: automatic extraction of protocol message format using dynamic binary analysis CCS 2007. In: Proceedings of the 14th ACM Conference on Computer and Communications Security, pp. 317–329. ACM, New York (2007)
3. Wondracek, G., Comparetti, P.M., Kruegel, C., Kirda, E.: Automatic network protocol analysis. In: Proceedings of the 15th Annual Network and Distributed System Security Symposium (NDSS 2008) (2008)
4. Cui, W., Peinado, M., Chen, K., Wang, H.J., Irun-Briz, L.: Tupni: automatic reverse engineering of input formats. In: Proceedings of the 15th ACM Conference on Computer and Communications Security, pp. 391–402 (2008)

5. Comparetti, P.M., Wondracek, G., Kruegel, C.: Prospex: protocol specification extraction. In: 30th IEEE Symposium on Security and Privacy, pp. 110–125 (2009)
6. Caballero, J., Poosankam, P., Kreibich, C., Song, D.: Dispatcher: enabling active Botnet infiltration using automatic protocol reverse-engineering. In: Proceedings of the 16th ACM Conference on Computer and Communications Security, Proceeding CCS 2009, pp. 621–634 (2009)
7. Caballero, J., Song, D.: Automatic protocol reverse-engineering: message format extraction and field semantics inference. Int. J. Comput. Telecommun. Netw. **57**(2), 451–474 (2012)
8. Lin, R., Li, O., Li, Q., Liu, Y.: Unknown network protocol classification method based on semi-supervised learning. In: Computer and Communications (ICCC), pp. 300–308 (2015)
9. McGregor, A., Hall, M., Lorier, P., Brunskill, J.: Flow clustering using machine learning techniques. In: Proceedings of the Passive and Active Measurement Workshop (PAM 2004), Antibes Juan-les-Pins, France, April 2004
10. Zander, S., Nguyen, T., Armitage, G.: Automated traffic classification and application identification using machine learning. In: IEEE 30th Conference on Local Computer Networks (LCN 2005), Sydney, Australia, November 2005
11. Moore, A., Zuev, D.: Internet traffic classification using Bayesian analysis techniques. In: ACM International Conference on Measurement and Modeling of Computer Systems (SIGMETRICS) 2005, Banff, Alberta, Canada, June 2005
12. Auld, T., Moore, A.W., Gull, S.F.: Bayesian neural networks for Internet traffic classification. IEEE Trans. Neural Netw. **18**(1), 223–239 (2007)
13. Williams, N., Zander, S., Armitage, G.: A preliminary performance comparison of five machine learning algorithms for practical IP traffic flow classification. Special Interest Group on Data Communication (SIGCOMM) Comput. Commun. Rev., **36**(5), 5–16 (2006)
14. Erman, J., Mahanti, A., Arlitt, M.: Internet traffic identification using machine learning techniques. In: Proceedings of 49th IEEE Global Telecommunications Conference (GLOBECOM 2006), San Francisco, USA, December 2006

Covert Timing Channel Design for Uniprocessor Real-Time Systems

Jaeheon Kwak and Jinkyu Lee(✉)

College of Software, Sungkyunkwan University, Suwon, Republic of Korea
{jaehunny, jinkyu.lee}@skku.edu

Abstract. Different from a general-purpose system, a real-time system requires stringent timing guarantees. While existing offline analysis techniques can provide timing guarantees using the worst-case execution time (WCET) of individual tasks, a variation of actual execution time makes it difficult to build covert timing channel. In this paper, we first present a novel covert timing channel, which considers actual execution time distribution of tasks and controls execution time to leak data between conspirators; we demonstrate that it is possible to leak data in real-time systems. Second, we suggest two enhancing techniques called S-R LCM (sender-receiver least common multiple) and noise area to reduce noise in communication. Through simulations, we demonstrate that our covert timing channel can serve trade-off between transmission speed and accuracy; that is, it shows average 50.2%, 54.6% and 51.3% accuracy for 100 test cases with thresholds 0, 1.4 and 2.8. Average 58.4% accuracy is accomplished with best threshold values for 100 test cases, and the maximum accuracy for a single test case is recorded 100.0%.

Keywords: Real-time systems · Covert channel · Timing inference attack Rate monotonic

1 Introduction

In a real-time system, a single failure of meeting deadlines may cause disastrous damage. Therefore, the system's temporal correctness should be verified, and the verification is based on the worst-case execution time (WCET) of individual tasks in the system. However, actual execution time differs greatly from WCET. Many researchers analyzed WCET [3–5], and they revealed that only few tasks run as much as WCET while most actual execution time is significantly shorter than its WCET. Therefore, it may be possible for some malicious tasks to utilize this gap.

Another problem is that scheduling algorithms commonly used in a real-time system such as RM (Rate Monotonic), EDF (Earliest Deadline First) or LLF (Leas Laxity First) [1, 2], are not affected by actual execution time when they determine priority of tasks. Most real-time systems are designed for work-conserving, and this point yields exposure to timing inference attack like side channel or covert timing channel. Several studies have already suggested that various attack methods [6–8] as well as defense methods [9–12].

J. H. Park et al. (Eds.): PDCAT 2018, CCIS 931, pp. 159–168, 2019.
https://doi.org/10.1007/978-981-13-5907-1_17

Among those studies, Son et al. (2006) proposed covert timing channel on a uniprocessor real-time system which uses RM scheduler [7]. They designed covert timing channel which is established by controlling execution time and categorized that sort of covert timing channels into four types. Our previous work [17] also proposed covert timing channel on a real-time system which utilizes execution time distribution. But, until now, no research has implemented covert timing channel or checked its accuracy.

In this paper, we propose advanced covert timing channel in a real-time system. Two tasks, a sender and a receiver utilize timing changes as bit data with conspiration. The sender controls its job's execution time to send data, and the receiver infers it by changes of start execution time of itself. We also devise some inevitable transmission fail scenarios and suggest two advanced methods (called S-R LCM and noise area) to solve those scenarios. We evaluate our covert timing channel by simulating the channel with its accuracy. Result shows average 50.2%, 54.6% and 51.3% accuracy for 100 test cases with thresholds 0, 1.4 and 2.8. With best threshold values for 100 test cases, average 58.4% accuracy is accomplished. The maximum accuracy for a single test case is 100.0%.

This paper makes following contributions:

1. We introduce a novel covert timing channel for real-time systems;
2. We suggest several scenarios which can preclude transmission of covert timing channel on real-time systems;
3. We propose two improving techniques, S-R LCM (Sender-Receiver Least Common Multiple) and noise area, to resolve limitation of 2; and
4. We demonstrate effectiveness of the proposed scheme via simulations.

The rest of this paper is organized as follows. In Sect. 2, we check background knowledge. In Sect. 3, we describe our basic covert timing channel model and its system model. In Sect. 4, we figure out accuracy-problematic scenarios and suggest techniques to relax the scenarios. In Sect. 5, we evaluate our model. Lastly, we conclude this paper in Sect. 6.

2 Background

2.1 Worst Case Execution Time (WCET) and Gumbel Distribution

Real-time system tasks are usually assumed to run as WCET, in the worst case to ensure operation and deadline compliance. But they are analyzed to have shorter actual execution times than WCET [3–5] in most cases. It is known that the actual execution time of the real-time system tasks has a probabilistic distribution of the shape shown in Fig. 1(a). The probability increases exponentially from the best-case execution time (BCET) and the probability decreases rapidly as it approaches the WCET, after its peak. According to Edgar and Burns, Gumbel distribution [15], which has probability distribution model as shown in Fig. 1(b), can be used to model the execution time probability distribution of real-time system tasks [4]. Although other more complex probability distribution models can be used, the covert timing channel covered in this paper does not have a large causal relationship with the complexity of the models or accuracy of the probability distribution model. Thus, in this paper, we assume that the execution time of all tasks follows the Gumbel distribution, which is representative and simple to implement.

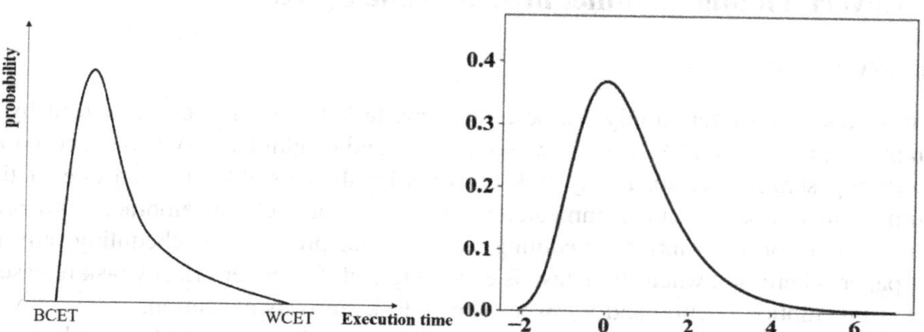

Fig. 1. (a) Typical execution time distribution of a real-time system task (b) Gumbel distribution (location = 0, scale = 1)

2.2 Covert Timing Channel

Covert timing channel is a kind of covert channel. Covert channel is a hidden channel that allows unauthorized information to be leaked between a sender and a receiver who colluded in advance [13].

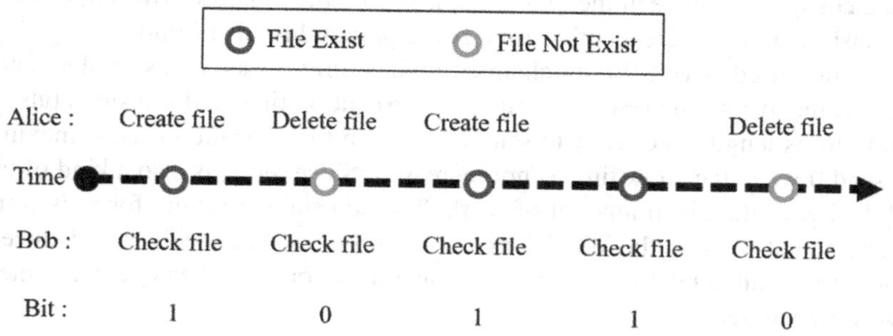

Fig. 2. Example of covert timing channel

A typical example of a covert channel is to transmit 0 and 1 bit depending on the presence of a file [14]. Figure 2 is a conceptual diagram of covert channel that can be implemented through Alice and Bob's conspiration. Suppose Alice can read, write, save, and delete files including a file "FileASDF", and Bob can only check whether some files exist or not. Alice can delete the file to send a 0 bit and send 1 bit by creating the file. Bob receives 1 bit if "FileASDF" is present, and 0 bit otherwise. This allows data leakage while checking and verifying the existence of the file every time, which is appointed between the two. As Alice and Bob send and receive data by the existence of the file via this covert channel, covert timing channel controls system resources to change timing to send and receive data.

3 Covert Timing Channel in Real-Time Systems

3.1 System Model

In this paper we target strictly periodic real-time task τ_i which is characterized by 3 parameters, period T_i, execution time (WCET) C_i, and deadline D_i. We consider work-conserving scheduling, so, if any task ends earlier than its WCET, then one of the waiting other tasks executes immediately. While our model can embrace both pre-emptive and non-preemptive scheduling, we consider preemptive scheduling only in this paper. Therefore, when some task is executing and if a higher priority task releases, then a preemption occurs, and a lower priority task stops its execution.

Let's denote τ_s as sender task, τ_r as receiver task and τ_{n} as noise tasks, which are any other tasks. About the tasks and process, we suppose one process has one task and there exist no methods to communicate with each other, except our covert timing channel. Our attack model's goal is sending secret data from sender τ_s to receiver τ_r.

3.2 Overall Design

In the RM scheduling algorithm, which is commonly used in real-time system, task's periods determine priority. Therefore, if two tasks know each other's period, they can inference which task between the two has higher priority than another. Furthermore, if there is a conspiracy between the two tasks, it is possible to know how much WCET is in each task and how the execution time has a probability distribution.

In our proposed covert timing channel, high priority tasks act as senders and low tasks act as receivers. The sender τ_s adjusts its execution time and transmits bits 0 or 1 according to its length. We set τ_s to send bit 1 when the execution time is maximized, and to send 0 when the work time is not increased. Since the τ_s is also a kind of task, it should deal with the given amount of work. The adjustable runtime for τ_s is from the end of the actual work to the WCET, as shown in Fig. 3. The receiver τ_r deduces that the sender transmitted bit 1 if its work starts later than normal, otherwise it assumes that bit 0 was transmitted.

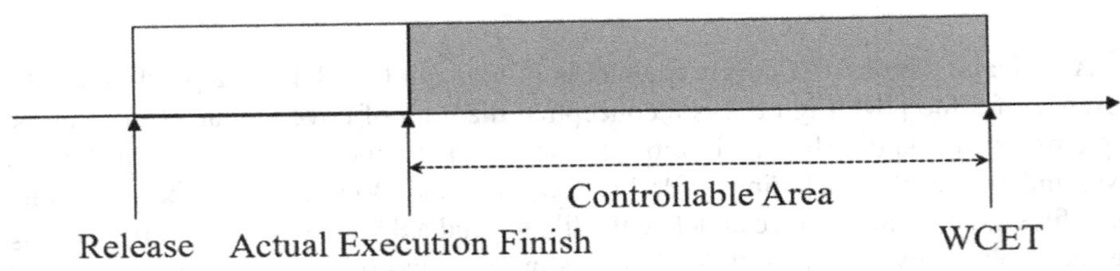

Fig. 3. Controllable area of sender task

3.3 Synchronization

The point to consider, when configuring the covert timing channel, is that other tasks between τ_s and τ_r are also performed. We denote those other tasks, noise τ_n. Figure 4 illustrates noise between τ_s and τ_r's job. τ_s and τ_r are interrupted by this τ_n. Therefore, τ_s and τ_r must be able to understand the effects of this noise. However, it is difficult to understand this because the types and numbers of incoming tasks between them are different depending on various intervals. Nevertheless, if they know the LCM (Least Common Multiple) of the entire tasks' periods, observation can be used to infer the average effect of τ_n on the interval.

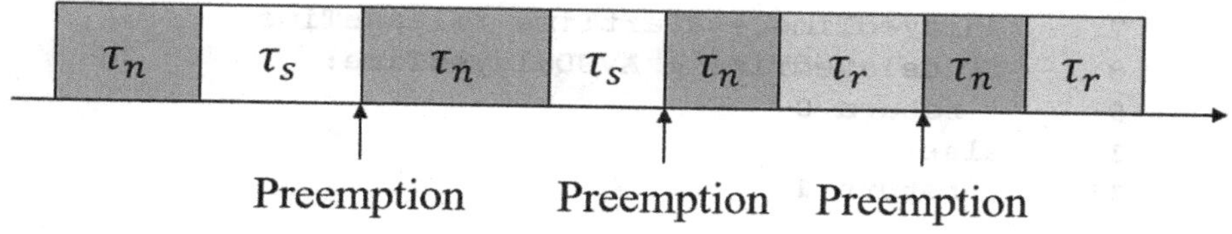

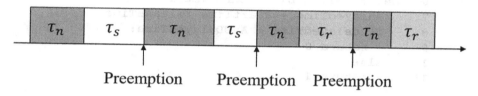

Fig. 4. Sender τ_s, receiver τ_r and noise τ_n

In Synchronization step, τ_s and the receiver observe other tasks without taking special action. They record delayed time by other tasks and find a pattern to inference whole tasks periods LCM. This process can be simply conducted. In a real-time system, task's execution time distribution is similar to that of the Gumbel distribution and the model's distribution is not so large. Therefore, if the sender and the receiver find a pattern that repeats the time delayed by another task, LCM is easy to infer. In this paper, we do not deal with the whole process of obtaining LCM in depth but assume that we have already found the LCM at the synchronization stage.

Once the LCM is obtained, the receiver calculates the average time delayed by different jobs in the interval. Because the sender does not have to consider other tasks, this process is only done by the receiver. The synchronization step ends when the calculation of the average delayed time $\left(L^k_{avg}\right)$, by other tasks for each interval (k), is completed during the period which is set by the pre-conspiration.

3.4 Sending and Receiving

Our covert timing channel consists of two tasks: the sender τ_s and the receiver τ_r. If τ_s wants to send bit 0 to τ_r, it executes as normal time; if τ_s wants to send bit 1, then it extends its work and executes as WCET. Because τ_r has lower priority than τ_s, τ_s's execution delays τ_r job's execution starts. τ_r can know whether τ_s extends its execution time or not at a specific interval k, by comparing with average delayed time L^k_{avg}. If τ_s extended its execution time, then τ_r's execution starts later than usual. τ_r infers those things by calculating observed delayed time at current L^k_{cur}, and subtracts with L^k_{avg}. If $L^k_{cur} - L^k_{avg}$ is bigger than 0, τ_r get 1 bit from τ_s; otherwise, it receives bit 0.

This process is done by Algorithm 1. The sender's work is described in "sendingBit", and the receiver's work is described in "receivingBit". The sender checks data to send (line 2 and 4) controls its execution time (line 3 and 5). The receiver calculates its delayed execution start time (line 7) and confirms the sender's data (line 8–11).

```
1   def sendingBit (actualExecutionTime, data):
2     if data == 0:
3       return actualExecutionTime
4     else:
5       return WCET

6   def receivingBit (startTime):
7     delayedTime = startTime-releaseTime
8     if delayedTime > AVGDelayedTime:
9       return 0
10    else:
11      return 1
```

Algorithm 1. Sender τ_s, and receiver τ_r work

4 Improving Techniques

4.1 S-R LCM (Sender-Receiver Least Common Multiple)

In some special cases, the covert timing channel may not work at all. Figure 5 describes that situation. If distance between τ_s and τ_r is too huge, then τ_s's execution time extension may not reach to τ_s. In worst scenario, idle time can exist between them. To resolve this problem, we propose a S-R LCM (Sender-Receiver Least Common Multiple) technique. The idea is very simple. Send and receive information only when τ_s and τ_r start at the same time.

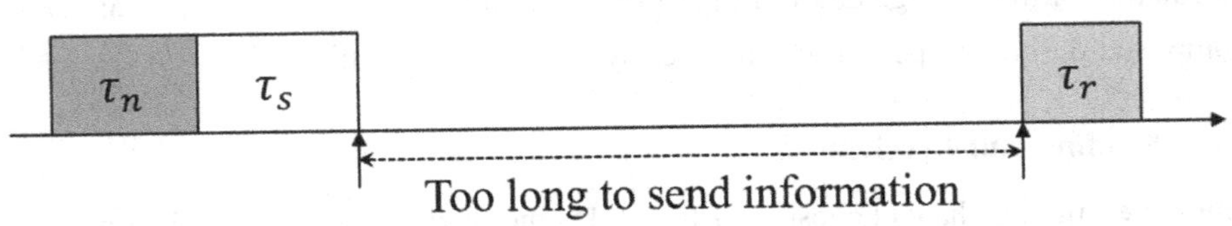

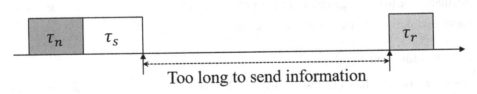

Fig. 5. Sender τ_s, receiver τ_r and noise τ_n

Because τ_s and τ_r know each other's period, they can infer another task's starting time. At the sender task and receiver task's period LCM, it is guaranteed that those tasks start at the same time. With this simple technique, covert timing channel can avoid such special case with sacrificing transmission speed. From Algorithm 2, line 2, 7, 10, 19 do this work by checking periods.

```
def sendingBit (actualExecutionTime, data):
  if releaseTime % SR_LCM == 0:
    if data == 0:
      return actualExecutionTime
    else:
      return WCET
  else:
    return Null (Pass)

def receivingBit (startTime):
  if releaseTime % SR_LCM == 0:
    delayedTime = startTime-releaseTime
    marginCheck = delayedTime - AVGDelayedTime
    if marginCheck > margin:
      return 0
    elif marginCheck < -margin:
      return 1
    else:
      return -1 (Noise)
  else:
    return Null (Pass)
```

Algorithm 2. Sender τ_s, and receiver τ_r work with SR-LCM and noise area

4.2 Noise Area

However, another problem still exists. Figure 6 describes the problem. If there is too huge τ_n exists between τ_s and τ_r, τ_s's influence is hard to reach to τ_r. To make this problem more difficult to solve, this problem cannot be avoided in the same way as S-R LCM.

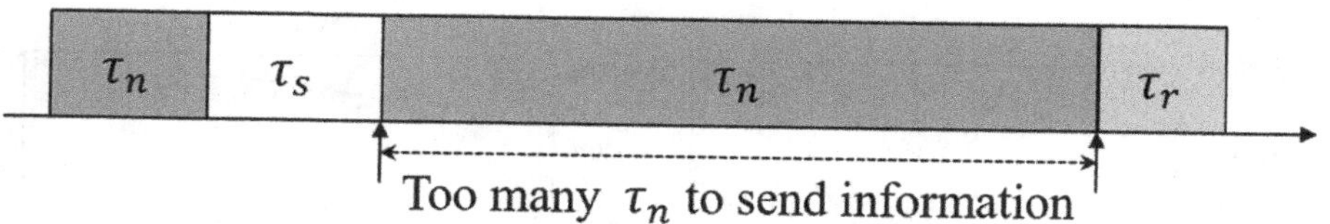

Fig. 6. Too many noises between sender and receiver for transmission to success

To relax above problem, we suggest noise area, which withdraws hard to infer bits information and retains relatively confident bits information. In our basic model, τ_r infers bit as whether $L^k_{cur} - L^k_{avg}$ is bigger than 0 or not. Noise area add margin to the value 0, and if $L^k_{cur} - L^k_{avg}$ cannot exceed the margin (x), regards that information as just noise and dump. So, when $-x < L_{cur}^k - L_{avg}^k < x$, τ_r just passes this calculated result and waits next bit. Like S-R LCM, it decreases transmission speed but improves accuracy. From Algorithm 2, only receiver does this process at line 11–18.

5 Evaluation

We evaluate the proposed covert timing channel by estimating its accuracy. To estimate its accuracy, we implement simulation of real-time system task and covert timing channel. RM scheduling is used, and all task generated to follow Gumbel distribution. It is well-known fact that RM scheduling ensures schedulability when whole system's usability is less than 0.69 [16]. We discard all scenarios when usability excesses 0.69. We target system which has 10 tasks and among those 10 tasks, τ_s is picked from top 20% highest priority tasks and τ_r is picked from top 20% to top 40% highest priority tasks. Gumbel distribution is used to generate task set, and its parameters are also selected randomly. Each task's usability is set from 0.1 to 0.05. For noise area, margin x is searched from 0 to 2.8 with 0.2 interval, and task's WCET is ranged from 0 to 10. Simulation conducts with S-R LCM methods and compares noise area margin's effectiveness with its margin value x. We give LCM observation for 20 times to calculate L_{avg}^k, and we check that 20 is a sufficient number because mean absolute error of L_{avg}^k is 7.92%.

We test the simulation with 100 tasks sets. The result shows that covert timing channel in real-time system is available but also faces limit. When we do not apply noise area technique, its accuracy is not reasonable to be used in practice. However, with sacrificing transmission speed, we can improve its accuracy up to from 50.2%, to 54.6% and 51.3% with x, 1.4 and 2.8. Figure 7(a) shows overall tendency accuracy versus x. Up to 58.4% accuracy can be obtained if channel can find optimal x. We check that too large x reduces accuracy. This is probably because when x is too large, then sender ignores the influence of the delay time given by the sender and to be more affected by noise. Moreover, increasing x results increasing of noise, as shown in Fig. 7(b), so channel should find proper x. For the best case, covert timing channel accuracy shows 100.00% accuracy even without noise area. It seems that if environment fits well with the channel, the accuracy of the channel increases substantially.

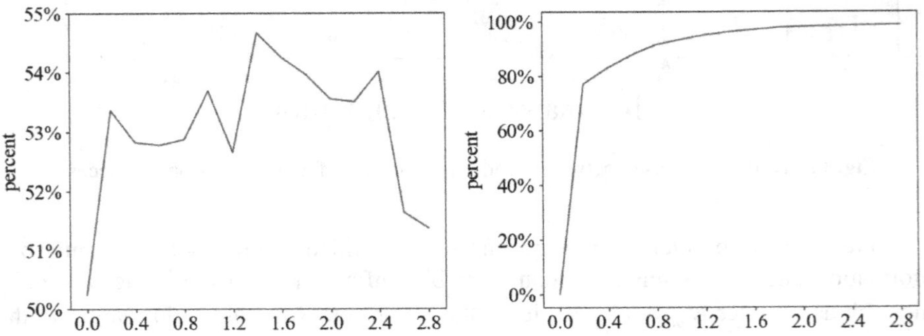

Fig. 7. (a) Accuracy versus noise area margin x (b) Noise portion versus noise area margin x

6 Conclusion

This research aims to design covert timing channel in uniprocessor real-time system. It proposes covert timing channel with handling execution time by specific conspired tasks. Also, the paper analyzes bad scenarios when the covert timing channel fails to communicate and proposes two methods, S-R LCM and noise area, to relax those problems.

Proposed covert timing channel is evaluated by simulation. The result shows 50.2%, 54.6% and 51.3% accuracy with margin 0, 1.4 and 2.8. The accuracy seems that proposed covert timing channel attack in real-time system needs improvement and lacks practicality. However, if channel can find suitable margin length its accuracy can be up to 58.4%, and even for best case, its accuracy reaches 100.0% without any our noise area techniques. This point sufficiently implies that system designers need to be careful with this attack method. Future work will study reasonable margin searching algorithm and other accuracy improving methods.

Acknowledgement. A short, earlier version of this paper has been presented as a domestic conference [17], which is 3-page-long.

This research was supported by the National Research Foundation of Korea (NRF) funded by the Ministry of Science and ICT (2017R1A2B2002458, 2017H1D8A2031628, 2017K2A9A1A01092689) and by the Ministry of Education (2018R1D1A1B07040321). This research was also supported by the IITP (Institute for Information & communications Technology Promotion) funded by the MSIT (Ministry of Science and ICT) (2015-0-00914, IITP-2017-2015-0-00742).

References

1. Liu, C.L., Layland, J.W.: Scheduling algorithms for multiprogramming in a hard-real-time environment. J. ACM (JACM) **20**(1), 46–61 (1973)
2. Liu, F., Narayanan, A., Bai, Q.: Real-time systems (2000)
3. Hansen, J., Hissam, S., Moreno, G.A.: Statistical-based WCET estimation and validation. In: OASIcs-OpenAccess Series in Informatics, vol. 10. Schloss Dagstuhl-Leibniz-Zentrum für Informatik (2009)
4. Bernat, G., Colin, A., Petters, S.M.: WCET analysis of probabilistic hard real-time systems. In: null, p. 279. IEEE, December 2002
5. Edgar, S., Burns, A.: Statistical analysis of WCET for scheduling. In: 2001 22nd IEEE Proceedings of Real-Time Systems Symposium, (RTSS 2001), pp. 215–224. IEEE, December 2001
6. Kocher, P.C.: Timing attacks on implementations of Diffie-Hellman, RSA, DSS, and other systems. In: Koblitz, N. (ed.) CRYPTO 1996. LNCS, vol. 1109, pp. 104–113. Springer, Heidelberg (1996). https://doi.org/10.1007/3-540-68697-5_9
7. Son, J.: Covert timing channel analysis of rate monotonic real-time scheduling algorithm in MLS systems. In: 2006 IEEE Information Assurance Workshop, pp. 361–368. IEEE, June 2006
8. Chen, C.Y., et al.: Schedule-based side-channel attack in fixed-priority real-time systems (2015)

9. Völp, M., Hamann, C.J., Härtig, H.: Avoiding timing channels in fixed-priority schedulers. In: Proceedings of the 2008 ACM Symposium on Information, Computer and Communications Security, pp. 44–55. ACM, March 2008
10. Yoon, M.K., Mohan, S., Chen, C.Y., Sha, L.: TaskShuffler: a schedule randomization protocol for obfuscation against timing inference attacks in real-time systems. In: 2016 IEEE Real-Time and Embedded Technology and Applications Symposium (RTAS), pp. 1–12. IEEE, April 2016
11. Pellizzoni, R., Paryab, N., Yoon, M.K., Bak, S., Mohan, S., Bobba, R.B.: A generalized model for preventing information leakage in hard real-time systems. In: 2015 IEEE Real-Time and Embedded Technology and Applications Symposium (RTAS), pp. 271–282. IEEE, April 2015
12. Mohan, S., Yoon, M.K., Pellizzoni, R., Bobba, R.B.: Integrating security constraints into fixed priority real-time schedulers. Real-Time Syst. **52**(5), 644–674 (2016)
13. Wray, J.C.: An analysis of covert timing channels. J. Comput. Secur. **1**(3–4), 219–232 (1992)
14. Stamp, M.: Information Security: Principles and Practice. Wiley, New York (2011)
15. Gumbel, E.J.: Statistics of Extremes. Courier Corporation, Mineola (2012)
16. Sha, L., Rajkumar, R., Sathaye, S.S.: Generalized rate-monotonic scheduling theory: a framework for developing real-time systems. Proc. IEEE **82**(1), 68–82 (1994)
17. Jaeheon, K., Jinkyu, L.: Covert timing channel considering execution time distribution in real-time systems. In: Korea Computer Congress (KCC), pp. 1920–1922 (2017)

Algorithms and Applications

Parallelization of the DIANA Algorithm in OpenMP

Hethini Ribeiro[1], Roberta Spolon[1(✉)], Aleardo Manacero Jr.[2], and Renata S. Lobato[2]

[1] Computer Department, Universidade Estadual Paulista "Júlio de Mesquita Filho" (UNESP), Bauru, SP, Brazil
hethini.ribeiro@outlook.com, roberta@fc.unesp.br
[2] Department of Computer Science and Statistics, Universidade Estadual Paulista "Júlio de Mesquita Filho" (UNESP), São José do Rio Preto, SP, Brazil
aleardo@sjrp.unesp.br, renata.spolon@unesp.br

Abstract. Global data production has been increasing by approximately 40% per year since the beginning of the last decade. These large datasets, also called Big Data, are posing great challenges in many areas and in particular in the Machine Learning (ML) field. Although ML algorithms are able to extract useful information from these large data repositories, they are computationally expensive such as AGNES and DIANA, which have O(n) and $O(2^n)$ complexity, respectively. Therefore, the big challenge is to process large amounts of data in a realistic time frame. In this context, this paper proposes the parallelization of the DIANA OpenMP algorithm. Initial tests with a database with 5000 elements presented a speed up of 5,2521. It is believed that, according to Gustafson's law, for a larger database the results will also be larger.

Keywords: Machine learning · Parallelization · DIANA · OpenMP

1 Introduction

Due to the diversity and large amount of data produced in the world, the analysis of these data is practically impossible to be carried out within a reasonable time. There is great demand to understand, synthesize, and make decisions on virtually all types of data [1]. With the production of data growing there is a need for machines capable of analyzing and processing the set of data. In this context machine learning algorithms (ML) play a central role in the analysis. The unprecedented production of these large amounts of data, greatly increases the analysis of these algorithms.

Another point to be taken into consideration is the computational complexity of ML methodologies. This is a limiting factor that can make the application of many algorithms impracticable, especially when they work with large datasets [2]. Thus, the parallelization extends the applicability of existing algorithms to larger and more complex datasets [2, 3].

The studies conducted by [4–6] showed the advantage of the combination of the two fields. For [4] the parallel version has reached 35x of speedup. In the experiment reported by [5] the OpenMP performed better with small and medium images while

J. H. Park et al. (Eds.): PDCAT 2018, CCIS 931, pp. 171–176, 2019.
https://doi.org/10.1007/978-981-13-5907-1_18

CUDA had better performance with the larger ones. In all cases the parallel processing performed better than the sequential one, reaching an acceleration of 30.93. In other work, the K-means parallelization with OpenMP, [6] also had interesting results by reducing the processing time.

According to [7], the high complexity of some algorithms implies their exclusion in some studies, such as the DIANA (divisive analysis) algorithm. For this reason, this algorithm was chosen for this study. In this paper we optimize the DIANA algorithm through the use of parallelism using OpenMP.

2 Parallelization of the DIANA Algorithm

The DIANA algorithm analyzes data and, through the patterns found using Euclidean distance, that means that the data must have coordinate values. It creates sub classifications for the objects, that is, the process starts with all the objects in a group, the algorithm at each iteration makes a segmentation of the cluster until there are N groups or only the elements [7]. This process divides larger groups into smaller groups. For this, the algorithm constructs an array of distances that contains the interval between all points (Fig. 1 [8]).

d	1	2	3	4	5	6
1	0	0.31	0.23	0.31	0.23	0.23
2	0.31	0	0.31	0.23	0.31	0.31
3	0.23	0.31	0	0.31	0.4	0.7
4	0.31	0.23	0.31	0	0.31	0.31
5	0.23	0.31	0.4	0.31	0	0.7
6	0.23	0.31	0.7	0.31	0.7	0

Fig. 1. An example of a distance matrix.

As shown in Fig. 1, the matrix formed is symmetric, that is, the opposing elements in relation to the main diagonal line are identical. With this feature, part of the analysis can be reduced because only one side - above or below the diagonal line - can be used in pattern recognition (Fig. 2).

d	1	2	3	4	5	6
1	0	0.31	0.23	0.31	0.23	0.23
2	0.31	0	0.31	0.23	0.31	0.31
3	0.23	0.31	0	0.31	0.4	0.7
4	0.31	0.23	0.31	0	0.31	0.31
5	0.23	0.31	0.4	0.31	0	0.7
6	0.23	0.31	0.7	0.31	0.7	0

Fig. 2. Demonstration of which side of the matrix will be used for the pattern recognition.

To distribute distances on one side of the matrix, it is necessary to determine how many distances are analyzed. To do this, the following formula (Eq. 1) is used, where "*numberPoints*" refers to the total number of data in the set.

$$Total = \frac{(numberPoints^2) - numberPoints}{2} \tag{1}$$

The distribution of the total amount of distances, for example in the Fig. 2, between three threads, is similar to Fig. 3, where each thread will choose the best local result.

d	1	2	3	4	5	6
1	0	0.31	0.23	0.31	0.23	0.23
2	0.31	0	0.31	0.23	0.31	0.31
3	0.23	0.31	0	0.31	0.4	0.7
4	0.31	0.23	0.31	0	0.31	0.31
5	0.23	0.31	0.4	0.31	0	0.7
6	0.23	0.31	0.7	0.31	0.7	0

Fig. 3. Example of distributing distances between three threads.

Once each thread has finished its calculation, an overall result is chosen among the finalists. In this case, the third thread is the winner with the longest distance (Fig. 4), that is, samples 3 and 6 can be updated (Fig. 5). The highest distance is the best global result once the that the pattern recognition model chosen was Complete Linkage, in which the highest result is always sought.

d	1	2	3	4	5	6
1	0	0.31	0.23	0.31	0.23	0.23
2	0.31	0	0.31	0.23	0.31	0.31
3	0.23	0.31	0	0.31	0.4	0.7
4	0.31	0.23	0.31	0	0.31	0.31
5	0.23	0.31	0.4	0.31	0	0.7
6	0.23	0.31	0.7	0.31	0.7	0

Fig. 4. Three final results are found by the three threads, in which the blue one is the winner with the highest result.

That is, with the determination of the total number of distances to be analyzed the set is divided between the processors. The following diagram (Fig. 6) illustrates the distribution of the dataset and the analysis of the set.

D	1	2	[3,6]	4	5
1	0	0.31	0.23	0.31	0.23
2	0.31	0	0.31	0.23	0.31
[3,6]	0.23	0.31	0	0.31	0.7
4	0.31	0.23	0.31	0	0.31
5	0.23	0.31	0.7	0.31	0

Fig. 5. The distance matrix updated after the first regrouping.

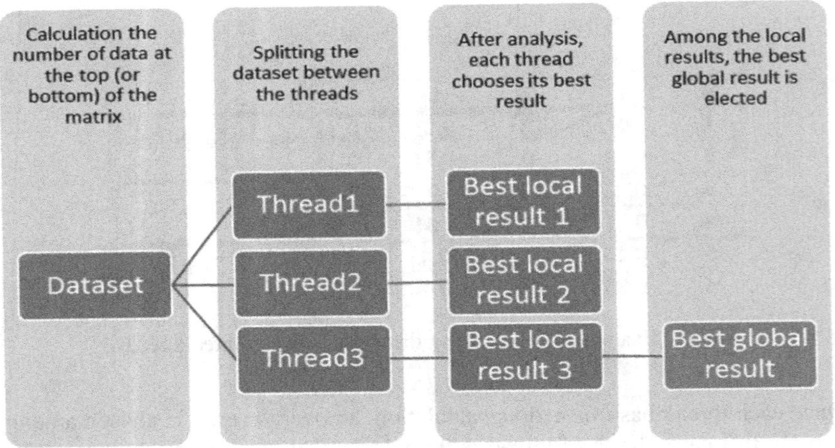

Fig. 6. Schematization of the parallelization for DIANA algorithm.

Once the amount of data is calculated by Eq. 1, it is divided by the number of threads. In a two-threaded environment, each thread will analyze half of the set. As OpenMP is relatively a high level extension for C, it can parallelize a loop with a single directive [3]. The distribution is done through a directive according to the example presented in Fig. 7, in which the variable "posA" is used to walk through the distances matrix. The loop iterations are distributed to the threads where MAX is the constant with the total number of points in the dataset.

```
#pragma omp for schedule (static)
for(posA=0; posA< MAX; posA++)
```

Fig. 7. Example of the use of the parallelism directive in OpenMP.

Therefore, the analysis of the distances in the matrix is carried out in parallel, accelerating the process.

3 Results

An OpenMP version of the DIANA algorithm has been developed using the Single Linkage method - it seeks to find patterns looking for the least Euclidean distance between the samples and tested on a machine with an AMD FX-8320 processor, 8 Gb RAM, and Ubuntu 16.04.2 x64 operating system. The execution time of the sequential part (corresponding to 19.65%) and the parallel part (corresponding to 80.35%) have been measured in a single thread. Using the percentages of the execution time of the sequential and parallel portions of the code, the gain in speed was calculated for a database with 5000 points. The first calculation refers to the law of Amdahl, in which performance is studied for problems of fixed size. This was determined by Eq. 2, where N is the number of processors:

$$Speedup = \frac{1}{0,1965 + (\frac{0,8035}{N})} \tag{2}$$

The second law applied was Gustafson's Law (Eq. 3). This law as well as the Amdahl Law, predicts the performance of the algorithm, but Gustafson's Law takes into consideration the variation of the problem size because "in practice, the problem size scales with the number of processors" [8].

$$Speedup = N - 0,1965.(N - 1) \tag{3}$$

The program developed for this study was tested three times for the same database with 5000 points. The time for the analysis and classification of the elements is described in Fig. 8.

Time in seconds to analyze 5000 elements			
Nº *threads*	Time	Time	Time
1	1412	1406	1415
2	423	429	426
4	314	310	311
8	299	299	299
16	299	300	299
32	298	299	297
64	297	296	296
128	296	297	295
256	291	292	293
512	283	285	283
1024	266	270	270

Fig. 8. Time to analyze 5000 points in relation to the increase of the amount of threads.

4 Conclusions

It is observed a transformer with 81% gain of time when comparing any parallel result with single thread, achieving an average speed up of 5,2521. Another view on Fig. 8 is the stabilization of the gain after 8 threads, that is, for a database with 5000 elements, the addition of more computational resources greater than 8 threads is no longer advantageous. This is due to the limitation of the sequential section in the code, update of the points effecting the relation with the clusters.

Under Amdahl's Law, a gain of 5.0694 was foreseen, i.e. the program developed outperformed the expectations. For Gustafson's Law, a speed up of 823,004 was expected. As Gustafson's Law results in greater gains for larger problems, it is believed that for databases with greater volumes of data the expected speed up will be achieved. To this end, a dataset with 13,467 samples provided by the Department of Informatics of the University of Eastern Finland was chosen. The database has 13,467 locations of MOPSI application users in Finland [10].

In summary, the use of parallelism in the DIANA data mining algorithm is advantageous, since the acceleration of the algorithm is higher than expected by the theoretical speedup.

References

1. Bell, J.: Machine Learning: Hands-On for Developers and Technical Professionals. Wiley, Hoboken (2015)
2. Lopes, N., Ribeiro, B., Machine learning for adaptive many-core machines: a practical approach (2015)
3. Pacheco, P.S.: An Introduction to Parallel Programming. Morgan Kaufmann Publishers, Burlington (2011)
4. Danalis, A., Mccurdy, C., Vetter, J.S.: Efficient Quality Threshold Clustering for Parallel Architectures (2012)
5. Bhimani, J., Leeser, M., Mi, N.: Accelerating K-means clustering with parallel implementations and GPU computing. In: High Performance Extreme Computing Conference (HPEC) (2015)
6. Naik, D.S.B., Kumar, S.D., Ramakrishna, S.V.: Parallel Processing of enhanced K-means using OpenMP (2014)
7. Kaufman, L., Rousseeuw, P.J.: Finding Groups in Data: An Introduction to Cluster Analysis. Wiley, Hoboken (1990)
8. Johnson, S.: Hierarchical clustering schemes. Psychometrika (1967)
9. Gustafson, J.L.: Reevaluating Amdahl's Law. Communications of the ACM, Technical Note (1988)
10. Fränti, P., Rezaei, M., Zhao, Q.: Centroid index: cluster level similarity measure. Patt. Recogn. **47**, 3034–3045 (2014)

Flash Animation Watermarking Algorithm Based on SWF Tag Attributes

YiAn Zuo, ZhiXun Zheng, and De Li(✉)

Department of Computer Science, Yanbian University, Yanji, China
245821459@qq.com, 2698817372@qq.com,
leader1223@ybu.edu.cn

Abstract. This paper presents a Flash animation watermarking algorithm based on SWF tag attributes. First, the SWF file is converted into an XML file, and the watermark image is converted into a binary data stream. Second, filter and categorize the tags and their attributes that have minimal effect on the animation in the XML file, and embed the binary data stream into the XML file according to a certain sequence of tag attributes. Finally, the XML file embedded in the watermark information is converted into a SWF file. Experiments show that the algorithm proposed in this paper has good transparency and invisibility, and it has the characteristics of high hiding capacity.

Keywords: Digital watermark · Flash animation · SWF · Tag attribute

1 Introduction

The development of computer technology and the popularization of the Internet have brought great convenience to people, but the information security issues ensued from such development and popularization should also receive sufficient attention. As a widely used form of application today, the issue of copyright infringement such as copying and dissemination of digital multimedia, is nothing new. In order to prevent such infringement from intensifying, a variety of related research on information hiding technologies have emerged. As the last line of defense for copyright protection, digital watermarking technology can embed relevant copyright logos of digital multimedia products into the carrier so as to play a key role in dealing with possible copyright disputes in the future. With the development of digital watermarking technology in the direction of carriers diversity, digital imaging, audio, video and other fields have made rapid progress. However, there is still few research on Flash animation.

Flash animation is a vector-based interactive movie format that integrates various media elements, dynamic effects, and user interactions. It expresses rich semantic information through internal objects and their attribute features. It is an efficient media format that enables the vector graphics, text, videos, and sound to be efficiently uploaded to the Internet [1]. In general, Flash files come in two formats, with the .fla being the source file, and the SWF files can interpret and display animations more efficiently, which is the main format of animation.

In the existing methods of protecting Flash animation, the traditional protection methods do not have high concealment. Through the understanding of the SWF File

J. H. Park et al. (Eds.): PDCAT 2018, CCIS 931, pp. 177–187, 2019.
https://doi.org/10.1007/978-981-13-5907-1_19

Format Specification [2] published by Adobe and the analysis of attributes in the SWF format file, the existing information hiding methods can be divided into four categories [3]. These techniques include appending data at the end of the SWF file, adding metadata tags, adding definition tag method, and replacing fill bits with hidden data. The literature [4] has carried on the experimental research to these four kinds of methods separately and obtained that these methods are usually successful. Zhang and Zhang [5] proposed a method of using user-defined tags to hide information. This method reads in the Flash file as an array of bytes, embedding and extracting based on the parity of the count of bits “1” in each byte of the byte array. The algorithm has no influence on the playback effect of Flash animation, can play a role in encryption, has the characteristics of high hiding capacity, good security, and has certain performance in the anti-attack. Ye [6] proposed a hidden method based on the SWF attributes, and achieved the hidden purpose by modifying the depth value of the object on the premise that the order of the depth between objects is the same. Experiments show that the method does not have visual effects before and after embedding hidden information, and it has higher invisibility. In literature [7], the author used the three attributes of coordinates, matrix, and color in the object properties of Flash to embed the secret information. All three methods use the LSB method to embed data to be hidden into Flash’s three attributes.

Based on previous research, this paper proposes a digital watermarking algorithm based on SWF tag attributes. The classical LSB method is applied in the process of watermark embedding and extraction. Experimental results show that the algorithm has good invisibility and good hiding capacity.

2 Related Theory and Technology

The algorithm in this paper mainly analyzes the animated SWF file, and then explains the format of the SWF file and its conversion to another file that is easy to read and write. The algorithm flow of the classical algorithm LSB is also described.

2.1 SWF File Structure Analysis

The need for interactive web pages makes Flash technology (SWF) a necessary tool, but its potential as a hidden information medium has been overlooked [7]. Analyzing SWF format specification [2] is very necessary to study the information hiding in the method of Flash. The SWF file format stands for “ShockWave Flash”. The SWF file can be played by Adobe Flash Player, either by a browser plug-in or by a stand-alone player. A SWF file is roughly composed of three parts, starting with a file header, followed by a file body containing a series of tags, and an ending tag.

The file header defines some basic properties of the SWF file, including the compression identifier, file version, file length, stage size, frame rate, and frame number. The compression identifiers are “FWS” and “CWS” respectively, “FWS” represents an uncompressed file, and “CWS” represents a compressed file. Here we discuss the uncompressed SWF file. The tags in the file body are divided into the definition tag and the control tag. The definition tag defines the attributes and feature parameters of all the constituent elements. The control tag controls the attribute changes, dynamic effects, and human-machine interaction of various components. The SWF tag format is shown in Fig. 1.

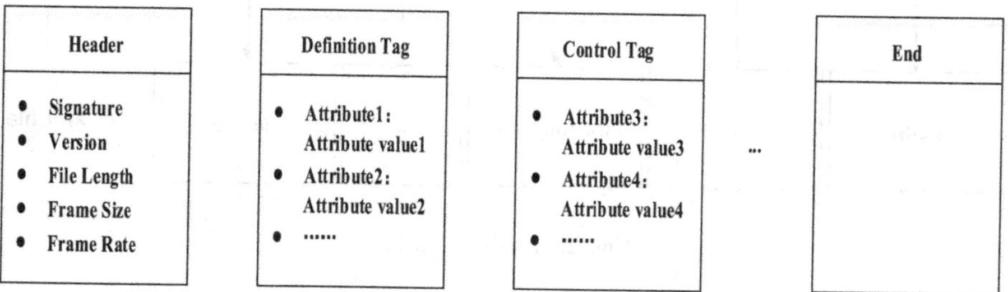

Fig. 1. SWF tag format

With the declaration in the SWF format specification document, this article uses the attributes of elements in Flash to achieve the purpose of embedding the watermark information. Since each element tag of the main part of the file is a separate entity, any tag and other tags have no data association. The modification of the value of any one element does not affect the playback of other elements. Therefore, under the premise of not affecting the playback effect of the entire Flash animation, it is feasible to appropriately modify the values of some element attributes to embed the watermark information. By collecting, analyzing and comparing Flash animations of different levels of complexity and attempting to modify the values of some element attributes, some element attributes that can be embedded in watermarks can be summarized. And these tag attributes are sorted according to the order of insensitivity from strong to weak, that is, the order of watermark information embedding is obtained.

SWF uses a series of binary tags to define various objects, events, and interactions. This has a similar internal structure to XML files, and XML files are well readable. Based on the in-depth research of the current Flash technology and the development of XML format technology, there are applications that can convert the Flash and XML formats to each other. The conversion of the formats can achieve a one-to-one correspondence between tags and attribute values, as shown in Fig. 2. This facilitates subsequent research.

2.2 LSB Algorithm

The LSB (Least Significant Bit) is a typical spatial domain information hiding algorithm based on bit planes. The LSB algorithm embeds the secret information into the least significant bit of the carrier attribute value, and changing this position has the least influence on the quality of the carrier.

The LSB embedding method of the stream carrier: select a subset of the carrier elements $\{j_1, j_2, \ldots j_{L(m)}\}$, in which there are altogether elements for $L(m)$ Bits which hide the information using information. Then perform a substitution operation this

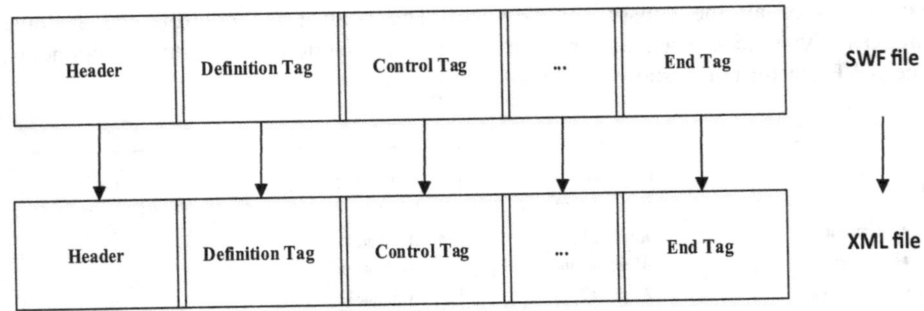

Fig. 2. Flash-to-XML

subset, replacing the lowest bit of each element with the watermark information. The LSB extraction method: It finds a subset of camouflaged elements that embed information $\{j_1, j_2, \cdots j_{L(m)}\}$, extracts the lowest bits from these disguised objects, and arranges and combines the secret information.

Since the watermark signal is hidden in the lowest position, it is equivalent to superimposing a signal with a weak energy, so it is difficult to detect visually and acoustically. LSB algorithm can hide more information, but it is less robust. However, as a large amount of information hiding method, LSB still occupies a very important position in covert communication [8].

3 Flash Animation Watermark Algorithm Design

The digital watermarking method based on tag attributes used in this paper makes use of the LSB algorithm. This method has the obvious disadvantage that the hidden capacity is limited by the number of tags with attributes in the carrier. So the effect on the hidden capacity is not ideal. This article summarizes the list of attributes that can be embedded, and sorts them according to insensitivity from strong to weak, so that it can have better hidden capacity and inherit the advantages of LSB without affecting the quality of the original carrier.

3.1 Watermark Embedding Algorithm

The watermark image in this paper is a 32 × 32 binary image. The process of embedding watermark information into Flash files is shown in Fig. 3, which is mainly divided into the following steps:

The first step: (1) Select a 32 × 32 pixel binary image as a watermark and convert it into a binary data stream; (2) Convert the SWF file to an XML file.

The second step: (1) Get the first tag attribute of the attribute sequence list; (2) Search from the <header> tag of the original carrier to find the same attribute, and convert the attribute value from decimal to binary; One-bit information of the binary secret information replaces the least significant bit of the attribute value; (3) Convert the binary attribute value containing the secret information to decimal; (4) Continue to search for the matching tag in the carrier in the order of the tag attribute list until Watermark information is all embedded.

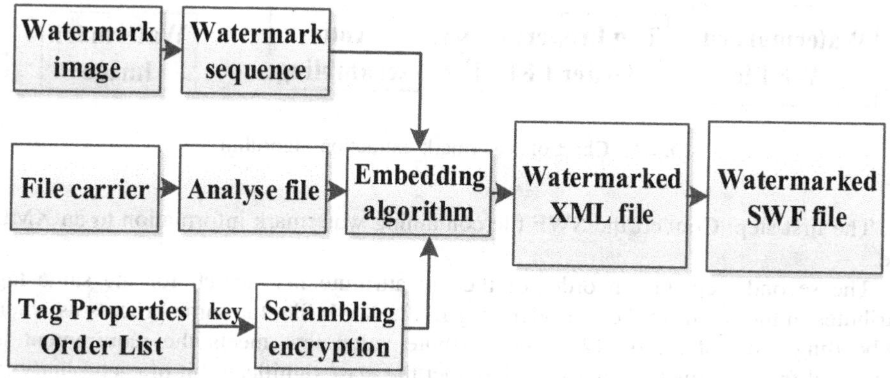

Fig. 3. Chart of watermark embedding algorithm

The third step: Stores the rewritten XML file and converts it to a SWF file, which results in a Flash file with secret information.

Based on the algorithm proposed above, an example is given to illustrate the above embedding idea.

1. Suppose there is an attribute sequence list $\{<Move>, <Bounds>, \cdots\}$, which is sorted by label insensitivity.
2. Transform the carrier SWF file into an XML file and search for the first label in the attribute sequence list, <Move>, starting from the <Header> tag. Get its decimal attribute value, such as <Move DX= "15" DY= "7">, and convert both the attribute value and the secret information into a binary stream.
3. Embed 1 bit of secret information into the end of attribute value. If the information to be hidden is "10010…", the binary attribute value to be embedded are "1111", "111", and the first bit "1" of the hidden information is embedded into "1111". Since the end of "1111" is also "1", no change is made. Embed the second bit "0" into the end of "111", it needs to be modified to "110". After converting the modified attribute value back to decimal, the embedding is complete.
4. When the attributes of the same tag in the carrier file are embedded with the hidden information, the next tag is searched in the order of attribute sequence list, embedded in the way of step 3, and so on, until the hidden information is embedded.

3.2 Watermark Extraction Algorithm

The process of watermark extraction is the inverse process of embedding watermark information. The algorithm is described as follows (Fig. 4):

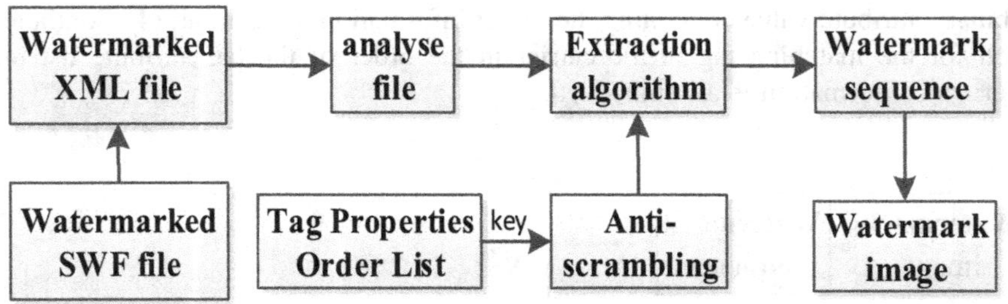

Fig. 4. Chart of watermark extraction algorithm

The first step: Convert the SWF file containing watermark information to an XML file.

The second step: (1) In order of the tag attribute list, search for the same tag attributes in the list from the <Header> tag in the XML file, the same process as when embedding the watermark; (2) Each attribute value that meets the requirement is converted from decimal to binary, and extract the least significant bit of each binary to compose the binary sequence until the watermark information is completely extracted.

The third step: Convert the generated binary sequence to a watermark image.

4 The Experimental Results and Analysis

This paper simulates the proposed watermark algorithm in MATLAB2014a. In this experiment, "simple", "medium" and "complex" animation in three different levels of complexity (mainly reflected in the number of animation frames, animation effects, scene changes, etc.) were tested. They are Christmas.swf, festival.swf, boy.swf respectively (the three files are all animations exported from the AS3 source files created by Flash CS6). They are 51 KB, 156 KB, and 391 KB in size. The watermark image is a binary image whose size is 32 × 32, written in "Y" of "Yanbian University", as shown in Fig. 5.

Fig. 5. Watermark image

4.1 Transparency Experiment

Figures 6, 7 and 8 show experimental charts of the test objects Christmas.swf, festival.swf, and boy.swf, respectively, in which figure (a) represents the display effect of the original carrier file and (b) represents the display effect of the generated hidden file. Figure (c) represents the watermark image generated after the extraction and restoration.

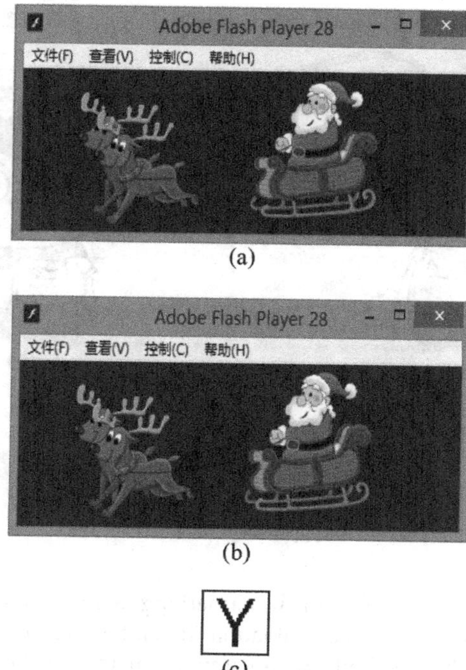

Fig. 6. Experimental charts (Christmas.swf)

Fig. 7. Experimental charts (festival.swf)

(a) (b)

(c)

Fig. 8. Experimental charts (boy.swf)

The imperceptible evaluation of watermarking algorithm can be divided into objective evaluation and subjective evaluation. It can be seen from the figures that the hiding techniques of this method has no impact on the display effect of the Flash animation, and it can be effectively hidden and highly concealed. And it can be seen from the figure that the extraction result is correct and effective.

The peak signal-to-noise ratio (PSNR) can quantitatively evaluate the invisibility of the watermark. By calculation, the PSNR values after embedding the watermark are 34.5213 dB, 36.7243 dB, and 36.9012 dB, respectively. Generally, the invisibility metric can be considered as meeting the requirement of invisibility as long as it is greater than a certain value. When the PSNR value is more than 28 dB, the difference in image quality is not significant. When it is higher than 35.4 dB, it is not recognized by the naked eyes, which ensures the invisibility of the watermark.

The effect on the file size before and after watermark embedding: The method proposed in this paper is mainly aimed at uncompressed SWF files. Since this algorithm only changes the attribute value and does not change the size of the storage space, it does not have any effect on the file size.

4.2 Embedded Capacity Experiment

In the algorithm proposed in this paper, the embedded capacity of the watermark information is related to the number of tag attributes that can be embedded in the carrier. Take a carrier file festival.swf as an example. The size of the original file is 156 KB and there are 3653 tags that can be embedded, but the number of attributes of each tag is different. For example, each <Bounds> tag can be embedded 4 bits of information. Each <scale> tag can be embedded 16 bits of information. Theoretically, the file carrier can be embedded with 44617 bits information, and embedding a

32 × 32 binary watermark image is sufficient. In addition, it was found in the experiment that the higher the complexity of the animation may have more object elements, more complex change effects, more background switching, etc., which shows that the more available embedded tag attributes for embedding, the greater the embedded capacity.

Another example. The carrier Christmas.swf is a relatively simple animation, only 12 frames, but because he has 4224 <Curveto> tags and 16933 <Lineto> tags, which makes the animation can be embedded into the more amount of information, can satisfy the embedded 32 × 32 binary watermark image in this article. But at the same time, this may be the reason why his PSNR is not ideal compared to other more complex animations.

4.3 Invisibility Experiments

Using this method, in addition to achieving hidden effects, there is no effect on the SWF animation display effect before and after embedding, so the hidden information is not visible to the naked eye. Even when the SWF file is parsed using decompile software, the SWF with hidden secret information is not easy to attract people's attention. Because the observer does not know whether the properties of the animation have been intentionally modified.

4.4 Algorithm Performance Comparison

Through the above analysis, we can select several existing algorithms: Replacing fill bits method [4], adding definition tag method, appending data method, adding Metadata tag method [5], depth replacement method [6], coordinate substitution method and color replacement method [7] are compared with the tag attribute list algorithm in this paper.

The following table compares the performance of this algorithm and the above algorithms.

Table 1. Algorithm performance comparison

Technique	File size change	Hiding capacity	Transparence	Robustness	Invisibility
Proposed method	No effect	Medium	Relatively strong	Relatively weak	Strong
Method [4]	No effect	Small	Strong	Relatively weak	Medium
Method [5]	Increase	Big	Strong	Relatively weak	Weak
Method [6]	No effect	Small	Strong	Medium	Strong
Method [7] (1)	No effect	Small	Relatively strong	Relatively weak	Strong
Method [7] (2)	No effect	Small	Medium	Relatively weak	Strong

From Table 1, we can see that with respect to the effect of embedded information on file size, the method based on tag attributes in this paper tends to outperform the appending data method, adding metadata tag method and adding definition tag method; in terms of the amount of concealment, this method optimizes algorithms that use tag attribute values, such as the depth substitution method, coordinate substitution method, color replacement method, and summarizes more tag attributes that can be used to embed information. Although it is still affected by the number of tag attributes, the number is greatly increased; the transparency of the algorithm is currently the part that can be optimized. By improving the list of tag attributes and further testing and adjusting the attribute values, there is room for improvement; robustness can be said to be a weakness of this method and several existing methods. If maliciously tampered with data embedded in the location, it would lead to the failure to extract secret information; in terms of invisibility, using third-party decompile software (such as Sothink SWP) is easier to find hidden secret information in the appending data method, adding the metadata tag method, and adding the hidden secret information in the definition tag method. Although the method of replacing the filling bits is less noticeable than the above methods, careful observation can still find that the filling bits are used. Even if the method of this article changes the value of the tag attribute, but if there is no comparison of the original file, it is not easy to arouse people's suspicion. Therefore, it obviously outperforms the above methods.

5 Conclusions and Next Steps

This article proposes a watermarking algorithm based on SWF file tag attributes. Experiments show that the algorithm proposed in this paper can hide the secret information in the Flash animation, and the effect of animation after embedding is not affected, and it will not affect the original playback process and design. The hidden capacity of this method will be limited by the number of carrier tag attributes, but to conclude, the method has a higher embedded capacity. However, there are still some shortcomings in the algorithm, especially in terms of robustness, and the list of tag attributes needs further research and optimization to improve the algorithm.

Acknowledgments. 1. This research project was supported by the Education Department of Jilin Province ("Thirteen Five" Scientific Planning Project, Grant No. 249 [2016]).

2. This research project was supported by the National Natural Science Foundation of China (Grant No. 61262090).

References

1. Liu, F., Meng, X.: Research on content feature analysis and image information extraction of flash animation. Mod. Educ. Technol. **19**, 91–94 (2009)
2. San Jose, Adobe Systems: Adobe Flash Video File Format Specification (Version 10.1), California (2010). www.adobe.com/devnet/f4v

3. Martini, A., Zaharis, A., Ilioudis, C.: Data hiding in the SWF format and spreading through social network services. In: Proceedings of the Fourth International Workshop on Digital Forensics and Incident Analysis, pp. 105–114 (2009)
4. Fouche, M.-A., Olivier, M.: Steganographic techniques for hiding data in SWF files. In: Peterson, G., Shenoi, S. (eds.) Advances in Digital Forensics VII. IFIP AICT, vol. 361, pp. 245–255. Springer, Heidelberg (2011). https://doi.org/10.1007/978-3-642-24212-0_19
5. Zhang, X., Zhang, X.: Information hiding algorithm based on flash animation. Comput. Eng. **36**(1), 181–183 (2010)
6. Ye, X.: Research on the Method of Hiding Information in FLASH Animation. Sun Yat-sen University, Guangzhou (2012)
7. Yu, J.: Hidden Message Technology Applied in Flash Files, pp. 1–39. Datong University, Datong (2010)
8. Wu, W., Zhang, Q.: Design of copyright protection system based on LSB digital watermarking technology. Autom. Instrum. **3**, 86–88 (2013)

Analysis of Massive E-learning Processes: An Approach Based on Big Association Rules Mining

Asma Hassani[1(✉)] and Sonia Ayachi Ghannouchi[2(✉)]

[1] ISITCom Hammam Sousse, RIADI Laboratory, ENSI Manouba, Manouba, Tunisia
Asmahassani08@yahoo.fr
[2] ISG Sousse, RIADI Laboratory, ENSI Manouba, Manouba, Tunisia
s.ayachi@coselearn.org

Abstract. In today's learning environments (MOOCs), process learning in highly dynamic. This dynamicity generates a large volume of data describing the handling of such processes. Mining that data is a powerful discipline for getting valuable insights. Thus big data mining becomes more and more an integral part of handling a business process management (BPM) project. Furthermore, using data mining outputs to extend process model with additional knowledge allows to provide more details for future process improvement. In this paper we aim to apply data mining techniques (Association rules mining) to recorded data in the context of massive learning process.

Keywords: Learning process · MOOCs · Analytics · BPM
Association rules mining · Big data

1 Introduction

Today, Business Process Management has emerged as a substantial approach that has been widely deployed in several areas (industry, commerce, healthcare, …) essentially to increase organization's performance. The educational field is not far from the power of BPM that leverages process orientation to analyze the effectiveness of learning approaches. Thus, by understanding, modeling and analyzing learning processes, tutors and learners can achieve a great visibility of their progression which will result in enhancing learning performance. The process-oriented approach is one of the prospective orientations of learning environments. Although, content-oriented, tool-oriented and task-oriented approaches are mainly neglected. The process aspect and the majority of E-learning systems do not deal with learning process per se [1]. On the other hand, dropout rate in MOOCs is considerable and is due to the lack of support [2]. An assistance is fundamental to help learners in performing their learning via a process-oriented approach. In the development of such learning, participants feel part of something to which they want to belong especially when we deal with collaborative MOOCs (CMOOCs). The objective of BPM in the field of learning, is to improve learning performance by highlighting interactions and data exchange between actors.

J. H. Park et al. (Eds.): PDCAT 2018, CCIS 931, pp. 188–199, 2019.
https://doi.org/10.1007/978-981-13-5907-1_20

Further, this approach aims to ensure efficiency and quality of the process by adapting to learner profile by regularly adapting processes to identified constraints such as behavior, level, pace, etc. Thus, all of these improvements are a direct result of process flexibility and continuous evolution. In this case, Business Process Analysis (BPA) [3, 4] aims to provide process participants with knowledge to understand how their processes are performed. These insights are used to detect possible improvements to align process model with learner profile.

Moreover, learning data stored in databases tables hide a lot of knowledge that describe learners performance. Fundamentally, data mining techniques deals with data processing and focus on data flow analysis. However process mining deals with event logs and focuses on control flow analysis. Data mining covers classification, clustering and association rules which concern two fundamental approaches: supervised and unsupervised. Association rules mining (ARM) is one of the unsupervised data mining methods. In the e-learning field this method allows to find rules between frequent itemsets in the dataset. The generated rules based on frequent data should be taken into account in order to enhance learning system by discovering behavioral patterns in the database and to group similar learners.

In this work we focus on the application of association rule mining as a data mining technique in the context of business process, especially learning process. This paper is organized as follow: Sect. 2 presents an overview of learning process in the context of MOOCs and their main characteristics. Section 3 introduces the concept of learning process analysis and its basic analysis techniques. Section 4 presents the idea of the applicability of BPM in the context of e-learning. Association rules technique and its algorithms are presented in Sect. 5. Section 6 describes the research goals and next our implementation contribution. Section 8 presents the application and results. Section 9 is dedicated to present the related work and our positioning. Finally, the paper concludes by the conclusion and future research directions.

2 Learning Process via MOOCs

In the majority of Learning Management Systems (LMS), environments are more oriented towards the management of the content than towards the learning process itself. So, in order to have a great visibility of learning activities, we need a perception of learning from a different angle, considering the process as an essential foundation for learning. Learning process can be defined as the series of tasks to be carried out by learner to achieve individual or collective objective and guarantee a good acquisition of knowledge. The interest is on interactions generated between learner and tutor, where each learner accomplishes a certain set of tasks representing the activities of the learning process [5, 6]. Thus, every learner carries on her/his own learning process, adapted to her/his profile and exchanges information with other learners and tutors.

Nowadays, learning environments, especially Massive Open Online Courses (MOOCs), known a rapid evolution and are more and more complex and generate more and more data. Modeling corresponding processes enables on one hand to understand the learning scenario and on another hand to analyze data in order to bring enhancement.

From this vision, modeling and analyzing process are the fundamental steps of BPM approach.

MOOCs environment and its corresponding process have particular characteristics related to the nature of the Massiveness, dynamism and online learning. Learning process has the following characteristics:

- Massiveness: the first letter of MOOC means massive. The first fundamental and revolutionary feature of the web is to be massive, and for learning this feature is fundamental. During learning process, it is not only about the number of enrolled persons but also the amount of data circulated in. This massiveness comes from the open and free registrations.
- Dynamism: MOOCs are characterized by their dynamic aspect. Sharing and discussion environments creates dynamism between learners essentially through the forum. In fact, in the collaborative learning process a great dynamism is deployed.
- Distribution: Learning processes are increasingly executed via a wide range of distributed activities by involving external and heterogeneous environments to accomplish learning and communication tasks. This distributed environment involves web services, social networks and other systems.
- Complexity: complexity arises from the complexity of large amounts of data to be exchanged. The amount of data that supports learning process via MOOCs is large and of various forms and sources. Furthermore, due to the variety of learner's profiles and behavior, multiple pathways can be followed and thus create complexity.

3 Learning Process Analysis

Understanding the learner's interactions can further enhance learning process after proceeding with the suitable analysis. Learning process analysis, has received a great attention and several mechanisms to analyze learner's learning process are adopted. Learning Analytics (LA) or Educational data mining (EDM) are defined as the analysis of learner's data to understand and optimize learning and the environments in which it occurs [7]. Deeper analysis of learning process would require selecting the appropriate feature. Web analytics and business activity monitoring (BAM) have inspired the works on learning analytics to improve learning processes. In the context of process, monitoring process models based on pathways and actors' behavior involve data mining techniques or/ and process mining approaches.

3.1 Data Mining

Data mining or Knowledge Discovery in Databases (KDD) is a powerful technology that helps organizations to get knowledge from data collected and analyzed. In the education field, data mining can help to predict learners performance, to detect learners behavior and many other objectives that data mining looks for solving [8]. Several researches have studied the corporation of data mining techniques into e-learning environments. Educational data mining provides a set of techniques such as clustering, classification, association rules, etc. [9] in order to discover useful knowledge from learning data. Analyzing mining results allows to provide more insights about the overall handling of learning in order to better organize and improve the learning process.

3.2 Process Mining

Process mining is a technique that allows the analysis of business process based on event logs. Process mining is considered as a set of techniques for discovering process, data and organizational structures from event logs [10]. Those techniques allow to detect bottlenecks and deviations and measure process performance by bringing out significant problems. Discovery, conformance, enhancement are the basic procedures of process mining [11, 12].

Learnflow mining [13], or educational process mining [14] combines process mining with technology-enhanced learning environments. Process-oriented knowledge discovery techniques in e-learning systems are a rich and fertile area of interest. In fact, learning systems generate a wide range of precious data: which tasks are performed, by whom, when and what it results in. This event data contains a history of what happened during process execution and can be analyzed using process mining techniques (discovery, conformance, enhancement).

4 BPM

Business Process Management (BPM) is one of the most expanded approaches in guiding organizations. The BPM approach is based on the concept of business process which is a fundamental element in its definition. BPM can be seen as successive steps that form the lifecycle of such approach. BPM lifecycle model systemizes the steps and activities that should be followed for conducting a BPM project. According to Netjes et al. [15], BPM lifecycle is composed of five steps: design, configuration, execution, control and diagnosis.

While BPM deals with the management of business processes, it also deals with the learning processes, indeed the strong and relative adjustment between the two [16]. First, both of them, business and learning processes, deal with activities structured in a certain way to achieve business or pedagogical goals. Business or learning process activities are executed by following business or pedagogical process rules. Then, both processes consist of fundamental steps: design, execution and control. Further, business organizations, even pedagogical institutions, are concerned in the improvement of their process.

Thereby, supporting the idea of process-oriented learning situations, the management of such processes includes their design, execution as well as their enactment. Such application of BPM in the field of education is significant for the management of such process by monitoring and analyzing learning performance through learning process analytics.

5 Association Rules Mining

In the field of data mining, applying association rules algorithms can help to find associated relationships between data items. The idea of association rules mining was presented by Agrawal et al. [17] since 1993 and was originally developed for market basket data. Since that time, this research area has captured more and more attention and has been widely studied and applied in many contexts.

Association rules were formally defined [17] as follows.

Let $I = \{i_1, i_2, \ldots, i_m\}$ be a set of items and $D = \{t_1, t_2, \ldots, t_n\}$ a set of n transactions in which t_j contains a subset of items. So, a rule can be defined as follows:

$$X \rightarrow Y, \text{where}\, X, Y \subseteq \text{and}\, X \cap Y = \varnothing$$

where X is referred to as the antecedent and Y is the consequent. Two thresholds of ARM are minimum support (min sup) and minimum confidence (min con). The support of a rule $x \rightarrow y$ is the probability of the Itemset {x, y} that means the relevance of the rule and the confidence of a rule $x \rightarrow y$ is the conditional probability of y given x that indicates the accuracy of the rule. Support and confidence are denoted as below:

$$\text{Support}(x \rightarrow y) = \frac{\text{support}(\{x, y\}}{\text{Total number of transaction in D}}$$

$$\text{Confidence}(x \rightarrow y) = \frac{\text{support}\{(x, y)\}}{\text{support}(x)}$$

$D = \{d1, d2, \ldots, dn\}$ each $di \subseteq I$: set of transactions.

Association rule mining identifies the associations and frequent patterns among a set of items in a given database. It is composed of two steps: (1) Discover frequent Itemset according to some support threshold; (2) Generate association rules satisfying the confidence measure.

In literature, there is a wide range of algorithms for extracting frequent patterns from a transactional database. These algorithms can be classified into two categories: those based on the test and generate approach (example: apriori) and those are based on the divide and conquer approach (example: frequent pattern growth). Fp-growth is more efficient than Apriori in fact it does not need to make successive consultations of dataset.

Association rules algorithms have known a great interest from researchers in last decades facing new implementations with big data techniques. Traditional algorithms does not offers significant performance when input data is more and more voluminous. Parallel computing offer a new potential for the growing amount of data. An efficient implementation and good performance results can be achieved by adopting the mapreduce model.

Various researchers have proposed approaches to parallelize Apriori on Hadoop distributed framework and mapreduce were presented in [18, 19]. Others, have proposed to parallelize the fp-growth exploiting the mapreduce paradigm [20, 21].

ARM has been successfully applied in educational domain. It can be used for finding interesting relations among learners data by discovering behavioral patterns to make suggestions to learners who share similar characteristics, to make changes in learning approaches and adapt content with learner types. In MOOCs environments, we can have voluminous data with different types and which is volatile in time (profile data, logging information, grades, ...). Based on this, various data can be obtained. Despite the importance of MOOC data, it has not been actively researched with Big data analytics paradigm; basically mapreduce.

6 Research Goals

Nowadays, there is a great interest on how data in e-learning environments can be used to improve learning experience. Therefore, the emergence of learning analytics has been inspired by many other analytics tools such as web analytics, business intelligence (BI) and business activity monitoring (BAM). These techniques have been applied in the educational field and captured by learning analytics and educational data mining.

The main goal of our work is to propose an approach based on data mining in order to support Business Process Analysis in massive learning process. In fact data mining is often integrated with business process to provide value. Adapting process and respective actions must be accomplished based on data mining results. The approach is based on association rules mining techniques that are especially useful in MOOCs environments.

ARM aims at discovering association rules and frequent itemsets in data. This technique allows to obtain behavioral patterns in learning data. It may also help to find frequent process path discovery. Thus, applying ARM may be an interesting technique to support the learning process analysis.

As the data grows, quantity and quality of discovered associations will be significant and change dynamically. Mapreduce computing is an optimized technique in time and in memory in extracting association rules. Generated rules are stored on a suitable dataset from which future process improvement will be based.

7 Our Contribution

BPM, data mining and big data are the basic components of our work. Data mining is an integral and basic part of executing business process and an important factor of the success of Business Process Management in order to improve process models.

Figure represents the general conceptual infrastructure that will be used to proceed with the improvement of learning process based on data mining techniques (ARM) (Fig. 1).

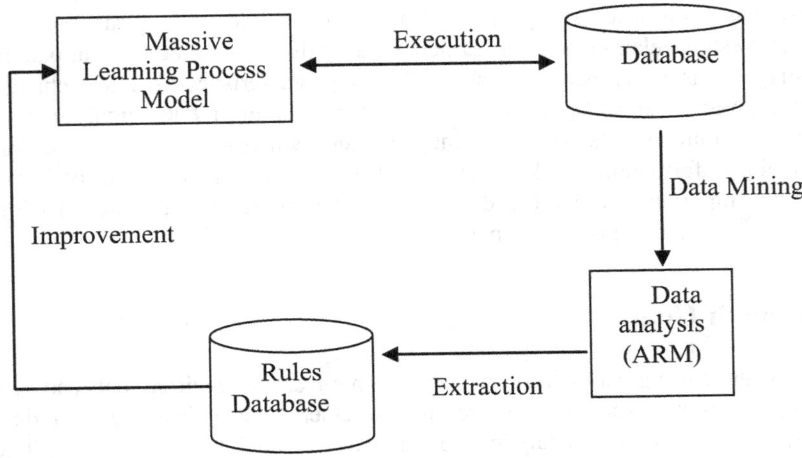

Fig. 1. Conceptual infrastructure

Massive learning models are developed and executed using a Business Process Management System (BPMS). Generated data is stored in a suitable big data base: we consider Mysql for structured data, CMS for course content and hive for big data warehousing. Suitable data mining technique (ARM) is applied. An ARM algorithm based on mapreduce is used to find out relations between data that describe the correlation of activities. Thus, the process can be mined from rules with appropriate thresholds (support and confidence). After finding rules, relevant rules that fulfill criteria values are stored. Through these rules, we can get patterns and information to facilitate learning decision and guide learner in an adaptive learning process.

The proposed model infrastructure will be simulated with different massive learning models to clearly understand the effective integration of data mining and learning process to sustain the project of BPM.

8 Application and Results

In the e-learning field, various available information is used to support learning process. In the case of MOOC environment, massive e-learning processes describe the general perspective of handling a scenario of learning.

Figure 2 presents the general model of massive e-learning process. We consider that it is composed of the following sub-processes:

- Course's launching
- Course's running
- Evaluation

Fig. 2. General model of massive e-learning process

In this paper we will focus on the second sub-process which model is shown in Fig. 2. This model presents the core of the model. It describes the different steps of the progress of the course. The process includes the various steps required during the learning by learners and teaching team (Fig. 3).

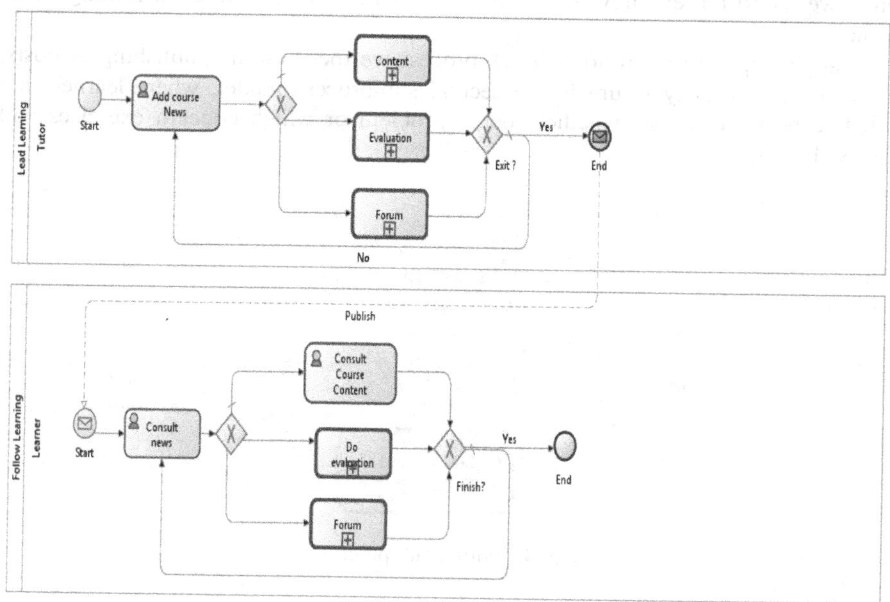

Fig. 3. Course's running process

Information learners, their enrolment courses, their grades and prerequisites make the difference between one scenario and another. This data can be analyzed from various perspectives and can help to give us more insights into the overall learning environment. Data mining techniques can be applied to find interesting patterns from large volumes of data accumulated in real time and in various types.

Furthermore, a learning process describes who is doing what, how and when. A BPMN learning process model can be represented by the following keys elements:

– roles: they identify who is involved in the process. In our case, tutor and learner are the basic roles (represented by lanes).

- events; represent any phenomenon that can take place and can start one or more activities (start, intermediate, end).
- activities: capture what will be done by actors.
- gateways: control the way of sequence flows in a process.

Exclusive gateways (Xor gateways) are used to model a choice in the process. When the execution arrives at this point, all outgoing sequence flows are evaluated in the order in which they have been defined. In our model, the learner will take exactly one path in the flow and thus describes his behavior in handling learning process; which is influenced by other information.

Our study attempts to identify factors influencing the learning behavior after each Xor gateway. We try to retrieve rules to describe choices in learning process. In other words, we try to retrieve how values of some attributes can affect the routing of an instance.

As an example, in the following sub process we focus on the publishing of posts, part of the call activity forum in the second sub-process model, where learners are invited to post their posts whether social, problem or which concern exercises and quizzes (Fig. 4).

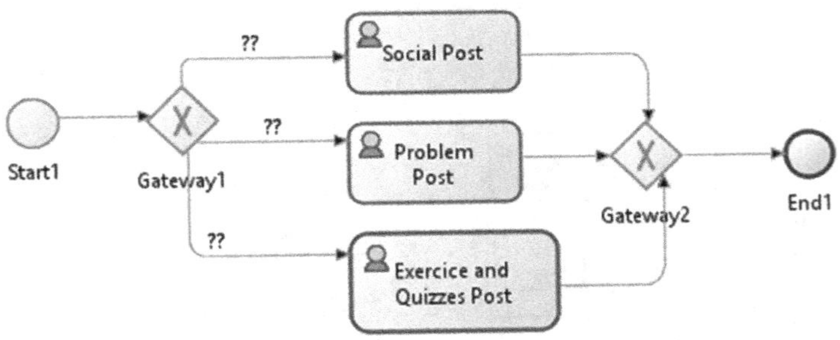

Fig. 4. Forum sub-process

We can take as hypothesis the fact that for each learner the taken path is influenced his profile, his progression and many other information. For example if the learner is a teenager, he usually publishes social posts. In contrast, old people face many problems when they deal with the platform. Those who have bad grades, they rely on forum to post their questions related to quizzes and evaluation exercises. We can formulate such rules as follows:

If age<30 THEN post=social
If age>30 THEN post=problem
If grade=bad THEN post=exercise

For this purpose, we use the ARM technique which aims to establish relationships between items. Thus, in order to analyze attributes that are generated in previous process executions, we need to find out, in real time, which rules qualify the corresponding taken path that describes choices made in the past and predict future behavior. Consequently, we provide for each learner the process that describes his attitude.

9 Related Work and Positioning

Data mining in the context of business process aims to discover various kinds of knowledge about process instances. In our work we refer to data mining as an integral and unbreakable part of the management of business process. Process mining and data mining are the fundamental two axis on which we concentrate our research literature.

Rozinat et al. have shown how a decision point analysis can be performed with process mining. Authors have considered each Xor split in a Petri net model as a classification problem, where according to values in prior task a particular path is taken. In this work [22], decision mining is based on event log enriched by data attributes.

Van der Aalst [11] shows the applicability of process mining by recording events, using techniques in the PROM framework. The objective is to enhance process model with much more information describing performance or organizational perspective.

According to Girald et al. [23], BPM adjusted to data mining allows to make decision easier. The application of Data Mining techniques in the context of BP is beneficial beyond the analysis of decision that have been made. Dennis and Stefan in their work [24] have discussed the integration of data mining into business process based on BPMN and CRISP-DM (CRoss-Industry Standard Process for Data Mining).

Furthermore, data mining can be applied on e-learning data to understand learners behavior by discovering association rules. These knowledge can be used to improve the performance of learner and enhance learning process by improving decision making process.

These are the point from which we start addressing our work. We propose to apply data mining techniques for several reasons. First, data of instances is recorded for a predefined period, with hidden knowledge about learners behavior. Consequently, knowledge can express path taken in choice point. Second, ARM allows to discover patterns from data by finding relationships among items in a given dataset. Antecedent is profile information or other data. Consequent has the value of attributes relative to tasks after the Xor split gateway in the process model.

The main difference between our proposition and existing work in literature is that we rely on Data Mining Techniques and not on Process Mining. But the objective is the same. Moreover, we consider in our approach threshold values (support and confidence), whereas the other works don't consider such limitations. The advantage of considering such limitations is that the analysis will be more efficient and results will probably be more precise.

10 Conclusion

Current MOOC platforms have generated valuable data which provided significant analysis and values to understand the way in which learning is conducted. BPM provides a continuous monitoring mechanism for process performance and improvement. Based on these capabilities, BPM can be applied for Massive learning processes in an efficiency manner to offer an adaptive and a flexible learning process.

In this paper, we have proposed a process improvement approach based on association rules mining to mine the association relationship between process's actor profile and information and process's activities. In our work, we aim to mine the potential associated relationship between activities which can help to improve process and offer different adaptive versions. Thus, Learning processes should be able to be executed in different ways; each learner follows the pathway which corresponds the best to his/her profile, learning styles, pace and behavior.

References

1. Helic, D.: Technology-supported management of collaborative learning processes. Int. J. Learn. Change **1**(3), 285 (2006)
2. Onah, D.F.O., Sinclair, J., Boyatt, R.: Dropout rates of massive open online courses: behavioural patterns. In: EDULEARN14 Proceedings, pp. 5825–5834 (2014)
3. Vera-Baquero, A., Colomo-Palacios, R., Molloy, O.: Business process analytics using a big data approach. IT Prof. **15**(6), 29–35 (2013)
4. Vergidis, K., Tiwari, A., Majeed, B.: Business process analysis and optimization: beyond reengineering. IEEE Trans. Syst. Man Cybern. Part C (Appl. Rev.) **38**(1), 69–82 (2008)
5. Cesarini, M., Monga, M., Tedesco, R.: Carrying on the e-learning process with a workflow management engine. In: Proceedings of the 2004 ACM Symposium on Applied Computing, New York, NY, USA, pp. 940–945 (2004)
6. Lin, J., Ho, C., Sadiq, W., Orlowska, M.E.: Using workflow technology to manage flexible e-learning services. J. Educ. Technol. Soc. **5**(4), 116–123 (2002)
7. Siemens, G., Baker, R.S.J.d.: Learning analytics and educational data mining: towards communication and collaboration. In: Proceedings of the 2nd International Conference on Learning Analytics and Knowledge, New York, NY, USA, pp. 252–254 (2012)
8. Ioniţă, I.: Data mining techniques for e-learning. J. Appl. Comput. Sci. Math. **10**, 26–31 (2016)
9. Castro, F., Vellido, A., Nebot, À., Mugica, F., México, H.: Applying Data Mining Techniques to E-learning Problems: A Survey and State of the Art: Evolution of Technology and Pedagogy. Springer, Heidelberg (2007)
10. Costa, C.J., Aparicio, M.: Analysis of e-learning processes. In: Proceedings of the 2011 Workshop on Open Source and Design of Communication, New York, NY, USA, pp. 37–40 (2011)
11. van der Aalst, W.: Process mining: overview and opportunities. ACM Trans. Manag. Inf. Syst. **3**(2), 7:1–7:17 (2012)
12. van der Aalst, W., et al.: Process mining manifesto. In: Daniel, F., Barkaoui, K., Dustdar, S. (eds.) BPM 2011, Part I. LNBIP, vol. 99, pp. 169–194. Springer, Heidelberg (2012). https://doi.org/10.1007/978-3-642-28108-2_19

13. Bergenthum, R., Desel, J., Harrer, A., Mauser, S.: Modeling and mining of learnflows. In: Jensen, K., Donatelli, S., Kleijn, J. (eds.) Transactions on Petri Nets and Other Models of Concurrency V. LNCS, vol. 6900, pp. 22–50. Springer, Heidelberg (2012). https://doi.org/10.1007/978-3-642-29072-5_2
14. Bogarín, A., Romero, C., Cerezo, R., Sánchez-Santillán, M.: Clustering for improving educational process mining. In: Proceedings of the Fourth International Conference on Learning Analytics and Knowledge, New York, NY, USA, pp. 11–15 (2014)
15. Netjes, M., Reijers, H.A.: Supporting the BPM life-cycle with FileNet. In: Proceedings of the CAISE, pp. 497–508 (2006)
16. Helic, D., Hrastnik, J., Maurer, H.: An analysis of application of business process management technology in e-learning systems. In: E-Learn: World Conference on E-Learning in Corporate, Government; Healthcare, and Higher Education, pp. 2937–2942 (2005)
17. Rakesh, A., Imielinski, T., Road, H.: Mining association rules between sets of items in large databases. ACM SIGMOD Rec. **22**(2), 207–216 (1993)
18. Li, N., Zeng, L., He, Q., Shi, Z.: Parallel implementation of apriori algorithm based on MapReduce (2012)
19. Singh, S., Garg, R., Mishra, P.K.: Review of Apriori Based Algorithms on MapReduce Framework (2017). arXiv:1702.06284 [cs]
20. Wang, J., Dai, Q.-H., Zeng, Y., Yang, D.-R.: Association rule mining with parallel frequent pattern growth algorithm on hadoop. Comput. Integr. Manuf. Syst. **18**(9), 2124–2129 (2012)
21. Wei, X., Ma, Y., Zhang, F.: Incremental FP-growth mining strategy for dynamic threshold value and database based on MapReduce. In: Computer Supported Cooperative Work in Design, CSCWD, pp. 271–276 (2014)
22. Rozinat, A., van der Aalst, W.M.P.: Decision mining in ProM. In: Dustdar, S., Fiadeiro, J.L., Sheth, A.P. (eds.) BPM 2006. LNCS, vol. 4102, pp. 420–425. Springer, Heidelberg (2006). https://doi.org/10.1007/11841760_33
23. Giraldo, J., Jiménez, J., Tabares, M.: Integrating business process management and data mining for organizational decision making (2015)
24. Wegener, D., Rüping, S.: On integrating data mining into business processes. In: Abramowicz, W., Tolksdorf, R. (eds.) BIS 2010. LNBIP, vol. 47, pp. 183–194. Springer, Heidelberg (2010). https://doi.org/10.1007/978-3-642-12814-1_16

SGNet: Design of Optimized DCNN for Real-Time Face Detection

Seunghyun Lee, Minseop Kim, and Inwhee Joe(✉)

Department of Computer Software, Hanyang University, Seoul, Korea
iwjoe@hanyang.ac.kr

Abstract. This paper proposes to optimize the deep convolution neural networks for real time video processing on detecting faces and facial landmarks. For that, we have to reduce the existing weight size and duplication of weight parameters. By utilizing the strengths of the two previous powerful algorithms which have shown the best performance, we overcome the weakness of the existing methods. Instead of using the old-fashioned searching method like sliding window, we propose our grid-based one-shot detection method. Furthermore, instead of forwarding one image frame through a very deep CNN, we divide the process into 3 stages for incremental detection improvements to overcome the existing limitation of grid-based detection. After lots of experiments with different frameworks, deep learning frameworks are chosen as the best for integration of 3-stage DCNN. By using transfer learning, we can remove the unnecessary convolution layers in the existing DCNN and retrain hidden layers repeatedly and finally succeed in obtaining the best speed and accuracy which can run on the embedded platform. The performance to find small sized faces is better than YOLO v2.

Keywords: DCNN · Scalable face detection · Transfer learning
Grid-based one-shot detection method

1 Introduction

Using the feature information of the human face, some algorithms have been developed that can detect 68 points called face landmarks. These face landmarks can be used to get a variety of facial information for human faces. As shown in Fig. 1, it is used not only in movie industry, various entertainment industries, but also in robotics and automobile industries.

As a traditional algorithm, SVM (SVM; Support Vector Machine) was used for face detection. Recently, Face detection algorithms have been developed using CNN (CNN; Convolutional Neural Network). Using Neural Network, face detection algorithms have greatly improved accuracy than traditional face detection algorithms.

The algorithms designed with Deep CNN like yolov2, Light-Head-R-CNN, Faster R-CNN have the detection performance of State of Art, but it is difficult to apply to mobile or embedded environment, because of the large amount of computation. In this paper, we designed an optimize network that can detect faces even in limited GPU environment.

J. H. Park et al. (Eds.): PDCAT 2018, CCIS 931, pp. 200–209, 2019.
https://doi.org/10.1007/978-981-13-5907-1_21

We designed a deep network that works on an embedded board or mobile by providing an architecture for detecting an accurate and 68 face landmarks. Rather than improve detection accuracy, we focus on detecting face and landmarks in real-time video and reducing network model size in a limited embedded board environment.

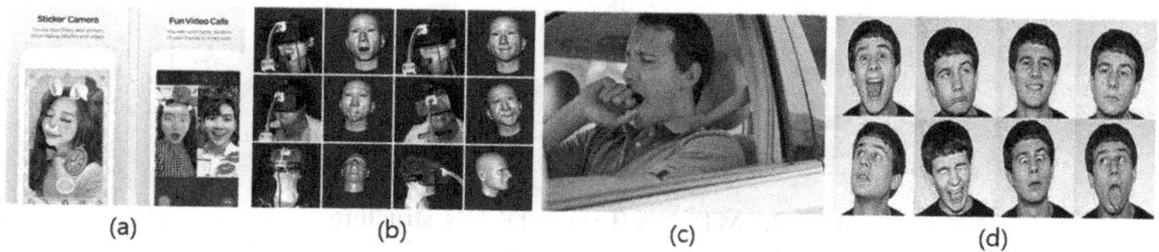

Fig. 1. Systems or applications using human face landmarks. (a) A camera application. (b) An application of Virtual Realization Avatar. (c) A system to estimate the driver's sleepiness by measuring the amount of blink in the eye and the number of yawning. (d) An emotional analysis system.

2 Related Works

Research on face detection and landmark detection has a long history, and accuracy has greatly improved recently using DCNN. In the case of face detection, the user moves through the entire image. Because the face can be larger than this box, it will increase the size of the box by a few pixels to slide the entire image from the beginning. This brutal force method analyzes the entire area of the image thoroughly, which can provide high detection accuracy but has a fatal disadvantage of very slow processing.

The whole image is divided into several Grids such as 7 × 7 or 13 × 13, and YOLO analyzes each cell at a time to estimate the probability of the object, the probability value of what class the object belongs to, and the position and size of the object. As the introduction, interest in Grid detection has increased. This method has the advantage that the detection speed can be increased up to 40 times (45–200 FPS) compared with the conventional than slide window method.

In the case of the face landmark, it is possible to detect 21 points, 36, 49, 68 points, etc. as the entire outline of the face is searched in detail, and the more difficult the detection, the more difficult it is. In addition, not only the front face but also the face may be rotated to the left, right, up and down, or some of the face may be covered or shadow may appear on the face, which makes the difficulty of landmark detection difficult.

2.1 MTCNN

In case of MTCNN, it increases efficiency by reducing number of ROIs while removing candidate groups of face boundary boxes through three stages. The uniqueness of MTCNN is that one small size network performs three tasks such as Bound Box and Landmark point regression and face classification as shown in Fig. 2. Training strategies are also unique. After P-Net study, this model collected hard images with low detection rates and recycled them to learn R-Net-Net.

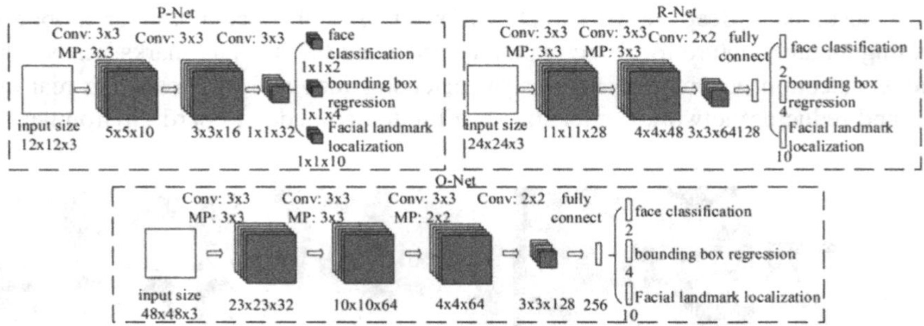

Fig. 2. MTCNN 3 stage DCNN structure

2.2 YOLO

YOLO v2 [17] and SSD (Single Shot Detection) [12] are typical DCNNs that perform Grid-based object detection. YOLO v2 is the algorithm that received the best paper award of CVPR 2017.

This Grid detection method has the advantage of high detection accuracy and especially a very fast processing speed (45–100 FPS). However, instead of searching through the whole image area in a sliding window sequentially, while it is forwarding to the network, the entire image area is transferred to the layer. As shown in Fig. 3, the whole image area is transferred to the hidden layer as it is, and the detection information is trained for each cell divided by the grid. So that objects smaller than a cell do not transfer the characteristics of the object to the latter part of the layer, distorted or crumpled and disappear.

Fig. 3. Image grid segmentation detection.

Yolo V2 is a reference because it performs better than other SSD model. In the case of the Yolo V2, the performance was improved by applying various techniques compared to the V1. Instead of using the VGG-19 as before, the features of the 1 × 1 Point-wise Convolution layer introduced in 2016 and the Fully Connected layer.

Instead, it replaced the 1 × 1 Point-wise Convolution layer with the same effect and Average Pooling with a 7 × 7 kernel size. The result is better speed and feature extraction, and this network is called Darknet-19.

2.3 Pruning

The pruning introduced in [2, 10, 11] is not that all the connections connecting each neuron contribute to activation, and the ratio varies from layer to layer, but generally 10% to 40% Value of the algorithm. Disconnecting is the same as setting these small parameters to zero. In fact, [19] was able to compress LeNet-5 by 12 times, AlexNet by 9 times, and VGG-16 by 13 times (Fig. 4).

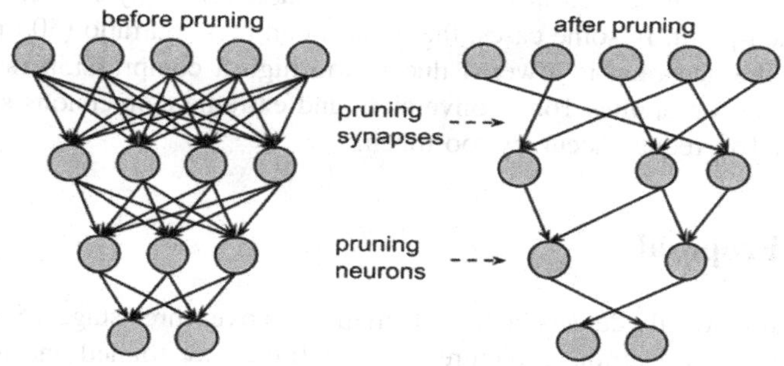

Fig. 4. Network pruning.

But basically, because DCNN has a connection sharing the weight, disconnecting the network connection and reducing the amount of annual traffic are not exactly proportional to each other. That is, as shown in [10], a reduction in the size of a model does not result in a 10-times faster computation, and sometimes the amount of annual output increases.

2.4 Separable Depth-Wise Convolution

Separable depth wise convolution in [4] and [5], a new convolution method is widely used. Convolution is performed by dividing it by channel (Depth-wise) as shown in Fig. 5. This is then combined with the 1 × 1 Point-wise Convolution feature to learn new features.

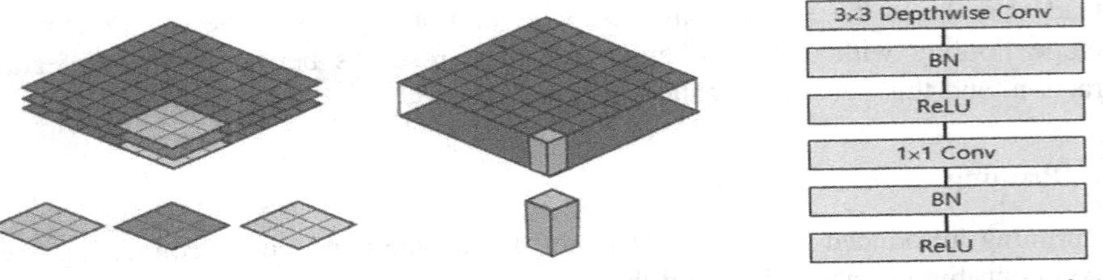

Fig. 5. Separable depth-wise convolution.

2.5 Quantization/Binarization

This technique was introduced in [6, 14], and is characterized by converting floating-points into fixed-point. In some cases, the highest compression ratio (50 times) can be obtained, but the accuracy is lowered due to too high a compression ratio. In most cases, Half Precision applies 16bit conversion, and extreme conversions such as 8bit, 4bit, 2bit, and 1bit reduce accuracy too much.

3 SGNet Proposal

In this paper, it gradually detects large and small faces over three stages. Since learning video data set is like learning a picture for each frame, we trained and tested model using some image. As shown in Fig. 6 two different network models detect faces and landmarks. One is BMSNet (Big Medium Small Network) that detects faces and the other is Landmark Network which detects landmarks.

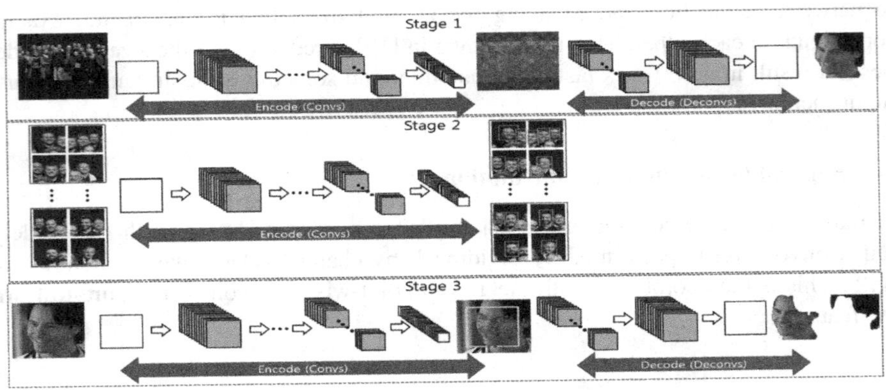

Fig. 6. The proposed DCNN structure

The first and second steps are the process of detecting the face, and the third step is the process of detecting the landmark. The first step is to detect areas that are expected to have large faces and small faces. The second step assumes that there are more faces

Table 1. Difference in MTCNN, YOLO v2, SGNet

	MTCNN	YOLO V2	SGNet
DCNN Architecture	3 Small Nets	DarkNet-19	BMS Net Landmark/Seg Net
Training DB	WIDER FACE/CeleA	VOC/COCO/ ImageNet	ImageNet/VOC/ WIDER FACE/300W/ IBUG/Helen/AFW/LFPW
FPS	2 ~ 4	190 ~ 200	60 ~ 70
Accuracy	Big face: more than 95% Small face: 60 ~70%	Big face: more than 95% Small face: 60 ~70%	Big face: more than 95% Small face: 90%
Input image	RGB	RGB	BMS Net: RGB L/S Net: Gray Scale
Detection method	Sliding window	Grid	Recursively Split Grid
Proposal ROI	N.A.	Direct initialization	Coarse to Fine
Task	Face Classification	20/80 Object Detection	Face Classification
	Boundary Box Regression		
	5 Landmark Regression		68 Landmark Regression
Loss function	$L_i^{box} = \left\| \hat{y}_i^{box} - y_i^{box} \right\|_2^2$ $L_i^{landmark} = \left\| \hat{y}_i^{landmark} - y_i^{landmark} \right\|_2^2$	$\lambda_{coord} \sum_{i=0}^{S^2} \sum_{j=0}^{B} \mathbb{1}_{ij}^{obj} \left[(x_i - \hat{x}_i)^2 + (y_i - \hat{y}_i)^2 \right] + \lambda_{coord} \sum_{i=0}^{S^2} \sum_{j=0}^{B} \mathbb{1}_{ij}^{obj} \left[\left(\sqrt{w_i} - \sqrt{\hat{w}_i} \right)^2 + \left(\sqrt{h_i} - \sqrt{\hat{h}_i} \right)^2 \right] + \sum_{i=0}^{S^2} \sum_{j=0}^{B} \mathbb{1}_{ij}^{obj} \left(C_i - \hat{C}_i \right)^2 + \lambda_{noobj} \sum_{i=0}^{S^2} \sum_{j=0}^{B} \mathbb{1}_{ij}^{noobj} \left(C_i - \hat{C}_i \right)^2 + \sum_{i=0}^{S^2} \mathbb{1}_{i}^{obj} \sum_{c \in classes} (p_i(c) - \hat{p}_i(c))^2$	
		N.A.	$\min_{\Delta S_t} \frac{\left\| T_t^{-1}(T_t(S_{t-1}) + \Delta S_t) - S^* \right\|}{d_{ipd}} + l(x;\theta) = \sum_{i,j} l(x_{i,j};\theta)$

between the small faces found in the first step and clustering these faces to redetect four regions into one image. In the third step, landmark detection and segmentation are performed on the faces found in the previous step. Encoders are responsible for landmarks and decoders are for segmentation. Landmark net had learned 68 landmark points of images consisting of about 3,000 images of 300 W, IBUG, Helen, AFW and LFPW based on Darknet-19. In the learning, instead of Euclidean distance, the value obtained by normalizing distance between eyes was the loss value.

The proposed method and the existing two methods showing the state of art performance as the present in face and object detection are as follows (Table 1).

BMSNet is transfer learning based on Yolo V2 network. The face detection process consists of two steps and Take the Coarse to Fine strategy based on the Grid. That is,

Fig. 7. Face detection algorithm

firstly a large face (96 * 96 or more), a medium size face (32 * 32–96 * 96), and a small face (32 * 32 or less) are detected first. Next, the area where the small faces are gathered is selected as a multiple of 4 samples, and these areas are upscale to the size of 208 * 208 and then made into one image again. This combined image is secondarily detected. Thus, all the small faces are detected in two steps.

The detailed face detection algorithm is performed as follows. First, through BMSNet, entire image area is analyzed with 13 * 13 Grid. Detects bounding boxes corresponding to a human class. Detect bounding boxes corresponding to large and small face classes, estimate small face areas. For a small face, 32 × 32 boxes including the face area and the surrounding area are learned, which is different from the existing

Table 2. BMS net architecture

			Kernel size/ stride	w_i		h_i		n_i		w_o		h_o		n_o
0	conv	32	3 × 3 / 1	416	×	416	×	3	→	416	×	416	×	32
1	max		2 × 2 / 2	416	×	416	×	32	→	208	×	208	×	32
2	conv	64	3 × 3 / 1	208	×	208	×	32	→	208	×	208	×	64
3	max		2 × 2 / 2	208	×	208	×	64	→	104	×	104	×	64
4	conv	128	3 × 3 / 1	104	×	104	×	64	→	104	×	104	×	128
5	conv	64		104	×	104	×	128	→	104	×	104	×	64
6	conv	128	3 × 3 / 1	104	×	104	×	64	→	104	×	104	×	128
7	max		2 × 2 / 2	104	×	104	×	128	→	52	×	52	×	128
8	conv	256	3 × 3 / 1	52	×	52	×	128	→	52	×	52	×	256
9	conv	128	1 × 1 / 1	52	×	52	×	256	→	52	×	52	×	128
10	conv	256	3 × 3 / 1	52	×	52	×	128	→	52	×	52	×	256
11	max		2 × 2 / 2	52	×	52	×	256	→	26	×	26	×	256
12	conv	512	3 × 3 / 1	26	×	26	×	256	→	26	×	26	×	512
13	conv	256	1 × 1 / 1	26	×	26	×	512	→	26	×	26	×	256
14	conv	512	3 × 3 / 1	26	×	26	×	256	→	26	×	26	×	512
15	max		2 × 2 / 2	26	×	26	×	512	→	13	×	13	×	512
16	conv	1024	3 × 3 / 1	13	×	13	×	512	→	13	×	13	×	1024
17	conv	512	1 × 1 / 1	13	×	13	×	1024	→	13	×	13	×	512
18	conv	1024	3 × 3 / 1	13	×	13	×	512	→	13	×	13	×	1024
19	conv	1024	3 × 3 / 1	13	×	13	×	1024	→	13	×	13	×	1024
20	conv	1024	3 × 3 / 1	13	×	13	×	1024	→	13	×	13	×	1024
21	route	14	-9		×		×		→		×		×	
22	reorg		/ 2	26	×	26	×	64	→	13	×	13	×	256
23	route	24 22	-1 -3		×		×		→		×		×	
24	conv	1024	3 × 3 / 1	13	×	13	×	1280	→	13	×	13	×	1024
25	conv	30	1 × 1 / 1	13	×	13	×	102	→	13	×	13	×	30

algorithm. And expand the selected small face areas to 416 * 416, and then configure the grid as one image (Fig. 7).

The Loss function of BMS Net is composed of the sum of squared error of Euclidian distance like Yolo V2 and Landmark Net's Loss function is the normalized Euclidean distance to the distance between the eyes Normalized Euclidean Distance was used (Table 2).

4 Performance and Conclusion

We were able to optimize the network by deleting or replacing some hidden layers. The graph that we learned about the basic Yolo V2 before optimization is as follows (Fig. 8).

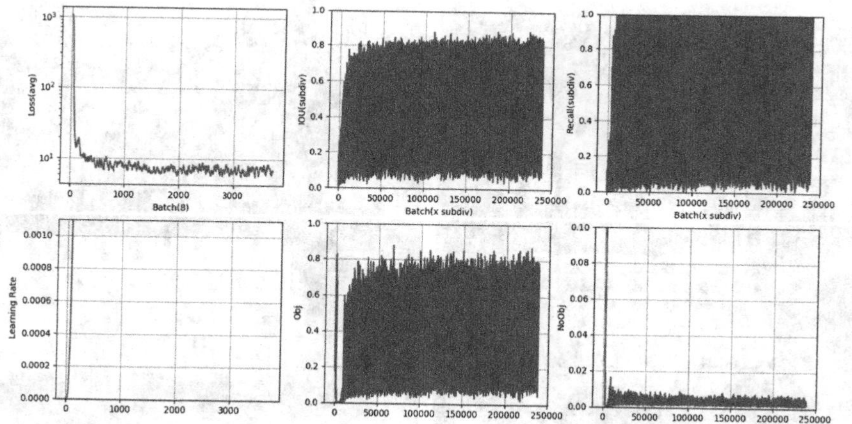

Fig. 8. BMS Network loss, IOU, Recall, Learning rate, object detection rate and Non-object detection rate graph

The first graph shows loss value per each batch, and it converges almost after 1000 iteration. The second graph is about IOU graphs. The third graph is about recall, the fifth and sixth graph are about interesting object or non-object detection rate.

When the proposed model was used, it shows the following performance. For optimization, layers 15, 16, 19 and 20 were removed and layers 23, 24 and 29 were replaced as separable depthwise Convolution. Although training required more Iteration to converge loss values than before based on larger faces, the accuracy is finally 2% to 4% lower performance (Fig. 9).

In face detection, yolov2 did not detect small faces well. We compared the face detection between BMSNet and MTCNN for small face (Fig. 10).

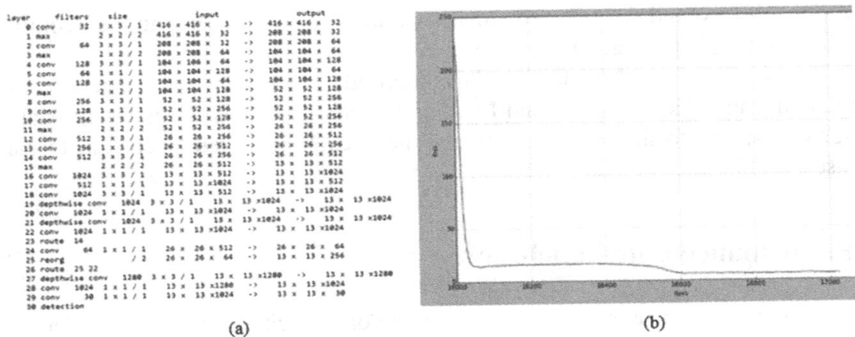

layer	filters	size	input		output
0 conv	32	3 x 3 / 1	416 x 416 x 3	->	416 x 416 x 32
1 max		2 x 2 / 2	416 x 416 x 32	->	208 x 208 x 32
2 conv	64	3 x 3 / 1	208 x 208 x 32	->	208 x 208 x 64
3 max		2 x 2 / 2	208 x 208 x 64	->	104 x 104 x 64
4 conv	128	3 x 3 / 1	104 x 104 x 64	->	104 x 104 x 128
5 conv	64	1 x 1 / 1	104 x 104 x 128	->	104 x 104 x 64
6 conv	128	3 x 3 / 1	104 x 104 x 64	->	104 x 104 x 128
7 max		2 x 2 / 2	104 x 104 x 128	->	52 x 52 x 128
8 conv	256	3 x 3 / 1	52 x 52 x 128	->	52 x 52 x 256
9 conv	128	1 x 1 / 1	52 x 52 x 256	->	52 x 52 x 128
10 conv	256	3 x 3 / 1	52 x 52 x 128	->	52 x 52 x 256
11 max		2 x 2 / 2	52 x 52 x 256	->	26 x 26 x 256
12 conv	512	3 x 3 / 1	26 x 26 x 256	->	26 x 26 x 512
13 conv	256	1 x 1 / 1	26 x 26 x 512	->	26 x 26 x 256
14 conv	512	3 x 3 / 1	26 x 26 x 256	->	26 x 26 x 512
15 max		2 x 2 / 2	26 x 26 x 512	->	13 x 13 x 512
16 conv	1024	3 x 3 / 1	13 x 13 x 512	->	13 x 13 x1024
17 conv	512	1 x 1 / 1	13 x 13 x1024	->	13 x 13 x 512
18 conv	1024	3 x 3 / 1	13 x 13 x 512	->	13 x 13 x1024
19 depthwise conv	1024	3 x 3 / 1	13 x 13 x1024	->	13 x 13 x1024
20 conv	1024	1 x 1 / 1	13 x 13 x1024	->	13 x 13 x1024
21 depthwise conv	1024	3 x 3 / 1	13 x 13 x1024	->	13 x 13 x1024
22 conv	1024	1 x 1 / 1	13 x 13 x1024	->	13 x 13 x1024
23 route 14					
24 conv	64	1 x 1 / 1	26 x 26 x 512	->	26 x 26 x 64
25 reorg		/ 2	26 x 26 x 64	->	13 x 13 x 256
26 route 25 22					
27 depthwise conv	1280	3 x 3 / 1	13 x 13 x1280	->	13 x 13 x1280
28 conv	1024	1 x 1 / 1	13 x 13 x1280	->	13 x 13 x1024
29 conv	30	1 x 1 / 1	13 x 13 x1024	->	13 x 13 x 30
30 detection					

(a) (b)

Fig. 9. (a) BMS Network, (b) Loss value of each iteration

(a) (b)

Fig. 10. Performance comparison between MTCNN and SGNet: (a) MTCNN (b) Proposed SGNet

In addition, after applying the three optimization techniques, the existing 200 MB model size could be reduced to 70 MB. The detection performance was approximately 1 s (1 FPS) for the MTPNN even if the threshold was lowered as low as 0.1 for the detection performance of SGNet. In case of SGNet, the performance was 30 FPS. However, the number of face detectors was significantly higher than that of MTPNN. Through these results, it could be able to verify that a network design that can be operated even in an embedded environment with GPGPU was possible.

In summary, we obtained the experimental results that were like the performance of two representative DCNNs showing the performance of the State of Art among the schemes introduced until 2017, while still being able to operate on the embedded system.

Acknowledgment. This work was supported by the Technology Development Program (S2521883) funded by the Ministry of SMEs and Startups (MSS, Korea).

References

1. Zhang, S., et al.: Faceboxes: a CPU real-time face detector with high accuracy. In: 2017 IEEE International Joint Conference on Biometrics (IJCB). IEEE (2017)
2. He, Y., Zhang, X., Sun, J.: Channel pruning for accelerating very deep neural networks. In: International Conference on Computer Vision (ICCV), vol. 2, no. 6 (2017)
3. Kowalski, M., Naruniec, J., Trzcinski, T.: Deep alignment network: a convolutional neural network for robust face alignment. In: Proceedings of the International Conference on Computer Vision & Pattern Recognition (CVPRW), Faces-in-the-wild Workshop/Challenge, vol. 3, no. 5 (2017)
4. Chollet, F.: Xception: deep learning with depthwise separable convolutions. arXiv preprint arXiv:1610-02357 (2017)
5. Howard, A.G., et al.: Mobilenets: efficient convolutional neural networks for mobile vision applications. arXiv preprint arXiv:1704.04861 (2017)
6. Courbariaux, M., Bengio, V.: Binarynet: training deep neural networks with weights and activations constrained to +1 or −1. arXiv preprint arXiv:1602.02830 (2016)
7. Wu, Y., Hassner, T.: Facial landmark detection with tweaked convolutional neural networks. arXiv preprint arXiv:1511.04031 (2015)
8. Zhang, K., et al.: Joint face detection and alignment using multitask cascaded convolutional networks. IEEE Signal Process. Lett. **23**(10), 1499–1503 (2016)
9. Redmon, J., Farhadi, A.: YOLO9000: better, faster, stronger. arXiv preprint (2017)
10. Han, S., Mao, H., Dally, W.J.: Deep compression: compressing deep neural networks with pruning, trained quantization and huffman coding. arXiv preprint arXiv:1510.00149 (2015)
11. Iandola, F.N., Moskewicz, M.W., Ashraf, K., Han, S., Dally, W.J., Keutzer, K.: Squeezenet: alexnet-level accuracy with 50x fewer parameters and <1 MB model size. arXiv preprint arXiv:1602.07360 (2016)
12. Liu, W., Anguelov, D., Erhan, D., Szegedy, C., Reed, S.: SSD: single shot multibox detector. arXiv preprint arXiv:1512.02325 (2015)
13. Szegedy, C., Ioffe, S., Vanhoucke, V.: Inception-v4, inception-resnet and the impact of residual connections on learning. arXiv preprint arXiv:1602.07261 (2016)
14. Wu, J., Leng, C., Wang, Y., Hu, Q., Cheng, J.: Quantized convolutional neural networks for mobile devices. arXiv preprint arXiv:1512.06473 (2015)
15. Yang, S., Luo, P., Loy, C.C., Tang, X.: WIDER FACE: a face detection benchmark. arXiv preprint arXiv:1511.06523
16. Deng, J., Dong, W., Socher, R., Li, L.-J., Li, K., FeiFei, L.: Imagenet: a large-scale hierarchical image database. In: 2009 IEEE Conference on Computer Vision and Pattern Recognition, CVPR 2009, pp. 248–255. IEEE (2009)
17. Redmon, J.: Darknet: open source neural networks in c (2013–2016). http://pjreddie.com/darknet
18. Jia, Y., et al.: Caffe: convolutional architecture for fast feature embedding. arXiv preprint arXiv:1408.5093 (2014)
19. Han, S., et al.: Learning both weights and connections for efficient neural network. In: Advances in Neural Information Processing Systems (2015)

A Study on L1 Data Cache Bypassing Methods for High-Performance GPUs

Cong Thuan Do[1], Min Goo Moon[2], Jong Myon Kim[3], and Cheol Hong Kim[2(✉)]

[1] Department of Computer Science, Korea University, Seoul, Korea
congthuan.hut@gmail.com
[2] School of Electronics and Computer Engineering, Chonnam National University, Gwangju, Korea
airnia41@gmail.com, chkim22@jnu.ac.kr
[3] School of Electrical Engineering, University of Ulsan, Ulsan, Korea
jmkim07@ulsan.ac.kr

Abstract. Graphics Processing Units (GPUs) with massive parallel architecture have been widely used to boost performance of both graphics and general-purpose programs. GPGPUs become one of the most attractive platforms in exploiting plentiful thread-level parallelism. In recent GPUs, cache hierarchies have been employed to deal with applications with irregular memory access patterns. Unfortunately, GPU caches exhibit poor efficiency due to arising many performance challenges such as cache contention and resource congestion caused by large number of active threads in GPUs. Cache bypassing can be a solution to reduce the impact of cache contention and resource congestion. In this paper, we introduce a new cache bypassing technique that is able to make effective bypassing decisions. In particular, the proposed mechanism employs a small memory, which can be accessed before actual cache access, to record the tag information of the L1 data cache. By using this information, the mechanism can know the status of the L1 data cache and use it as a bypassing hint to make the cache bypassing decision close to optimal. Our experimental results based on a modern GPU platform reveal that our proposed cache bypassing technique achieves up to 10.4% of IPC improvement on average.

Keywords: GPU · CPU · Bypassing · Cache · Performance

1 Introduction

In recent years, GPUs have been employed for handling general purpose computation and become one of the most attractive computing platforms for executing general purpose applications. Such GPUs are known as General Purpose Graphics Processing Units (GPGPUs). In fact, for parallel applications, it was proven that GPUs achieve much higher performance than CPUs due to the exploitation of a large number of streaming multiprocessors (SMs) and gigabytes of high memory bandwidth [1]. To deal with wide range of workloads, caches have been used in conjunction with scratchpad memory as on-chip memory in the new generations of GPUs like NVIDIA Kepler architecture. Unfortunately, GPU caches have to confront with many challenges

J. H. Park et al. (Eds.): PDCAT 2018, CCIS 931, pp. 210–219, 2019.
https://doi.org/10.1007/978-981-13-5907-1_22

caused by the limitation of memory subsystem and the design of GPU caches, meanwhile the number of incoming requests issued by GPU applications is much larger compared to CPU applications. Consequently, the low per-thread capacity, associativity, etc. are easily overwhelmed; making GPU caches a system bottleneck and causing performance unpredictability. Due to the large number of active threads in GPUs, significantly increasing per-thread capacity by increasing the physical size of a cache is difficult, hence, we have to consider new thought about cache management. Cache bypassing, where memory requests are selectively bypassed a particular cache, is a popular technique that has been widely applied on CPUs. However, the number of studies in cache bypassing techniques for GPUs is modest. In this paper, we propose a cache bypassing technique that employs a small memory, which can be accessed before actual cache accesses, to record the tag information of the L1 data cache (L1-D) to reveal the status of the L1-D. Compared to always turning cache on, our cache bypassing design achieves up to 10.4% performance speedup on average for a 16 KB L1-D. The rest of this paper is organized as follows. Section 2 discusses related work and briefly describes baseline GPU architecture. Section 3 presents our proposed cache bypassing architecture and Sect. 4 describes experimental methodology and results. Section 5 concludes the paper.

2 Background

2.1 Related Work

To fully utilize the GPU hardware resources, a large number of thread scheduling policies were proposed [2–5]. Remarkably, Narasiman et al. [3] proposed a two-level warp scheduler that classified warps into active set and pending set to improve the performance of GPUs. A subset of warps (called a fetch group) is run as long as possible until these warps all hit long-latency operations. Rogers et al. [4] proposed a family of static and dynamic greedy scheduling approaches to improve intra-warp data locality. Their work can improve the L1 hit rates for cache-sensitive applications. Control flow divergence is another critical that constrains the performance potential of GPUs. Furthermore, there is number of studies on optimizing the effect of control flow divergence issue [6–8]. Modern GPUs are equipped on-chip cache memories to deal with high frequency of diverse and irregular memory access patterns in general purpose applications, as a result, memory system optimizations are increasingly important. In fact, there are many techniques being applied on CPUs, for example complicated cache replacement polices [9–11], cannot be applied on GPUs effectively due to the fundamental difference on computing architectures. Recently, cache bypassing has been considered as on effective solution to the problem of cache contention and cache resource congestion. Jia et al. [12] proposed a hardware structure called memory request prioritization buffer (MRPB), which employs request reordering and cache bypassing, to avoid a system bottleneck in GPU caches. Chen et al. [13] proposed an adaptive cache management technique by combining the protection distance [14] and cache bypassing to improve the cache performance. Other recent cache bypassing techniques for GPUs were proposed in [15–17].

2.2 Baseline GPU Architecture

In this section, we describe briefly a typical modern GPU design of NVIDIA. Further details on these architectures can be found in [4, 18, 19]. Baseline GPU architecture consists of many streaming multiprocessors (SMs) also called shader cores, with each typically having single-instruction, multiple-threads (SIMT) lanes of 8 to 32. The target GPU architecture used in this work consists of 15 SMs and they are connected each other by an interconnection network (NoC). Each SM is equipped various types of private caches, including a read-only texture, constant cache, low-latency shared memory and L1-D. For supporting parallel computation, GPUs are supported vast on-chip register file resources per SM in order to accommodate a large number of set of threads. The memory hierarchy of the GPU includes register files, L1 memories (scratchpad and L1-D), special-purpose caches (such as texture, read-only, etc.), shared L2 cache, and off-chip GDDR DRAM. The private L1-D has short latency and is accessible from all of the threads within a thread block. It also works as the central point of coherency. Each memory controller is associated with a slice of shared L2 cache bank. An L2 cache bank with a memory controller is defined as a memory partition. On NVIDIA GPUs, the programmers use CUDA to parallelize the applications. A typical CUDA application consists of many kernels. Execution on GPUs starts with the launch of kernel. Each kernel includes groups of threads called cooperative thread arrays (CTAs). Within a CTA, groups of threads (usually 32 threads) called warps are executed in lockstep fashion. A CTA is an abstraction that encapsulates all synchronization and barrier primitives among warps [20], helping CTAs to be executed on an SM in any order. This leads to an increase of available parallelism since there are no restrictions on CTAs and any SM is free to schedule any CTAs. Modern GPUs can provide better performance by effectively utilizing the computational resources since they can exploit FGMT which allows the assignment of one or multiple CTAs to one SM [21]. Thread scheduling on the GPU is performed as a three-step process. In the first step, a kernel is launched on the GPU. After launching the kernel, the global CTA scheduler (e.g. GigaThread [22]) will assign CTAs to all available SMs. After this assignment, if an SM is capable of executing multiple CTAs and there are enough CTAs, a second round of assignment starts. This process (second step) continues until all CTAs are assigned their maximum limit of CTAs (limited only by the hardware resources) [23]. After the CTA assignment, warp schedulers are responsible for scheduling warps associated with the launched CTAs to SIMT lanes of the corresponding SM (third step).

3 Proposed Cache Bypassing Architecture

The resources such as a cache line, an MSHR entry, a miss queue entry, etc. must be allocated by the missed request before the miss can be processed. If any of these resources are unavailable, the pipeline will be stalled. When the pipeline is stalled, this request must be retried in subsequent cycles until all resources are available to process it. However, with the L1-D bypassing turned on, such read requests can be immediately sent to main memory as in multi-core processors, therefore, avoiding congesting the pipeline. In the case of GPUs, the request will be forwarded to the L2 data cache through the NoC. Afterwards, the returned data is written directly to the registers without filling the cache.

One of the main requirements for a good cache bypassing technique is the logic must be simple to make bypass decision in terms of hardware implementation. In addition to this, when a memory request is bypassed is an important design decision that must be concerned. Optimally, if an incoming memory request is generated by a mostly-miss or all-miss warp, this request should be bypassed to the L2 data cache. When a bypassing decision is known, it can simply use a 2-bit multiplexer to determine whether the incoming request is sent to the L1-D or to the L2 bank arbiter. Our cache bypassing technique is based on the memory address of the block access as the input of the mechanism to make the bypassing decision. This is because an address-based scheme would be preferred when the same memory address has similar amounts of locality in each phase of the program execution, especially when the code around the load instructions does not dictate the data locality. More specifically, the proposed bypassing scheme uses the memory address with the aim to probe the status of the L1-D before the real cache access. The lookup result is used as a hint to decide whether the request can bypass the L1-D and need not be allocated in the L1-D. When the L1-D tag array is accessed, it is either a tag hit or a tag miss. A bypassing decision will be make if the lookup result is a tag miss. The incoming request then is forwarded to the L2 data cache through the NoC. When a memory request is bypassed, its bypass-bit, which is added to every memory request, is set to keep track of the returned data. In our proposed technique, the returned data requested from lower-level memories will be sent directly to processors without filling the L1-D. This is due to the low data locality of GPGPU applications, which represents through very high cache miss rates at the L1-D.

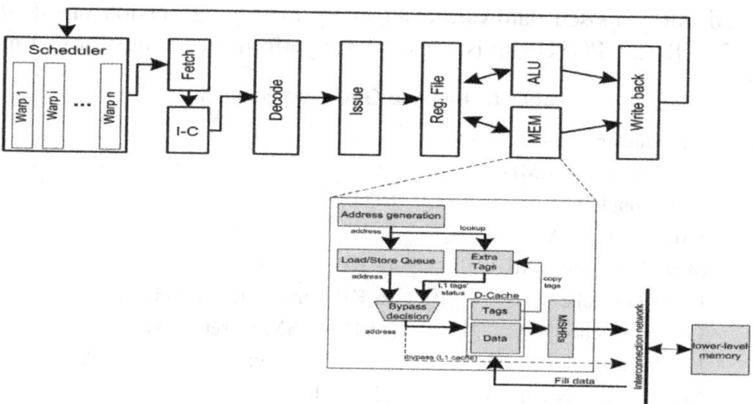

Fig. 1. Modified pipeline between the address generation and the memory stage

The difficulty here is how to implement early tag-array access in hardware. We observe that a block's memory address can be generated one clock cycle before the Load/Store Queue (LSQ) stage (as seen in the bottom part of Fig. 1). This means that the address is available one clock cycle before L1 cache access stage. With the aim at knowing the status of the L1-D before the memory request decides to either access or bypass the L1-D, we only need to examine the tag array of the L1-D instead of the whole cache. Furthermore, the latency of a typical L1-D access is one clock cycle and could be longer in deeply pipelined processors, whereas the access to the tag array can be finished in at most one cycle [24, 25]. Since a load/store instruction is sent to the LSQ before being issued to the data cache, the instruction is compared with the existing ones in the LSQ to determine whether the instruction can be issued to the data cache at the next clock cycle. If not, the instruction will stay in the LSQ stage for more than one cycle. Therefore, the memory address of a load/store is available in the LSQ stage for at least one clock cycle before being issued to the data cache and the status of the L1-D can be known before accessing the L1-D. We employ an additional memory, which is organized as the L1-D tag array, to record the tag information and the status of cache lines at the LSQ stage. This memory will be searched prior to a cache access, during the LSQ stage. By adding this memory, it can avoid the data contention with the L1-D. Each time a memory instruction is sent into the LSQ, an access to the corresponding set of extra tag array is performed (as shown in the bottom part in Fig. 1). Hence, the status of the L1 cache can be determined before the L1-D access and used as a hint for the cache bypassing decision. Every time a new cache line is filled to the L1-D, the corresponding tag in the extra tag-array will be updated. Note that the tag array will not be updated until a cache miss occurs due to the temporal/spatial locality.

4 Experiments

4.1 Experimental Methodology

We modeled our proposed hardware changes by modifying version of GPGPU-sim (version 3.2.2) [23]. GPGPU-sim is a cycle-level performance simulator that models a

Table 1. Baseline GPU configuration

# of shader cores	15
Warp size and SIMD width	32
# of threads/SM	1024
# of registers/SM	32768
Shared memory/SM	48 KB
L1 Data Cache	16 KB per SM (32-sets/4-ways)
L1 Inst Cache	2 KB per SM (4-sets/4-ways)
L2 Cache	768 KB (64-sets/16-ways/6-banks)
Min. L2 access latency	120 cycles
Min. DRAM access latency	220 cycles
# of memory controllers	12
# of memory chips/controllers	2
Memory channel bandwidth	4 KB
GDDR5	tCL = 12, tRP = 12, tRC = 40, tRAS = 28, tRCD = 12, tRRD = 6

general-purpose GPU architecture supporting NVIDIA CUDA [1] and its PTX ISA. The details of the baseline platform configuration used in this work are presented in Table 1. The benchmarks used to evaluate the performance of our proposed technique are selected from ISPASS [23] and CUDA SDK [26]. The selected programs are EstimatePiInlineP, FastWalshTransform, Histogram, Merge-Sort, Reduction, Scalar-Prod, SimpleMultiGPU, SortingNetworks, BFS, LIB.

4.2 Experimental Results

This section shows the impact of the proposed cache bypassing technique on GPGPU applications. Note that the proposed technique is applied for global loads since the L1-Ds in modern GPUs do not cache global stores. Figure 2 plots the speedup of the proposed technique normalized to the baseline. As we can see, the average IPC improvement is 10.4% without slowing down any applications. Remarkably, the IPC improvement is up to 55% on scalarProd, 11.7% on EstimatePiLineP and 12.6% on BFS. Figure 3 shows that the number of L1-D accesses is reduced by 6.6% compared to the baseline when applying the proposed technique. From this figure, we can see that scalarProd has the largest L1-D access reduction, up to 43%. As this application is unfriendly to the L1-D, hence, bypassing a majority of cache requests alleviates the negative impacts of the L1-D on performance. Figure 3 also represents the L1-D cache miss rate reduction of the proposed technique. On average, the cache miss rate is improved slightly, about 1% due to very low data locality of GPGPU applications. Notably, in the cases of EstimatePiLineP and LIB, though the miss rate is reduced by 3%, their performance improvement is still considerable. These results point out that the performance of GPUs depends on various factors rather than only L1-D cache miss rate.

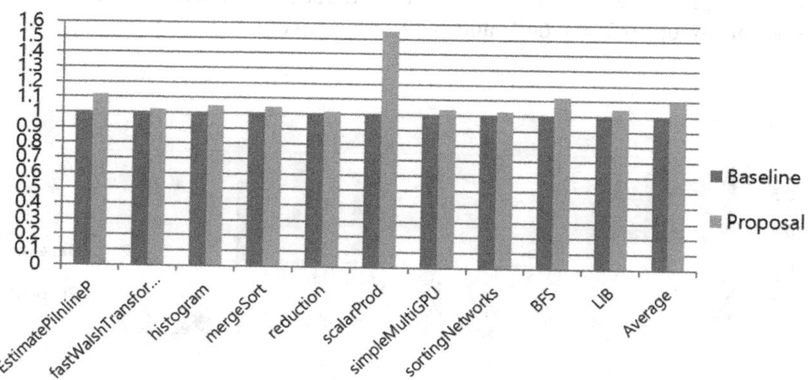

Fig. 2. Normalized IPC of Bypass compared to the baseline

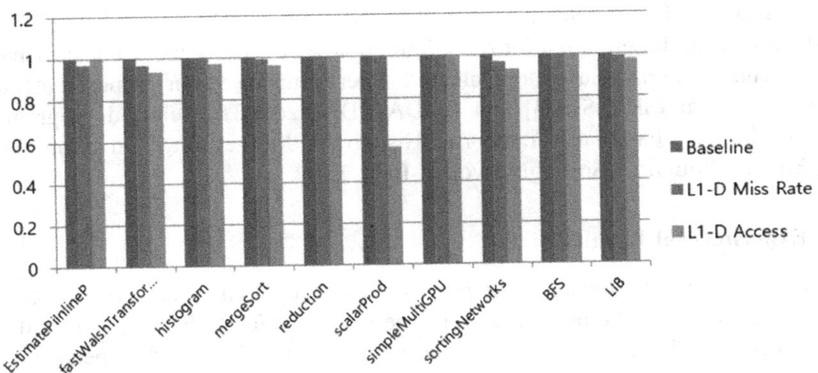

Fig. 3. Normalized L1-D miss rate and number of cache

There are several main reasons behind the performance improvement when the proposed technique is applied. First of all, the system can get negligible advantage even there are a few L1-D hits while bypassing these requests can provide much more benefits. On the other hand, if these requests miss in the L1-D, they have to suffer longer queuing latencies at both L1 and L2 caches. In that case, bypassing these requests can reduce contention at the L1-D. These benefits are especially clearer for applications that are sensitive to cache. Figure 4 shows the number of pipeline stall cycles at the memory stage, which are mainly caused by (1) shared memory bank conflict, (2) non-coalesced memory access and (3) serialized constant access. As shown in the figure, the degree of pipeline stall at the memory stage is reduced by 27.4% across all applications compared to the baseline. In general, the reduction of this pipeline stall has a strong impact on overall performance since GPUs use in-order pipelines. The proposed technique significantly reduces such stall for BFS, scalarProd and reduction by up to 70%, 64% and 62%, respectively.

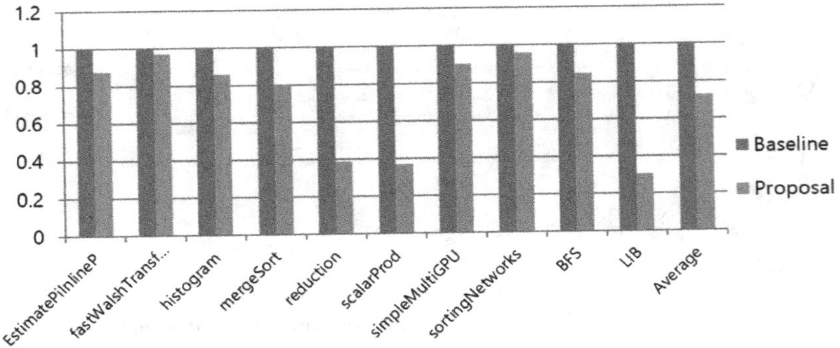

Fig. 4. Normalized number of pipeline stall cycles at the memory stage

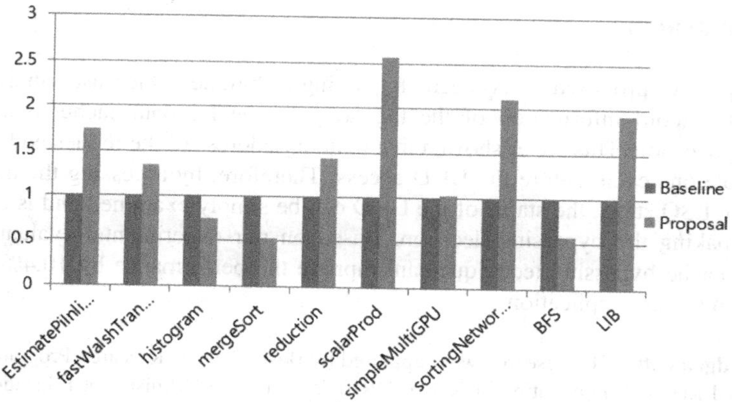

Fig. 5. Normalized number of cycles that the dram channels are stalled due to the interconnection congestion with the proposed technique compared to the baseline

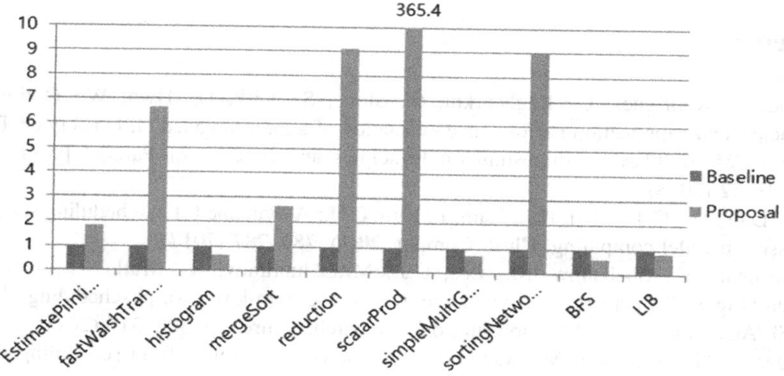

Fig. 6. Normalized number of cycles that the interconnection outputs to dram channels is stalled with proposed technique compared to the baseline

We also study the impact of the proposed technique on other factors of the GPU. As the L2 cache is off-chip and is associated to the shared cores via the NoC, therefore, the proposed cache bypassing technique can impact the NoC. Figure 5 shows the number of stall cycles that the dram channels are stalled due to the NoC congestion (Stall_Inct_to_Core). As expected, the proposed technique increases Stall_Inct_to_Core for most of applications. The reason is the increasing number of memory requests to the NoC that causes higher stall probability of the NoC. Figure 6 shows the number of cycles that the NoC outputs to dram channel is stalled (Stall_DRAM_Full). Similarly, the proposed technique also results in an increase in Stall_DRAM_Full for most of applications. Note that there is no performance degradation is observed.

5 Conclusions

In this paper, we proposed a new cache bypassing technique, which uses an additional memory to record information of the tag array of the L1 data cache to make the bypassing decision. This paper showed that memory address of the requested block can be available one cycle before the L1-D access. Therefore, by accessing the additional memory at LSQ stage, the status of the L1-D can be simply examined and is used as a hint for making the bypassing decision. Based on our experimental evaluation, our proposed cache bypassing technique can improve the performance by 10.4% without slowing down any applications.

Acknowledgements. This research was supported by Basic Science Research Program through the National Research Foundation of Korea (NRF) funded by the Ministry of Education (NRF-2018R1A2B6005740), and it was also supported by the MSIT (Ministry of Science and ICT), Korea, under the ITRC (Information Technology Research Center) support program (IITP-2018-2016-0-00314) supervised by the IITP (Institute for Information & communications Technology Promotion).

References

1. Ryoo, S., Rodrigues, C., Baghsorkhi, S., Stone, S., Kirk, D., Hwu, W.: Optimization principles and application performance evaluation of a multithreaded GPU using CUDA. In: The ACM SIGPLAN Symposium on Principles and Practice of Parallel Programming, pp. 73–82 (2008)
2. Son, D.O., Do, C.T., Choi, H.J., Nam, J., Kim, C.H.: A dynamic CTA scheduling scheme for massive parallel computing. Clust. Comput. **20**(1), 781–787 (2017)
3. Narasiman, V., Shebanow, M., Lee, C.J., Miftakhutdinov, R., Mutlu, O., Patt, Y.N.: Improving GPU performance via large warps and two-level warp scheduling. In: The IEEE/ACM International Symposium on Microarchitecture, pp. 308–317 (2011)
4. Rogers, T.G., O'Connor, M., Aamodt, T.: Cache-conscious wavefront scheduling. In: The International Symposium on Microarchitecture, pp. 72–83 (2012)
5. Lee, S.-Y., Arunkumar, A., Wu, C.-J.: CAWA: coordinated warp scheduling and cache prioritization for critical warp acceleration of GPGPU workloads. In: The International Symposium on Computer Architecture, pp. 515–527 (2015)
6. Meng, J., Tarjan, D., Skadron, K.: Dynamic warp subdivision for integrated branch and memory divergence tolerance. In: The International Symposium on Computer Architecture, pp. 235–246 (2010)
7. Park, Y., Park, J.J.K., Park, H., Mahlke, S.: Libra: tailoring SIMD execution using heterogeneous hardware and dynamic configurability. In: The IEEE/ACM International Symposium on Microarchitecture, pp. 84–95 (2012)
8. Rhu, M., Erez, M.: Maximizing SIMD resource utilization in GPGPUs with SIMD lane permutation. In: The International Symposium on Computer Architecture, pp. 356–367 (2013)
9. Do, C.T., Choi, H.J., Kim, J.M., Kim, C.H.: A new cache replacement algorithm for last-level caches by exploiting tag-distance correlation of cache lines. Microprocess. Microsyst. **39**(4–5), 286–295 (2015)

10. Jaleel, A., Theobald, K.B., Steely, S.C., Emer, J.: High performance cache replacement using re-reference interval prediction (RRIP). In: The International Symposium on Computer Architecture, pp. 60–71 (2010)
11. Qureshi, M.K., Jaleel, A., Patt, Y.N., Steely, S.C., Emer, J.: Adaptive insertion policies for high performance caching. In: The International Symposium on Computer Architecture, pp. 381–391(2007)
12. Jia, W., Shaw, K., Martonosi, M.: MRPB: memory request prioritization for massively parallel processors. In: The IEEE International Symposium on High Performance Computer Architecture, pp. 272–283 (2014)
13. Chen, X., Chang, L.-W. Rodrigues, C.I., Lv, J., Wang, Z., Hwu, W.-M.W.: Adaptive cache bypass and insertion for many-core accelerators. In: The International Workshop on Manycore Embedded Systems, p. 1 (2014)
14. Duong, N., Zhao, D., Kim, T., Cammarota, R., Valero, M., Veidenbaum, A.V.: Improving cache management policies using dynamic reuse distances. In: The IEEE/ACM International Symposium on Microarchitecture, pp. 389–400 (2012)
15. Do, C.T., Kim, J.M., Kim, C.H.: Early miss prediction based periodic cache bypassing for high performance GPUs. Microprocess. Microsyst. **55**, 44–54 (2017)
16. Xie, X., Liang, Y., Wang, Y., Sun, G., Wang, T.: Coordinated static and dynamic cache bypassing for GPUs. In: The IEEE International Symposium on High Performance Computer Architecture, pp. 76–88 (2015)
17. Xie, X., Liang, Y., Sun, G., Chen, D.: An efficient compiler framework for cache bypassing on GPUs. In: The International Conference on Computer-Aided Design, pp. 516–523 (2013)
18. Krewell, K.: AMD's Fusion Finally Arrives, Microprocessor Report (2011)
19. Krewell, K.: NVIDIA Lowers the Heat on Kepler, Microprocessor Report (2012)
20. Kirk, D., Hwu, W.: Programming Massively Parallel Processors. Elsevier, London (2010)
21. Abdalla, K.M., et al.: Scheduling and Execution of Compute Task. US Patent US20130185725 (2013)
22. NVIDIA. NVIDIA Tegra Multiprocessor Architecture (2010)
23. Bakhola, A., Yuan, G., Fung, W., Wong, H., Aamodt, T.: Analyzing CUDA workloads using a detailed GPU simulator. In: The International Symposium on Analysis of Systems and Software, pp. 163–174 (2009)
24. Delano, E., Mulla, D.: Data cache design considerations for the Itanium2 processor. In the International Conference on Computer Design, pp. 356–362 (2002)
25. Brock, B., Exerman, M.: Cache Latencies of the PowerPC MPC7451. Freescale Semiconductor, Inc., Austin, TX, USA (2006)
26. NVIDA. CUDA SDK. http://developer.nvidia.com/gpu-computing-sdk

Memory Contention Aware Power Management for High Performance GPUs

Hong Jun Choi[1], Dong Oh Son[2], and Cheol Hong Kim[3(✉)]

[1] The Attached Institute of ETRI, Daejeon, Korea
chj6083@nsr.re.kr

[2] Avionics R&D Lab, LIG Nex1, Daejeon, Korea
dongoh.son@lignex1.com

[3] School of Electronics and Computer Engineering, Chonnam National University, Gwangju, Korea
chkim22@jnu.ac.kr

Abstract. To improve the performance of the GPU, more parallelism should be exploited and the GPU should be operated at higher clock frequency. However, high parallelism and high clock frequency cause serious memory contention problems, resulting in significant power consumption and increased idle cycles in the GPU. This paper proposes a new memory contention aware (MC-aware) power management scheme to reduce the power consumption of the GPU with little impact on the performance. When serious memory contention problems occur in the GPU, the proposed MC-aware scheme changes the mode of the SM (Streaming Multiprocessor) to power saving mode with little performance degradation. The proposed scheme monitors the degree of memory contention, since severe memory contention causes serious performance degradation. The proposed GPU architecture includes SM management unit that generates the control signals based on the estimated degree of memory contention. According to our simulation results, the proposed MC-aware scheme can increase the power efficiency, IPC per watt, by up to 31.4% compared to the conventional architecture.

Keywords: GPU · Performance · Memory contention · Power efficiency Streaming multiprocessor

1 Introduction

Graphics processing unit (GPU), which is originally designed to run graphics-related workloads, can provide high performance by employing thousands of threads in parallel. The GPU uses a single instruction multiple threads (SIMT) model that can execute the same instruction but with different data simultaneously to support concurrent execution at low cost. Moreover, GPUs have become more flexible by including programmable processors to execute diverse graphics operations [1, 2]. Recent GPUs, which are named as general purpose GPUs (GPGPUs), can execute both

This study was financially supported by Chonnam National University. (Grant number: 2017-2727).

J. H. Park et al. (Eds.): PDCAT 2018, CCIS 931, pp. 220–229, 2019.
https://doi.org/10.1007/978-981-13-5907-1_23

general purpose applications and graphic applications with the help of convenient APIs such as CUDA and OpenCL. Therefore, up-to-date GPUs have been widely used as a solution to improve the system performance in various application domains [3]. To provide high performance, GPUs are operated at high clock frequency and include massively parallel computing resources. For instance, the GTX480 GPU contains 16 streaming multiprocessors (SMs) and one SM includes 32 processing elements, 16 load store units, 4 special function units for special-purpose arithmetic computations, and it is operated at 700 MHz [4]. However, the power consumption in the GPU is increased dramatically as the GPU provides high performance. Consequently, the power efficiency has become a growing concern in designing the GPU [5]. Unfortunately, reducing the power consumption inevitably sacrifices the performance [6]. Therefore, the performance should be considered more carefully when low-power techniques are applied to the GPU architecture. In this work, our goal is to reduce the power consumption in the GPU with little impact on the performance. We focus on the most power-hungry component in the GPU, the SMs, which consume at least 85% of the total dynamic power of the GPU in executing compute-intensive applications [7]. To reduce the power consumption of the GPU, we propose the technique to manage the number of active SMs dynamically, since reducing the number of active SMs can save the power consumption. Decreasing the number of active SMs may cause serious performance degradation, resulting in significantly increased execution time. In this case, total power consumption required for completing the workload may be increased due to increased execution time. To prevent serious performance degradation, the proposed technique manages the number of SMs while keeping the GPU at a comparable performance level. To reduce the power consumption in the GPU with little impact on the performance, we propose a memory contention aware streaming multiprocessor management technique which exploits the memory contention information. Memory contention occurs when multiple accesses require the same memory block at the same time, which causes performance degradation due to pipeline stalling. Therefore, reducing the number of active SMs when memory contention is serious has little impact on the performance. The rest of this paper is organized as follows. Section 2 presents the baseline GPU architecture and Sect. 3 explains the proposed memory contention aware power management technique and the architectural support required for the proposed scheme. The experimental environment and detailed results are discussed in Sect. 4. Finally, we conclude this paper in Sect. 5.

2 Background

In GPU architectures, multiple threads are grouped into batches known as warps to support single instruction multiple threads (SIMT). A warp, which usually contains 32 threads, is processed in lock-step [8]. GPUs can provide high performance; however, they also suffer from performance degradation due to the long memory access latencies. To alleviate this problem, GPUs employ fine-grained multithreading (FGMT) which allows executing multiple warps to hide the long memory access latencies [9]. Therefore, warps with the same instructions are grouped together, called a cooperative thread array (CTA). The baseline GPU is composed of GPU clusters, interconnection network and memory partitions. A GPU cluster consists of multiple SMs that are connected to memory partitions through an on-chip interconnection network [10]. Every SM clock cycle, the CTA scheduler selects a SM of a GPU cluster to assign a CTA that is issued from the launch kernel. The process is continued in a round-robin order. Modern GPU architectures can launch multiple kernels simultaneously, however, an SM can execute only one CTA at one time. Each kernel can assign its CTAs to the SM independently to support multiple kernel execution. The CTA scheduler examines available computational resources (available threads per SM, register file, shared memory, etc.) to determine the maximum number of CTAs per SM [11].

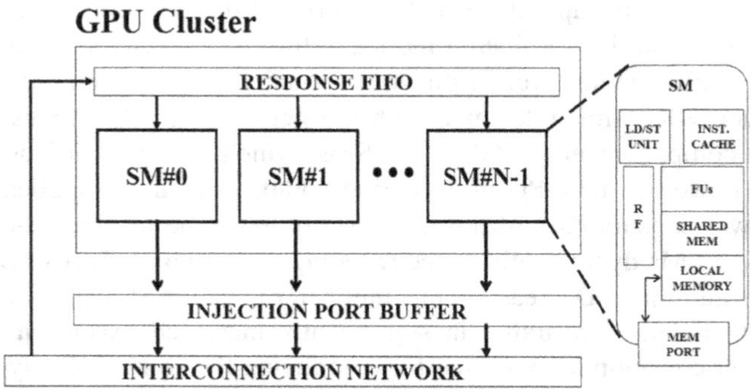

Fig. 1. Overview of GPU cluster

Figure 1 presents an overview of a GPU cluster. Response FIFO stores the data obtained from the interconnection network. If the data is an instruction, Response FIFO will forward the instruction to the caches of SMs. On the other hand, if the data is a memory request, Response FIFO will send it to the load/store unit of SMs. When all SMs are busy, Response FIFO will be stalled. The rightmost diagram in Fig. 1 shows the components inside an SM. To be able to execute multiple threads in parallel, each SM is provided a load/store unit (LSU), an instruction cache, a huge register file, and various functional units (FUs) such as 32 processing elements (PEs), a multiply-add (MAD), and special function unit (SFU). Furthermore, each SM includes various local memories (shared memory, texture cache, constant cache, and L1 data cache). The memory data is transferred to the interconnection network via an injection port buffer [11]. In order to issue the warp efficiently, SMs monitor the status of all the warps by keeping their own scoreboard and warp scheduler (not seen in Fig. 1) [12]. The scoreboard keeps track of the execution status of prior instructions of each warp to preserve data hazards such as RAW and WAW. Because modern GPUs execute instructions in an in-order fashion, the scoreboard is responsible for reserving the destination registers until warp execution is completed to avoid data hazards among the threads within a warp. In other words, the scoreboard can know when the execution of each warp is completed. A warp scheduler is employed to check the ready warps in the warp pool. In this work, we use a two-level scheduler as the warp scheduler of the baseline GPU architecture. In such warp scheduler, a warp within the warp pool has to wait for operands until all the operands for the warp are ready. When the warp becomes ready, it is moved into the active warp queue by the level-one scheduler. Then, the level-two scheduler will select a ready warp to issue.

3 Memory Contention Aware Streaming Multiprocessor Management

In this paper, we propose a dynamic memory contention aware streaming multiprocessor management (that we called MC-aware) scheme to improve the power efficiency of the GPU while not seriously degrading the performance. With our MC-aware, each SM can operate in two modes: normal mode and sleep mode. C-states technique has already been applied for improving the power efficiency of GPUs [13]. In our MC-aware, normal mode corresponds to C0 state for supporting high performance, whereas sleep mode corresponds to C3 state for power-saving. In order to manage the operation modes of SMs, the proposed MC-aware scheme selects one SM in normal mode when a serious memory contention occurs. Then, the CTA scheduler will stop issuing any CTA to the selected SM. After the selected SM completes all of the operations, the MC-aware scheme will change the mode of the SM from normal mode to sleep mode. By contrast, in case memory contention does not occur, the MC-aware scheme will select an SM in sleep mode. Afterwards, the CTA scheduler assigns CTAs to selected SM. In the next subsections, we describe the proposed architecture in detail, especially the SM management unit.

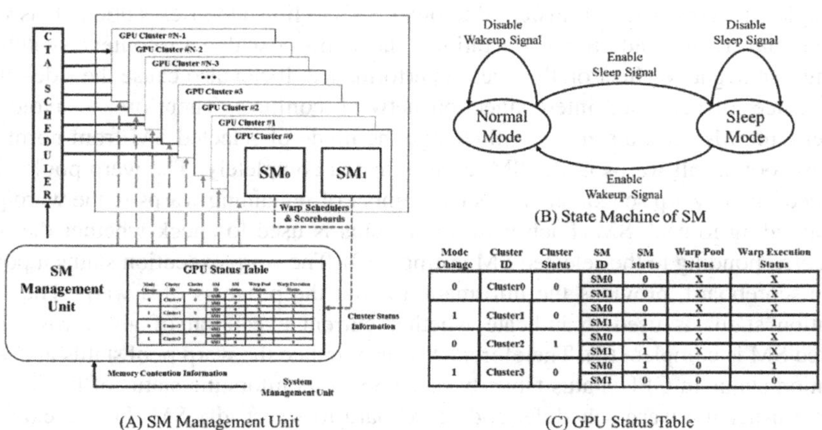

Mode Change	Cluster ID	Cluster Status	SM ID	SM status	Warp Pool Status	Warp Execution Status
0	Cluster0	0	SM0	0	X	X
			SM1	0	X	X
1	Cluster1	0	SM0	0	X	X
			SM1	1	1	1
0	Cluster2	1	SM0	1	X	X
			SM1	1	X	X
1	Cluster3	0	SM0	1	0	1
			SM1	0	0	0

(A) SM Management Unit

(B) State Machine of SM

(C) GPU Status Table

Fig. 2. Proposed architecture for SMU

An In the MC-aware scheme, to easily manage the operation modes of the SMs, we propose an SM management Unit (SMU) including a GPU status table. SMU exploits the information about memory contention for estimating the degree of memory contention. In addition, SMU exploits the information about SM status to update the GPU status table. GPU status table is used to determine whether the SM can be selected or not.

Figure 2(A) depicts the proposed SMU including the GPU status table. SMU uses the estimated the information about memory contention obtained from GPU's system management unit to determine the degree of memory contention. After calculating the degree of memory contention, SMU will generate the output control signals based on an algorithm. The details of the proposed algorithm for SMU are explained as follows. These control signals are used to change the mode of the SMs. Figure 2(B) shows the state machine of SM. Initially, all SMs are in the normal mode that aims at high performance. As shown in Fig. 2(B), an SM will not change its mode until the control signal is enabled. For example, when sleep signal comes to an SM in a normal mode, the SM's mode will be changed to sleep mode and when the SM receives the wake-up signal later, its mode will return to normal mode. SMU wakes up an idle SM and turns a busy SM into sleep mode so that MC-aware can dynamically manage SM's operation mode. To generate the control signal properly, SMU employs a GPU status table. An entry in the GPU status table consists of 7 fields: cluster ID, cluster status, SM ID, SM status, warp pool status, warp execution status, and mode change as shown in Fig. 2 (C). The number of entries in the GPU status table is the same as the number of SMs. The GPU status table of SMU stores the SM status and cluster status. The cluster status of a GPU cluster depends on the status of its SMs. When all SMs are in sleep mode (0), the GPU cluster will be in sleep mode (0). If at least one SM has the status to be normal mode, the cluster status will be normal mode (1). This can be implemented by simply using OR gate. Each SM status is updated based on warp pool status and warp execution status that are tracked by scoreboard. In MC-aware scheme, SMU can change the mode of an idle SM to reduce the cost caused by the context switch. If there are uncompleted warps when the SM is changed to sleep mode, the context related to these uncompleted warps will be transferred to other SMs to finish their execution. This warp transfer may require additional operations and extra cost due to context switching, leading to a negative effect on the overall performance. It can also cause the side-effect, for instance, of increased interconnection network conflicts. Therefore, to avoid this problem, the MC-aware scheme will change the mode of selected SM from normal to sleep as soon as all warps in the SM are terminated completely. The warp pool status, managed by the warp scheduler shows the number of remaining warps in the warp pool corresponding to each SM. Each warp pool status is used to check whether the warp pool corresponding to the selected SM is empty (0). The warp execution status, updated by the scoreboard, provides the information about the progress of a warp. The warp execution status is used to indicate whether current execution of all warps within selected SM is complete (0). Therefore, SMU can update the warp pool status and warp execution status in GPU status table, because SMU monitors the status of the selected SMs by using the warp scheduler and scoreboard to select idle SM. For an example, when warp pool status and warp execution status corresponding to the selected SM are both '0', the SM can be selected to change mode from normal to sleep since it is idle. Lastly, the mode change field is used to check whether the mode of an SM was changed (1) or not (0). This is to prevent SMU to choose an SM that has been already selected. To support this, SMU uses the mode change field in the GPU status table. Note that the SMU receives memory contention information every clock cycle. However, to reduce power consumption, the information about SM status is sent to the SMU only when GPU status table receives the enabled signal from SMU. In the MC-aware scheme, the SM status information is only required when necessary.

The flowchart in Fig. 3 shows the process stages of the proposed SMU in Fig. 2A. Firstly, SMU receives the memory contention information to calculate the degree of memory contention. Then, the MC-aware mechanism determines the control signal. If the control signal is null, there will be no extra operation like conventional mechanism; otherwise, the control signal will be changed. Depending on the contents of the control signal the MC-aware scheme will perform the corresponding operations. SMU looks up the GPGU status table to properly select the SMs whose mode was changed. In case the control signal is sleep, SMU will stop distributing CTAs to the GPU cluster in normal mode and then wait for the idle status of the selected SM before turning it into sleep mode. To further reduce power consumption, the mode of a GPU cluster will be changed to sleep when all SMs in the cluster are in sleep mode. On the other hand, the GPU cluster can be waked up earlier than the selected SM since GPU cluster involves the hardware components to support execution on SMs. After changing the mode of the selected SM mode from sleep to normal, SMU again allows distributing CTAs to the SM. In this work, CTAs are assigned to GPU clusters in a round-robin fashion. In summary, unlike the baseline GPU architecture in which CTAs are assigned to all GPU clusters, our proposed MC-aware scheme is able to assign CTAs to the particular GPU clusters that are permitted by SMU. As a result, this can improve power efficiency of the GPU. The key idea of this mechanism is to change the mode of a GPU cluster and its SMs.

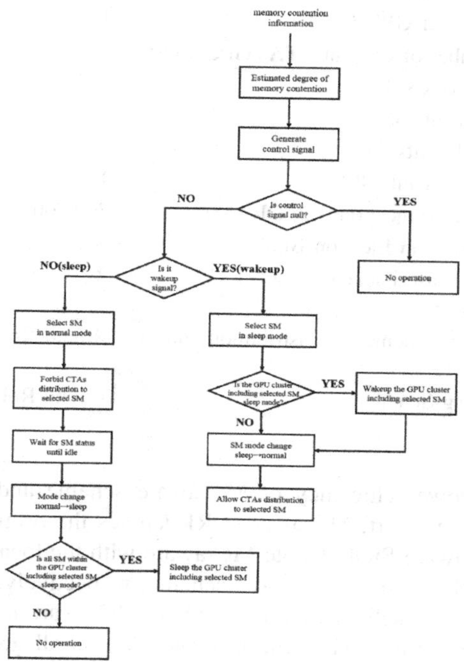

Fig. 3. Flowchart for MC-aware mechanism

4 Evaluation

We modified GPUWattch [14] constructed by GPGPU-Sim and McPAT to evaluate the performance and power savings of our proposed techniques. GPGPU-Sim [15] was developed to model the various aspects of parallel architectures such as the GPUs. An architectural modeling tool, McPAT by Sheng Li et al. [16], is used to evaluate the power consumption on the components within the GPU. In this work, we use a wide range of diverse domains, various benchmarks from NVIDIA, SDK [17], ISPASS [15], and Rodinia [18]. Table 1 shows the system configurations used in this work. Our target GPU is modeled based on an NVIDIA GTX480-like configuration which is the state-of-the-art GPU architecture [12]. The baseline GPU is operated at 700 MHz and composed of 15 SMs, each with 2 SP units and separate L1 instruction and data caches. Each SP unit contains 16 double-frequency CUDA cores, which execute integer and floating point pipelines independently, resulting in 480 ($15 \times 2 \times 16$) GPU CUDA cores in total. Round robin fashion is employed as the default scheduling scheme for selecting the CTAs or SMs. In the next subsections, we will explain how to determine the parameters values in the Signal Control Algorithm. After that, the simulation results and analysis will be presented in detail.

Table 1. System configuration

Parameter	Value
Number of GPU Clusters	15
Number of SMs per GPU Cluster	1
Maximum Number of Current CTAs/GPU Cluster	8
Number of Threads/SM	1536
Number of SP Units/SM	2
Number of SFU Units/SM	1
Number of LSDT Units/SM	1
Clock(Core; Interconnection; L2; DRAM)	700;700;700;924 (MHz)
Number of Packets in Ejection Buffer	8
Number of Response Packets in LDST Unit of Ejection Buffer	2
CTA Scheduling Scheme for Distribution and GPU Status Table	Round Robin Policy
Warp Scheduling Scheme	Round Robin Policy

We analyze the power efficiency of MC-aware scheme and compared it to the baseline. Note that in this part, MC-aware-ORI denotes the original MC-aware, MC-aware-LoSe and MC-aware-SiOb denote MC-aware with the locality-sensitive scheme and MC-aware with the signal-obstructive scheme, respectively. Figure 4 shows the power efficiency of the baseline, MC-aware-ORI, MC-aware-LoSe, and MC-aware-SiOb. The results are normalized to the baseline. Across all groups of benchmarks, MC-aware-ORI results in a power efficiency improvement of 12.7% on average (17.5%

for API, 31.4% for SQRNG, 21.8% for VA, 21.0% for BFS, 25.5% for MUM, 9.1% for GE, 5.5% for DCT, 16.0% for SP, 7.9% for BP, 15.6% for LPS, 9.8% for WP, 1.0% for HW, 2.4% for NW, 2.5% for SRAD, 3.0% for STO). In comparison with the other groups, MC-aware-ORI is able to improve the power efficiency by 21.1% on average for the Group I benchmarks, by 11.0% on average for the Group II benchmarks, and by 2.2% on average for the Group III benchmarks.

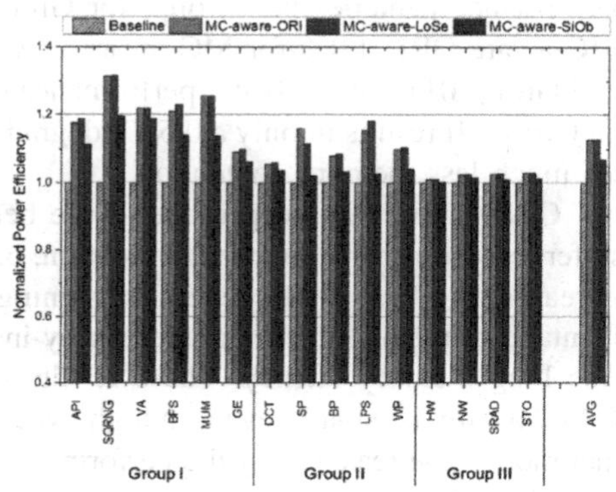

Fig. 4. Power efficiency according to GPU architecture

MC-aware-LoSe and MC-aware-SiOb also improve the power efficiency by 12.9% and 6.9% on average, respectively, and up to 31.4% and 19.4% for SQRNG. Compared to MC-aware-ORI, the results show that MC-aware-LoSe significantly improves the power efficiency, whereas MC-aware-SiOb does not provide comparable benefit. However, in terms of the performance, MC-aware-SiOb is the most effective technique among the three proposed MC-aware techniques, because it has minimal impact on the overall performance. To analyze the power efficiency of the proposed MC-aware, we show the power consumption and IPC separately.

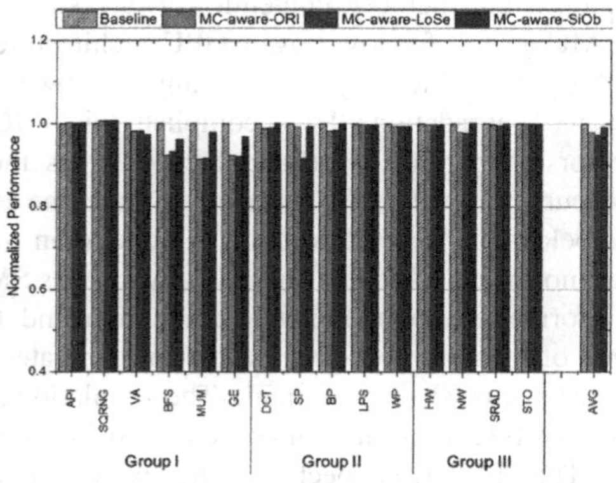

Fig. 5. Performance comparison

Figure 5 shows the performance for the baseline, MC-aware-ORI, MC-aware-LoSe, and MC-aware-SiOb. The results, IPC is normalized to that of the baseline. As shown in Fig. 5, the performance of MC-aware-ORI and MC-aware-LoSe is decreased considerably compared to that of MC-aware-SiOb, because the increased number of GPU clusters in sleep mode causes significant performance degradation. However, the performance degradation of the proposed MC-aware compared to the baseline is quite small (2.2% for MC-aware-ORI, 2.6% for MC-aware-LoSe, 0.9% for MC-aware-SiOb). According to performance penalties per Group, for Group I, the performance overhead is 4.2% for MC-aware-ORI, 4.0% for MC-aware-LoSe, and 1.8% for MC-aware-SiOb. Group II and Group III provides better performance compared to Group I. Especially, MC-aware in Group III results in only a 0.51% degradation in performance, as Group III experiences much less memory contention.

Although the ratio of GPU clusters in sleep mode of the benchmarks, especially Group I, is high, they suffer less from performance degradation, because the number of memory contentions decreases significantly. In general, assigning more CTAs to each core can take better advantage of the additional parallelism by increasing the resource utilization and hiding the long memory latency. These results show that the performance degradation of MC-aware is small, even though MC-aware decreases the number of SMs in normal mode. The reason why the performance penalty is low, even though decreasing the number of GPU clusters in normal mode reduces the performance considerably, is because the number of memory contentions is decreased. As reducing the memory contention diminishes the number of stall cycles, MC-aware alleviates the performance degradation caused by decreasing the number of GPU clusters in normal mode. In other words, even though MC-aware decreases the number of SMs in normal mode, it can improve the power efficiency without inducing any serious performance degradation, because the benefit to the performance obtained by reducing the memory conflicts compensates for the performance loss by increasing the number of GPU clusters in sleep mode.

5 Conclusion

In this work, we proposed the memory contention aware streaming multiprocessor management scheme, MC-aware, for low power GPU architecture. The goal of MC-aware is to improve the power efficiency by reducing the power consumption while having little performance degradation. To accomplish this, MC-aware selects the streaming multiprocessors in normal mode and then changes their mode when the memory contention occurs seriously. By contrast, in case the estimated degree of memory contention is below the lower threshold, the operation mode of the selected SMs is turned back into normal. To support this, MC-aware uses SM management unit, SMU, that contains information about memory contention and GPU cluster status. SMU judges the degree of memory contention and then generates appropriate control signals to manage the SM status (sleep or normal). The simulation results show that the power efficiency in units of IPC per watt is improved up to 31.4% while inducing little performance overhead. Therefore, we expect that MC-aware can be a good solution to improve the power efficiency.

Acknowledgements. This study was financially supported by Chonnam National University (Grant number: 2017-2727).

References

1. Luebke, D., Humphreys, G.: How GPUs work. J. Comput. **40**, 96–100 (2007)
2. Buck, I., et al.: Brook for GPUs: stream computing on graphics hardware. ACM Trans. Graph. **23**, 777–786 (2004)
3. General-purpose computation on graphics hardware. http://www.gpgpu.org/
4. GTX480 NVIDIA. http://www.geforce.com/hardware/desktop-gpus/geforce-gtx-480
5. Jing, N., et al.: An energy-efficient and scalable eDRAM-Based register file architecture for GPGPU. ACM SIGARCH Comput. Arch. News **41**, 344–355 (2013)
6. Rhu, M., Erez, M.: The dual-path execution model for efficient GPU Control Flow. In: High Performance Computer Architecture, pp. 591–602 (2013)
7. Gilani, S.Z., Kim, N.S., Schulte, M.J.: Power-efficient computing for compute-intensive GPGPU applications. In: High Performance Computer Architecture, pp. 330–341 (2013)
8. Fung, W.W.L., Sham, I., Yuan, G., Aamodt, T.M.: Dynamic warp formation and scheduling for efficient GPU Control Flow. In: Proceedings of the 40th Annual IEEE/ACM International Symposium on Microarchitecture, pp. 407–420 (2007)
9. Thornton, J.E.: Parallel operation in the control data 6600. In: Fall Joint Computer Conference, Part II: Very High Speed Computer Systems, AMC (1964)
10. CUDA Programming Guide Version 3.0. https://developer.nvidia.com/cuda-toolkit-30-downloads/
11. Abdalla, K.M., et al.: US Patent US20130185725: Scheduling and Execution of Compute Tasks (2013)
12. Abdel-Majeed, M., et al.: Gating aware scheduling and power gating for GPGPUs. In: Proceedings of the 46th Annual IEEE/ACM International Symposium on Microarchitecture, pp. 111–122 (2013)
13. Wang, P.-H., Yang, C.-L., Chen, Y.-M., Cheng, Y.-J.: Power gating strategies on GPUs. ACM Trans. Arch. Code Optim., **8**, (2011)
14. Leng, J., et al.: GPUWattch: enabling energy optimizations in GPGPUs. In: Proceedings of the International Symposium Computer Architecture, pp. 487–498 (2013)
15. Bakhoda, A., Yuan, G.L., Fung, W.W.L., Wong, H., Aamodt, T.M.: Analyzing CUDA workloads using a detailed GPU simulator. In: Performance Analysis of Systems and Software, pp. 163–174 (2009)
16. Li, S., Ahn, J.H., Strong, R.D., Brockman, J.B., Tullsen, D.M., Jouppi, N.P.: McPAT: an integrated power, area, and timing modeling framework for multicore and manycore architectures. In: Microarchitecture MICRO-42, pp. 469–480 (2009)
17. SDK CUDA SDK. http://developer.download.nvidia.com/compute/cuda/sdk/
18. Goodrum, M.A., Trotter, M.J., Aksel, A., Acton, S.T., Skadron, K.: Parallelization of particle filter algorithms. In: Varbanescu, A.L., Molnos, A., van Nieuwpoort, R. (eds.) ISCA 2010. LNCS, vol. 6161, pp. 139–149. Springer, Heidelberg (2011). https://doi.org/10.1007/978-3-642-24322-6_12

Dynamic Selective Warp Scheduling for GPUs Using L1 Data Cache Locality Information

Gwang Bok Kim[1], Jong Myon Kim[2], and Cheol Hong Kim[1(✉)]

[1] School of Electronics and Computer Engineering, Chonnam National University, Gwangju, Korea
loopaz63@gmail.com, chkim22@jnu.ac.kr

[2] School of Electronical Engineering, University of Ulsan, Ulsan, Korea
jongmyon.kim@gmail.com

Abstract. Warp scheduling policy for GPUs has significant impact on performance since the order of executed warps determines the degree of data cache locality. Greedy warp scheduling policy such as GTO shows better performance than fair scheduling policy for numerous applications. However, cache locality by multiple warps is underutilized when the GTO is adopted, resulting in overall performance degradation. In this paper, we propose a dynamic selective warp scheduling exploiting data locality of workload. Inter-warp locality and intra-warp locality are determined based on the access history information of the L1 data cache. By adjusting scheduling policy dynamically, the performance and cache efficiency are improved compared LRR and GTO significantly. According to our experimental results, the proposed technique provides IPC improvement by 19% and 3.8% over LRR and GTO, respectively.

Keywords: GPU · Warp scheduling · Cache · Data locality · Access history

1 Introduction

In general, the Graphics Processing Units (GPUs) show higher throughput than CPU by supporting parallel processing with powerful hardware resources. However, the GPU doesn't achieve expected throughput due to cache inefficiency. The cache inefficiency makes frequently long stall cycle which brings data from off-chip memory to register file. Then, resources for parallel processing are underutilized. The threads are grouped into the warp in order to process multiple data per an instruction within SIMD (Single Instruction Multiple Data) structure. The Cooperative Thread Arrays (CTAs) are assigned to multiple Streaming Multiprocessors (SMs) from a kernel. And, these are divided into the warp in order to be executed in parallel. Generally, warp scheduler impacts on GPU performance considerably since ordering warps leads to request order. Therefore, CTA/warp scheduling techniques for improving GPU performance have been proposed [1–4]. These researches focus on utilizing several resources or reduce unnecessary waiting cycles for memory. In terms of cache, cache structure of CPU optimizes memory latency, but thrashing problem by massive multi-thread is the main factor of GPU cache efficiency. For GPU, the residence time in cache gets shorter

J. H. Park et al. (Eds.): PDCAT 2018, CCIS 931, pp. 230–239, 2019.
https://doi.org/10.1007/978-981-13-5907-1_24

relatively than CPU since GPU cache allows multi-threads to access in parallel. Cache contention by several warps hamper utilization of data locality, hence overall performance is decreased. Therefore, this problem cannot be solved by studies related to cache management for CPU. CCWS [16] throttles the number of active warp to be executed in parallel. The CCWS limits the number of warp to improve cache efficiency of warps if the intra-warp locality is underutilized in L1 data cache. To capture the loss locality information in L1 data cache, the CCWS introduces the victim tag array (VTA). However, this unit stores 512 entries to track victim tags for all warps. Recently, iPAWS [7] proposed the warp scheduler selecting issue policy based on warp issue pattern. Pipeline stall occurrence and locality characteristics feed into warp issue pattern. Therefore, the scheduler predicts suitable warp scheduling policy after estimating warp issue pattern for previous period. They select scheduling policy between LRR and GTO to utilize locality of warp. However, information about the cache is not used directly.

In this paper, we categorize data locality of GPU workload in terms of warp. The data characteristic, which tends to be referenced by multiple warps is categorized as inter-warp locality. On the other hand, if data reused by same warp more than the case of locality by multiple warp, we define this characteristic as intra-warp locality. The proposed warp scheduler selects a policy either LRR or GTO dynamically after determining locality type of current workload. Cache requests are stored in the proposed unit when a new kernel is launched. Then, favorable scheduling policy is applied from next cycle. The localities from different workload are utilized efficiently by our proposed warp scheduler. The rest of paper is organized as follows. The background of GPU architecture is introduced in Sect. 2. Section 3 describes the proposed dynamic warp scheduling technique and architecture to support it. The evaluation methodology and experimental results are presented in Sect. 4. In the final Sect. 5, we make a conclusion about this paper.

2 GPU Architecture

GPU consists of multiple SMs for parallel processing. The modern GPU pipeline such as NVIDIA Fermi architecture is shown in Fig. 1. The fetch and decode stages communicate with warp scheduler (issue stage) by using instruction buffer. A warp-instruction is fetched with PC (Program Counter), then decoded instruction is stored in the instruction buffer. Each warp has two entries in instruction buffer to store instructions. Each entry of the instruction buffer has a valid bit, which indicates whether there is non-issued instruction for an entry. Each entry also updates a ready bit when the warp is ready to be issued by warp scheduler. Then, the warp scheduler selects a warp and feed it to corresponding execution unit after verifying the ready status of warp. The ready bit is updated if data dependency is free and operands are buffered in operand collector unit from the multi-banked register file.

A warp grouped by 32 threads executes an instruction following lock-step fashion. Every cycle, the scheduler selects a warp and feeds to a required execution unit by priority policy. When an SM proceeds memory instruction, the Load Store Unit (LSU) accepts it and processes it in order of application. In the NVIDIA Fermi architecture, an SM has 16 LSUs and totally GPU has 256 LSUs. The memory requests from Load/Store instruction are sent to the L1 data cache, shared memory, constant or texture cache. These caches are private to each SM, on the other hand, an L2 cache is shared across all SMs. The structure of L1 data cache on GPU is revealed by prior studies [14, 15]. The L1 data cache is structured as 128 block size, 32-set, and 4-way for the 16 KB configuration. The several memory requests are merged by coalescing unit before being sent to L1 data cache. For a memory instruction forwarded to the L1 data cache, multiple requests are generated if required memory addresses from a warp are not coalesced into one request. Meanwhile, a cache line can be referenced in parallel in a single transaction if requests belonging to a warp is merged by coalescing unit. If there is a cache hit, the required data are transferred to registers directly. Meanwhile, for cache miss, the information of miss request is stored in MSHR entry. If there is a no resource for handing current miss, the LSU is stalled. Since the L1 data cache suffers from contention of massive threads on GPU, the cache would be inefficient. For example, we can simply divide 16 KB L1 data cache by the number of threads. The space available per thread would be very small since the data stored in L1 data cache are evicted frequently in a short time.

Warp scheduling policies used in our baseline architecture are LRR (Loose Round Robin) and GTO (Greedy Then Oldest). The LRR policy issues warps in a round robin manner basically. Therefore, the equal priority assigned to all warps enhances utilization of data reuse by multiple warps. Meanwhile, the GTO policy can utilize intra-warp locality since it gives a priory to single warp until it has stalled. However, preference for a few warps should hamper inter-warp locality, which data is referenced by multiple warps. These characteristics shown differently in various workloads. Since the LSU accepts warps in order of application, the warp scheduling policy also determines memory access order. Furthermore, the warp scheduler easily improves overall GPU performance by utilizing locality or latency hiding. Especially, the LRR policy easily makes long latency stall across warps for irregular applications. Hence, there is no opportunity to hide latency by warp scheduler. Hence, warp scheduler which adapts to various workload characteristics can improve overall performance.

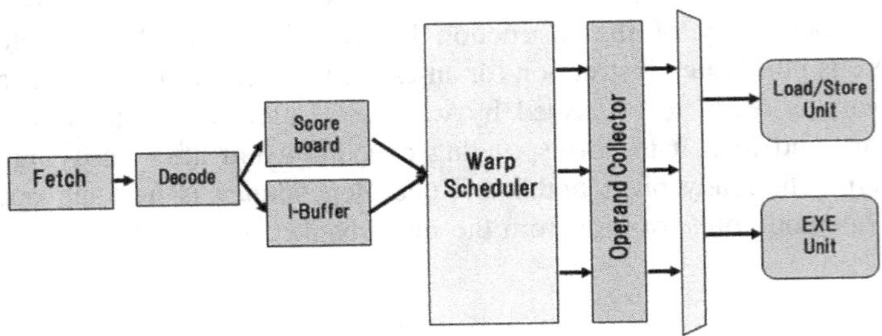

Fig. 1. Modern GPU pipeline

3 Dynamic Selective Warp Scheduling

The proposed warp scheduler requires three phases for each kernel. For the first phase, the information of L1 data cache accesses are stored in proposed unit. And the data locality type of workload is determined in terms of warp. In order to record references for few cache blocks, we propose locality decision unit (LDU), which consists of entries based on the requests to L1 data cache. For a request, the LDU is accessed to lookup all entries. The additional hardware space of proposed unit is limited since the cache of GPU is accessed massively in short time. Therefore, partial cache sets are pre-reserved to limit the number of accesses in LDU. Thus, we use cache set sampling technique that few sets of L1 data cache are used for collecting accesses information. In our baseline, only one set is used for sampling. If several sets are pre-reserved for sampling, the bits which indicates set number are required. This approach is to approximate the behavior of cache technique by using small hardware overhead.

The Fig. 2 shows structure of LDU with warp scheduler in detail. The warp scheduler selects a warp to execute an instruction. In this paper, we focus on only the request to L1 data cache since it is accessed with frequency and effects on performance significantly. To create entry of new access for L1 data cache and LDU, corresponding requested address and warp ID are delivered to LDU. If there is a request to the tag stored in LDU, the bit of corresponding entry sets. Each entry has 96 bits counter to trace each access by warp ID. Basically, 48 warps can be issued in parallel for a SM, and each warp has 2-bits counter in LDU. Therefore, the access to the address from same warp updates first bit of corresponding warp ID position. From this method, the access information for intra-warp locality is collected. On the other hand, a request from different warp ID increase counter bit located in the second bit of corresponding warp. Then, the proposed warp scheduler detects intra-warp locality and inter-warp locality by using the LDU. access to L1 data cache information for each warp ID is collected until the predefined period is finished. To consider the tradeoff between hardware overhead and accuracy of locality decision, we allow 16 entries for LDU in our experiments.

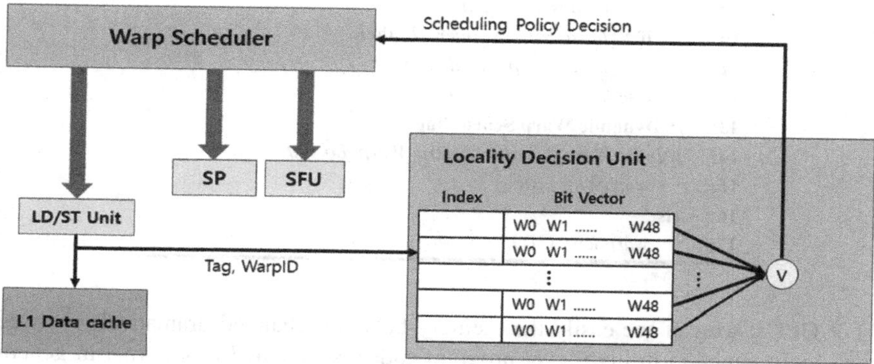

Fig. 2. Locality decision unit

The proposed warp scheduling technique selects relatively favorable policy after determining locality type of current workload. The proposed scheduler has 3 stages totally, the sampling phase for collecting memory access information of current workload, the locality type decision phase for determining major locality type and the running phase for applying scheduling policy dynamically. The Algorithm 1 shows how to determine locality type of current workload as pseudo-code after sampling phase. If the number of accumulated cycles from kernel launch is bigger than sampling phase cycles, then decision stage is started. The number of bits which indicate intra-warp accesses and inter-warp accesses are accumulated respectively. After accumulating the number of bits for each locality type, the entry concludes the locality type by comparing two accumulated degrees. We present degree of locality type of each entry as *Intra_Warp_Locality* and *Inter_Warp_Locality* in Fig. 2. Thus, we add up the degrees of locality about all entries and compare them by voting. The voting method make final decision based on major results from all entries although each entry has different value or direction with the major result. This method is a well-known technique which seen easily in research related to hardware even software [5, 6]. The final locality type determined by voting is applied to the scheduler for running phase. If the intra-warp locality is strongly presented for sampling phase, GTO is selected for scheduler in order to enhance intra-warp locality of workload. On the other hand, the LRR policy for warp scheduling is applied in order to allow all warps issue with fair priority. Note that allocation time information of each warp is required to support GTO with LRR.

Algorithm 1. Proposed warp scheduling algorithm

```
if Kernel_Cycle > SAMPLING_PERIOD then
/* Voting and Determining Locality Type */
for Idx=1,2.., MAX_TABLE_ENTRIES do
    Current_Entry=Table[Idx];
    for W_Idx=1,3.., MAX_WARPS X 2-1 do
      if Currrent_Entry[W_Idx] = 1 then
          N_Intra_Hit++;
      if Currrent_Entry[W_Idx+1] = 1
          then N_Inter_Hit++;
    if N_Intra_Hit > N_Inter_Hit then
          Intra_Warp_Locality++;
    else Inter_Warp_Locality++;
/* Dynamic Warp Scheduling */
if Inter_Warp_Locality > Intra_Warp_Locality
    warpSch= LRR;
else
    warpSch= GTO;
```

For GPUs, we can see easily the memory behavior changed dramatically between kernels since function in high-level program codes is executed as a kernel in general. Furthermore, according to previous researches, cache shows similar behavior within a kernel execution [7, 8]. However, longer sampling phase leads to shorter running phase with fixed the number of instruction of a kernel. Therefore, we minimize sampling phase by performing offline-training with several benchmarks, thus we apply fixed the number of phase cycles to our scheduler.

4 Experiments

4.1 Experimental Methodology

In this section, we present and discuss the impact of our warp scheduler on GPU performance. We use GPGPU-SIM v3.2 [9], a cycle-accurate GPU architecture simulator, to evaluate our proposed warp scheduler. The baseline GPU architectural parameters are configured as summarized in Table 1. To evaluate the performance of our proposed technique we chose benchmarks from the Rodinia [10], ISPASS [11], Polybench [12], CUDA SDK [13] as shown in Table 2. In our experiments, we categorize a wide range of GPU benchmarks into LRR-friendly (RF), GTO-friendly (GF) and scheduler insensitive (SI) type. If a warp scheduling policy outperforms another policy for a benchmark by more than 1.1x, then we categorize the benchmark as a friendly benchmark for favorable policy.

Table 1. Baseline configurations

Parameter	Value
GPU core	15 SMs, 1.4 GHz
Warp size and SIMD width	32
Max. number of warps/SM	48
Max. number of blocks/SM	8
Warp scheduling	LRR/GTO
L1 Inst. cache	2 KB per SM, 4-ways
L1 data cache	16 KB per SM, 4-ways
L2 data cache	64 KB, 8-way, 768 KB
DRAM	440 cycles latency, 924 MHz

Table 2. Benchmarks

Application	Description	Locality type
3MM [12]	3 Matrix multiplications	RF
GEMM [12]	Matrix-multiply	RF
SC [10]	Solving the online clustering problem	RF
MVT [12]	Matrix vector product and transpose	GF
BICG [12]	BiCG sub kernel of BiCGStab linear solver	GF
ATAX [12]	Matrix transpose and vetor multiplication	GF
MUM [11]	High-throughput DNA exact sequence alignment	SI
NW [10]	Global optimization for DNA sequence alignment	SI
TF [13]	Reduction operation on an array of values	SI

4.2 Experimental Results

Figure 3 shows the miss rate of L1 data cache access with our scheduler compared with LRR and GTO. The miss rate of our proposed scheduler shows almost same with GTO on average. For RF type, the GTO scheduler shows high miss-rate, and benchmarks of GF type shows high miss-rate. TF benchmark shows 100% miss rate due to zero-reuse of data in cache even with our proposed scheduler. Our proposed scheduler shows the lowest miss-rate in our experiments, lower miss-rate than LRR policy by 11.6%. Therefore, our scheduler achieves high efficiency of L1 data cache on average by utilizing data locality. The GTO scheduler and our scheduler reduces similar L1 data cache miss rate in our experiments.

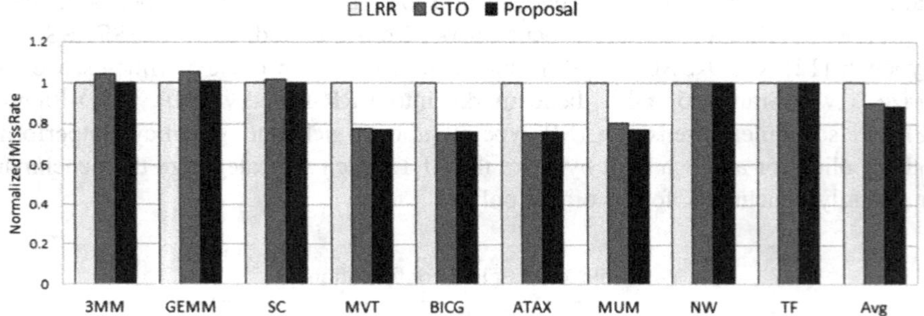

Fig. 3. L1 data cache miss rate comparison

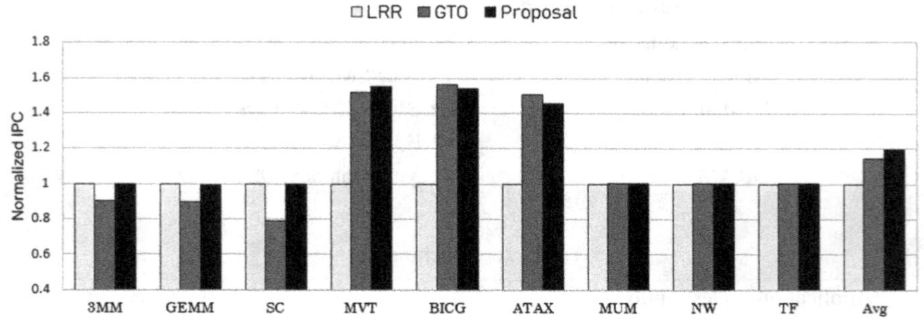

Fig. 4. Performance improvement with proposed warp scheduler

Figure 4 shows the IPC performance of proposed scheduler and GTO normalized to LRR by the metric, Instructions Per Cycle (IPC). As a result, proposed scheduler shows similar performance with LRR, and better performance than GTO for RF type. In case of GEMM, the IPC is higher than GTO by 11.3% due to reduced stall cycles for issue stage. Performance of SC benchmark by the proposed scheduler is improved by 26% compared with GTO since the GTO decrease the inter-warp locality. MVT of GF type with our scheduler shows slightly higher IPC than GTO. However, it improves IPC by 55% compared with LRR. Our scheduler shows much better performance than LRR for MVT, BICG and ATAX benchmarks as shown in Fig. 4.

For MUM of SI type, cache miss-rate is reduced, however performance is almost same with other schedulers. The cache efficiency is not affected due to relatively few cache accesses. Therefore, it is difficult to improve IPC by utilizing data locality only. The performance of NW and TF is not improved by our scheduler. This is because the data locality tends not to be captured due to limited cache capacity and working set size. As a result, proposed scheduler shows similar performance with the best policy for each benchmark. And, there is no performance degradation for SI type benchmarks.

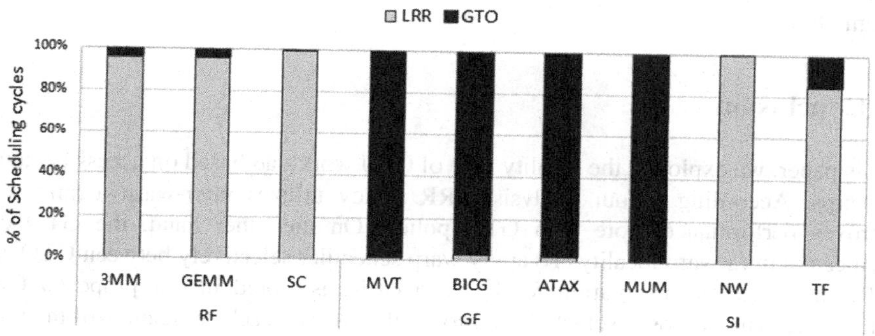

Fig. 5. Cycle ratio of applied scheduling policy

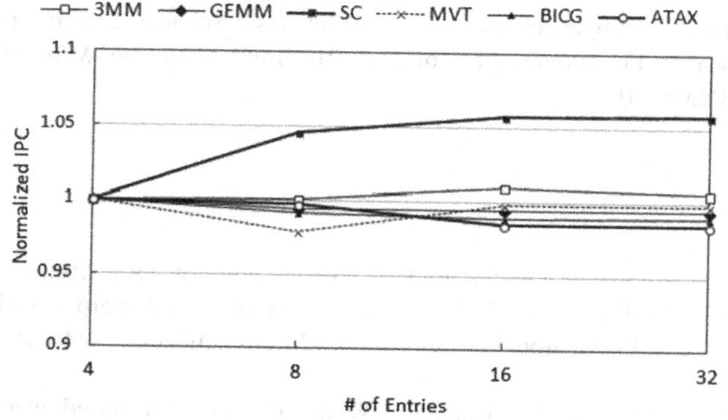

Fig. 6. IPC comparison according to LDU entries

Figure 5 shows the cycle ratio of applied scheduling policies with proposed warp scheduler for running phase. The LRR and GTO are applied across the benchmarks dynamically in our evaluation. We accumulate cycles for each warp scheduler policy when a warp is issued. The benchmarks of RF and GTO types show that inter-warp locality and intra-warp locality are detected suitably, and preferable policies are applied to each benchmark. For RF type, LRR policy accounts for 97% on average of whole running phase. Meanwhile, LRR policy occupies only 3% of GF benchmarks for running time. This is because intra-warp locality is detected more than inter-warp locality, then GTO takes advantages from cache efficiency. In case of SI type, there is no performance degradation although LRR and GTO ratio vary widely, shown in Fig. 5. The MVT, BICG and ATAX consist of 2 kernels each. With our proposed scheduler, we can see that warps are issued and access to cache based on locality favor.

The additional hardware to support our proposed LDU has limited entries. These are used for storing partial tags of L1 data cache. Figure 6 shows the performance impact of using different the number of LDU entries from 4 to 32. In our experiment, the optimal numbers for each benchmark is different each other. SC benchmark shows improvement up to 5.8% with 32 entries. On average, 16 entries for LDU is adequate design choice.

5 Conclusion

In this paper, we exploited the locality type of GPU workload based on access behavior of warps. According to our analysis, LRR policy utilizes inter-warp locality and improves performance more than GTO policy. On the other hand, the GTO can enhance the intra-warp locality. To apply warp scheduler selectively between GTO and LRR, cache access information for L1 data cache is stored in our proposed GPU architecture. The proposed scheduler improves the L1 data cache efficiency of the GPU by dynamically selecting better scheduling policy between GTO and LRR. According to our simulation results, the proposed scheduler improves the overall performance by 19% over LRR and by 3.8% over GTO, respectively.

Acknowledgements. This research was supported by Basic Science Research Program through the National Research Foundation of Korea (NRF) funded by the Ministry of Education (NRF2018R1A2B6005740).

References

1. Narasiman, V., Shebanow, M., Lee, C.J., Miftakhutdinov, R., Mutlu, O., Patt, Y.N.: Improving GPU performance via large warps and two-level warp scheduling. In: 44th Annual IEEE/ACM International Symposium on Microarchitecture (MICRO), pp. 308–317. IEEE (2011)
2. Zhang, Y., Xing, Z., Liu, C., Tang, C., Wang, Q.: Locality based warp scheduling in GPGPUs, Futur. Gener. Comput. Syst. (2017)
3. Wang, B., Zhu, Y., Yu, W.: OAWS: memory occlusion aware warp scheduling. In: International Conference on Parallel Architecture and Compilation Techniques, pp. 45–55. IEEE (2016)
4. Wang, J., Rubin, N., Sidelnik, A., Yalamanchili, S.: LaPerm: locality aware scheduler for dynamic parallelism on GPUs. ACM SIGARCH Comput. Arch. News **44**(3), 583–595 (2016)

5. Zhang, W.: Enhancing data cache reliability by the addition of a small fully-associative replication cache. In: Proceedings of the 18th Annual International Conference on Supercomputing, pp. 12–19 (2004)
6. Sato, M., Egawa, R., Takizawa, H., Kobayashi, H.: A voting-based working set assessment scheme for dynamic cache resizing mechanisms. In: IEEE International Conference on Computer Design (ICCD), pp. 98–105. IEEE (2010)
7. Lee, M., Kim, G., Kim, J., Seo, W., Cho, Y., Ryu, S.: iPAWS: instruction-issue pattern-based adaptive warp scheduling for GPGPUs. In: IEEE International Symposium on High Performance Computer Architecture (HPCA), pp. 370–381. IEEE (2016)
8. Oh, Y., Kim, K., Yoon, M.K., Park, J.H., Ro, W.W., Annavaram, M.: APRES: improving cache efficiency by exploiting load characteristics on GPUs. ACM SIGARCH Comput. Arch. News **44**(3), 191–203 (2016)
9. Aamodt, T.M., Fung, W.W.L.: GPGPU-Sim 3.x Manual (2014). http://gpgpu-sim.org/manual/index.php/GPGPU-Sim 3.x Manual
10. Che, S., et al.: Rodinia: a benchmark suite for heterogeneous computing, workload characterization. In: IEEE International Symposium on IISWC 2009, pp. 44–54 (2009)
11. Bakhoda, A., Yuan, G.L., Fung, W.W.L., Wong, H., Aamodt, T.M.: Analyzing CUDA workloads using a detailed GPU simulator. In: Performance Analysis of Systems and Software, pp. 163–174 (2009)
12. Grauer-Gray, S., Xu, L., Searles, R., Ayalasomayajula, S., Cavazos, J.: Auto-tuning a high-level language targeted to GPU Codes. In: Innovative Parallel Computing, pp. 1–10 (2012)
13. NVIDIA, NVIDIA CUDA C programming guide v4.2, April 2012. http://developer.nvidia.com/nvidia-gpu-computing-documentation
14. Nugteren, C., van den Braak, G.-J., Corporaal, H., Bal, H.: A detailed GPU cache model based on reuse distance theory. In: High Performance Computer Architecture, pp. 37–48 (2014)
15. Wong, H., Papadopoulou, M.-M., Sadooghi-Alvandi, M., Moshovos, A.: Demystifying GPU microarchitecture through microbenchmarking. In: Performance Analysis of Systems & Software, pp. 235–246 (2010)
16. Rogers, T.G., O'Connor, M., Aamodt, T.M.: Cache-conscious wavefront scheduling. In: Proceedings of the IEEE/ACM International Symposium on Microarchitecture, pp. 72–83 (2012)

An Efficient Model and Algorithm for Privacy-Preserving Trajectory Data Publishing

Songyuan Li(✉), Hong Shen, and Yingpeng Sang

School of Data and Computer Science, Sun Yat-sen University, Guangzhou, China
lisy36@mail2.sysu.edu.cn, {shenh3, sangyp}@mail.sysu.edu.cn

Abstract. Since Abul et al. first proposed the *k*-anonymity based privacy protection for trajectory data, the researchers have proposed a variety of trajectory privacy-preserving methods, these methods mainly adopt the static anonymity algorithm, which directly anonymize processing and data publishing after initialization. They do not take into account the real application scenarios of moving trajectory data. The objective of this paper is to realize the dynamic data publishing of high dimensional vehicle trajectory data privacy protection under (k, δ) security constraints. First of all, we propose the partition storage and calculation for trajectory data. According to the spatial and temporal characteristics of vehicle trajectory data, we choose the sample point (x_2, y_2, t) at the time t_i as partition fields, partition storage of the trajectory data according to the time sequence and the location of the running vehicle is $Region(m, n)\_(x_i, y_i, t_i)$. The computation of data scanning in trajectory data clustering and privacy processing is reduced greatly through this method. Secondly, the dynamic clustering method is used to cluster the regional data. According to the characteristics of the vehicle trajectory data, (x_i, y_i, t_{m-n}) as the release data identifier, trajectory attributes of the vehicle as the sensitive attributes, we use Data Partitioning and Cartesian Product (*DPCP*) method to cluster trajectory data under the (k, δ) security constraints. Thirdly, the anonymization function f_{DPCP} is used to preserve the privacy of clustering trajectory data. In each sampling time slice, f_{DPCP} function is used to generalize the location data in the grouping. Through the continuous algorithm optimization and the experimental verification of real trajectory data, this model and algorithm can effectively protect privacy under the security constraint of (k, δ). By means of data simulation and data availability evaluation, the data processed by the anonymization method has a certain usability under the threshold of δ. At the same time, the experimental results are compared with the classical *NWA* algorithm, and *DLBG*, the method in this paper have been proved to be advanced in time cost and data availability evaluation.

1 Introduction

Public service vehicle's trajectory need to be opened to the public because of their public nature. At the same time, because of the sensitivity of public service vehicle's trajectory data, it is the data that many attackers want to obtain, which is used for

J. H. Park et al. (Eds.): PDCAT 2018, CCIS 931, pp. 240–249, 2019.
https://doi.org/10.1007/978-981-13-5907-1_25

malicious attacks. However, public service entities or government departments need to publish relevant data on a regular basis in accordance with marketing or policy requirements. How to effectively protect the trajectory privacy data of a public service vehicle and avoid the target of being attacked, and also meet the requirements of data publication or data publicity, is an urgent problem to be solved.

For vehicle trajectory data publishing, some issues need to be addressed:

(1) For publishing public service vehicle trajectory data, it is necessary to preserve the location data of some sensitive targets, such as public passenger vehicles and freight cars. For example, a malicious attacker should be prevented from finding a vehicle for bank service through the release of these tracks from the trajectory data, and then inferring its destination in the next time, or inferring which bank outlets it is to service.
(2) How to protect high dimensional data privacy in data publishing. In order to protect the high dimensional data, it is necessary to keep the data availability and integrity as far as possible, and to protect the trajectory privacy of each moving target. Unlike the two dimensional data privacy preserving, the trajectory data of moving object belongs to three-dimensional spatial and temporal data. So far there is no effective means to dealing with the problems on how to preserve a series of moving tracks in a three-dimensional space, and then publish it to the outside.
(3) How to store, calculate and protect privacy data for mass trajectory data. Trajectory data is the time series of position coordinates of moving targets. With the increasing of moving targets and the increasing of position coverage area, the volume of trajectory data of moving objects is very large. According to estimations, for 100 thousand mobile targets, by o sampling interval of *10* s, *24* h of trajectory data will be more than *200 GB*.

For the above questions, the innovative work of this paper includes: (1) We first propose the partition storage and calculation for trajectory data. According to the spatial and temporal characteristics of vehicle trajectory data. The computation of data scanning in trajectory data clustering and privacy processing is reduced greatly through this method. (2) According to the characteristics of the vehicle trajectory data, (x_i, y_i, t_{m-n}) as the release data identifier, trajectory attributes of the vehicle as the sensitive attributes, we use dynamic clustering method Data Partitioning and Cartesian Product *(DPCP)* to cluster trajectory data under the (k, δ) security constraints. (3) The anonymization function f_{DPCP} is used to preserve the privacy of clustering trajectory data. In each sampling time slice, f_{DPCP} function is used to generalize the location data in the grouping. (4) Through the continuous algorithm optimization and the experimental verification of real trajectory data, this model and algorithm can effectively protect privacy under the security constraint of (k, δ), and the l-diversity processing [1–3] can also effectively prevent the background attack and association attack.

2 Related Work

Most existing privacy-preserving trajectory data publishing techniques use generalization or disruption method to deal with the published trajectory to meet the k-anonymity model. Samarati and Sweeny [14] first proposed a privacy preserving k-anonymity model for relational data publishing. The k-anonymity model requires that each record in the table should be at least the same as the value of the other k-1 record about the quasi-identifier (QI). After many years of research, k-anonymity model has become increasingly mature.

Through research, it is found that in some cases, two types of privacy leak attacks may still exist in k-anonymity data: homogeneous attack and background knowledge attack. Machanavajjhala et al. proposed an enhanced k-anonymity model, l-diversity model. The l-diversity principle requires that each k-anonymity group in a data table contain at least l different sensitive attribute values. The attacker inferred that the probability of a recorded privacy message would be less than 1/l [16].

Abul et al. proposed (k, δ)-anonymity model [6] based on the uncertainty of moving trajectory data. On the basis of the model, the problem of trajectory anonymity was treated by clustering. However, by analyzing the protection degree of (k, δ)-anonymity model, the model can only realize the k-anonymity of the trajectory just under the condition of $\delta = 0$ [1, 8, 16].

At the same time, they also put forward the use of Euclidean distance as a measure function of trajectory clustering, requiring that the trajectory has the same time interval, for different time interval trajectory data can not be clustered.

Aiming at the problem that the trajectory data of different time intervals can not be clustered, the real sequence edit distance is proposed as the metric function of trajectory clustering.

Shin et al. proposed an algorithm based on trajectory partitioning in document [18] to ensure privacy security. The algorithm will be the future trajectory is divided into several sections, anonymous and ensure all the time in the area of the trajectory and the minimum section to anonymous as a unit, in order to ensure privacy needs as much as possible to improve the quality of service.

Similar to the k-model in the application of anonymous structured data, we used the same core idea of anonymous, anonymize the quasi identifiers of anonymous trajectory data, in each anonymous group, k-1 data and records about the quasi-identifier (QI) value of the same.

3 Privacy-Preserving Trajectory Data Publishing

3.1 Definition

Trajectories are based on positional data, adding temporal dimensions to high-dimensional data. Because of the location irregularity of trajectory data and time characteristics, it is difficult to protect the trajectory data by using the existing two dimensional anonymous privacy preserving algorithm.

Table 1. Symbols in this paper

Symbol	Meaning
Tr	Trajectory
p_k,	The position of *Tr* at t_k
$Dist(tr_1[t], tr_2[t])$	The distance between tr_1 and tr_2 and time t
δ	Threshold of trajectories similarity
$\prod_{QI}^{(T)}$	The projection of *Tr(d)* on the attribute *QI*
$[t_{min}, t_{max}]$	Time interval of trajectory
$T_{[min,max]}$	The trajectory at $[t_{min}, t_{max}]$
π	$t_{max} - t_{min}$
heap	Position point heap of the trajectory

In this paper, according to the characteristics of trajectory data, the temporal characteristics of trajectory are used as the quasi-identifier*(QI)*, and the position coordinates of t_m at some point are defined as sensitive attributes of trajectory. *k*-anonymity *l*-diversity method is used to preserve the trajectory data privacy and prevent malicious attack [4, 5].

There are many mathematical symbols are shown in this paper, and Table 1 shows the notations which have been used.

Definition 1. *Distance Between Trajectories.* Euclidean distance [6, 7] is used to measure the distance between trajectories. Define two trajectories tr_1 and tr_2, at the time t, the distance is:

$$Dist(tr_1, tr_2) = \sum_{t=1}^{m} Dist(tr_1[t], tr_2[t])$$

Definition 2. *Discriminant the Similarity of Two Trajectories.* Within the time range $[t_1, t_m]$, two trajectories Tr_1 and Tr_2. are called similar trajectories, only if the distance between points (x_1, y_1, t) on $Tr_1(x, y, t)$ and points (x_2, y_2, t) on $Tr_2(x, y, t)$ at any point in time $\mathrm{Dist}(x_1, y_1)((x_2, y_2) \leq \delta$ is denoted as $\mathrm{Coloc}_\delta(\mathrm{tr}_1, \mathrm{tr}_2)$, δ is the threshold of trajectory similarity.

$$Coloc_\sigma(tr_1, tr_2) = \begin{cases} \sum_{p_i \in tr_1} \sigma(p_i, \perp), & |tr_2| = 0 \\ \sum_{p_i \in tr_2} \sigma(p_i, \perp), & |tr_1| = 0 \\ \min \begin{array}{c} \{Coloc_\delta(tr_1 - tr_1.p_1, tr_2 - tr_2.p_1) + \sigma(tr_1.p_1, tr_2.p_1), \\ Coloc_\delta(tr_1, tr_2 - tr_2.p_1) + \sigma(tr_2.p_1, \perp), \\ Coloc_\delta(tr_1 - tr_1.p_1, p_1) + \sigma(tr_1.p_1, \perp)\} \end{array} & |tr_1|, |tr_2| > 0 \end{cases}$$

At present, the trajectory data publishing as a common form of non-relational data released, its privacy problem has become a research hotspot of privacy protection data, attracted many researchers to study. Since Abul et al. proposed the k-anonymous trajectory data privacy-preserving model based on clustering for the first time [8, 9], some research achievements have been made in the anonymous trajectory data protection technology based on clustering.

The trajectory data privacy-preserving technology based on clustering is divided into three steps:

(1) Trajectory data preprocessing. According to the spatial and temporal characteristics of vehicle trajectory data, choose the sample point (x_1, y_1) coordinates and time t_i as partition fields, the trajectory data according to the time sequence and the location of the running vehicle partition storage trajectory data in $Region(m, n)\_(x_i, y_i, t_i)$.
(2) The trajectory data to be published are initialized and grouped according to the trajectory similarity. According to the characteristics of the vehicle trajectory data, (x_i, y_i, t_{m-n}) as the release data identifier, trajectory attribute information of the vehicle as the sensitive attributes, using Cartesian Product and Euclidean Distance method to cluster trajectory data grouping trajectory data clustering under the (k, δ) security constraints.
(3) Using some method to anonymous each cluster group. The anonymous function f_{DPCP} is used to preserve the privacy of clustering trajectory data. In each sampling time slice, f_{DPCP} function is used to generalize the location data in the grouping. After anonymization, the clustering group data can be published directly.

3.2 Trajectory Data Preprocessing

The partition field is determined and the partition step size needs to be considered. The partition step size determines the amount of partition data. Choose a reasonable amount of data, combined with the real life situation, the last partition based on the following: since the trajectory for anonymity from vehicles, so the reality is that general motors is within *0* to *120 km/h*. We have set a district cap of *200 km/h*, which is based on the speed of the high-speed train, that is to say, the maximum speed of the vehicle is supposed to catch up with the high-speed train. Then, the speed of the vehicle is converted into x. latitude and y longitude (Figs. 1 and 2).

When a point is located at the edge of the partition, then the position of the point within δ radius will exceed the partition size. At this point, take *2* or *4* adjacent partitions with the location as the center, δ as the radius, to be calculated. The value of δ does not necessarily equal to the partition size. The value of δ, on the one hand, based on the maximum speed of the vehicle to divide the area, on the other hand, based on the size of the position point grouping.

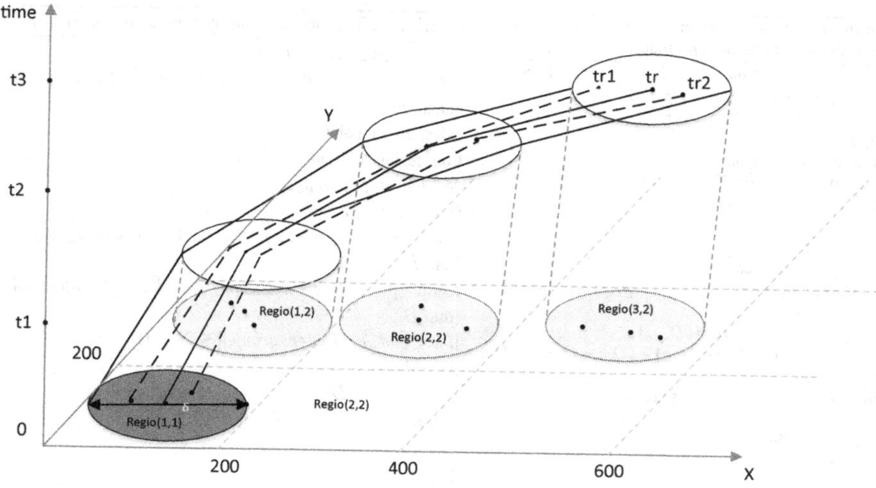

Fig. 1. Trajectory data preprocessing

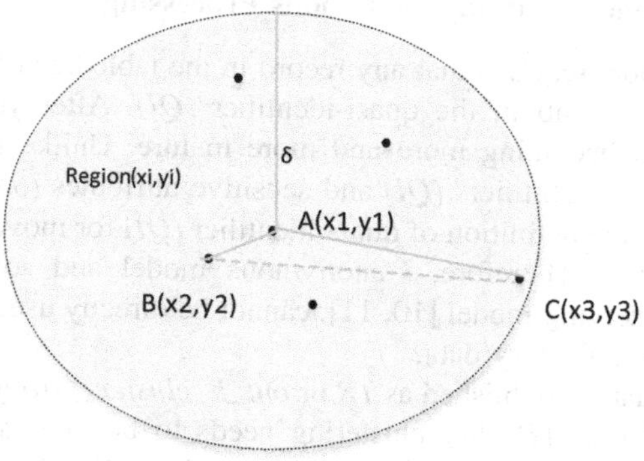

Fig. 2. Trajectory anonymous processing

3.3 *k*-anonymous Algorithm Based on Trajectory Similarity Grouping

According to the uncertainty of trajectory, the distance between trajectory and trajectory distance, the delta is used as trajectory similarity judgment threshold, and the trajectory set T is grouped into different subsets, denoted as m, $T_2, \ldots, T_q$, q is the number of locus subsets.

Given the trajectory data $T(d)$, the privacy security protection mechanism P, the information loss metric, and the chi square $Tr(d)$ map, it can cover the minimum space for all records at the time t as the identifier. According to the characteristics of trajectory data, dynamic programming algorithm Data Partitioning and Cartesian Product *(DPCP)* are adopted to make trajectory data cluster and *k*-anonymous.

Algorithm 1 Trajectory data clustering algorithm based on Data Partitioning and Cartesian Product ($DPCP$)

```
InPut: Tr = Region = (m,n); tr_i, tr_j, GPS_time, steps_length, steps_interval, k, Tr_i
OutPut: return the out_k_cluster_Group
initialization.
Select all the points at time t_0 and
from region table(m,n). Tr = (tr_i, tr_j) ; step = 1
do while step < steps_length
do {
copy D to D';
D × D';//Cartesian product
select ,SQRT((p1.x − p2.x) × (p1.x − p2.x) + (p1.y − p2.y) × (p1.y − p2.y)) from D';
IF D.ObjectID!= D'.ObjectID and
sqrt((D.x − D'.x) × (D.x − D'.x) + (D.y − D'.y) × (D.y − D'.y)) < δ and
k < |out_k_cluster_group_ti| ≤ 2k − 1;
if step! = 0
Do Get D^ti × D^ti−1;
End if
}
End while
```

Algorithm 2 Trajectory data anonymious algorithm f_{DPCP}

```
InPut: out_k_cluster_Group
OutPut: return the out_k − anonymious_cluster_Group
initialization.
Select all the points at time t_i and
from region table(m,n). TR = (tr_i, tr_j) ; step = 1
do while step < steps_length
do {
return ((p1.x + p2.x + ... + pm.x)/|out_k_cluster_group_ti|
from D';
return ((p1.y + p2.y + ... + pm.y)/|out_k_cluster_group_ti|
from D';
if k < |out_k_cluster_group_ti| ≤ 2k − 1;
if step! = 0
Do Get D^ti × D^ti−1;
End if
End if
}
End while
```

3.4 Trajectory Generalization Anonymous Processing

The *k*-anonymity model requires that any record in the table be at least the same value as any other *k-1* record about the quasi-identifier *(QI)*. After years of research, *k*-anonymous model is becoming more and more mature. Unlike relational data, it is difficult to define quasi-identifiers *(QI)* and sensitive attributes *(SA)* in moving trajectory data. At present, the definition of quasi-identifier *(QI)* for moving trajectory data is still an open question. Therefore, *k*-anonymous model and some other relational database privacy preserving model [10, 11], cannot be directly used to prevent the high dimension of mobile trajectory data.

If the trajectory data is published as *TR* or *out_k_cluster_Group*, the location of the individual is leaked out. *TR* after clustering needs to be treated anonymously with certain anonymous methods. Therefore, a generalized function f_{DPCP} is designed in this paper, represented by *DPCP* function, which can be used for further anonymous processing of clustering data.

f_{DPCP} function principle: the number of clustering points in the group and the size of the generalized regional clustering has a negative effect on information loss, because the location within the clustering of the clustering was anonymous in the later have the opportunity to be published; At the same time, the greater the region, the greater the uncertainty. According to this principle, the generalized function is designed based on the method of dynamic generalization of the position points in the cluster.

The generalization function of the clustering group at sampling time t_i is as follows:

$$f_{DPCP} = \left(\frac{p_1x + p_2x + \ldots, p_mx}{|out\_k\_cluster\_Group_{t_i}|}, \frac{p_1y + p_2y + \ldots, p_my}{|out\_k\_cluster\_Group_{t_i}|} \right)$$

Among them, $out\_k\_cluster\_Group_{t_i}$ is an anonymous group after clustering at time t_i.

4 Experiment and Analysis

The experimental data using the two control groups [12]. The first data group is generated by the Brinkhoff network based on mobile object data generator, generating data represented by the German city of Oldenburg the road network condition day mobile data. The trajectory target is *10 thousand* moving targets (http://www.fh-oow.de/institute/iapg/personen/brinkhoff/generator). The other is based on the actual moving track data of *100 thousand* vehicles in an area in one day (Table 2).

Table 2. Data sets used in experiments

Dataset	MBB radius(D)	|D|	π	$\|D^{ec}\|$	π	Size/Mb
OLDEN	35779.3	10 k	5	435	141	350
VichelTraj	743400	100 k	10	7	3062	32200

MBBradius(D) represents half the diagonal of the minimum coverage rectangle in the dataset D, and |D| represents the number of tracks, $|D^{ec}|$ said after the pretreatment of equivalence class number, MaxTime said the data set maximum time interval, Size said the size of the data. It should be noted that the actual track data VehicleTraj set extracted from this experiment is larger than the Oldenburg dataset, and the coverage is also greater.

In the given data table *Tr*, the final publishing data is assumed to be processed $Tr'(p_1, p_2, \ldots, p_m)$, $p_1, p_2, \ldots, p_m$ is *m* anonymous sets. $P_1, P_2, \ldots, P_m$ is *m* anonymous grouping. Then, according to the Discernibility Metric *(DM)* measure function, the table is punished by the following values [14] (Figs. 3, 4, 5 and 6).

$$\mathrm{DM}(\mathrm{Tr}') = \sum_{1 \le i \le m} |P_i|^2$$

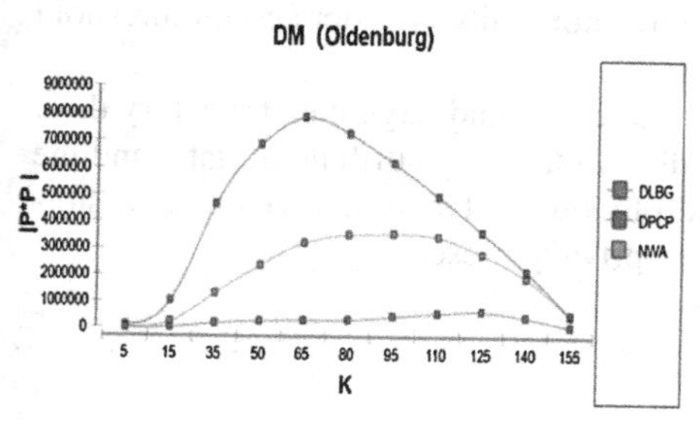

Fig. 3. DM in OldenBurg (t_2–t_1 = 1, *heap* = 7)

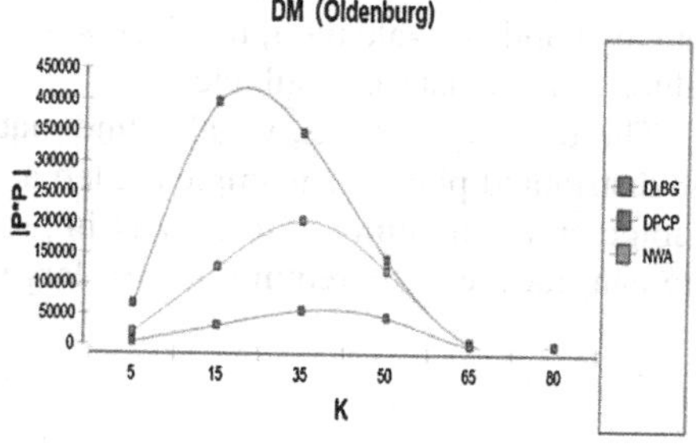

Fig. 4. DM in OldenBurg (t_2–t_1 = 2, *heap* = 14)

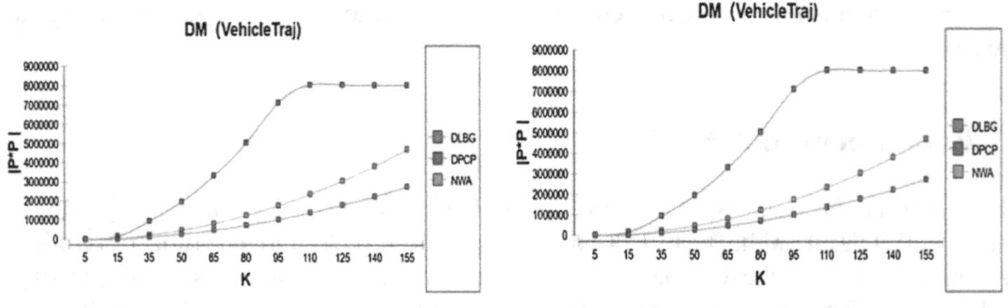

Fig. 5. DM in VehicleTraj (t_2–t_1 = 3, *heap* = 15)

Fig. 6. DM in VehicleTraj (t_2–t_1 = 3, *heap* = 15)

5 Conclusion

According to the trajectory data release, first proposed by time division based on t (x_i, y_i, t_{m-n}) to be released as a quasi identifier data, using *k*-model to trajectory data segmentation of anonymous and anonymous publishing.

Because of the characteristic of trajectory data with time correlation, position correlation, large scale and high dimension, based on the traditional relational data released by the *k*-anonymity and privacy protection model proposed in recent years for some other relational database privacy protection model cannot be used for privacy issue in mobile trajectory data of high dimension direct prevention. At present on the mobile trajectory data research, it is difficult to define the identifier and the sensitive attribute. The definition of identifiers for moving trace data is still an open question. According to the characteristic of the public vehicle trajectory data, t partition based on (x_i, y_i, t_{m-n}) to be released as a *QI* data, carrying information for sensitive attributes to the trajectory of the target, using *k-l*-anonymity ideological diversity data to trajectory data respectively by *NWA(Greedy_algorithm)* and *DLBG_algorithm* generalization, and compare all kinds of anonymous methods on data availability evaluation. Through the continuous optimization algorithm, the trajectory data can be effectively protected, and the *l*-diversity processing can also effectively prevent the background attack and association attack of the trajectory data. On the other hand, through data simulation and data availability evaluation, the data processed by anonymity is under certain threshold value, and the data is available.

Through experiments, we also find that for large size and large data trajectory data, the distributed planning is implemented according to the time attribute of data, and the storage and computing efficiency is improved obviously. Traditional centralized storage can not meet the requirement of large data privacy-preserving.

References

1. Sweeney, L.: k-Anonymity: a model for protecting privacy. Int. J. Uncertain. Fuzziness and Knowl.-Based Syst. **10**(5), 1–14 (2002)
2. Machanavajjhala, A., et al.: L-diversity: privacy beyond k-anonymity. ACM Trans. Knowl. Discov. Data **1**(1), 3 (2007)
3. Li, N., Li, T., Venkatasubramanian, S.: t-Closeness: privacy beyond k-Anonymity and l-Diversity. In: IEEE, International Conference on Data Engineering, pp. 106–115. IEEE (2007)
4. Tramp, S., Frischmuth, P., Arndt, N., Ermilov, T., Auer, S.: Weaving a distributed, semantic social network for mobile users. In: Antoniou, G., Grobelnik, M., Simperl, E., Parsia, B., Plexousakis, D., De Leenheer, P., Pan, J. (eds.) ESWC 2011, Part I. LNCS, vol. 6643, pp. 200–214. Springer, Heidelberg (2011). https://doi.org/10.1007/978-3-642-21034-1_14
5. Li, F., Gao, F., Yao, L., Pan, Yu.: Privacy preserving in the publication of large-scale trajectory databases. In: Wang, Yu., Yu, G., Zhang, Y., Han, Z., Wang, G. (eds.) BigCom 2016. LNCS, vol. 9784, pp. 367–376. Springer, Cham (2016). https://doi.org/10.1007/978-3-319-42553-5_31
6. Liberti, L., et al.: Euclidean distance geometry and applications. Quant. Biol. **56**(1), 3–69 (2012)
7. Zhu, T., et al.: Faces of the cone of Euclidean distance matrices: characterizations, structure and induced geometry. Linear Algebr. Its Appl. **408**(1), 1–13 (2005)
8. Abul, O., Bonchi, F., Nanni, M.: Never walk alone: uncertainty for anonymity in moving objects databases. In: IEEE, International Conference on Data Engineering. IEEE Computer Society, pp. 376–385 (2008)
9. Basu, A., et al.: A privacy risk model for trajectory data. In: Zhou, J., Gal-Oz, N., Zhang, J., Gudes, E. (eds.) IFIPTM 2014. IAICT, vol. 430, pp. 125–140. Springer, Heidelberg (2014). https://doi.org/10.1007/978-3-662-43813-8_9
10. Sun, X., Sun, L., Wang, H.: Extended k-anonymity models against sensitive attribute disclosure. Comput. Commun. **34**(4), 526–535 (2011)
11. Poulis, G., et al.: Distance-based k^m-anonymization of trajectory data. In: IEEE, International Conference on Mobile Data Management, pp. 57–62. IEEE (2013)
12. Xin, Y., Xie, Z.Q., Yang, J.: The privacy preserving method for dynamic trajectory releasing based on adaptive clustering. Inf. Sci. **378**, 131–143 (2017)
13. Geometry and applications. Quant. Biol. **56**(1), 3–69 (2012)
14. Kiran, P., Kavya, N.P.: A survey on methods, attacks and metric for privacy preserving data publishing. Int. J. Comput. Appl. **53**(18), 20–28 (2013)
15. Samarati, P.: Protecting respondents' identities in microdata release. IEEE Trans. Knowl. Data Eng. **13**(6), 1010–1027 (2001)
16. Gehrke, J., Kifer, D., Machanavajjhala, A.: ℓ-Diversity. In: van Tilborg H.C.A., Jajodia S. (eds.) Encyclopedia of Cryptography and Security, pp. 707–709. Springer, Boston (2011). https://doi.org/10.1007/978-1-4419-5906-5
17. Bonchi, F., Lakshmanan, L.V.S., Wang, H.: Trajectory anonymity in publishing personal mobility data. ACM Sigkdd Explor. Newsl. **13**(1), 30–42 (2011)
18. Shin, H., et al. Ensuring Privacy and Security for LBS through Trajectory Partitioning. In: Eleventh International Conference on Mobile Data Management IEEE Computer Society, pp. 224–226 (2010)

What Makes Charitable Crowdfunding Projects Successful: A Research Based on Data Mining and Social Capital Theory

Xizi Wang(✉) and Li Wang

Shanghai University, Shanghai, China
katewang91@163.com, 1015811070@qq.com

Abstract. The charitable crowdfunding platforms, as a new model for donation, enable fund seekers to solicit funds from the public over the Internet. Despite of the rapid development of crowdfunding platforms, knowledge about charitable crowdfunding remains obscure. Few studies have investigated the determinants of charitable crowdfunding projects' success. In this paper, we adopt data mining techniques to collect data from ZhongChou platform, an important crowdfunding website in China. The theoretical foundation of our research model is social capital theory. Our question is what factors can affect the success of charitable crowdfunding projects. The findings show that the more structural social capital the fundraisers have, the high success rate the project will achieve. Besides, frequent updates on project progress can attract more supporters and more donation. More comments and more followers can attract more supporters, but have no significant impact on attracting more donation.

Keywords: Social capital · Charitable crowdfunding · Data mining

1 Introduction

The maturity of Web 2.0 technologies and the success of crowdsourcing had resulted in the rapid development of crowdfunding platforms [1–3]. As an innovative business model under the umbrella of micro-finance, crowdfunding platforms provide entrepreneurs with a viable alternative to raise funds from the crowd via Internet, facilitating information flow and transactions between fundraisers and investors [1, 4]. Crowdfunding can offer support for both commercial and charitable purposes [5]. In this study, we focus on charitable crowdfunding platforms, where individuals and charity organizations raise monetary contribution for people in developing or impoverished areas. From an economical perspective, online charitable crowdfunding projects can reduce transaction and coordination costs, connecting donors and fundraisers regardless of geographical distance. Given the fact that most impoverished areas scattered in middle and western provinces of China, the success of charitable crowdfunding have significant implications for society and regional development. According to the report from zhongchoujia [6], over 9500 charitable projects had successfully raised 400 million yuan in 2017. However, little is known about factors that affect the success of charitable crowdfunding. This paper aims to understand determinants of charitable

J. H. Park et al. (Eds.): PDCAT 2018, CCIS 931, pp. 250–260, 2019.
https://doi.org/10.1007/978-981-13-5907-1_26

crowdfunding projects' success by introducing social capital theory. Indeed, social capital theory has been employed in financial development [7] and reward-based crowdfunding [8] settings, yet it hasn't been studied in the case of charitable crowdfunding platforms. In particular, we examined the fundraiser's social capital that was embedded in the platform, and the influence of which on the success of crowdfunding projects. Data used in this paper was crawled from ZhongChou website, one of the most popular crowdfunding platforms in China. This paper extends prior studies on charitable crowdfunding in two ways. First of all, most of prior literature that examined the determinants of crowdfunding performance collected data by questionnaires or surveys. While we wrote a program via Python to collect project information from the website. Moreover, our findings shed light on how to implement successful charitable crowdfunding projects. This paper extends the contemporary knowledge about charitable crowdfunding platforms, and lead both fundraisers and industry to improve the way in which they design platforms and campaigns.

The rest of the paper is organized as follows: Sect. 2 reviewed the existing literature in donation-based crowdfunding platforms, and we subsequently present previous arguments of social capital. In Sect. 3, we propose our research model and hypotheses in this specific context. Section 4 describes how we design our research and collect data. The results of research are also discussed in this section. We conclude the paper by the end of this paper.

2 Literature Review

2.1 Related Studies on Crowdfunding

In fact, considerable research findings have been yield in the crowdfunding domain. First of all, the definition and some concepts in this domain have been well-elaborated by prior literature. According to the nature of the reward, crowdfunding projects can be classified into commerce-oriented and charity-oriented. Crowdfunding projects with commercial goals can be further divided into reward-based crowdfunding, equity-based crowdfunding, and lending-based crowdfunding. It has also been found that there were three primary reasons for companies to raise fund through crowdfunding, including raising capital, attracting public's attention, and receiving feedback for their products [4]. Secondly, for either commerce-oriented crowdfunding or charity-oriented crowdfunding, it's crucial to understand the factors associated with projects' success. Factors that have been proved that had impact on commercial-oriented crowdfunding projects success included social information such as others' funding decisions [9], project specified characteristics such as description and images [10], and individuals' social capital [11]. However, little is known about what affect the success of charitable crowdfunding projects.

2.2 Charitable Crowdfunding

Charitable crowdfunding platforms have been considered as the intersection between charitable giving and IT-enabled crowdfunding, transform the way in which charity has traditionally operated [5]. Explaining online users' motivations and behavior intention has always been highlighted in the IS field. Thus, a majority of studies have studied the dynamic behaviors on the platform along with motivations of charitable giving. Smith [12] found that there is a positive peer effect in donation-based crowdfunding. Donors use donation information from other donors to decide which projects they should donate. Koning's study [13] proved that the amount of donations received at the beginning of the project can affect the subsequent donation behavior of donors, which have a great influence on the success of the project. In [9], it has been found that the donor's support of the project is in line with the U-shaped model, which means that the project fundraiser will receive more support at the beginning and end of the project. The results of [14] proved the influence of the word of mouth effect on crowdfunding projects. Supporters actively promote the donation campaign via their social networks, attracting more people to support the project.

2.3 Social Capital Theory

In a complex system of action, people provide new resources for individuals through social networks formed by social relationships—social capital [15]. It's a key factor affecting the success of crowdfunding projects. For instance, Giudici [3] divided social capital into individual social capital and regional social capital from the perspective of individual and regional sharing. Their study found that individuals' social capital and crowdfunding success were significantly correlated. [16] studied the relationship between crowdfunding success and geographical locations. It has been proved that the donors of crowdfunding projects mainly come from the geographical area where the fundraisers live in. They suggested that regional social capital was positively related to the success of crowdfunding project. Mollick's study [10] suggested that fundraisers' personal social networks and project quality had positive influence on crowdfunding success. Informed by previous literature, this paper analyzed the determinants of crowdfunding projects' success based on the foundation of social capital theory.

3 Research Framework and Hypotheses

3.1 Research Framework

There are three dimensions of social capital [17]: structural social capital, relational social capital, and cognitive social capital. Structural social capital examines the role of individual-owned social capital in the acquisition of external resources from the perspective of embedded social relations. While relational social capital focuses on the exchange of knowledge in the social network and emphasizes cooperation with others. Intellectual capital refers to the beliefs, rules, and paradigms shared among members in the network. Usually, donors have limited knowledge and it's difficult to establish cognitive social capital. Therefore, this paper only analyzes the impact of structural social capital and relational social capital on the crowdfunding projects. The research model is shown in Fig. 1.

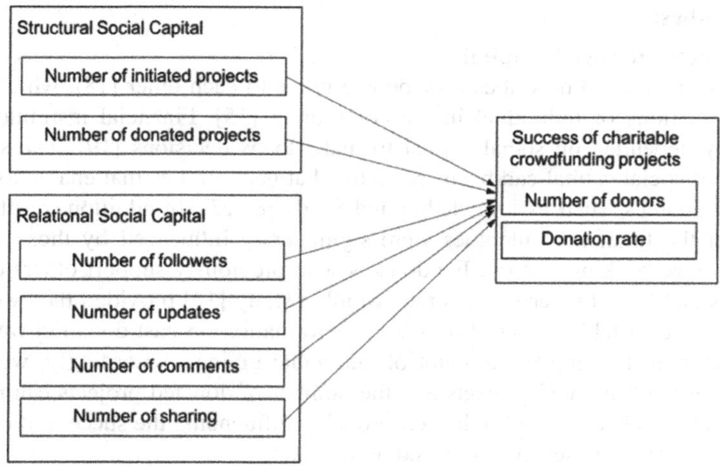

Fig. 1. Research model

The number of initiated projects and the number of donated projects of fundraisers on the crowdfunding platform were used to represent their structural social capital. While we adopted the number of project followers, the number of updates, the number of comments, and the number of sharing to measure their relational social capital. The crowdfunding performance has two measurements which are the number of donors and the donation rate. Target amount serves as the control variable. Measurements are described in Table 1.

Table 1. Descriptions of measurements

Variables	Measurements	Descriptions
Structural social capital	Number of initiated projects (NIP)	Number of projects that fundraiser used to initiate
	Number of donated projects (NDP)	Number of projects that fundraiser used to donate
Relational social capital	Number of followers (NF)	Number of project followers
	Number of updates (NU)	Number of project updates
	Number of comments (NC)	Number of project comment
	Number of sharing (NS)	Number of project that have been forwarded via social networking sites
Control variable	Target amount (TA)	Target amount of charitable crowdfunding projects
Success of charitable crowdfunding	Donation rate (DR)	Rate of the donated amount to the target amount
	Number of donors (ND)	Numbers of donors that participated in crowdfunding

3.2 Hypotheses

3.2.1 Structural Social Capital

Structural social capital describes how people relate to each other [18], which reflects the social relations of individual in the community [15]. Financial institutions have increasingly depended on social capital to make loans decisions [19]. The structural dimension of social capital can been see as the that network ties that enable to provide access to resources, such as knowledge and funds [8, 17]. In addition, results of [1] proved that the financing outcomes were significantly influenced by those entrepreneurs who have backing-history. Fundraisers who previously support others can raise more funds and have higher success rate. Smith's study [12] provided the evidence of peer effects in charitable giving. For donors, information on past donation was served as benchmark in deciding the amount of charitable giving. Accordingly, we suggest that the number of initiated projects and the number of donated projects can represent the structural social capital of fundraiser, probably influencing the success of charitable crowdfunding. Hypotheses are proposed as follows:

Hypothesis 1a: the number of initiated projects is positively associated with project's donation rate.
Hypothesis 1b: the number of initiated projects is positively associated with the number of donors.
Hypothesis 2a: the number of donated projects is positively associated with project's donation rate.
Hypothesis 2b: the number of donated projects is positively associated with the number of donors.

3.2.2 Relational Social Capital

The relational aspect of social capital is the degree of closeness between individuals' interactions [17]. It refers to the strength or quality of the relationship and evaluated by trust, norms, obligations, and identification [17]. In the crowdfunding community, the number of followers means how many people give attention to the projects. With more attention, the high recognition the fundraiser will probably achieve. The strength of relationship might strength as well. Therefore, hypotheses are proposed as follows:

Hypothesis 3a: the number of followers is positively associated with project's donation rate.
Hypothesis 3b: the number of followers is positively associated with the number of donors.

Prior literature argued that the high frequency of interaction is conductive to forming the social relationship, thus increasing the probability of crowdfunding success [20]. The significant role of frequent project updates has already been verified [21]. Frequent updates reflect the fundraiser's ability of implementation and help build trust between donors and fundraisers. We therefore assume that more updates can contribute to the success of projects, implying the following hypotheses.

Hypothesis 4a: the number of updates is positively associated with project's donation rate.

Hypothesis 4b: the number of updates is positively associated with the number of donors.

De Choudhury [22] analyzed the blogs' characteristics among four US technology companies from January to November in 2007. Their investigation showed that the number of blogs, the number of comments, the average length of comments, and the average time of responses are all significantly related to change ratio of company's stock price. These factors can effectively predict the trends and scales of stock price movement, with 87% and 78% accuracy respectively. Evers [23] found that the number of comments on crowdfunding projects is related to the reputation and recognition of fundraisers. In the case of crowdfunding platform, commenting on project can be regarded as a kind of interaction. Thus, our hypotheses are proposed as follows:

Hypothesis 5a: the number of comments is positively associated with project's donation rate.

Hypothesis 5b: the number of comments is positively associated with the number of donors.

ZhongChou platform allow donors to forward the crowdfunding project to WeChat, QQ and Sina Weibo, sharing the donation information with their friends or acquaintances. This help more people participate in the donation project. The probability of crowdfunding success increases as well. Hypotheses are proposed as follows:

Hypothesis 6a: the number of sharing is positively associated with project's donation rate.

Hypothesis 6b: the number of sharing is positively associated with the number of donors.

4 Research Design and Discussion

4.1 Research Design

ZhongChou (www.zhongchou.com) is one of the most popular crowdfunding platforms in China. People can initiate product crowdfunding, charitable crowdfunding and equity crowdfunding on the platform. This paper crawled public data on the charitable crowdfunding website by programming techniques. The information we targeted consists of basic project information, project funding information, and participants' interaction information. We collected 636 samples which were projects had completed from June 2016 to December 2016. There were 412 projects had successfully raised funds, with 64.78% success rate. Table 2 presents the descriptive statistics of variables.

Table 2. Descriptive statistics of variables

Variables	Minimum	Maximum	Mean	Standard Variation
NIP	0	29	2.14	5.96
NDP	0	10	1.21	1.80
NF	0	1673	61.89	131.84
NU	0	43	4.24	3.93
NC	0	224	10.69	22.82
NS	0	14	0.21	0.98
TM	500	1200000	27082.69	93699.49
ND	0	2803	102.47	195.04
DR	0	1586	92.11	112.28

Figures 2 and 3 show the donation distribution among failed projects and successful projects respectively.

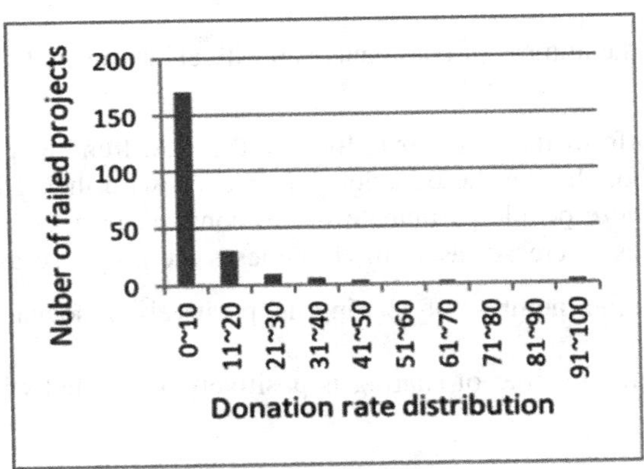

Fig. 2. Donation rate distribution among failed projects

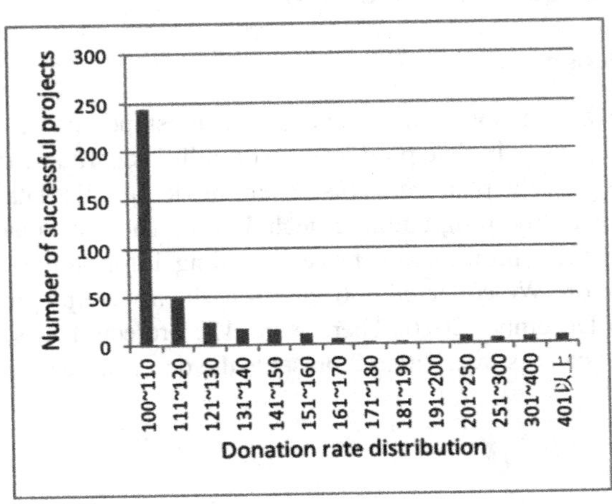

Fig. 3. Donation rate distribution among successful projects

Table 3. Correlations between different variables

	NIP	NDP	NF	NU	NC	NS	TA	ND	DR
NIP	1								
NDP	−.028	1							
NF	−.053	.111**	1						
NU	.023	.219**	.090*	1					
NC	−.071	.159**	.311**	.247**	1				
NS	−.048	.258	.076	.131**	.258**	1			
TA	−.075	−.030	.036	.009	.128**	.266**	1		
ND	−.060	.204**	.212**	.249**	.553**	.272**	.263**	1	
DR	.485**	.149**	.124**	.219**	.100*	−.007	−.089*	.147**	1

N = 636 significance level *P < 0.1, **P < 0.05, ***P < 0.01

Table 3 shows the correlation coefficient and significance level of each variable. Obviously, the correlation coefficients between the independent variables are all less than 0.7, indicating that there is no multicollinearity problem between independent variables.

4.2 Discussion

We applied multiple regression analysis on the donation rate and the number of donors respectively. For the donation rate variable, the coefficient determination (R^2) is 31.3%, and the adjusted R^2 is 30.5%. For the number of donors, R^2 is 37.7% and the adjusted R^2 equals 37.0%.

Based on the results in Table 4, numbers of initiated projects and numbers of donated projects have significantly impacts on donation rate, supporting the hypotheses H1a and H2a. Numbers of initiated projects, however, are not significantly related to numbers of donors, rejecting H1b. Numbers of donated projects are positively related to numbers of donors, supporting H2b.

Table 4. Multiple regression analyses on donation rate and number of donors

Variables	Donation rate			Number of donors		
	Coefficient	P value	T value	Coefficient	P value	T value
NIP	0.489	0.000***	14.672	−0.007	0.832	−.212
NDP	0.105	0.002***	3.062	0.108	0.001***	3.309
NF	0.109	0.002***	3.119	0.037	0.264	1.119
NU	0.164	0.000***	4.708	0.097	0.004***	2.923
NC	0.056	0.131	1.512	0.454	0.000***	12.887
NS	−0.017	0.628	−.485	0.086	0.011**	2.543
TA	−0.057	1.101	−1.642	0.182	0.000***	5.550

N = 636 significance level *P < 0.1, **P < 0.05, ***P < 0.01

In the relational social capital construct, numbers of followers and frequency of projects updating have influence on donation rate, which support the hypotheses H3a, H4a. While H5a and H6a have been proved insignificant. Regarding to the number of donors, frequent updates, comments and sharing are positively related to the dependent variable, supporting H4b, H5b and H6b. While the H3b has not been verified.

For the control variables, we found that the target amount has no significant effect on donation rate, which is inconsistent with prior findings from reward-based crowdfunding and equity-based crowdfunding [10, 24], in which target amount had positive effects on investors' perceived risk and uncertainty. One explanation might be that charitable crowdfunding projects usually request a small amount of contribution. Donors' decisions are mainly affected by their perception of trust rather than the rewards.

5 Conclusion

This work extends the knowledge of social capital in explaining charitable crowdfunding success, providing reference for future studies. The primary major contribution is evidence of empirically analyzing determinants of charitable projects success in electronic crowdfunding context. Our research throws light on the how to use crowdfunding technologies impacts the nature of charitable giving and solicit more funding. For a more practical purpose, China particularly need charitable crowdfunding as supplementary measure to alleviate poverty. Currently, impoverished areas are scattered throughout the remote rural areas. With the help of crowdfunding platforms, Individuals and charities can use these platforms to raise money directly despite of their geographical locations. The second contribution lies in the evidence suggesting that interaction and recognition play an important role for charitable projects success. We thus contribute to the bedrock of applying social capital theory to charitable crowdfunding.

This study also has practical implications for both fundraisers and crowdfunding platforms. For fundraisers: (1) more interactions and activities in crowdfunding communities are encouraged. For example, initiating more innovative and valuable crowdfunding projects. Donating more charitable crowdfunding projects in the community is also a useful way to improve their social capital. (2) Projects should be updated frequently, then donors can follow the project progress in time. This will help stimulate the donors' enthusiasm to participate in crowdfunding and solicit target funds. For crowdfunding platform: (1) The fundraiser credit rating system (structured social capital) should be continuously improved, revealing more information such as the success rate and failure rate of initiated projects. Fundraiser's ability of project implementation is supposed to displayed meanwhile. (2) The crowdfunding platform should constantly improve the crowdfunding community by enhancing interactions and facilitating communications in the community.

References

1. Zvilichovsky, D., Inbar, Y., Barzilay, O.: Playing both sides of the market: success and reciprocity on crowdfunding platforms (2015)
2. Kleemann, F., Voß, G.G., Rieder, K.: Un (der) paid innovators: the commercial utilization of consumer work through crowdsourcing. Sci. Technol. Innov. Stud. **4**(1), 5 (2008)
3. Giudici, G., Nava, R., Rossi Lamastra, C., Verecondo, C.: Crowdfunding: the new frontier for financing entrepreneurship? (2012)
4. Belleflamme, P., Lambert, T., Schwienbacher, A.: Individual crowdfunding practices. Venture Cap. **15**(4), 313–333 (2013)
5. Choy, K., Schlagwein, D.: Crowdsourcing for a better world: on the relation between IT affordances and donor motivations in charitable crowdfunding. Inf. Technol. People **29**(1), 221–247 (2016)
6. Information. http://www.zhongchoujia.com/data/30373.html
7. Guiso, L., Sapienza, P., Zingales, L.: The role of social capital in financial development. Am. Econ. Rev. **94**(3), 526–556 (2004)
8. Zheng, H., Li, D., Wu, J., Xu, Y.: The role of multidimensional social capital in crowdfunding: a comparative study in China and US. Inf. Manag. **51**(4), 488–496 (2014)
9. Kuppuswamy, V., Bayus, B.L.: Crowdfunding creative ideas: the dynamics of project backers. In: Cumming, D., Hornuf, L. (eds.) The Economics of Crowdfunding, pp. 151–182. Springer, Cham (2018). https://doi.org/10.1007/978-3-319-66119-3_8
10. Mollick, E.: The dynamics of crowdfunding: an exploratory study. J. Bus. Ventur. **29**(1), 1–16 (2014)
11. Giudici, G., Guerini, M., Rossi Lamastra, C.: Why crowdfunding projects can succeed: the role of proponents' individual and territorial social capital (2013)
12. Smith, S., Windmeijer, F., Wright, E.: Peer effects in charitable giving: evidence from the (running) field. Econ. J. **125**(585), 1053–1071 (2015)
13. Koning, R., Model, J.: Experimental study of crowdfunding cascades: when nothing is better than something (2013)
14. Choy, K., Schlagwein, D.: IT affordances and donor motivations in charitable crowdfunding: the "earthship kapita" case. In: ECIS (2015)
15. Coleman, J.S.: Social capital in the creation of human capital. Am. J. Sociol. **94**, S95–S120 (1988)
16. Agrawal, A.K., Catalini, C., Goldfarb, A.: The geography of crowdfunding. National Bureau of Economic Research, No. w16820 (2011)
17. Nahapiet, J., Ghoshal, S.: Social capital, intellectual capital, and the organizational advantage. In: Knowledge and Social Capital, pp. 119–157 (2000)
18. Burt, R.S.: Structural Holes: The Social Structure of Competition. Harvard University Press, Cambridge (2009)
19. Everett, C.R.: Group membership, relationship banking and loan default risk: the case of online social lending. Bank. Finance Rev. **7**(2) (2015)
20. Beier, M., Wagner, K.: Crowdfunding success: a perspective from social media and e-commerce (2015)
21. Koch, J.-A., Siering, M.: Crowdfunding success factors: the characteristics of successfully funded projects on crowdfunding platforms (2015)

22. De Choudhury, M., Sundaram, H., John, A., Seligmann, D.D.: Can blog communication dynamics be correlated with stock market activity? In: Proceedings of the Nineteenth ACM Conference on Hypertext and Hypermedia, pp. 55–60. ACM (2008)
23. Evers, M.W., Lourenço, C., Beije, P.: Main Drivers of Crowdfunding Success: A Conceptual Framework and Empirical Analysis. Erasmus Universiteit, Rotterdam (2012)
24. Ahlers, G.K.C., Cumming, D., Günther, C., Schweizer, D.: Signaling in equity crowdfunding. Entrepreneurship Theory Pract. **39**(4), 955–980 (2015)

A *SwarmESB* Based Architecture for an European Healthcare Insurance System in Compliance with GDPR

Cristina Georgiana Calancea, Lenuţa Alboaie, and Andrei Panu(✉)

Faculty of Computer Science, Alexandru Ioan Cuza University of Iasi, Iaşi, Romania
{cristina.calancea, adria, andrei.panu}@info.uaic.ro

Abstract. With the everlasting development of technology and society, data privacy has proven to grow into a pressing issue. The bureaucratic state system seems to expand the number of personal documents required for any kind of request. Therefore, it becomes obvious that the number of people having access to information that should be private is on the rise as well. This paper offers an alternative cloud integration solution centered on user data privacy, its main purpose being to help software services providers and public institutions to comply with the General Data Protection Regulation. Throughout this proposal we describe how data confidentiality can be achieved by transitioning complex human procedures into a coordinated and decoupled swarm system, whose core lies within the "Privacy by Design" principles.

Keywords: Privacy · GDPR · SwarmESB · Integration

1 Introduction

Alongside with the evolution of society, people have become more prone to understanding how digitized systems could improve the everyday life experience. Moreover, the need for automation has surged from the countless irregularities regarding privacy discovered in the legal procedures that require public employees to go through citizen's personal data in order to validate it. The issue demanding a solution is the lack of coordination among governmental institutions and therefore, the failure to sustain the right to privacy.

General Data Protection Regulation [1], which took effect on May 2018, states that collecting, processing and storing user data without explicit consent is a punishable offense. In order to comply with regulations stipulated in the GDPR [1], many companies have sought ways to improve their systems as to support monitoring of the data access and information removal mechanisms. For instance, after the Cambridge Analytica scandal [2], Facebook is said to have improved its confidentiality policy by presenting the users choices regarding what they want to share and offering them the option to view and delete the data they store [3, 4]. Even though GDPR is a regulation promoted by the EU, Google is also subjected to these laws when providing any of its

J. H. Park et al. (Eds.): PDCAT 2018, CCIS 931, pp. 261–267, 2019.
https://doi.org/10.1007/978-981-13-5907-1_27

services. Besides the explicit consent agreements, the enterprise has also worked on its ads mechanism by making targeting less aggressive [5].

Unlike private enterprises, governmental institutions find it difficult to cope with all the GDPR changes, since in most of them, the legal procedures are still executed by using pen and paper. A relevant example is Romania, an European country, where the health insurance system relies on excessive interaction between citizen personal data and human resources. The latter are necessary to complete almost any task, which often leads to privacy breaches. The obvious solution is the automation of the whole process, as a distributed system, which offers the same services, whilst shielding user privacy. According to the "Privacy by Design" principles [6], our digital processing unit must be built in order to foresee data breaches and integrate confidentiality in its components from the beginning, not as a last minute extension. This paper promotes an alternative cloud integration technology, Swarm ESB [7], which centers on protecting data confidentiality, while decoupling complex systems in small entities that can coordinate themselves in order to work with as little user data as possible.

2 Swarm Architecture as a Viable Alternative for Integration and Privacy

There are a few Enterprise Service Bus implementations that offer Integration Platform as a Service solutions, oriented on privacy: Mule ESB [8], WSO2 ESB [9] and Swarm ESB [7]. Mule ESB offers resource access constrained by several filters and policies, while preventing sensible data exposure by using encryption, digital signatures and access control techniques for APIs usage. WSO2 ESB integration solution implements sixteen security scenarios inspired from the web services security policies. Some examples are the Integrity, Confidentiality and Kerberos Token-based Security scenarios [12]. Each one of them uses either digital certificates or keys to verify the identity of the sender and the authenticity of the message.

In order to understand the way Swarm ESB approaches privacy and how it is modeled in Healthfuse (described in Sect. 3), we will briefly present the concepts behind it - for a detailed perspective on SwarmESB, see [7, 10]. Swarm communication is a pattern of sending and processing messages between adapters. An adapter is a server side software node that offers a functionality of the system, which can be used only through a swarm. Usually, communication in an ESB implementation takes place between complex entities that process simple messages. Swarm ESB pictures messages as "smart" entities, capable of taking over some of the workload, by being routed between specialized components, which helps to reduce the complexity of distributed systems, offering scalability, availability, decoupling and parallel use of resources [10].

The integration strategy proposed by Swarm ESB is one of the few that cover all privacy principles, by introducing the usage of executable choreographies. The implementation of this concept turns formal contracts between organizations in code executed by every communication participant. The standard classification that helps us model various processes of integration and privacy assurance contains three categories, according to [11]. A privacy advantage of using verified choreographies is the capability of monitoring the data stream directly in the integration layer, which is logically separated by the processing code, found in adapters. Encrypted Choreographies use various control access mechanisms, with the purpose of identifying and authenticating the key entities that communicate through swarms, while serverless choreographies are appropriate for deployment in a public cloud, which offers monitoring and full automation of processes' capabilities. Executable choreographies are the key principle of the "Privacy-Integration" model Swarm ESB proposes.

3 Healthfuse - Swarm Based Architecture in Compliance with GDPR

3.1 Romanian Health Insurance System

The outline in Fig. 1 summarizes the tedious experiment of obtaining a European Insurance Health Card for a person who is an employee and also has a business. The six steps that must be performed in order to obtain an insurance are: bringing the identity card and a copy of it, supplying an employment proof, providing an income declaration and its confirmation from the National Fiscal Administration Agency and showing the receipt which proves all taxes were paid. Having an extra income earned from

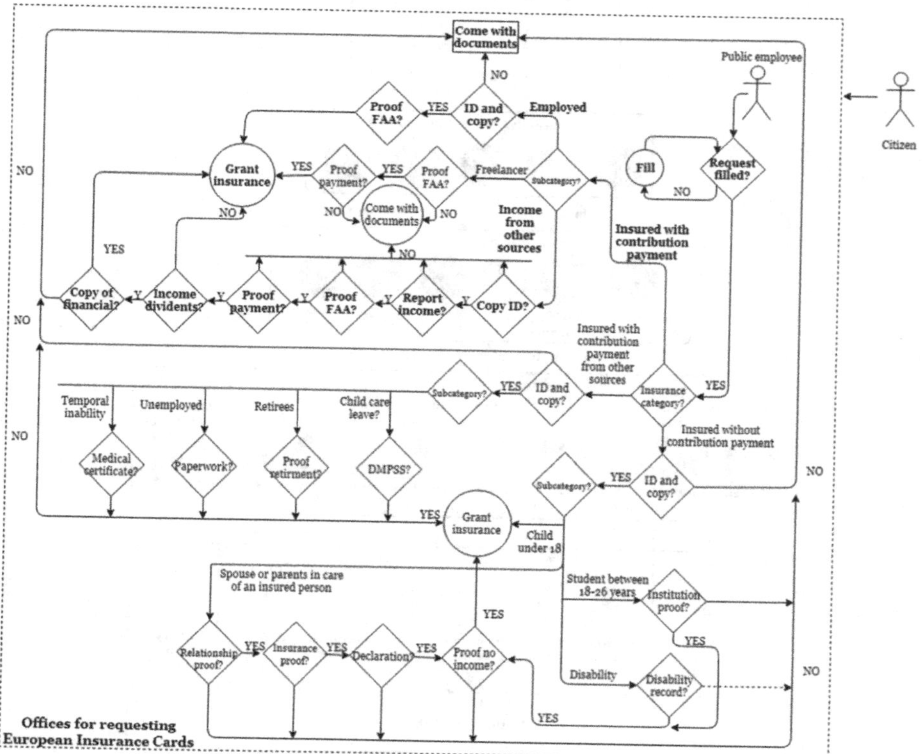

Fig. 1. Flow of interaction when requesting an European health insurance

dividends results in the need of bringing additional financial statistics from the Ministry of Finance. The total amount of time spent by going through all this flow revolves around ten hours, since we had to visit three public institutions in order to gather private paperwork and show it to at least five different people with the purpose of fulfilling procedures, while completely dismissing citizen's privacy.

3.2 Healthfuse - An Alternative Approach

In order to turn privacy into a key player, the flow shown in Fig. 1 is automated to reduce human interaction with user data, as it is outlined in the proposed Healthfuse architecture, described in Fig. 2. It reduces the time consuming six steps process to a three steps flow: log in, select the type of insurance and then wait for the request to be validated by a swarm that executes the automated version of the process presented in Fig. 1, synthesized in choreography. Since the suggested architecture is a distributed

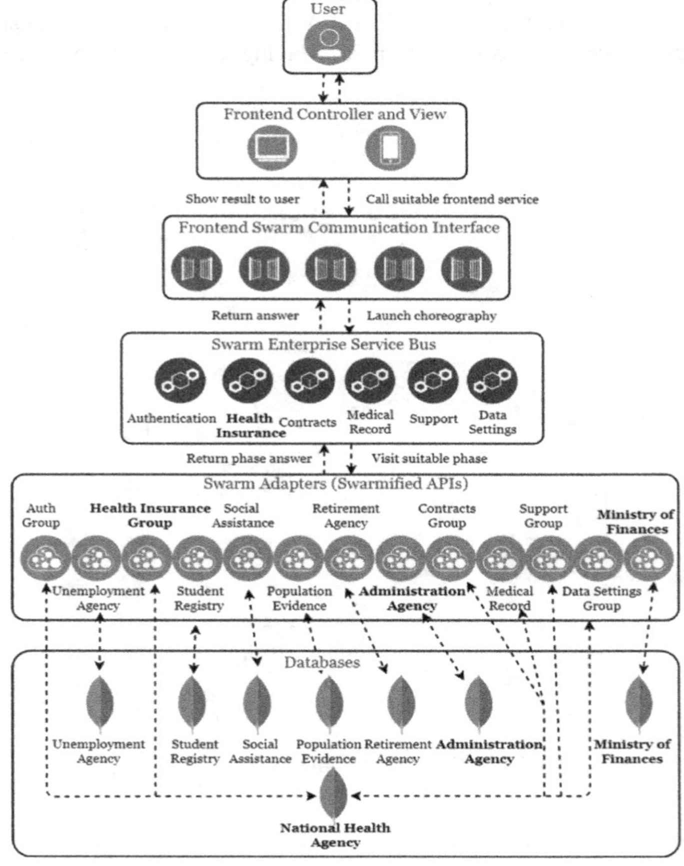

Fig. 2. HealthFuse swarm application architecture

one, it enhances the benefits of using a cloud system to coordinate all entities and keep their stored data confidentially. Therefore, our alternative approach provides the assurance that any shared data is kept private and used only for automated processing requested by the user. The only scenario that leads to data disclosure to the technical team is when an inquiry is submitted by the user directly through the integrated support system.

Since the significant amount of information needed to validate the insurance flows cannot be found in the resources belonging to the National Health Insurance Agency, we need to create additional adapters used to extract data from entities like the Administration Agency, described on the fourth level of Fig. 2. Adapters supply data to swarms - outlined on the third level of the architecture - in order to transport it between entities, following various choreographies. This process spares citizens from going between public institutions to gather their personal data and from handing it to strangers. The architecture is designed by separating entities and allowing them to exchange data on authenticated channels only when necessary. That way, the relevant data is distributed to the relevant entity and accessible to the others only on valid premises. The main features Healthfuse offers to its users are the ability of getting the appropriate insurance policy from just a few easy steps and opening a medical record

Table 1. Implementation of “Privacy by Design” principles in healthfuse

Privacy by design principles	Healthfuse - a SwarmESB based solution
1. Proactive not reactive; Preventative not remedial	Our solution offers an access monitoring mechanism, which makes it formally verifiable, since it can accurately state which entities has accessed the private data
2. Privacy as the default setting 3. Privacy embedded into design	Privacy concerns are treated by using verified choreographies in the swarm’s implementation. This way, adapters only process data, leaving the confidentiality checks to the integration layer, composed by swarms
4. Full functionality; Positive - sum, not zero-sum	The business model can be altered at any time, without producing any impact over the current components and flows; instead we would just develop choreography to encapsulate the new model
5. Visibility and transparency – keep it open	The implemented choreographies target communication between many organizations, all of them being known to the others. Private data can be accessed only by entities that have the privileges to run the proper choreography
6. End-to-end security – full lifecycle protection	Data is protected by the SCRAM-SHA1 encryption protocol applied to the authentication step of the storage unit. End-to-end security is guaranteed by using communication through WSS Protocol on server side and HTTPS Protocol on client side
7. Respect for user privacy – keep it user-centric	Access to private data is done only with the consent of the user when reporting an incident in the support system. The user is prioritized over commercial interest by offering the option to delete any collected data

by uploading, deleting and downloading files they may need to have closed. In Table 1, we describe the way our proposal is designed and implemented in respect to the "Privacy by Design" Principles.

4 Conclusion

This paper presents the impact GDPR has on public institutions, the difficult situation in which they found themselves while trying to comply with the new requirements regarding data privacy and a solution to most of their problems, based on Swarm architecture.

Throughout this paper, we propose a way to stop the ascending tendency of losing the control of our personal data to various governmental agencies. Since the key principle in creating Healthfuse was to protect data, we remodeled the health insurance system from its roots and built it as an independent and specialized entity that only exchanges data with other software entities built on the same principles.

As for the future, this project can be looked at as a continuously developing system [13], which implies collaboration from several other public institutions like the Financial Administration Agency and Retirement Agency. The current prototype can be developed into a public administration tool which works relying on controlled collected data distribution, avoiding at the same time its centralization. This leads to the usage of a coordinated system which doesn't keep the data it uses secret.

Revolutionizing the healthcare insurance procedures by installing such a system would most likely reduce data leaks and would most certainly offer a better user experience in terms of time and resources.

Acknowledgements. The dissemination of this work is partly funded by the European Union's Horizon 2020 research and innovation programme under grant agreement No 692178. It is also partially supported by the Private Sky Project, under the POC-A1-A1.2.3-G-2015 Programme (Grant Agreement no. P 40 371).

References

1. EU General Data Protection Regulation (2018). https://www.eugdpr.org/
2. Cambridge Analytica (2018). https://cambridgeanalytica.org/
3. What is General Data Protection Regulation? (2018). https://www.facebook.com/business/gdpr
4. Constine, J.: A flaw-by-flaw guide to Facebook's new GDPR privacy changes (2018). https://techcrunch.com/2018/04/17/facebook-gdpr-changes/
5. Frey, S., Crandall, M.: Google Cloud: Ready for the GDPR (2018). https://www.blog.google/products/google-cloud/google-cloud-ready-for-gdpr/

6. Cavoukian, A.: Privacy by design: origins, meaning, and prospects for assuring privacy and trust in the information era. In: Privacy Protection Measures and Technologies in Business Organizations: Aspects and Standards, pp. 170–208. IGI Global (2012)
7. Alboaie, L., Alboaie, S., Panu, A.: Swarm communication – a messaging pattern proposal for dynamic scalability in cloud. In: Proceedings of the 15th IEEE International Conference on High Performance Computing and Communications – IEEE HPCC, China, pp. 1930–1937 (2013). https://doi.org/10.1109/hpcc.and.euc.2013.277
8. MuleSoft Inc.: What is Mule ESB? (2018). https://www.mulesoft.com/resources/esb/what-mule-esb
9. WSO2 Inc.: WSO2 Enterprise Service Bus (2018). http://wso2.com/products/enterprise-service-bus/
10. Alboaie, S., Alboaie, L., Panu, A.: Levels of privacy for eHealth systems in the cloud era. In: 24th International Conference on Information Systems Development, Harbin, China, pp. 243–252 (2015)
11. Alboaie, L.: Towards a smart society through personal assistants employing executable choreographies. In: Proceedings of the 26th International Conference on Information Systems Development, ISD 2017, Larnaca, Cyprus, 6–8 September 2017
12. WSO2 Inc.: Security Implementation (2018). https://docs.wso2.com/display/DSS322/Security+Implementation
13. Calancea, C.G.: Healthfuse Presentation (2018). https://bitbucket.org/meoweh/healthfuse/src/HealthFusePresentation/

A Study on Deriving and Simulating Pre-risk on Complex Gas Facilities for Preventing Accidents

Jeong Seok Oh(✉)

Institute of Gas Safety R&D, Korea Gas Safety Corporation, 1390, Wonjung-ro, Maengdong-myeon, Eumseong-gun, Chungcheongbuk-do, Korea
jsoh90@gmail.com

Abstract. For preventing accident, the risk analysis about gas facilities has been more important since many gas facilities be superannuated. Especially, deriving and simulating risk is very important for preventing and corresponding accidents by means of specific analysis method in complex gas facilities. However, many studies have been not enough not yet in order to derive and simulate risk considering various situations. This paper aims to propose deriving and simulating risk method around limited area of complex gas facilities. Our study proposes total risk analysis that is composed four methods with individual point of view. Furthermore, this research is able to grant the independent grade of risk per logical zone area for simulating and predicting risk to complex gas faculties.

Keywords: Simulation of risk · Grade of risk · Zone-based risk analysis

1 Introduction

For preventing accident, the risk analysis about gas facilities has been more important since many gas facilities be superannuated. An accident of industrial can generate damage to person and facilities, because most industrials use noxious or combustible materials. Especially, deriving and simulating risk is very important for preventing and corresponding accidents by means of specific analysis method in complex gas facilities. However, many studies have been not enough not yet in order to derive and simulate risk considering various situations [1–4].

This paper aims to propose deriving and simulating risk method around limited area of complex gas facilities. Because the core of our method is risk analysis, our study proposes total risk analysis that is composed four methods with individual point of view [5]. Each of four methods are reflected by accident probability, work risk, unexpected peculiar situation, and condition deterioration respectively in order to

This research was financially supported by the Ministry of Trade, Industry, and Energy (MOTIE), Korea, under the "Regional Specialized Industry Development Program (R&D, P0002072)" supervised by the Korea Institute for Advancement of Technology (KIAT).

J. H. Park et al. (Eds.): PDCAT 2018, CCIS 931, pp. 268–273, 2019.
https://doi.org/10.1007/978-981-13-5907-1_28

consider various situations. Furthermore, this research is able to grant the independent grade of risk per logical zone area for simulating and predicting risk to complex gas faculties.

2 Design Method of Deriving and Simulating Pre-risk

Pre-risk simulation model is made up four kinds of risk-driven method. The first method is accident probability method by chemical engineering point of view. This method computes accident possibility with accident probability and damage seriousness about gas leak, fire, and explosion. This method has been developed six components as show Table 1 and Fig. 1.

Table 1. Component of accidents probability method

Component	Description
Material component	Derives material characteristics factor using material database
Vessel component	Converts related parameters such as temperature and pressure into a mass unit in target facilities
Scenario component	Calculate an accurate mass value by collecting status parameters such as crack and hole
Source component	Computes the velocity of leak by acquiring atmospheric factor such as wind speed and direction
Probability component	Leak: computes density Fire: computes thermal radiation Explosion: computes overpressure
Rendering component	Display fatality by Leak, fire, explosion

The second method is work risk analysis method in order to digitize work risk of worker. This method adds work error rate (or human error rate) to fault tree of facilities. A work error rate is decided by the work failure substance and frequency of instrument. The work failure substance and frequency is selected by human error probabilities derived from experience and generic rates for operations. This method has been developed six components as show Table 2 and Fig. 2.

The third method can recognize unexpected peculiar situation through deriving similarity among sensing data. A risk in this method is driven by comparing likeness between risk standard data and real current ambient data of facility. The method uses Euclidean method or variation coefficient for comparing likeness. Especially, our study must be adopted data normalization in order to prevent data bias and distortion. This method has been developed five components as show Table 3 and Fig. 3.

The fourth method can derive risk by predicting condition deterioration to gas facilities. The degree of condition deterioration is decided by thinning model and crack

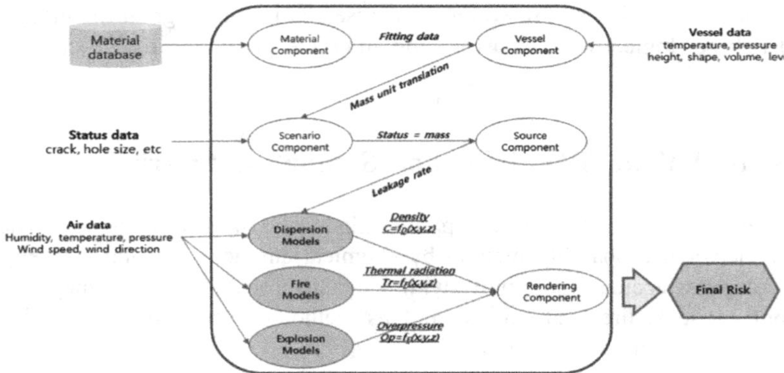

Fig. 1. Data flow of accidents probability method

Table 2. Component of work risk analysis method

Component	Description
Equipment component	Associates related information to facilities and devices
FTA component	Derives risk on facility by constructing fault tree based on fault rate of facility
Work component	Calculates work error rate by reading work information of the relevant zone and facility, and the adds fault tree
Combine component	Computes the combined risk using fault rate of facility and error rate of work
Decision component	Predicts risk using job schedules based on zone area
Rendering component	Display risk

severity index. A thinning model is created through standard data set in API RP 581. The crack severity index is able to reveal the degree of risk about stress spectrum of crack growth. Effective stress is given by subtracting maximum stress from opening stress. This method has been developed six components as show Table 4 and Fig. 4.

At last, four different four risks combine to a total risk in our study as shown Eq. 1. x_1, x_2, x_3, x_4 mean risk of accident probability method, risk of work analysis method, risk of intelligent cognition method and risk condition deterioration method respectively. w_n is weighted value in each method and selected differently by user requirements such as accident frequency.

$$\frac{x_{1w_1} + x_{2w_2} + x_{3w_3} + x_{4w_4}}{w_1 + w_2 + w_3 + w_4}$$

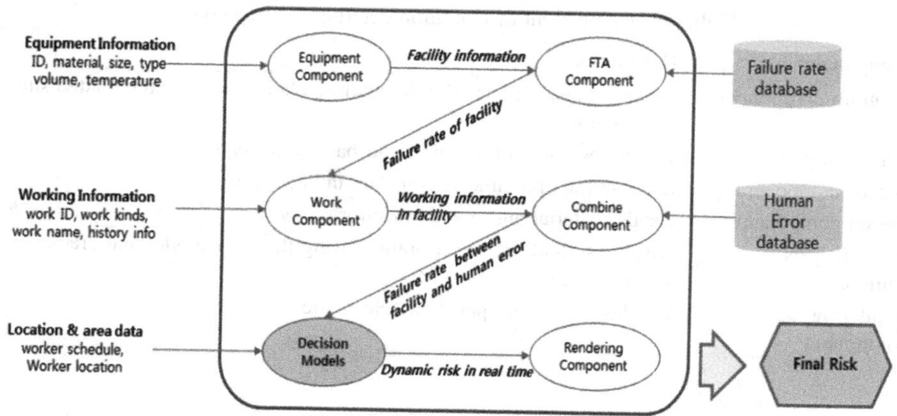

Fig. 2. Data flow of work analysis method

Table 3. Component of intelligent cognition method

Component	Description
Diagram component	Provides basic information by reading plan of site and the related site information
Zone component	Combines zone information and basic information
Sensor component	Sens various possible sensor data in zone area
Risk analysis component	Carries out intelligent risk analysis through similarity method
Rendering component	Display risk grade per zone rea in site

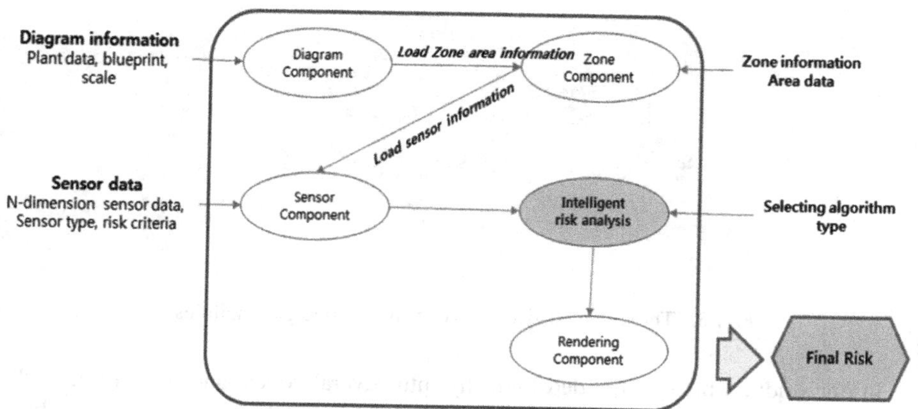

Fig. 3. Data flow of intelligent cognition method

Table 4. Component of condition deterioration method

Component	Description
Diagram component	Provides basic information by reading plan of site and the related site information
Zone component	Combines zone information and basic information
Sensor component	Sens various possible sensor data in zone area
Vessel component	Reads material and operating information
Risk analysis component	Analyzes condition deterioration using thinning model and crack severity index
Rendering component	Display risk grade per zone rea in site

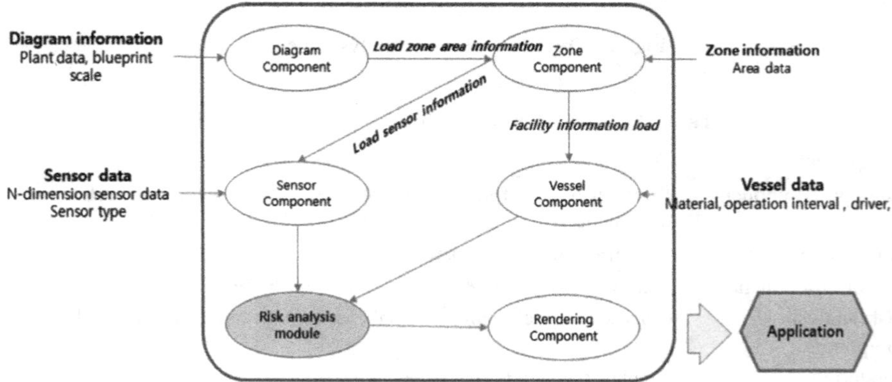

Fig. 4. Data flow of condition deterioration method

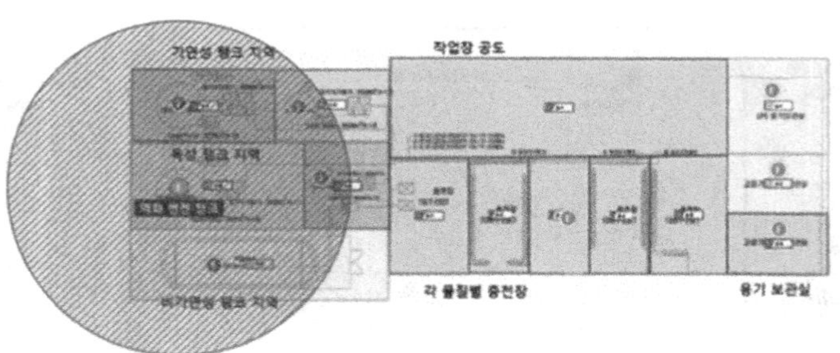

Fig. 5. The assigned risk per zone in multiple gas facilities

In our study, a plant is divided logically into several zones, and then a total risk, which is combined with our four different four risks, would give per a zone. Furthermore, the grade of risk is classified into five such as very high, high, normal, low, and very low. This grade of risk is expressed as color referencing API 581 standard code. For example, risk is expressed differently by zone area in Fig. 5.

3 Conclusion

This paper aims to propose deriving and simulating risk method around limited area of complex gas facilities. Because the core of our method is risk analysis, our study proposes total risk analysis that is composed four methods with individual point of view. Each of four methods are reflected by accident probability, work risk, unexpected peculiar situation, and condition deterioration respectively in order to consider various situations. The first method can analyze method accident possibility with probability and damage seriousness. The second method is able to digitize work risk of worker. The third method can recognize unexpected peculiar situation through deriving similarity among sensing data. The fourth method is able to decide the degree of condition deterioration by thinning model and crack severity index. Furthermore, this research is able to grant the independent grade of risk per logical zone area for simulating and predicting risk to complex gas faculties.

References

1. Moussa, M.I., Eid, T.: Risk management for chlorine producing factory in Egypt. Aust. J. Basic Appl. Sci. **1**, 239–248 (2007)
2. Daniel, A., Louvar, J.F.: Chemical Risk Analysis: Fundamentals with Applications. Prentice Hall, Englewood Cliffs (1990)
3. Zangle, C., Nannerer, J.: Data Mining Applications in the Petroleum Industry. IBM Round Oak Publishing, Thousand Oaks (2003)
4. Jhang, J.: Designing a cost effective and reliable pipeline leak detection system. Pipes Pipeline Int. **42**, 20–26 (1997)
5. Oh, J.S., Bang, H.J., Ko, H.: An empirical study on smart safety management architecture for gas facilities in Korea. Inf. Int. Interdisc. J. **15**, 1107–1122 (2012)
6. Oh, J.S.: Safety monitoring system design for LPG supply facilities. Asian Int. J. Life Sci. **12**, 167–177 (2015)

Body Gesture Modeling for Psychology Analysis in Job Interview Based on Deep Spatio-Temporal Approach

Intissar Khalifa(✉), Ridha Ejbali, and Mourad Zaied

Research Team in Intelligent Machines, National Engineering School of Gabes, Gabes, Tunisia
{intissar.khalifa.tn, ridha_ejbali, mourad.zaied}@ieee.org

Abstract. Social psychologists have long studied job interviews with the aim of knowing the relationships between behaviors, interview outcomes, and job performance. Several companies give great importance to psycho-test based on observation of the candidate is behavior more than the answers they even especially in sensitive positions like trade, marketing, investigation, etc. Our work will be a combination between two interesting topics of research in the last decades which are social psychology and affective computing. Some techniques were proposed until today to analyze automatically the candidate is non verbal behavior. This paper concentrates in body gestures which is an important non-verbal expression channel during affective communication that is not very studied in comparison to facial expressions. We proposed in this work a deep Spatio-temporal approach, it merges the temporal normalization method which is the energy binary motion information (EBMI) with deep learning based on stacked auto-encoder (SAE) for emotional body gesture recognition in job interview and the results prove the efficiency of our proposed approach.

Keywords: Non-verbal behavior · Body gestures · Deep learning EBMI · SAE

1 Introduction

Interpersonal Communication does not just mean words, on the contrary, thus the postures, the gestures, facial expressions [1], gaze direction and the intonation say many more than long speeches. Non-verbal communication experts (such as Mehrabian [2]) claimed that only 7% of the communication is verbal, 38% Para-verbal and 55% is non-verbal. There are four main categories of non-verbal behaviors [3] which are vocalics, proxemics, kinesics and haptics. In this paper we will concentrate in kinesics associated to upper body gesture [4] which is a real indicator for hiring decision because the negative emotions can be easily hidden facially by keeping a fake smile during an interview session but some errors in body language makes the candidate untrustworthy and unreliable, and haptics which is referred to the way that human communicates and interacts through using touching sense. Some self-touch behaviors are referred to as adaptors and they are generally related to negative feelings such as

J. H. Park et al. (Eds.): PDCAT 2018, CCIS 931, pp. 274–284, 2019.
https://doi.org/10.1007/978-981-13-5907-1_29

stress, psychological discomfort and anxiety that make a candidate less hirable than others.

1.1 Body Gesture Categories

Gestures are an essential component of interpersonal communication as they are used to interpret the vocal content and aid observer comprehension by reinforcing the attention, activating images in the listener's mind, and increasing the recall of what is being said. The gesture can be considered both from a physiological signal (result of reflex or voluntary muscle contractions), or from an interpersonal interaction (non-verbal, semiotic communication). We can for example classify the gestures according to the parts of the body involved [5]. There are generally three types of gestures which are: Head posture and facial expressions, gestures involving the whole body and gestures of the hand and arms: they form the main category of interactive gestures (semiotic function) and the main research work in this area is related to the recognition of hand positions [6–8], the interpretation of sign language and the development of human-machine interactions.

According to Gelder [9] 95% of the research was focused on facial expressions. Despite that body expressions have been relatively neglected in the field of affective computing, social studies have shown that body postures and gestures are significant sources for transmitting emotional information when other channels are inaccessible and inconsistent to the observers. Different classifications of gestures have been proposed by several researchers, (Kipp 2004) [10] used a set of six classes of gestures detailed in the work of (Ekman and Friesen 1969) [11] and (McNeill 2005) [12] which are: emblem, deictic, iconic, metaphoric, beat and the last one is the adaptor which is a non-communicative and spontaneous gesture such as touching an object or the own body (scratching head, touching nose, etc.) that can inform about the state of the speaker.

1.2 Body Gesture Cues

Body (arm, hand, shoulders…) has been used for gesturing to express the feelings and notify the thoughts. As a result of the feedback obtained from the work of Ekman and Friesen [11], Burgoon [13] and Coulson [14], Gunes [15] in his work identified the correlation between body gestures and the emotional state categories as presented in Table 1 and he took into consideration the relationship between the body parts (hands, hands and face, hands face and shoulders). Furthermore he added to the six basic emotions (happiness, surprise, fear, anger, sadness and disgust) other secondary emotions categories which are boredom, anxiety and uncertainty that can be real indicators and useful in our case which is the decisional review in job interview.

Table 1. Body movements related to emotions

Emotions	Body gesture
Happiness	Hands made into fists and kept high, hands clapping
Sadness	Body shift- forward leaning trunk, covering the face with two hands, dropped shoulders
Anger	Crossing the arms, finger point with right/left hand, lift the right/left hand up
Disgust	Hands covering the neck, backing, right/left hand on the mouth
Surprise	Self-touch, two hands covering the cheeks or the mouth
Fear	Covering the body parts, arms around the body/shoulders
Boredom	Move to the right/left, hands below the chin, elbow on the table
Anxiety	Tapping the tips of the fingers on the table, hands pressed together in a moving sequence, biting the nails
Uncertainty	Palms up and shoulder shrug, right/left hand touching the chin, touching the neck, touching the forehead

2 Related Work

For decades the gesture recognition is a major problem in the computer vision studied by several researchers and each proposed his own approach in order to apply it in a field such as robotics, remote control, interpretation of feelings, recognition of sign language, manipulation of virtual objects, etc. The techniques used are very numerous, we cannot limit them and each author can intervene in a specific stage: detection, tracking, classification, analysis and interpretation.

The most existing methods for emotion recognition were presented in two main steps: Firstly, how can we extract the features related to specific non-verbal cues? Secondly, the application of methods and algorithms related to machine learning like the use of Hidden Markov Model (HMM), Support Vector Machine (SVM), probabilistic graphical models or topic models for training and then for classification.

With the big progress of automatic markerless, motion capture, Kinect sensor and from RGB-D images the parts of body can be classified, these techniques were used in gaming applications when the full body is visible, the gestures are very clear and the behaviors of users were adapted to these applications. We talk for example about the body-controlled console games that were made by Savva et al. [16]. However, Kinect can fail in the registration of naturalistic and spontaneous body gesture that express emotions.

Chen et al. [17] in their work described the dynamics of body gesture so they applied like a first way the Image-HOG for feature extraction and then the combination between Histogram of Oriented Gradients (HOG) and the Motion History Image (MHI), and for the classification SVM was used.

Gunes and Piccardi [18] proposed the technique of Maximum voting of Apex frames for candidate key frame extraction then the features were extracted through optical flow, edge, geometry and the comparison to the neutral frame, then for classification they used HMM, SVM and Random Forest (RF).

In their another work [19]: the body motion modeling is a combination between color and silhouette based body models for upper body location, then a background subtraction for segmentation and Camshift technique for hands tracking were applied and for the best recognition results they used BayesNet classification algorithm.

The rest of the paper is presented as follows: In Sect. 3 we will describe in detail the architecture of our approach and then we move in Sect. 4 to present our experimental results while indicating the strong and weak points of this approach to finish in Sect. 5 with a conclusion and ideas for future work.

3 Proposed Deep Spatio-Temporal Approach

3.1 Overview of Our Method

In this paper we propose a concept of representation to construct an informative and discriminative semantic overview for body gesture recognition system. By exploiting the fact that hands are the part with more energy in the candidate is upper body in job interview video sequence and based in the strong relation between hands and head to express the spontaneous gestures, we will move after key frame extraction to a pretreatment which is the skin color segmentation binary image that will be after that accumulated by EBMI to obtain a coherent local motion descriptor for each body gesture, and then these images results will be the input of SAE that allows learning high level information from a large number of unlabeled image patch and it is basically different from a number of current methods like KNN, SVM and NN that need before the use of it the extraction of features (edge, texture, color, etc.), we talk about handcrafted features. An overview of our architecture is presented in Fig. 1:

3.2 Apex Frame Extraction

Video summarization is an interesting research domain in computer vision for video browsing and content retrieval. The summarized video increases the efficiency in extracting the key frames from each video. The techniques used for extraction are various and based in the features, they are divided into two categories:

- High level features we talk for example about Pair wise pixel comparison, Likelihood ratio or Motion Capture Data by Curve Saliency [20].
- Low level features which are the color, edge or block correlation [20].

In a long video sequence related to job interview, we needed unless the candidate key frames which are the most informative and meaningful for our psychology recognition. So, in this step we adopted to use the Color histogram difference method [21] which is a simple and robust method for key frame extraction. The idea behind this method is that two frames with unchanging background and unchanging moving objects will have little difference in their histograms. The color histogram difference D (i, i + 1) between two consecutive frames X_i and X_{i+1} can be calculated using:

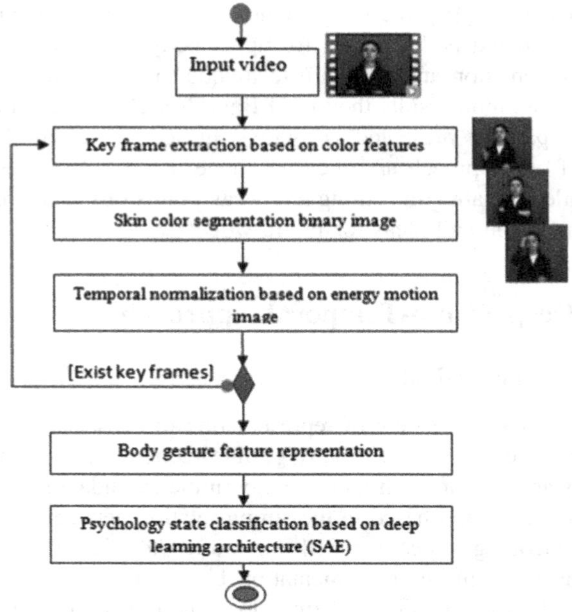

Fig. 1. Overview of our architecture

$$D(i, i+1) = \sum_{k=1}^{n} \frac{|H_i(k) - H_{i+1}(k)|^2}{H_{i+1}(k)} \tag{1}$$

Where H_i and H_{i+1} stand for the histogram of X_i and X_{i+1}. A shot transition occurred when the difference is bigger than a given threshold which is the key of this method, in our case it was the sum between mean (X) and standard deviation std (X).

3.3 Temporal Normalization for Body Gesture Representation

By exploiting the fact that hands and head have more energy and they are the rapidest part in job interview video sequence in a seating position of a candidate, we adopted to use our adaptive energy motion information which is a commonly used spatio-temporal representation of movement: it converts the 3D space-time information into 2D intensity image, so it is more robust than other information because of the use of depth information and the modeling of body gesture made it possible. This method cumulated the motion energy images M which were obtained through calculating the difference between consecutive key frames ($F_{j+1} - F_j$) [22] into a single binary image using the Eq. (2) and the result will be as shown the Fig. 2.

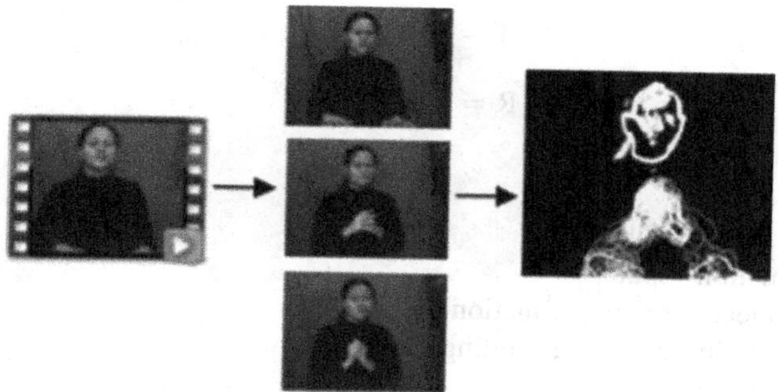

Fig. 2. The result of application of EBMI in video sequence for anxious candidate

$$\mathrm{EBMI}(x, y) = \sum_{i=1}^{n} M_{xy}(i) \tag{2}$$

Where:

EBMI (x, y): The energy of binary image result.
$M_{xy}(i)$: The energy motion image at time i.
n: Total number of key frames.

3.4 Psychology State Classification Based on Deep Learning Architecture

We used in our work SSAE (Stacked Sparse Auto-encoder) [23] that benefits from all the advantages of a deep network. The input layer of our SAE consists in transforming the input image X_n into the corresponding representation $\hat{X}_n$ and the hidden layer H presents the new presentation of the different functionalities of the input figure patch. SAE as shown Fig. 3 minimizes the gap between the input images and its reconstruction in order to find the optimal parameters. For each image input x from the dataset we have:

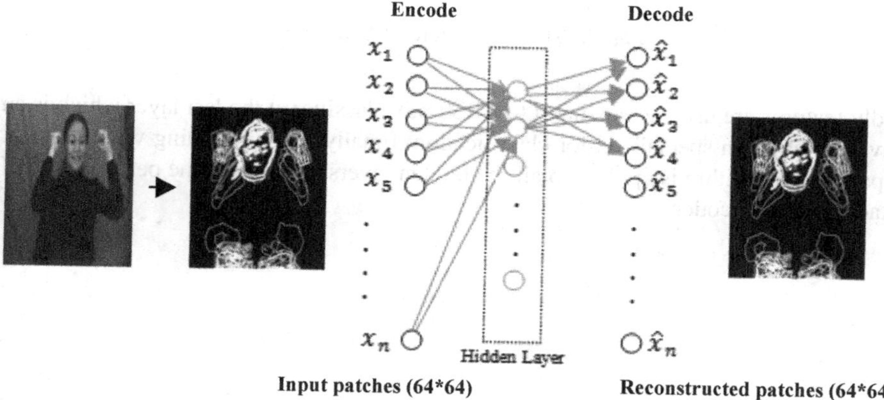

Fig. 3. The principle of auto-encoder

$$T = f_e(xW_e + a) \tag{3}$$

$$R = f_d(TW_d + b) \tag{4}$$

Where:

T: transformation of the input
R: reconstruction input
f_d, f_e: nonlinear activation function
W_e, W_d: encoding and the decoding weight
a, b: decoding bias.

We used a SSAE [24] with 3 hidden layers as illustrated in Fig. 4. First, the raw input is entered into the stacked trained auto-encoder to obtain the primary feature activations $h^{(1)}(x)$ for each of the input images x. Then, these primary features are used as the input to another auto-encoder to learn secondary features $h^{(2)}(x)$ of these primary features. After that, the secondary features are entered into the third auto-encoder to get the third feature activations $h^{(3)}(x)$ for each of the secondary features $h^{(2)}(x)$. Those

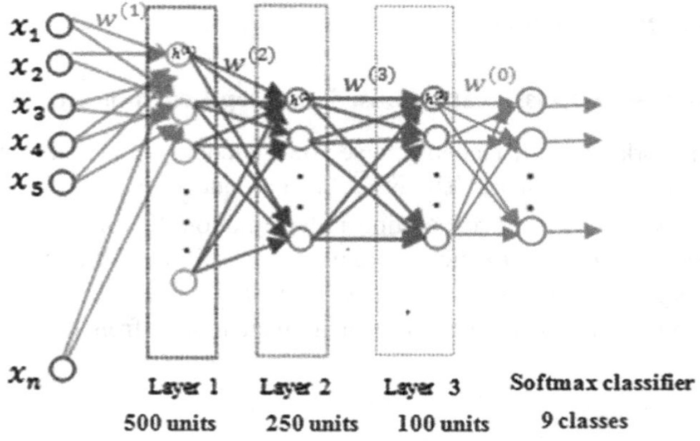

Fig. 4. The architecture of SAE used

thirdly features are used as raw input to a Softmax classifier at the last layer which is an activation function specialized for classification. Finally, the fine-tuning with the back propagation algorithm is applied to all the hidden layers to improve the performance of a stacked auto-encoder.

4 Experimental Results

In order to evaluate the efficiency of emotional body gesture recognition approach, a public dataset was used which is FABO dataset [15]: Bimodal face and body gesture database suitable for use in automatic analysis of human non verbal affective behavior

Fig. 5. Body gesture expressions of uncertainty person

because of its variety and richness of content. It is not limited in basic emotions but there are other non basic emotions like uncertainty, anxiety that express the discomfort in job interview. As indicated in Table 1 the same emotion can be expressed differently with body gestures. In our case we have 9 classes of emotions for 10 persons and each class contains subclasses (2 to 5 expressions of body gesture) in general we have 120

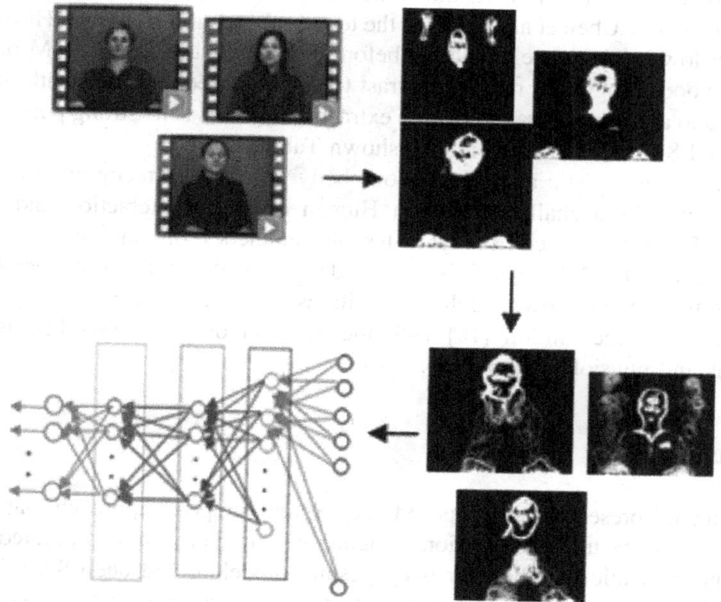

Fig. 6. Illustration of our body gesture modeling and recognition

videos (80 videos for train and 40 for test) and each video contains 70-300 frames. The following Fig. 5 represented an example of "uncertainty emotion" that is expressed differently:

The result of our approach's implementation can be summarized in the Fig. 6. We began with a key frame extraction, then skin color segmentation binary image to eliminate the parts that disrupt the good classification, next we moved to the representation of body gesture with EBMI that will be the input of SSAE for emotional body gesture classification.

Table 2. Comparison between our method and other methods for monomodal affect recognition based on body gesture

Methods	Emotions	Classification rate%
Chen et al. [17]: MHI + HOG + SVM	**9**	67%
Chen et al. [17]: HOG + SVM	**9**	64%
Gunes [18]: BayesNet	**12**	73.2%
Gunes [18]: Random Forest	**12**	76.87%
Wang et al. [4]: RF	**4**	72%
Wang et al. [4]: Adaptive RF	**4**	77.33%
Our method	**9**	81%

In comparison to the results found in other works for monomodal affect recognition based on body gestures [4, 17, 18] the video based accuracy is more performed than frame based accuracy that used candidate key frames without expression of dynamics, and even others like Chen et al. [17] used the temporal normalization with HMI, the use of HOG for low level feature extraction before the classification with SVM or RF did not lead to good recognition rate; in contrast to our proposed approach with SSAE we did not need to apply hand crafted feature extraction and this is its strong point and with it we reached 81% classification rate as shown Table 2:

Because of naturalistic and spontaneous body gesture, the recognition of the real meaning of it stills a challenge task in Human computer interaction and affective computing. So that, we need firstly to test the efficiency of our approach in other datasets like EMO FBVP which is a new dataset made in the last year but it is unavailable for use until now, and then we will test the performance of our approach in multimodal affect recognition [18] with the addition of other cues like the facial expressions and physiological signals.

5 Conclusion

In this paper we presented our proposed deep Spatio-temporal approach that could be useful in activity, gesture and emotion recognition. Our approach incorporated both the temporal normalization method for body gesture modeling and deep SAE for classification and the results obtained were important in comparison to previous works and it could be in our next work tested by adding other cues like the vocalic and physiological signals.

Acknowledgment. The authors would like to acknowledge the financial support of this work by grants from General Direction of Scientific Research (DGRST), Tunisia, under the ARUB program.

References

1. Afdhal, R., Ejbali, R., Zaied, M.: Emotion recognition using the shapes of the wrinkles. In: The 19th international Conference on Computer and Information Technology ICCIT (2016)
2. Mehrabian, A.: Communication without words. Psychol. Today **4**, 53–56 (1968)
3. Nguyen, L.S., Frauendorfer, D., Mast, M., Gatica-Perez, D.: Computational inference of hirability in employment interviews based on non verbal behavior. IEEE Trans. Multimed. **16**, 1018–1031 (2014)
4. Wang, W., Enescu, V., Sahli, H.: Adaptive real-time emotion recognition from body movements. ACM Trans. Interact. Intell. Syst. **5**, 18 (2015)
5. Abrilian, S.: Représentation de Comportements Emotionnels Multimodaux Spontanés: Perception, Annotation et Synthès., Thèse en informatique de l'Université Paris (2007)
6. Liang, H., Zhao, Y., Wei, J., Quan, D., Cheng, R., Wei, Y.: Robust hand detection and tracking based on monocular vision. In: IEEE International Conference on Intelligent Human-Machine Systems and Cybernetics (2014)
7. Bouchrika, T., Zaied, M., Jemai, O., Amar, C.B.: Neural solutions to interact with computers by hand gesture recognition. Multimed. Tools Appl. **72**, 2949–2975 (2014)
8. Khalifa, I., Ejbali, R., Zaied, M.: Hand motion modeling for psychology analysis in job interview using optical flow-history motion image (OF-HMI). In: The 10th International Conference on Machine Vision ICMV (2018)
9. de Gelder, B.: Why bodies? twelve reasons for including bodily expressions in affective neuroscience. Philos. Trans. Roy. Soc. B: Biol. Sci. **364**, 3475–3484 (2009)
10. Kipp, M.: Gesture Generation by Imitation From Human Behavior to Computer Character Animation. Dissertation.com, Boca Raton (2004)
11. Ekman, P., Friesen, W.V.: The repertoire of nonverbal behavior: categories, origins, and coding. Semiotica **1**, 49–98 (1969)
12. McNeill, D.: Gesture and Thought. The university of Chicago Press books, Chicago (2005)
13. Burgoon, J.K., Jensen, M.L., Meservy, T.O., Kruse, J., Nunamaker, J.F.: Augmenting human identification of emotional states in video. In: International Conference on Intelligent Data Analysis (2005)
14. Coulson, M.: Attributing emotion to static body postures: recognition accuracy, confusions, and viewpoint dependence. J. Nonverbal Behav. **39**, 117–139 (1992)
15. Gunes, H., Piccardi, M.: A bimodal face and body gesture database for automatic analysis of human nonverbal affective behavior. In: The 18th International Conference on Pattern Recognition ICPR (2006)
16. Savva, N., Bianchi-Berthouze, N.: Automatic recognition of affective body movement in a video game scenario. In: Camurri, A., Costa, C. (eds.) INTETAIN 2011. LNICST, vol. 78, pp. 149–159. Springer, Heidelberg (2012). https://doi.org/10.1007/978-3-642-30214-5_17
17. Chen, S., Tian, Y., Liu, Q., Metaxas, D.N.: Recognizing expressions from face and body gesture by temporal normalized motion and appearance features. J. Image Vis. Comput. **3**, 175–185 (2013)

18. Gunes, H., Piccardi, M.: Automatic temporal segment detection and affect recognition from face and body display. IEEE Trans. Syst. Man Cybern. **39**, 64–84 (2009)
19. Gunes, H., Piccardi, M.: Bi-modal emotion recognition from expressive face and body gestures. J. Netw. Comput. Appl. **30**, 1334–1345 (2007)
20. Kumthekar, A.V., Patil, J.K.: Key frame extraction using color histogram method. Int. J. Sci. Res. Eng. Technol. **2**, 207–214 (2013)
21. Shi, Y., Yang, H., Gong, M., Liu, X., Xia, Y.: Fast and robust key frame extraction method for video copyright protection. J. Electr. Comput. Eng. **3**, 1–17 (2017)
22. Liang, B., Zheng, L.: Gesture recognition from one example using depth images. J. Lect. Notes Softw. Eng. **1**, 339–343 (2013)
23. Hassairi, S., Ejbali, R., Zaied, M.: Supervised image classification using deep convolutional wavelets network. In: 27th International Conference on Tools with Artificial Intelligence ICTAI (2016)
24. Hassairi, S., Ejbal, R., Zaied, M.: A deep convolutional neural wavelet network to supervised Arabic letter image classification. In: 15th International Conference on Intelligent Systems Design and Applications ISDA (2015)

An Optimized Regularization Method to Enhance Low-Resource MT

Yatu Ji, Hongxu Hou(✉), Ying Lei, and Zhong Ren

College of Computer Science Inner, Mongolia University, Hohhot 010021, China
cshhx@imu.edu.cn

Abstract. Overfitting caused by scarce parallel corpus is a serious problem in low-resource machine translation task, resulting in the weak generalization ability of translation models. Dropout and Dropconnect can address this issue by reducing training neurons or weights randomly with increasing the generalization ability. In this paper, we optimize Dropconnect by adopting Gaussian approximation in the Bernoulli distribution in low-resource machine translation tasks, and make an integration to alleviate the uneven sampling effect in Dropout and Dropconnect, especially the inadequate training problem. It is an effective approach to approximate mask calculations to linear operations while being fully trained. An interesting finding is that the adhesive language is more sensitive to our regular methods. Our approach outperforms the Dropout and Dropconnect for low-resource translation tasks.

Keywords: Low-resource machine translation · Over-fitting
Uneven sampling · Regularization method

1 Introduction

Neural Machine Translation (NMT) has replaced Statistical Machine Translation (SMT) as the main model for commercial translation systems [1] due to its controllable architecture and the certain advantage of dealing with the complex data processing tasks. However, with the complication of task such as the ever-expanding network and parameter scale have become the issue that researchers have to consider, and data sparsity can result in overfitting problems for low-resource tasks easily [2, 3]. The usual practice is to use regularization method, such as L1, L2 and Bayesian [4]. Dropout [3] and Dropconnect [5] was proposed by randomly selecting neurons or weights in the network to participate in each epoch of training by binomial probability distribution, effectively solves the overfitting in model training and increases the generalization ability through a randomly distributed mask matrix. However, Srivastava et al. [3] and Wan et al. [5] proved unreasonable random sampling problem has important influence, which leads to uneven distribution of the mask. Insufficient mask operation makes training lack of randomness. As a result, the model cannot be fully trained, obviously in the task with data sparsity.

In this paper, Dropconnect is used for machine translation tasks rather than image recognition. We optimize Dropconnect (marked as G～Dropconnect, G refers to

J. H. Park et al. (Eds.): PDCAT 2018, CCIS 931, pp. 285–295, 2019.
https://doi.org/10.1007/978-981-13-5907-1_30

Gaussian), and according to the proportions randomizes the weights and nodes, integrate Dropout and G ~ Dropconnect (we call it *integration*) with the feature of optimized distribution to achieve better generalization. The dimensionality of the optimization calculation and the *integration* is the main problem that we encountered during the experiment. Based on the central limit theorem, we optimize the dimension and vector calculations of the matrix computation space for the sampling and masking process. This method can reduce the complexity associated with sparse data in vector calculations effectively.

In the rest of the paper, we first review Dropconnect in Sect. 2 and introduce the method architecture in detail in Sect. 3. In Sect. 4, we introduce the optimization method. Experimental setup, results and implementation tricks are discussed in Sect. 5. We conclude the whole paper and look forward to the future research in Sect. 6.

2 Dropconnect

Dropconnect (see Fig. 1) is a regularization method similar to Dropout in that it differs in the area of action that Dropconnect acts on weights.

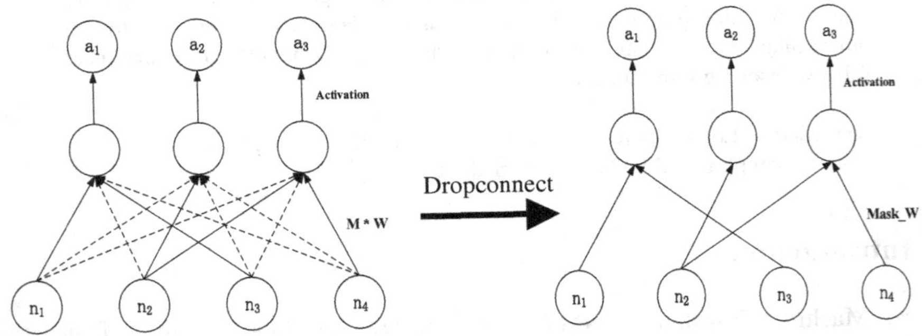

Fig. 1. Dropconnect is similar to Dropout as it introduces dynamic sparsity within the model, but differs in that the sparsity is on the weights W (denote by M*W), rather than the output vectors of a layer. Mask_W represents the weight matrix after the mask operation, where Mask is a binary matrix encoding the connection information and Maskij ~ Bernoulli(p).

The strategy randomly selects weights to generate different subnets to participate in the training. Due to masking weights is a Bernoulli stochastic process, it can give a strong ability to generalize which by training different subnets. For a Dropconnect layer, the output is given as:

$$Mask_i^l \sim Bernoulli(p) \tag{1}$$

$$r_i^l = \left(Mask_i^l * W\right) * X + b_i^l \tag{2}$$

$$output_i^l = a(r_i^l) \tag{3}$$

Where a() denotes the activation function (we use Relu) and * denotes element wise product and each element of the mask M is drawn independently for each example during training. Additionally, the biases (denote by b_i^l) are also masked out during training.

Dropconnect has some improvement in sampling relative to Dropout. However, insufficient training affects the quality of translation for low-resource task directly. Based on the above, we have integrated the optimized methods.

3 Model Description

We consider a standard model architecture composed of four basic components (see Fig. 2):

(1) Embedding Layer: we use Word2Vec to vectorize the input to 512 dimensions.
(2) Regularization Layer:
  (a) G ~ Dropconnect: output (w) = a(r) = a ((Mask * W) * X + b) where X is the output of the embedding layer, W is a fully connected weight matrix, a() is a non-linear activation function which used Relu and M is the binary mask matrix.
  (b) Dropout: output(x) = a(r) = a (W * (Mask * X)), the main definition is the same as G ~ Dropconnect layer, except that Mask acts on X and a differ of Weight matrix W and X comes from G ~ Dropconnect.
(3) Softmax Layer: prediction = s (output(x); W_s) takes as input output(x) and uses parameters Ws to map this to a k dimensional output (k being the number of classes).
(4) Cross Entropy Loss: A(y, prediction) = $-\sum y \log(prediction)$ takes probabilities prediction and the ground truth labels y as input.

Different from Dropconnect in image classification task, the *integration* can apply their respective advantages in regular process effectively by adjusting the probability. Here, we select the probability p randomly in the Bernoulli distribution as the threshold, the choice of p depends on the corpus scale and complexity of network [6, 7]. Among them, data sparsity is more significant [8, 9], it is the most important factor causing overfitting of low-resource tasks, which can be mitigated by random nodes and weights [10, 11]. Of course, we can shield certain strategies by controlling p.

By sharing parameter, Dropout can make the calculations between neurons independent, which is more suitable for large networks, while G ~ Dropconnect can train the network more comprehensive. Then we try to integrate them with the regularization constraint, but require increasing the corresponding p value respectively.

We note that Bernoulli sampling of the weights between neurons in the hidden layer is calculated by the dot multiplication of the mask matrix. Since the vectorize input has changed (dimension of the word vector is usually proportional to the vocabulary size, denoted as d), Mask is (m*d) and the weight matrix dimension is

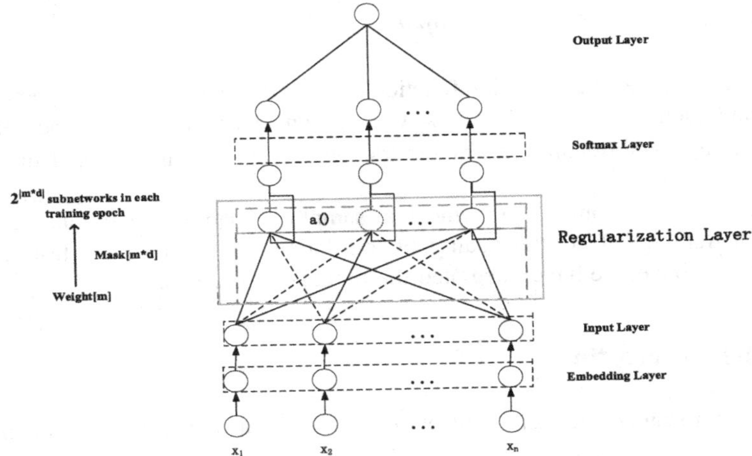

Fig. 2. To facilitate these connections, all sub-layers in the model, as well as the embedding layers, produce outputs of the same dimension. In particular, dimensionality is unified through the Mask between the Dropout and G ~ Dropconnect.

denoted as m. The output is a mixture of $2^{|m*d|}$ different networks. Therefore, this method enhances the generalization ability of the model greatly.

When the mask operation is completed, the activation value is obtained through the linear and non-linear mapping. If a further constraint is required, the mask calculation of the Dropout layer will perform on the output of G ~ Dropconnect to prevent over-fitting. When back propagating gradients, the weight matrix and nodes which undergone mask operation in the feedforward also need to perform mask calculations on the corresponding weights and nodes, so as to ensure that the same subnets can participate in training and the node's random distribution. A summary of these steps is provided in Algorithm.

Algorithm Training process

Input: Select sample $x_i(i \in (1,n))$ according to mini-batch, weight $W(W_s, W_d)$, learning rate r, target $y_i \in Y$

Output: Updated Mask_W , A'
Feedforward calculation

Embedding : Sample vectorization

Weight mask : $M \sim Bernoulli(p)$, $Mask\_W = M * W$

Nonlinear mapping : $O_a = a(Mask\_W * X) || O_a = a(W * X)$

Neuron mask : $O_a = M * a(W * X)$

Softmax : $O = s(O_a, W_s)$,s = softmax()

Loss : $A(y, O) = -\sum_{i=1}^{k} y_i * \log(o_i)$

Gradient update : $A' \sim (A'_S, A'_w)$

Backward propagation
Softmax Layer: $W_s = W_{s-r} * A'_S$

Weight mask : $Mask\_W = M * W$

---

4 Matrix Calculation Optimization

Considering the computational complexity during the training, model needs to be simplified to some extent. Therefore, an attempt is to calculate the mask by computationally much easier sampling methods. At the same time, the complexity of the matrix is also an important factor which should be considered [12].

The Sect. 2 mentioned that if the dimension of mask matrix is (m*d), the corresponding random subnet number is $2^{|m*d|}$ species. The usual practice in matrix calculations is to amplify the input by (1 − p) magnification when training the model or to reducing the corresponding weight by the factor of p in the test [5, 13]. It means that when model masks the weights with p, the weight is amplified by $1/(1-p)$ times, or it is traversed by the mask and then averaged, which is called rescale in the image recognition and classification task [14, 15]. Such a calculation not only makes the matrix very large, but also loses its original purpose of simplifying the model.

The Central Limit Theorem [16] shows that we can approximate the joint distribution results by combining the mean and variance of each dimension with a normal distribution. Specifically, if the sampling size is n and the prediction output denotes by o, the Eq. 4 can be obtained according to Eqs. 2 and 3.

$$o \sim y_i = \sum\nolimits_j Mask\_w_{ij} * x_j \sim N\left(\mu, \sigma^2\right) \tag{4}$$

$$\mu \sim E_m(o) = p * Mask\_w * x \tag{5}$$

$$\sigma^2 \sim V_m(o) = p(1-p) * (Mask\_w * Mask\_w)(x * x) \tag{6}$$

Therefore, after n times of sampling, the mask calculation of the prediction output is to approximate by using Eqs. 4, 5 and 6, which simplifies the calculation of the mask greatly and ensures the rationality of probability distribution. The complexity of the G~Dropconnect under Gaussian sampling is shown in Eq. 7. It can be seen a significant improvement over Dropconnect is that the complexity after distribution is approximated to a linear function about p, where n and d indicate the input and output dimensions, V_s and V_d indicate the maximum number of weights involved respectively in training, $V_{embedding}$ represents the dimension of the word vector.

$$complexity_{Gaussion} = p\left(2dV_s n\sqrt{d}V_d\right) V_{embedding} \tag{7}$$

5 Experiments

5.1 Evaluation Process

The evaluation criteria of the experiment is training error (cross-entropy algorithm, cost function as shown in Eq. 8) and BLEU (Bilingual Evaluation Understudy) score, which are used to measure the quality of the translation model [17, 18]. Moreover, training error is obtained based on the loss function, and a corresponding parameter file generated during the training process. The BLEU score is evaluated using the IBM model.

$$c = -\frac{1}{n}\sum\nolimits_x [y \ln a + (1 - y)\ln(1 - a)] \quad (8)$$

$$bp = \begin{Bmatrix} 1 & if\ c > r \\ e^{1-r/c} & f\ c \leq r \end{Bmatrix} \quad (9)$$

$$BLEU = bp * \exp\left(\sum\nolimits_{n=1}^{n} w_n \log p_n\right) \quad (10)$$

Where y is the target output, a is the actual output of the neuron, and n is the number of samples which in sample capacity z. The two values which dependent on the calculation of the BLEU score are the precision p_n and the penalty term bp. If the length of the translation to be evaluated is denoted by c and r is the length of the reference, the corresponding formula is shown in Eqs. 9 and 10.

The effectiveness and robustness of the experiment are verified in the comparison between the three low-resource language translation tasks of the two models. The model uses an Encoder-Decoder architecture to reduce dependence on semantic relationship in the sequence [19] which can reduce the obstacle and workload of preprocessing effectively. We validate our proposed strategy on LSTM (Long Short-Term Memory) and Transformer, the gate settings can solve the long-distance dependence problem effectively in sparse corpus [20]. Additionally, attaching more attention to the model can pass a richer context from the encoder to decoder, especially for the adhesion language has a better semantic transmission. Transformer [21] gets rid of the limitations of network architecture and realizes parallel work to some extent, it also solves the semantic transfer in the sequence by capturing the position information. Meanwhile, it is more worthwhile to try for low-resource sequence tasks due to its better generalization [21].

We choose GPU to perform matrix operations, since the RNN is usually implemented through multiple loops with large dimension, and slower response time.

5.2 Dataset

The alignment corpus from the Mongolian-Chinese (M-C), Tibetan-Chinese (T-C) and Uighur-Chinese (U-C) machine translation track of the CWMT2018 evaluation campaign (Table 1). The corpus consists of sentence-aligned subtitles of daily language, news, government literature and some literary works. We pre-process the training data using the tokenizer of the Moses toolkit and remove sentences longer than 50 words as well as casing.

We also processed the case of Mongolian corpus. The advantage of doing so is to reduce the proportion of words with a frequency of 1(more in Mongolian-Chinese vocabulary) and alleviate the sparsity of corpus. Specifically and simply, the case with additional elements and word stem are segmented during encoding. The form of the case in the Mongolian corpus we identified and processed is shown in the Table 2.

Table 1. Statistical analysis of experimental corpus

	Training	dev	Test
M-C	201643	1001	1000
T-C	136580	729	1000
U-C	352523	1000	1000

Table 2. Additional elements of case in Mongolian

1	2	3	4	5	6	7
Attribute case	Position case	Object case	Rely case	From case	Add case	Joint case
[illegible]	[illegible]	[illegible]	[illegible]	[illegible]	[illegible]	[illegible]

5.3 Results

We choose the BLEU score corresponding to different regularization in 50 epochs. Dropout and G∼Dropconnect in the initial training can make the model get higher learning efficiency, while the BLEU of the *integration* is relatively low. With constant training, the BLEU score of the *integration* is improved significantly because it is more balanced sampled at the beginning of the training, meaning that only the original weights are involved, but with the number of epoch increased, the model gains a better generalization, as a result is a more accurate prediction. See Table 3 for BLEU score obtained by different regularization in LSTM and Transformer.

Table 3. The performance of different regularization in the LSTM*(Transformer)* model

LSTM*(Transformer)*	No-drop	Dropout	G∼Dropconnect	*integration*
M-C	24.8*(26.1)*	26.7*(27.2)*	27.2*(27.7)*	**28.1*(28.9)***
T-C	23.3*(23.9)*	24.1***(24.5)***	24.1***(24.5)***	**24.8***(24.3)*
U-C	21.7*(22.4)*	22.6*(22.2)*	22.4*(23.3)*	**22.7*(23.5)***

We used three different regularization methods, which is worth mentioning is the *integration* method, it will reduce the probability p value(we verified an optimal *integration*, Dropout:0.2 and G∼Dropconnect:0.1), but reason for this is that Dropout is superior in complexity, while G∼Dropconnect is more uniform and reasonable in the distribution of choices.

5.4 Analysis

We observed that the addition of regularization method obviously improves the overfitting problem of the model, as a result, there is an average of 1–2 improvements on the BLEU score in each language.

The convergence of dev error can usually reflect learning ability intuitively. Surprisingly, during this process, we find that the model has advance efficiency and a faster convergence rate in the previous iteration. It is not easy to fall into the local optimal and the same at inference time. We select the optimal error which the model with Dropout and Dropconnect in the three languages as experimental basis (the thin line in right figure) to verify the effect of the regular methods we optimized, as shown in Fig. 3.

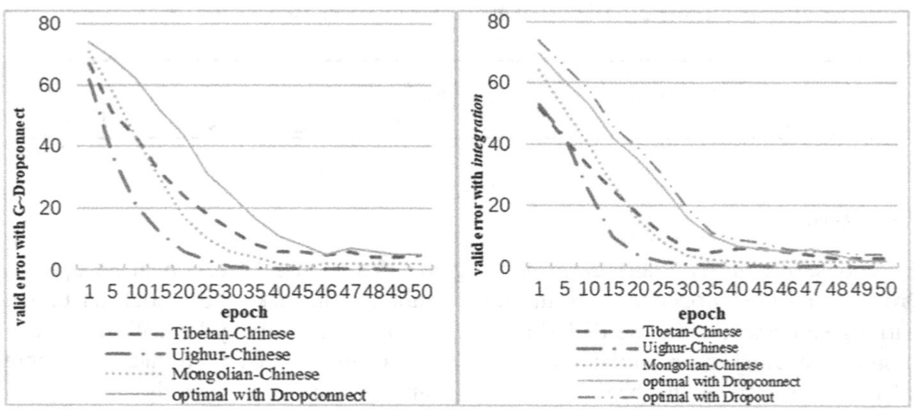

Fig. 3. Different regularization on the Transformer (We observed both models and found that Transformer is more sensitizer to regular methods, so more convincing.) valid error, the changes in the previous 40 epoch are noticeable and tend to be flat later.

Observed of convergence, it can be clearly seen that the regularization method in three tasks is effective for the baseline system, regardless of convergence speed or error value.

What's more interesting, it is more sensitive to the regular method and more obvious in the adhesive language (such as Mongolian and Uighur) task. The basis for this conclusion is that, in the convergence state of the model, the convergence speed of Mongolian-Chinese and Uighur-Chinese is better and the error rate is lower (see Fig. 3). In the final translation evaluation index, the corresponding BLEU value has been greatly improved (see Table 3). Due to the strong dependence of adhesive language on the relationship between words, the vocabulary is relatively sparse and the problem of overfitting is more prominent.

We also extracted a sentence in different language respectively to illustrate the influence of the regular method on the translation as shown in Table 4.

Table 4. Performance of translation under different regularization

	sentence	Mongolian-Chinese	Tibetan-Chinese	Uighur-Chinese
	source	[illegible]	[illegible]	شى جىنپىڭ مۇھىم سۆز قىلدى.
	target	在英国，来往车辆都靠左行驶，对吗？	她刚刚起床，头发乱七八糟的。	习近平发表重要讲话。
got	Non-regularization	在英国，都去左边车厢，应该去吗？	她刚刚才早上，头发七八个。	习近平提出了重要演讲。
	Dropout	在英国，都跟随左边车厢，应该跟随吗？	她刚刚才起来，头发乱了七八个。	习近平发表重要的演讲。
	G~Dropconnect	在英国，来回在左边的驾车，应该吗？	**她刚刚才起来，头发很了乱七八糟的。**	**习近平发表了重要的演讲。**
	Integration	**在英国，来回都在左边驾车，应该吗？**	她刚才起来，头发很乱了。	**习近平发表了重要的演讲。**

6 Conclusion

Our work is motivated by two major deficiencies in model generalization for low-resource machine translation: overfitting and insufficient training. However, regularization strategy can address these issues potentially by training the model with different subnets to get better generalization ability. Towards that end, we validated Dropout, G$\sim$Dropconnect and *integration* in LSTM and Transformer models, compared the regular effects in different low-resource language translation tasks. We also apply a variety of regularization in a more computationally intensive Gaussian distribution to approximate the matrix calculation that satisfies the Bernoulli distribution, and to perform sampling approximations based on the calculated mean and variance. Above process alleviates the computational complexity and overfitting caused by corpus sparsity.

Our contribution is a thorough empirical evaluation on the proposed strategies, we compare against the strong baseline. The results show that two models with the optimized strategy achieve much better accuracy compared to the baseline on the test set. The *integration* we construct has reduced the distribution probability p value respectively and used the same distribution approximation to calculate the mask of the network. It has obvious advantages in low-resource translation task with good computational complexity, and the algorithm is relatively stable. We also found that the adhesive language is more sensitive to the regular methods we optimized. There is reason to believe that such conclusions can provide reference for other similar adhesive language.

Finally, the computational efficiency and the optimization of the mask calculation are worth discussing. Although compared to the overfitting problem and the generalization ability of the model, the above two issues are not the primary to be solved for the Low-resource machine translation, it still imposes a burden on the training of the model. And we wish to approximate the results with a simpler calculation in matrix operations to achieve a simplified model in the future work.

Acknowledgments. We thank PDCAT-18 reviewers. This work is supported by Natural Science Foundation of Inner Mongolia (No. 2018MS06005), Mongolian Language Information Special Support Project of Inner Mongolia (No. MW-2018-MGYWXXH-302) and the Postgraduate Scientific Research Innovation Foundation of Inner Mongolia (No. 10000-16010109-14).

References

1. Liu, Y.: Advances in neural machine translation. J. Comput. Res. Dev. **54**(6), 1144–1149 (2017)
2. Lü, G., Luo, S., Huang, Y., et al.: A novel regularization method based on convolution neural network. J. Comput. Res. Dev. **51**(9), 1891–1900 (2014)
3. Srivastava, N., Hinton, G., Krizhevsky, A., et al.: Dropout: a simple way to prevent neural networks from overfitting. J. Mach. Learn. Res. **15**(1), 1929–1958 (2014)
4. Mackay, D.J.C.: Probable networks and plausible predictions—a review of practical Bayesian methods for supervised neural networks. Netw. Comput. Neural Syst. **6**(3), 469–505 (1995)
5. Wan, L., Zeiler, M., Zhang, S., et al.: Regularization of neural networks using dropconnect. In: International Conference on Machine Learning, pp. 1058–1066 (2013)
6. Zhou, W.H., Wang, A.H.: Discrete-time queue with Bernoulli bursty source arrival and generally distributed service times. Appl. Math. Model. **32**(11), 2233–2240 (2008)
7. Gal, Y., Ghahramani, Z.: Dropout as a Bayesian approximation: representing model uncertainty in deep learning. In: International Conference on International Conference on Machine Learning, pp. 1050–1059. JMLR.org (2016)
8. Shekhar, S., Xiong, H.: Model generalization. Encyclopedia of Gis, 682 (2013)
9. Xu, P., Jelinek, F.: Random forests and the data sparseness problem in language modeling. Comput. Speech Lang. **21**(1), 105–152 (2007)
10. Lecun, Y., Bengio, Y., Hinton, G.: Deep learning. Nature **521**(7553), 436 (2015)
11. Mitchell, T., Buchanan, B., Dejong, G., et al.: Machine Learning. McGraw-Hill, New York (2003)
12. Bahdanau D, Cho K, Bengio Y. Neural Machine Translation by Jointly Learning to Align and Translate [J]. Computer Science, 2014
13. Bianchini, M., Scarselli, F.: On the Complexity of Neural Network Classifiers: A Comparison Between Shallow and Deep Architectures [J]. IEEE Transactions on Neural Networks & Learning Systems **25**(8), 1553–1565 (2014)
14. Levy, O., Goldberg, Y.: Neural word embedding as implicit matrix factorization. Adv. Neural. Inf. Process. Syst. **3**, 2177–2185 (2014)
15. Dahl, G.E., Sainath, T.N., Hinton, G.E.: Improving deep neural networks for LVCSR using rectified linear units and dropout. In: IEEE International Conference on Acoustics, Speech and Signal Processing, pp. 8609–8613. IEEE (2013)
16. Ozonat, K.M., Gray, R.M.: Fast Gauss mixture image classification based on the central limit theorem. In: 2004 IEEE Workshop on Multimedia Signal Processing, pp. 446–449. IEEE (2005)
17. Kline, D.M., Berardi, V.L.: Revisiting squared-error and cross-entropy functions for training neural network classifiers. Neural Comput. Appl. **14**(4), 310–318 (2005)
18. Callison-Burch, C., Osborne, M., Koehn, P.: Re-evaluation the role of bleu in machine translation research. In: Proceedings of the Conference Eacl 2006, Conference of the European Chapter of the Association for Computational Linguistics, 3–7 Apr 2006, Trento, Italy, pp. 249–256. DBLP (2006)

19. Cho, K., Van Merrienboer, B., Gulcehre, C., et al.: Learning phrase representations using RNN encoder-decoder for statistical machine translation. Comput. Sci. (2014)
20. Dey, R., Salem, F.M.: Gate-variants of gated recurrent unit (LSTM) neural networks (2017)
21. Vaswani, A., Shazeer, N., Parmar, N., et al.: Attention is all you need (2017)

A Development of Architecture for Risk Area Prediction System

Ugiwyeon Lee and Jeongseok Oh(✉)

Korea Gas Safety Corporation, Eumseong, Republic of Korea
lyullee@gmail.com, jsoh90@gmail.com

Abstract. Risk area prediction system is a kind of safety management system that suggesting risk level to worker or humans resideing in chemical plant. This system is calculating safety distances based event consequence analysis according to API RP 581, EIGA Doc 75 standards, reference manual Bevi risk assessments. Safety distance could be estimated by calculating in sequential order of release module, dispersion module, and fire and explosion module. So, it needs to consider how to connect between each modules. Also, additional modules have to be developed to help derive safety distance and risk area from calculated results of fire or explosion effect modules. This paper introduce consequence analysis modules for estimating risk area from chemical plant data and chemical properties and architecture of connecting each modules. And risk gradation as a new risk expression concept will be provided for offering risk level to workers in chemical plant.

Keywords: Smart safety · Risk assessment · Risk prediction system

1 Introduction

In chemical plants, most chemical accidents begin with an incident, which is usually a leakage in hazard materials use facilities. And, gas leakage result in the loss of containment of the materials from the chemical process. If the material has hazard properties, which could increase a risk as fire accident, explosion, or toxic gas dispersion [1]. Risk assessment systems for analysis these chemical accident effects plays important rules to insure workspace safety and improve chemical process with respect to yield and quality [2]. That is why chemical accident analysis methods, as called consequence analysis, have been developed by many risk management institutions as Det Norske Veritas (DNV), Toegepast- Natuurwetenschappelijk Onderzoek (TNO), and Center for Chemical Process Safety (CCPS - AIChE) [1, 2].

This paper suggests what modules need to and how to organize and integrate the modules. To predict risk area, release rate or amount of a risk material has to be calculated firstly [1]. Discharge module plays a role in calculating leak rate or amount, and determining whether the phase of released material is vapor or liquid. Dispersion module estimates concentration of the released material in the air as distance from release point. Finally, fire and explosion module calculated the effects of fire or explosion generated from dispersed material cloud [1, 2]. Additionally, chemical database and safety distance module is required for providing basic chemical properties

J. H. Park et al. (Eds.): PDCAT 2018, CCIS 931, pp. 296–303, 2019.
https://doi.org/10.1007/978-981-13-5907-1_31

and deriving safety distance from fire or explosion effects, respectively. Also, event probability module is key requirement for total risk area calculation [1, 2].

Risk area can be calculated from area of circular formula with using safety distances as radius. If a user adopted fire and explosion thresholds from EIGA Doc 71 standard, the system could generate risk gradation as surface contours. In risk gradation, workers are possible to recognize their risk level through color density of their place [1, 2].

2 Architecture for Risk Area Prediction System

2.1 Input Values

Input values are categorized into two parts, "process & risk condition" and "chemical properties". Process & risk condition could be obtained from plant design data (plot plan, PFD, P&ID, HAZOP data, etc.) or experts [7]. Chemical properties depend on the kind of chemicals using in the process. Chemical database such as NIST chemistry webbook, OSHA Occupational Chemical Database, or yaws chemical handbook can provide required values in the system.

2.2 Material Discharge Section

In gas phase, discharged material is will dispersed into the air [7]. Dispersed flammable gas or toxic gas has a possible to cause major accidents [7]. In other case, when the phase of the material is liquid, the discharged material spill on the ground near leak point. If the material is being held in liquid phase, there could not be major accidents. But, liquefied gas in liquid phase is leaked, even if the discharged liquid keeps in form of liquid pool, the liquefied material will be vaporized to gas phase soon. Eventually, liquefied gases leakage is similar to gas leakage case. However, liquefied gas leakage is much more dangerous than gas leakage because evaporation rate generally is much more than gas leak rate [1, 2].

Thermodynamics module could calculate vapor-liquid ration and each mass in chemical process. Safety distance module estimates an enough distance for workers to be unaffected by fire or explosion hazards. On the other hand, workers could be in danger when they are located within the safety distance from hazard facility. In the same concept, risk area could be calculated from safety distance [1, 2].

Materials discharge section is comprised of flash module, vapor release module, liquid release module, pool spread diameter module, evaporation pool module, and boiling pool vaporization module as Fig. 1. Flash module analysis input data and predicts the phase of released materials. If the phase is predicted to vapor, the vapor release module will work. In opposite case, the liquid release module will work [1, 2].

All of equation in discharge section is referred guidelines for consequence analysis of chemical releases of center for chemical process safety (CCPS) in American institute of chemical engineers (AICHE).

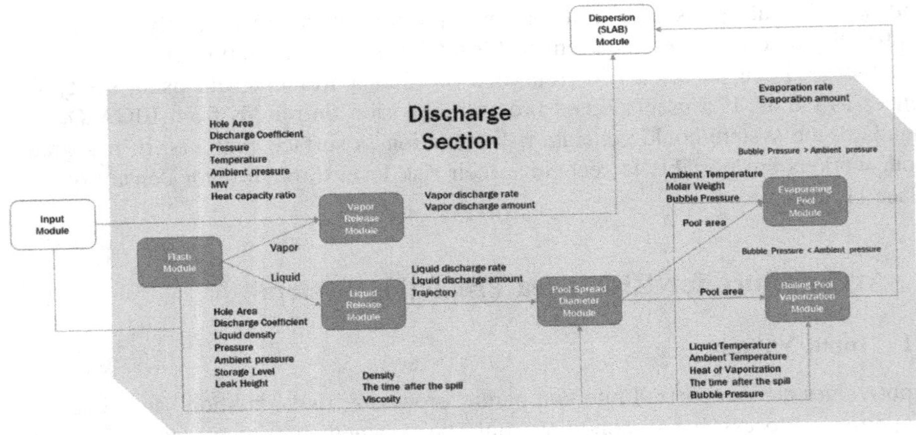

Fig. 1. The architecture of discharge section

2.3 Material Dispersion Section

Material dispersion section simulates the dispersion of flammable and toxic dense gas clouds [7]. This section receives and analysis discharge data from material discharge section and analysis, and estimate the dispersing area of released gas, concentration in the area, and dispersion rate. The output of this section will be used for calculating fire or explosion effects of dispersed material [5].

The core of this section is SLAB computer model. SLAB computer model is able to estimate dense gases dispersion mechanisms [5]. A dense gas is defined as a gas whose density is greater than that of the ambient air. Most gases that were used for chemical plants are belonging to dense gas [5]. When dense gases are initially discharged, these gases slump toward the ground and disperse both upwind and downwind. Moreover, dense gas mechanisms for mixing with air are markedly different from neutrally buoyant releases. SLAB model is one of the most commonly used for dense gases dispersion models and is core engine of typical risk assessment software, EFFECTS, made by TNO which is the Netherlands Organization for applied scientific research [5].

2.4 Fire and Explosion Module

Explosion section is comprised of 5 modules, VCE module, jet fire module, flash fire module, fireball module, and pool fire module. If a large amount of flammable vaporizing liquid or gas is instantaneously discharged, a vapor cloud is generated and disperses into the air. If the flammable cloud is ignited before the cloud is diluted below its lower flammability limit (LFL), a VCE or flash fire will occur. For this reason, VCE module and flash fire module required cloud dispersion data from dispersion section [1, 2] (Fig. 2).

Jet fire is a turbulent spread resulting from the combustion of a flammable material continuously released in a particular direction. Jet fire frame is not influenced by dispersion pressure, but discharge. Fire ball and pool fire is also directly affected by

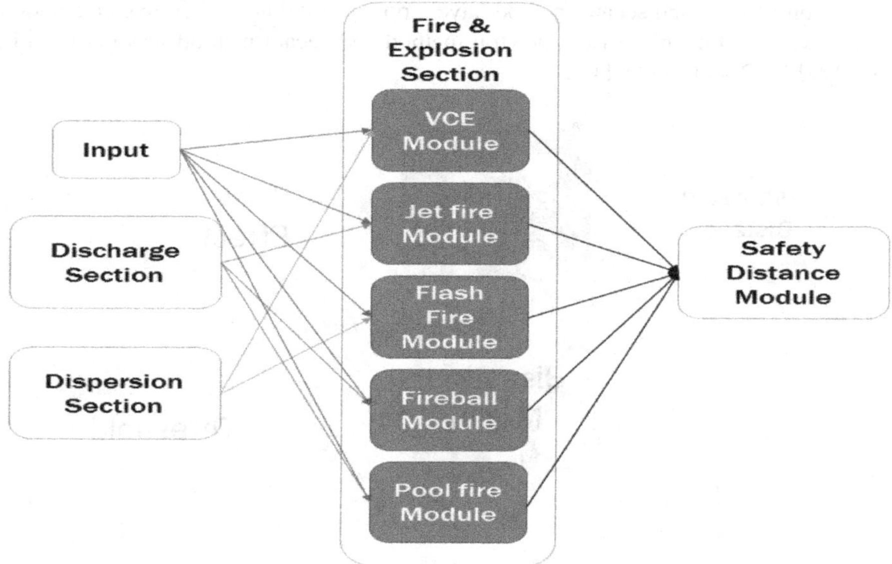

Fig. 2. The architecture of fire & explosion section

discharge section. Fire ball occur released gas cloud to ignite before dispersing into the air. Pool fire is generated from a flammable liquid pool. When a flammable liquid is released from process equipment, some of liquid may rainout onto the ground. The released materials are likely to form a liquid pool which will continue to evaporate or be fired. Pool fire or fire ball is not influenced by dispersion section because two events are occurred before the material is spread into the air. Results of this section are offered in the form of explosion pressure or radiant heat. But, risk area calculating needs a risk distance (or safety distance) which means the enough distance that humans or workers are not affected by explosion or fire from a specific location (loss of containment incident point in the chemical process). That is why additional module as Safety distance analysis module to estimate an appropriate distances from fire & explosion module results [1, 2].

2.5 Safety Distance Analysis Module

Safety distance calculating requires thresholds to find a proper distance to avoid damage from fire or explosion effects. As mentioned above, the result type of fire & explosion is explosion pressure or radiation [3]. Safety distance analysis module estimate safety distance corresponding with thresholds by using numerical (bisection) method [3].

Newton method could save more resources if it is possible to get the differential equation of the fire & explosion modules. However, newton method is not suitable in this system because most equations is to get differential equations too complex, and safety distance finding requires iterative calculation on too many equations [3]. Also, both newton method and secant method have a possibility that the iterative calculation is not converged. For this reason, newton method and secant method was not used in his study (Fig. 3 and Table 1).

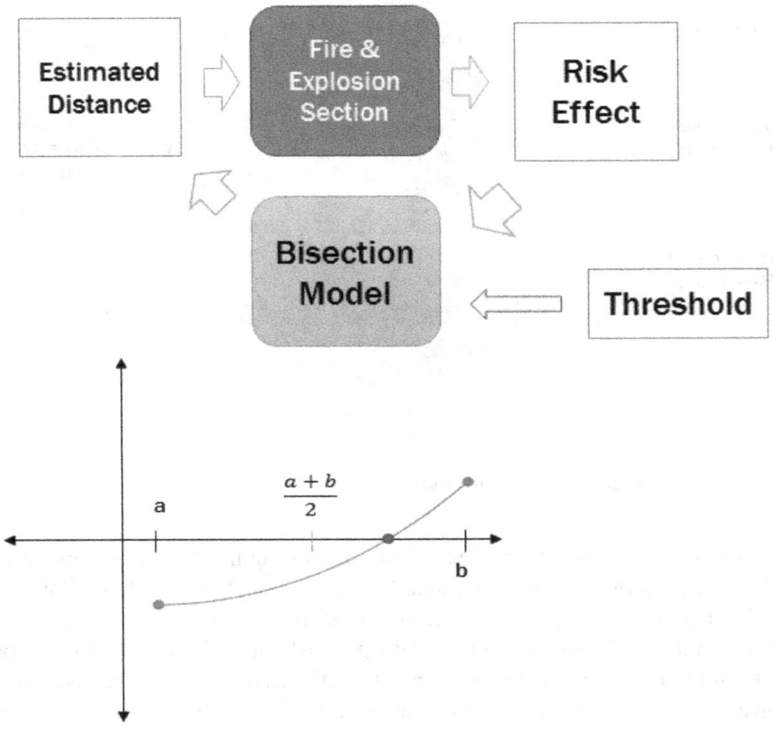

Fig. 3. The architecture of safety distance analysis module (left) and bisection method (right)

Table 1. Threshold limits for safety distance calculation (Ref. EIGA Doc. 75)

Thermal effects	Harm	No harm
Fire		
Heat radiation (Human)	9.5 kW/m2	1.6 kW/m2
Heat radiation (Machine)	37.5 kW/m2	15.8 kW/m2
Lower flammable limit	LFL 100%	LFL 50%
Explosion		
Pressure (Human)	7 kPa	2 kPa
Pressure (Machine)	30 kPa	20 kPa

For the purpose of estimating a safety distance, the severity can be defined for workers at two levels, harm and no harm. According to EIGA Doc 075, harm criteria is one that would cause serious distress, a high probability of a need for medical attention, probability of severe or fatal injury. No harm criteria is one that nearly all individuals could be exposed to without experiencing or developing irreversible or other serious health effects, or symptoms that could impair their abilities to take protective action.

2.6 Risk Area Calculation Module

Risk area could be calculated by using safety distance as radius with area of circular formation [6]. But, this system generates six risk areas of each risk type, so, there is necessary for six risk area to combine into one risk area as total risk area with using event probability (or incident occurrence rate) [6]. Total risk area could be estimated as the sum of safety distance multiplied by event probability values as weight value [6].

Event probability values could be gained from Reference Manual Bevi Risk Assessments introduction (RIVM 2009), or Guidelines for Determining the Probability of Ignition of a Released Flammable Mass (CCPS 2014).

Finally, Architecture of all modules integration is established as Fig. 4.

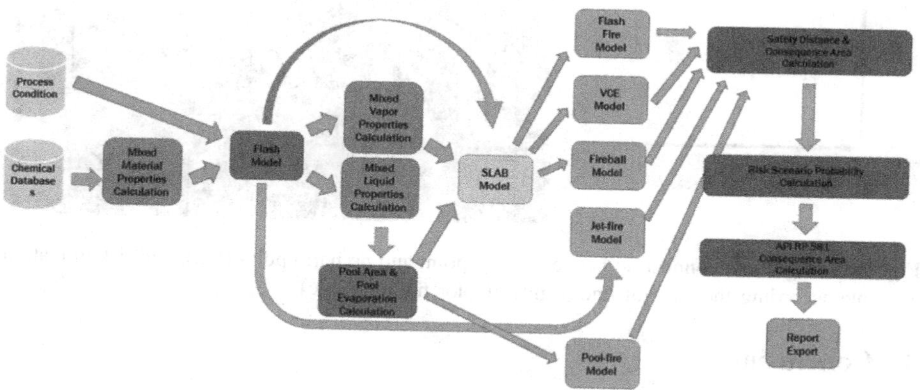

Fig. 4. The architecture of connected modules

3 Risk Area Prediction System Results

3.1 Risk Assessment Modules Connection

This system is possible to automatically generate risk area prediction report in Microsoft Excel form. The report contains release rate, total release amount by gas or liquid phase. Also, safety distances, risk areas, event probabilities are provided for each risk scenario as flash fire, vapor cloud explosion, jet fire, and fire ball. Additionally, user could check dispersed material concentrations in the air as distances from leak point.

3.2 Risk Area Gradation

According to EIGA Doc 75, this system is possible to determine two kinds of risk area, harm and no harm. Harm risk area is corresponding to approximation value of 1% probability of fatality to a general population and no harm risk area is corresponding to approximation value of 0.1 probabilities of that. If two thresholds are connected as straight line, other threshold values are approximately presumed corresponding to other probability of fatality. These thresholds could generate risk gradation as Fig. 5 right. The deeper red color the more dangerous a point in the risk gradation has.

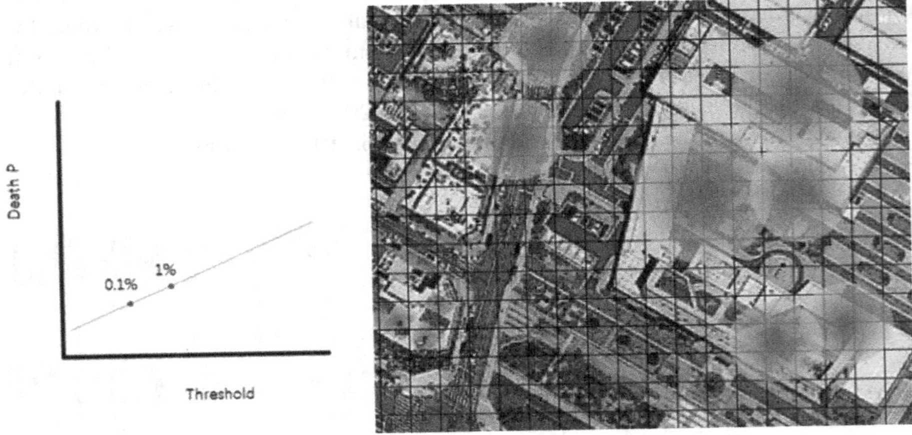

Fig. 5. Straight line connecting between harm point and no harm point (left), and risk gradation example according the straight line (right). (Color figure online)

4 Conclusion

Korea gas Safety Corporation has performed risk-based assessment for Korean chemical plants. Risk area prediction is core technology in risk-based assessment. Risk area can be calculated according to API (American Petroleum Institute) RP 581 standard. Risk area is the key to determine risk level for worker or human residing on chemical plant. When a human stays in a risk area, which is generated by safety distance from a risk facility, the human will get a risk corresponding to approximation value of 1% (harm distance case) or 0.1% (no harm distance case) probability of fatality. If safety management system provides workers this risk area information, worker would recognize their risk on their work place. Also, administrators are able to restrict a risk area and come up with safety measures efficiently to reduce the risk.

Korea Gas Safety Corporation has researched real-time risk assessment technologies. This system is also developing and established as a part of the research [4]. The result of this system depends on the pressure and temperature in chemical process. Pressure and temperature are time-sensitive, so, risk level for the worker's place may also be changeable over time. That is why real-time system is so important for risk area prediction system.

References

1. Guidelines for Analysis of Chemical Releases. CCPS (1999)
2. Methods for the Calculation of Physical Effects. TNO Yellow Book (2005)
3. Chapra, S.C., Canale, R.P.: Numerical Methods for Engineers, 6th edn. McGrawHill, New York City (2009)
4. Oh, J.S.: A study on developing IGAS for analyzing the status of gas facility. Int. J. U- E-Serv. Sci. Technol. **10**(7) (2017). ISSN 2015-4246
5. Deaves, D.M.: Dense gas dispersion modeling. J. Loss Prev. Process Ind. **5**(4), 219–227 (1992)
6. InfoNORMA: Reference Manual Bevi Risk Assessments Introduction (2009)
7. API RP 581 - Risk-Based Inspection Methodology, 3rd edn. (2018)

Design of SNS-Based English Word Learning System for Daily Study

Chungin Lee, Doyeon Kim, Sujeong Kim, and Yunsick Sung(✉)

Department of Multimedia Engineering, Dongguk University, Seoul 04620, Republic of Korea
{chungin1801, ggg5483, suju0328}@naver.com, sung@dongguk.edu

Abstract. It is possible to design an English vocabulary learning system using a text to speech (TTS) engine and a module which creates push alarms for new words or sentences at a specific time every day. The function of selecting new words and sentences and showing them to the user is a core technology in vocabulary learning systems, and it has a great effect on the systems' learning effectiveness. This paper proposes a method for creating a system which selects words and sentences. The system collects English words and sentences via social networks, refines them so that they are in a form which is appropriate for user learning, and allows the user to learn words which are related to their current location and weather.

Keywords: Listening system · SNS crawling · Data refine Android alarm system

1 Introduction

As global interaction becomes more common, English language education is moving away from methods which focus on academic grammar and vocabulary and shifting toward forms of learning which focus on the English conversation skills needed to communicate in daily life. Despite this, the most popular method of teaching English to children and adolescents in South Korea is still focused on memorizing English vocabulary and grammar. This is not effective for cultivating conversational skills. As such, there is a need for acquiring the varied lexicon that is required in conversation rather than memorizing the lexicon needed for academic grammar and vocabulary. There is also a need for methods which can cultivate the skills for listening to and understanding English in actual daily life. The listening skills needed in daily life are improved through listening training, and the most important research results in real-life vocabulary learning have been in the area of listening training [1].

The smartphones that we often use in everyday life can be incorporated into English learning so that learners can become familiar with English vocabulary from the time they are in early elementary school. For example, English learning can be implemented and provided as a smart phone application. Smart phones can provide an effective English learning method in which the parents of early elementary school students choose a time of day for learning and new English words or sentences are

J. H. Park et al. (Eds.): PDCAT 2018, CCIS 931, pp. 304–312, 2019.
https://doi.org/10.1007/978-981-13-5907-1_32

listened to every day at that time. This can be considered an ideal format that is suitable for a society which has a high level of interest in English learning.

The trends in English learning smartphone applications and related technologies are as follows. There are cases where a text to speech (TTS) engine was used to design systems for English education [2]. In the case of countries where English is not a native language, it is difficult to naturally acquire a lexicon during daily life, and methods for becoming familiar with new vocabulary are limited, so there is a need for learning methods that are just right for the individual. Increasingly, education systems are being developed based on web databases and social data collection systems that use web crawlers [3–5]. "Push" technology, which regularly sends information to users, is actively being used on mobile learning sites due to the growth of wireless LAN services and the inflow of large amounts of internet data as communications evolve. This paper focuses on a system which provides custom weather services using an application web server for English learning and the user's location information.

The present paper proposes a system for learning English conversational skills which are needed to communicate in daily life. It allows vocabulary to be learned easily during daily life through TTS technology and push technology.

This paper is organized as follows. Section 2 provides the system's overall design layout as well as a detailed description of the web server and the learning application system, which are the most important parts of the overall design. Section 3 presents conclusions about the method described in Sect. 2 and describes future research tasks.

2 Daily Learning System

2.1 Overall System Design

The social network-based English vocabulary alarm system for daily learning is divided into three systems which exchange data with each other: the learning application, the web server, and the database. The overall structure of this system can be seen in Fig. 1.

The proposed system's overall goal is to receive data that has been classified on the server and provide the optimal word or sentence to the user. In the learning application, the word or sentence is provided to the user through the TTS engine. The alarm time when the word or sentence can be heard is determined beforehand by the alarm module, and it has been implemented so that a fixed number of English words or sentences can be heard at the specified time. The TTS engine and alarm module are part of the learning application which runs on a smartphone owned by the user. Adjustments are made in the learning application so that a fixed number of words or sentences are heard by the user. In addition, the user's location data is collected for a feature which displays weather data based on the user's location and also recommends weather-related words and sentences.

The web server is divided into modules for collecting, refining, and transmitting words and sentences. The collection module collects English words and sentences from social network pages. The refinement module performs a sequential refinement process on data from the social network pages which is suitable for the user to learn.

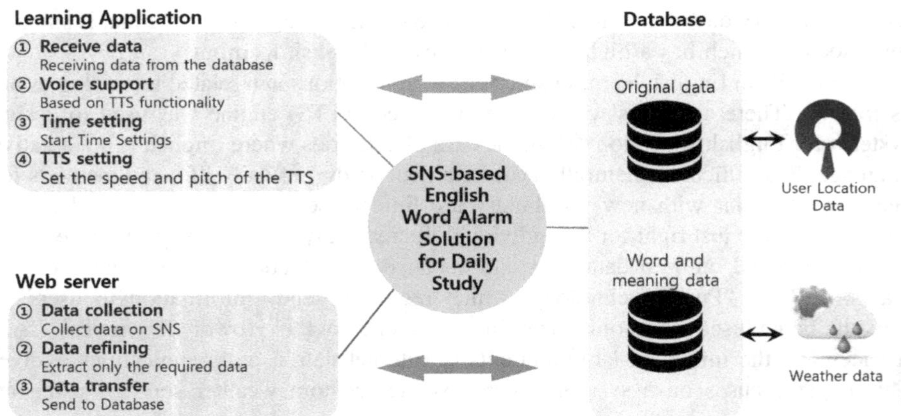

Fig. 1. Structure of social network-based English vocabulary alarm solution for daily learning

The transmission module performs a classification process and transmits the refined words and sentences to the learning application.

The database has two storage spaces. The storage spaces store the original data and the refined data, respectively. In addition, the meaning of each English word is also stored.

2.2 Web Server System

The web server collects words and sentences from social networks with a focus on Twitter. Rich data cannot be collected by using the Twitter REST API, which only collects data from certain Twitter account pages as shown in Fig. 2. This API is not suitable for a learning application which teaches the English words that various people use during daily life. Therefore, data is collected by using the Twitter streaming API which collects data from all twitter account pages via the specific keywords that will be learned. Of the data collected from the Twitter accounts, only the Tweet_ID, Screen_name, Created_at, and Text data are saved in the database as original data. The original database is shown in Fig. 3.

The Tweet_ID is the database column which stores the Twitter ID in the overall account, not the Twitter user account's ID. Screen_name is the database column which stores the Twitter user account's name. Created_at is the database column which stores the collected dates. Text is the database column which stores the content of the tweets. As can be seen in Fig. 3, the text data in the original database is saved in a form that is difficult to use for daily learning.

In the refinement module, the text data from the original database table is opened, and a refinement process is performed to extract only words. The meaning of these words is found and stored with the words. Tags such as HTML tags and unnecessary RT tags are removed, and then the punctuation which is often used in social networks is tokenized. Words are extracted according to spaces, and chains of single letters, such as vowels being repeated for emphasis, are removed from within words. Duplicate data is

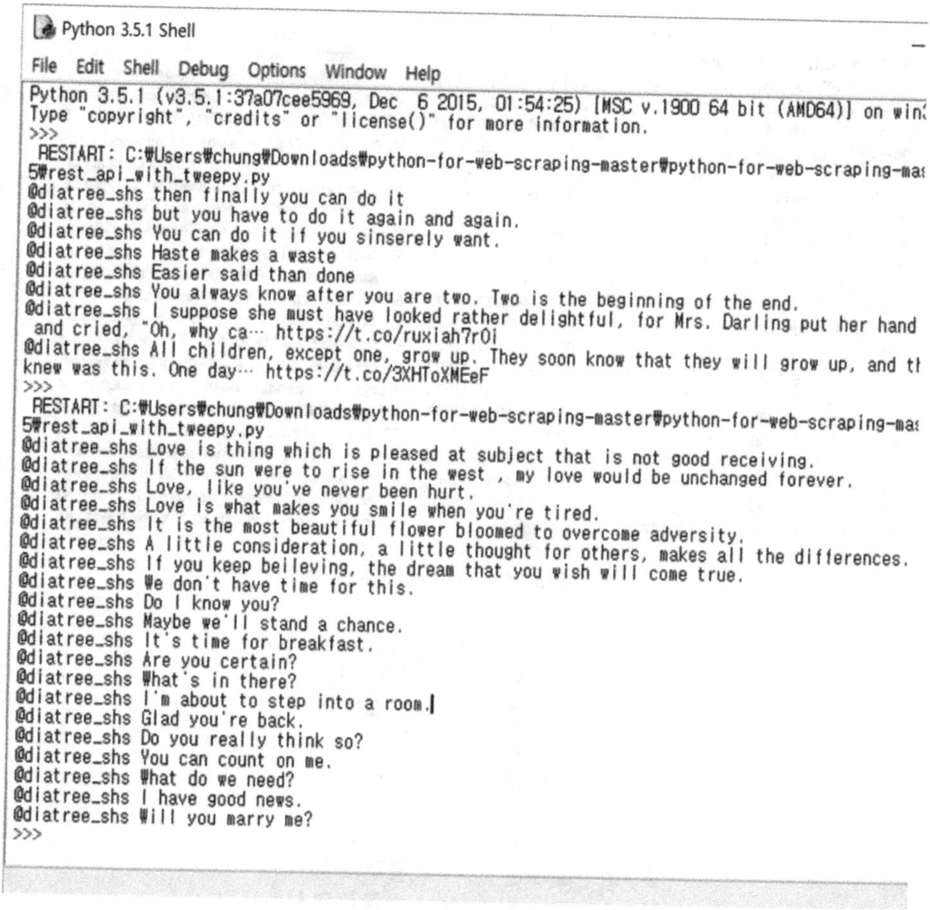

```
Python 3.5.1 Shell
File Edit Shell Debug Options Window Help
Python 3.5.1 (v3.5.1:37a07cee5969, Dec  6 2015, 01:54:25) [MSC v.1900 64 bit (AMD64)] on win
Type "copyright", "credits" or "license()" for more information.
>>>
 RESTART: C:₩Users₩chung₩Downloads₩python-for-web-scraping-master₩python-for-web-scraping-ma
5₩rest_api_with_tweepy.py
@diatree_shs then finally you can do it
@diatree_shs but you have to do it again and again.
@diatree_shs You can do it if you sinserely want.
@diatree_shs Haste makes a waste
@diatree_shs Easier said than done
@diatree_shs You always know after you are two. Two is the beginning of the end.
@diatree_shs I suppose she must have looked rather delightful, for Mrs. Darling put her hand
 and cried, "Oh, why ca… https://t.co/ruxiah7r0i
@diatree_shs All children, except one, grow up. They soon know that they will grow up, and t
knew was this. One day… https://t.co/3XHToXMEeF
>>>
 RESTART: C:₩Users₩chung₩Downloads₩python-for-web-scraping-master₩python-for-web-scraping-ma
5₩rest_api_with_tweepy.py
@diatree_shs Love is thing which is pleased at subject that is not good receiving.
@diatree_shs If the sun were to rise in the west , my love would be unchanged forever.
@diatree_shs Love, like you've never been hurt.
@diatree_shs Love is what makes you smile when you're tired.
@diatree_shs It is the most beautiful flower bloomed to overcome adversity.
@diatree_shs A little consideration, a little thought for others, makes all the differences.
@diatree_shs If you keep beileving, the dream that you wish will come true.
@diatree_shs We don't have time for this.
@diatree_shs Do I know you?
@diatree_shs Maybe we'll stand a chance.
@diatree_shs It's time for breakfast.
@diatree_shs Are you certain?
@diatree_shs What's in there?
@diatree_shs I'm about to step into a room.
@diatree_shs Glad you're back.
@diatree_shs Do you really think so?
@diatree_shs You can count on me.
@diatree_shs What do we need?
@diatree_shs I have good news.
@diatree_shs Will you marry me?
>>>
```

Fig. 2. Twitter REST API

removed, and stop words which have no particular meaning and are not needed for learning are removed. Finally, the refined words' roots are extracted. When the meanings that correspond to the extracted words are saved, the extracted words' roots are found in a web page dictionary via the web server, and the search results pages are accessed via HTML parsing. The part of the accessed data which corresponds to the word's meaning is extracted, and only the first listed meaning is saved. The database which stores the extracted words and meaning is shown in Fig. 4.

The id is an automatically increasing number column for the purpose of avoiding data duplication and organization. Word is the database column which stores the refined word or root. Means is the database column which stores the word's meaning. The data transmission module is implemented in PHP so that data can be sent and received between Android and the database, and the data can be saved as a file in Android.

tweet_id	screen_name	created_at	text
990566158511243264	Fabriziobustama	2018-04-29 12:19:10	1) Precision vs. Sionificance. 2) Accuracv vs. Pr...
990566164685295616	stirrinabewitch	2018-04-29 12:19:12	RT @ Peculium: If vou want to ioin our arowina...
990566171870085120	iustttxnwik1571	2018-04-29 12:19:13	RT @UserExperienceU: #Gettina into #User #E...
990566178346037248	strout retha	2018-04-29 12:19:15	RT @UserExperienceU: #Gettina into #User #E...
990566178635444224	machinelearn d	2018-04-29 12:19:15	RT @Pea21240: What is aoal for aoolvina machi...
990566179268956166	GameDev RT	2018-04-29 12:19:15	RT @xilinx9959: https://t.co/LuvDViLvB2 is For ...
990566183865905152	PleasureEthics	2018-04-29 12:19:16	Now I onlv wonder on who the Italian Maffia re...
990566190350249985	WIOMAX VA	2018-04-29 12:19:18	RT @ContractNetLabs: Measures for #IoTdata ...
990566196465602560	WIOMAX VA	2018-04-29 12:19:19	RT @Soohie Sander: #DDoS Attacks on Mobile/...
990566202111033344	PeterRecollet	2018-04-29 12:19:21	RT @iblefevre60: In 4 vears. vour #smartohon...
990566205042774016	sitisitiii7	2018-04-29 12:19:21	RT @AitheonOfficial: Reaardless of manufactur...
990566205097500672	111Lvnsev	2018-04-29 12:19:21	RT @TechforLifeUK: We're thrilled to be workin...
990566209723686914	machinelearn d	2018-04-29 12:19:22	RT @TechMeToHeaven: Let's all trv! #aoole #i...
990566217873281024	bdt svstems	2018-04-29 12:19:24	RT @whatsthebiodata: #2018 Guide to #BiaDa...
990566230531694593	wiomax cn	2018-04-29 12:19:27	RT @wiomax: Whv the #transoortation sector ...
990566232515534850	nafisalam	2018-04-29 12:19:28	A aood #comoendium on how #AI is sourrina #...
990566233509695488	JueraUnaer	2018-04-29 12:19:28	RT @coledhill: Artificial intelliaence intearated in...
990566242414100480	davcrv9	2018-04-29 12:19:30	RT @oroarammingurus: "What reallv killed the d...
990566243915780096	bdt svstems	2018-04-29 12:19:31	RT @aianlucacarrera: Most businesses are alre...
990829352697778181	aibson6171ic	2018-04-30 05:45:01	RT @iustinsuntron: (1/3) #TRON Weeklv Reoor...
990829362214768640	enaov	2018-04-30 05:45:03	BLOW OUT ->: Defeat decisivelv #ohrasal #e...
990829372742434816	UweMueaae	2018-04-30 05:45:05	#Amolexor seek a #German / #Enalish >: #S...

Fig. 3. Original database

id	word	means
1	aoostore	앱스토어
2	olease	제발
3	bouaht	사다
4	whv	왜
5	suoer	대단한
6	share	함께쓰다
7	banaladesh	방글라데시
8	maffia	마피아
9	music	음악
10	machine	기계
11	craft	공예
12	host	주인
13	robotics	로봇공학
14	user	이용자

Fig. 4. Word and meaning storage database

2.3 Learning Application

The learning application is composed of a weather data collection and storage module, a TT engine module, and a time setting alarm module. The TTS engine module parses the database of all refined words and meanings from the web server page. When the user presses the pink play button at the bottom center of the word list view in the main screen, the TTS engine reads the words in order. The words are displayed in the single list item view in a large font size so that the user can focus on words one at a time, and the screen UI is constructed so that the user can scroll downward. Young children and elementary school students are sensitive and change their minds quickly, so they can easily become bored if there are too many words in the list view or the words remain the same, so the number of words in the list view is set at around 10, and the words change to new words each time the play button is pressed so that children don't lose interest. A TTS settings feature is implemented so that the user can press the mint-colored settings button at the bottom right of the word list view on the main screen for additional TTS learning settings. The TTS engine settings feature can be considered a major feature in the learning application which allows the user to better adapt to actual circumstances and improve their listening skills more effectively because it considers the fact that each person who is using English in real-life has a different tone of voice and rate of speaking so they can be heard differently even if they are saying the same words. Figure 5 below shows the welcome screen, main screen, and TTS settings screen.

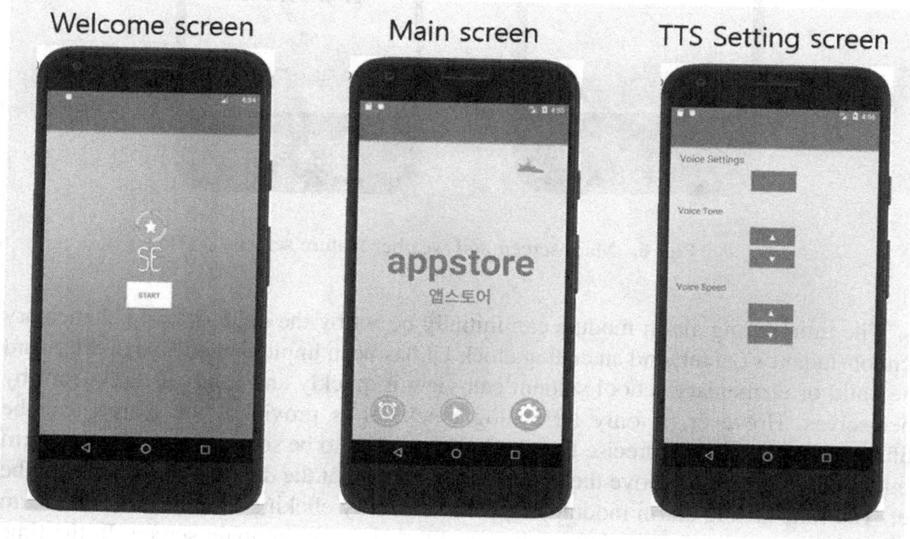

Fig. 5. Start and main screens and the TTS settings screen

The weather data collection module extracts only the weather data for the GPS location received from the application and saves it in the daily weather database. Images and words which describe the corresponding location and weather are provided to the user along with the humidity, temperature, etc. The learning application has been developed to be appropriate for daily learning by providing words and sentences based on weather data, which is checked daily by most users. The weather data collection module is more of a secondary feature in the learning application than the TTS engine feature or the TTS engine feature settings. Therefore, the size of the button on the main screen is smaller than those of other main features. It can be accessed by clicking the moon icon at the top right of the word list view. Figure 6 shows the aforementioned weather feature screen and the main screen.

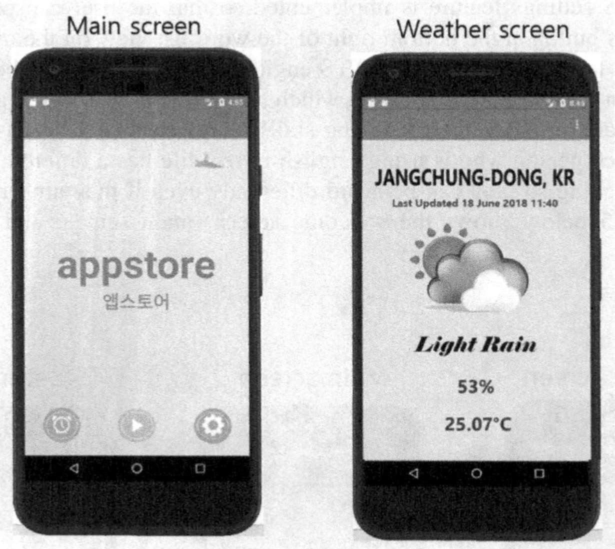

Fig. 6. Main screen and weather feature screen

The time setting alarm module can initially be set by the child or early elementary school student's parent, and an analog clock UI has been implemented so that afterward the child or elementary school student can view it quickly and easily as they learn by themselves. However, if only an analog clock UI is provided, the feature can be difficult to use when the precise hour and minute need to be set, so a smartphone alarm setting box is provided above the analog clock UI so that the detailed alarm time can be set. The time setting alarm module can be accessed by clicking the mint-colored alarm button located at the bottom left of the word list view on the main screen. In the time setting alarm screen, the user sets the time for the alarm to go off and presses the On button to set it. When the alarm goes off, the user clicks the Off button to turn the alarm off. The small triangle button below the alarm On and Off buttons can be clicked to set the alarm sound according to the user's preference. Figure 7 below shows the aforementioned time setting alarm feature. Figure 8 below shows the push alarm.

Fig. 7. Main screen and alarm feature

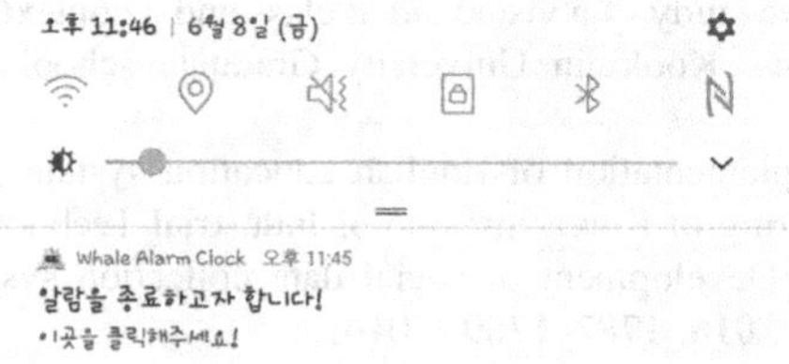

Fig. 8. Push alarm

3 Conclusions and Future Work

The present paper has proposed methods for collecting and refining words and sentences to use in a language study application which can help early elementary school students learn English effectively. The user is provided with a database of new words and meanings which is created daily from a web server, and the web server pages are parsed in the learning application. The learning application increases the user's interest and at the same time offers features which allow it to be used on a daily basis by providing location, temperature, humidity, and weather status data based on the user's location.

Finally, the paper has devised and proposed design methods for the web server and database, which are the core technologies of the social network-based English vocabulary alarm system for daily learning that provides these services. By proposing methods for designing the web server's core modules and the database, as well as methods for extracting and refining words and sentences, this paper can provide real help in implementing English learning applications which diverge from conventional English vocabulary learning methods that use simple memorization.

Future work will include making it possible to select words more precisely and accurately and classify their level of difficult by employing machine learning in the data collection approach which uses the Twitter streaming API. Currently, there are cases where words are not perfectly refined, and words are stored without their meaning, and this approach is a way of resolving these problems.

The level of difficulty can be set according to the learning level of the application user. Methods which have been considered for classifying the level of difficulty include methods which check the frequency of each word in the refined database of words and meanings that was created from the original database's text data, as well as methods which classify words by considering various factors such as length, number of vowels and parts of speech tags, etc.

Acknowledgments. This research was supported by the MSIT (Ministry of Science and ICT), Korea, under the National Program for Excellence in SW supervised by the IITP (Institute for Information & communications Technology Promotion).

References

1. Park, S.J.: A comparative study of a visual aided class and a context aided class in vocabulary learning. Master's thesis, Kookmin University Graduate school of education, pp. 14–67 (2009)
2. Youn, Y.H., et al.: Implementation of English education system using TTS. Collection of Academic Papers in Spring of Korea Institute of Industrial Technology, pp. 892–896 (2011)
3. Lee, C.E., Jang, J.W.: Development of social data collection system using web crawling. Korea Comput. Congr. **2016**, 1787–1789 (2016)
4. Na, S.W., Shin, S.M., Lee, S.A., Cho, Y.H.: Cyber learning system design and implementation based on web database. J. Korean Soc. Content **2**(1), 1–6 (2002)
5. Lee, S.J.: Design and implementation of English learning system in web and mobile environment. Master's thesis, Silla University of education (2004)

Study of Real Toy Tank-Based Mixed Reality Contents

Eunhee Park, Namgwang Ryu, Jisun Lee, and Yunsick Sung(✉)

Department of Multimedia Engineering, Dongguk University-Seoul, Seoul 04620, Republic of Korea
peh2017@gmail.com, pieisland@naver.com, wltjs95@daum.net, sung@dongguk.edu

Abstract. Mixed Reality (MR) has advantages of a high immersion from virtual reality (VR) and a high sense of reality from augmented reality (AR). MR enables users to have various experiences by providing an environment in which real and virtual worlds are interconnected. The MR technologies are advancing by combining the advantages of VR and AR. This study presents the elements and techniques required to efficiently implement contents in MR and creates MR contents based on them.

Keywords: MR · Marker · Mobile · Toy

1 Introduction

The mixed reality (MR) technology, which enables users to exchange information in real time within an integrated space that properly combines a sense of immersion from virtual reality (VR) and a sense of reality from augmented reality (AR), is developing. Conventional MR contents either use a newly developed device for interaction between virtual and real environments or simply provide the contents on the screen of a smartphone [1, 2]. However, to improve accessibility to MR contents, it is required to actively utilize the devices and things that already exist rather than developing new devices.

This paper introduces elements and techniques for implementing MR and creating MR contents. This paper is structured as follows. Section 2 reviews related works and Sect. 2.1 describes a definition of MR. Section 3 describes the design and creation of MR contents, and introduces the elements of MR contents and actual MR contents that utilize toys. Section 3.2.4 proposes the development direction of MR contents. Lastly, Sect. 4 presents conclusions.

2 Related Works

2.1 What is MR?

VR and AR are closely related to MR. VR is an interface between a user and a computer that makes users feel as if they are interacting with the actual surrounding

J. H. Park et al. (Eds.): PDCAT 2018, CCIS 931, pp. 313–321, 2019.
https://doi.org/10.1007/978-981-13-5907-1_33

situation or environment, which are, in fact, created using a computer [1]. It isolates users from reality and makes them completely immersed in a 'virtual' world that has been synthesized by computers. AR is an environment that assists users to perform jobs more conveniently by providing additional information (VR) in real time to the reality perceived by the users [5].

MR, which combines the advantages of both VR and AR, creates a new environment or information such as visualization by combining virtual and real worlds. MR refers to the interaction in real time among things that exist in real and virtual worlds.

Examples include applications using Microsoft HoloLens. 'RoboRaid' is a game to crash a robot that destroys the walls to come out of an actual house. 'Fragments' is a game in which users investigate traces of a crime scattered in their house, which has turned into a murder scene, to find a suspect. Both games allow users to get deeply immersed and interact with the real world by setting a house which is a space of reality as a background.

2.2 Studies Related to MR

On top of VR and AR, studies related to MR contents have been actively conducted. There is a study on MR contents that used NFC, smartphone and toy cars. A smartphone is loaded on a toy car and a user moves the toy car [1]. In this study, the location of a toy car was recognized by installing NFC chips in a map mat which can interact with the NFC feature in smartphones and recognize the location of the smartphone.

For studies related to tracing movements in MR, there is a study on spatial paratexts [3]. This study identifies spatiotemporal traces with texture mapping by tracking physical motions and movements of users based on the Kinect coordinate system when implementing a mobile MR. There is a study on a MR exhibition guide system that uses a transparent display [4]. In this study on contents, transparent displays are placed above real exhibits and the information related to the exhibits is provided on the display. The screes on the transparent displays can be removed if needed.

2.3 Development Direction of MR Contents

Currently, many studies on MR are being actively conducted, and some of them have been commercialized. NASA used MR devices in 'Project SideKick' which is an astronaut education project to save education time, and Ford, a global automobile manufacturer, used Microsoft HoloLens to check the design of a car under development in advance and reduced the time required for car design. Though MR has not been commercialized as much as VR or AR, it is highly likely that MR contents will be developed soon in various fields since both VR and AR are being developed and commercialized rapidly. VR and AR are diversely utilized in the game and education fields. MR not only covers the area that VR and AR are being used, but also further shows the potential for commercialization and advancement in actual business operations or medical treatment.

3 Design and Creation of MR Contents

3.1 Elements of MR Contents

This study attempts to create a MR mobile content enjoyable at home by using a mini-beam projector and a smartphone. It is a game in which elements, such as RC tanks in real world and missiles in virtual world, interact with each other. In order to implement this MR environment, the objects in real world need to be recognized and devices and an event processing system for interaction between the real and virtual worlds are used.

3.1.1 Recognition Method for Objects in Real World

HoloLens is mainly used to implement MR contents. However, it is not the goal of this study to show the real world and virtual world and make them interact with each other through a virtual display. The goal of this study is to watch the virtual world from the real world visually without a display and make the elements interact with one another. A method to recognize the actual objects is required since both real and virtual worlds must interact with each other without using HoloLens.

Two methods, marker and OpenCV, are applied to recognize the actual objects. In this study, mobile devices are used. When using OpenCV in the mobile environment, there is a problem of broken screen because of numerous operations. Thus, a marker is used to recognize the objects in real world.

This study uses the image marker recognition and tracking techniques of Vuforia to recognize the objects in real world. The features of a registered image marker are compared with those of the object that is reflected through a camera. If the two features match, the object is continuously traced as longs it stays in the camera angle. This marker allows to identify the location of actual object and enables the interaction among objects.

To improve the recognition rate of the marker, similar images having different distribution or different number of the features are used. The number of markers that can be stably recognized in a mobile device is limited. Thus, the number of markers used as virtual buttons can be reduced by making several markers that play different roles as one marker.

3.1.2 Interaction Between Real World and Virtual World

The most obvious difference of MR compared to VR or AR is an interaction between real world and virtual world. This study uses markers to recognize actual objects in real world so that they can interact with virtual world. In order to enable the interactions of the two worlds, the marker objects in real world are used to control the game without touching the screen of a mobile device and this can be achieved by using the virtual button function that is activated when a part of marker's features are blocked. The game screen of a virtual world is projected by a projector onto the game board on the floor, which is a real world, to create a mixed space of real and virtual worlds. Actual objects are recognized in real time and their motions are traced to create the event within a game by manipulation of a user. Such an event is again projected onto the real world to let both real and virtual worlds visually interact with each other. Since this study aims to create the MR mobile contents that can be used at home as well, the method using a mini-beam projector is selected to project the virtual world onto the real world.

3.2 MR Mobile Contents Utilizing Toys

This content is a game that actual toy RC tank in real world and missiles in virtual world interact with each other. The motions of tank are recognized by a smartphone, and various effects such as launching bombs and applying game items, which cannot be seen in real world, appear. This is the mobile MR contents using a smartphone and a projector, and allows users to enjoy the game relatively easily at home without watching the smartphone screen. This is a one-to-one fighting game using an RC tank. The tank used in the content is shown below (Fig. 1).

Fig. 1. RC Tank in size of 7×15 cm and its controller

3.2.1 MR Contents Design

The process of contents design is as follows. A user directly manipulates a RC tank with a controller. A tank object in real world is recognized using a marker and the information of movement is delivered through the motion recognition module. The event process module delivers the events according to motions of RC tank to the graphic module to create corresponding graphics and then the graphics are output to the real world through a smartphone and a projector. As an example, the motion recognition module recognizes a motion of firing a bomb in RC tank and makes the event process module issue a command to fire a bomb (Fig. 2).

3.2.2 Basic Settings of Contents

To create the contents, the following items are required: a game board onto which the virtual world is to be projected, a mobile device, a mini-beam projector, RC tanks and controllers, 5 markers in total consisting of two button markers, two tank markers, and one center marker. The layout of the markers is shown in Fig. 3.

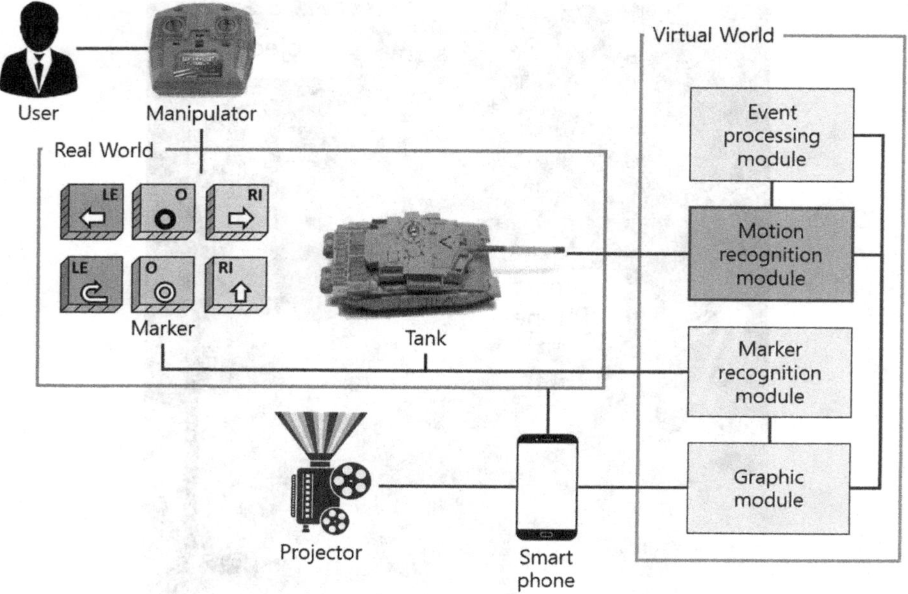

Fig. 2. Contents design configuration

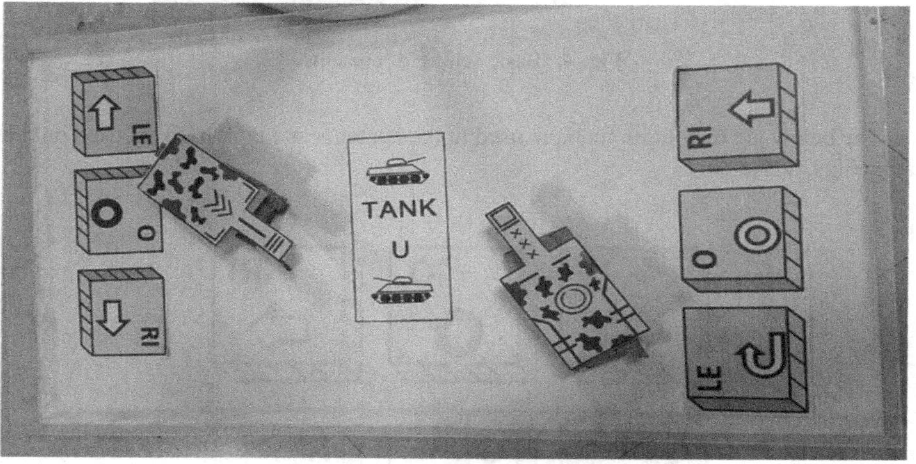

Fig. 3. Game board and marker layout

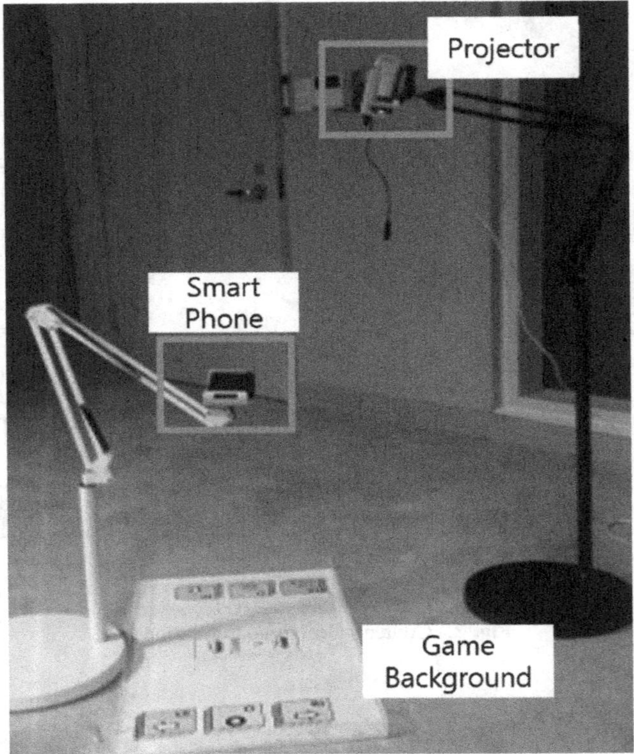

Fig. 4. Basic setting of contents

The below are the button markers used in the contents and tank markers used on the tank (Figs. 5 and 6).

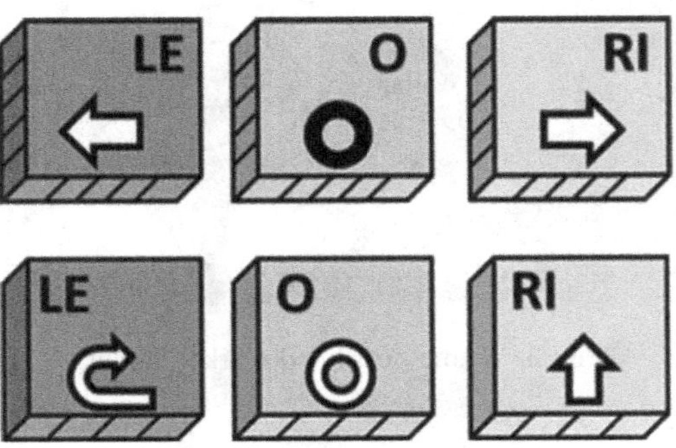

Fig. 5. Button marker

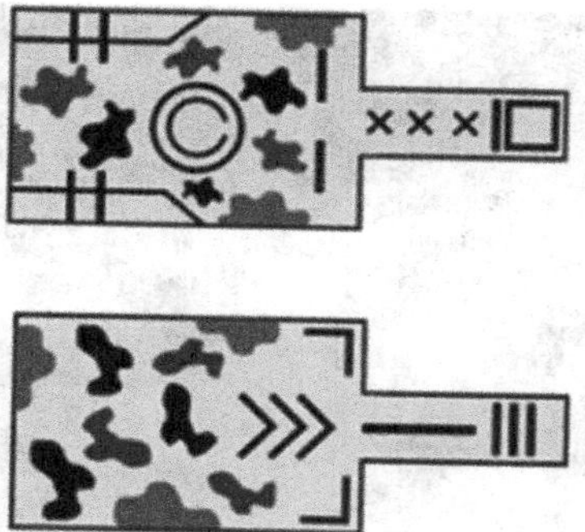

Fig. 6. Tank marker

3.2.3 Contents Scenario

This is a one-to-one fighting game content. Two users play together. Users are required to conduct the basic setting as described in Figs. 3 and 4 above prior to playing the game and take a seat. Since a projector is used, the game is played in a dark place. Once the application is executed, players perform all controls directly in the real space and not in the mobile space. If any selection is required in the game, the selection function is performed by blocking a part of the button marker, which serves as a virtual button, with a foot or hand (Figs. 7 and 8).

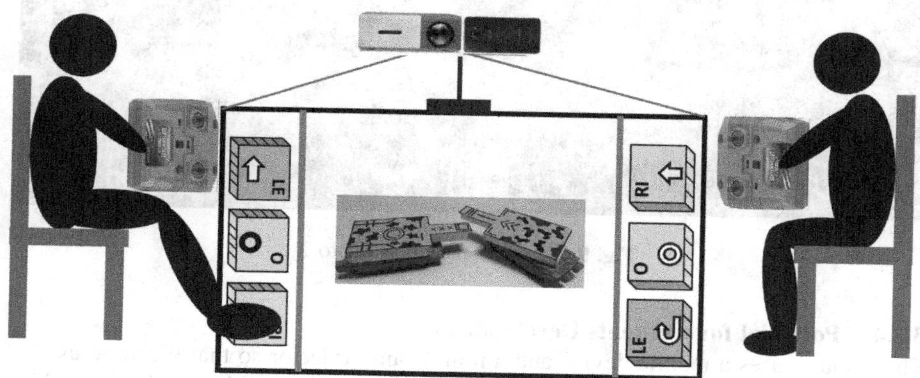

Fig. 7. Full view in content execution

RC tanks and controllers are used in the game. Users obtain the virtual game items or fire bombs by manipulating the actual RC tank with a controller. Items can be used by blocking the button marker that serves as a virtual button. If a bomb in virtual world hits a RC tank in real world, the HP decreases. Making the HP of the opponent zero by firing bombs to the opponent's tank wins the game (Fig. 9).

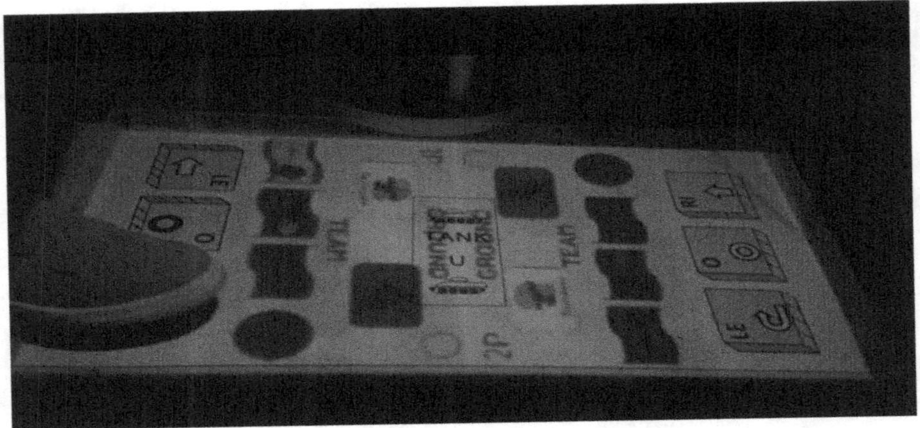

Fig. 8. Content execution photo 1

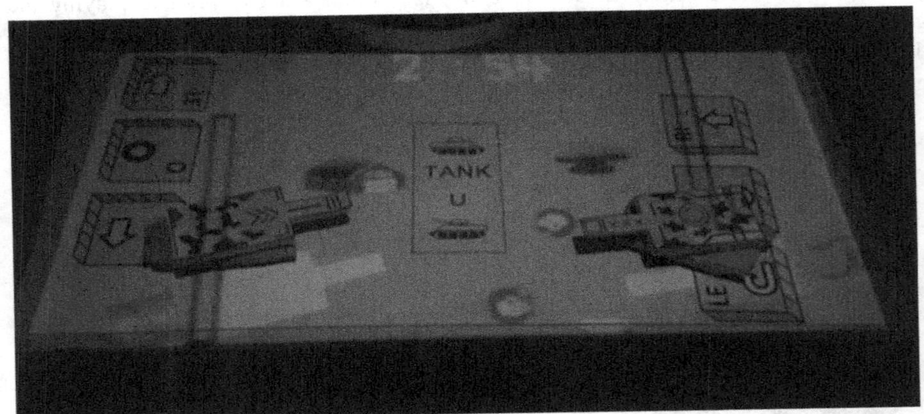

Fig. 9. Content execution photo 2

3.2.4 Potential for Contents Development

This content uses a mobile device and a mini-beam projector so that it can be used at home as well. It is difficult to visually synchronize the firing positions of bombs, which are virtual objects, because of a small movement range of RC tanks as well as the positions and shadows of the mobile device and projector. Due to the limited performance of the mini-beam projector, this game can be played only in a sufficiently dark place. If a high-performance projector is installed on the ceiling to project the virtual world onto the floor, the synchronization can be improved.

Since actual objects and markers are recognized through a mobile device, the coverage of the camera was narrow in this study. If a camera or lens that covers a wider range is used instead of a mobile device, the movement range of RC tank can be increased.

For actual object recognition, a marker was fabricated to cover the entire gun barrel of the RC tank in this study. If this is improved by either making an actual RC tank to be a marker or attaching a sticker that can be used as a marker to an RC tank, it will be possible to recognize the RC tank itself, which is an actual object.

Since this mobile MR content was created using actual toys, the target users can be expanded from children to adults and various functions can be added without concerns about safety accidents. Therefore, it is expected that the toy industry will be also developed together since the contents having different functions or stories can be created using one toy.

4 Conclusion

This study defined MR and demonstrated the elements of MR contents and created an example content by utilizing those elements. Furthermore, the development direction of MR contents was proposed. As a measure to improve accessibility to MR contents, this study developed an MR content that uses markers, a projector, mobile devices and toys without using a hologram, and suggested the potential for development. The MR contents field is expected to grow in the future together with the stagnant toy industry. Currently, a number of studies on MR technology are being conducted actively, and it is expected that MR will provide a variety of experiences and enable more convenient life in the future through the commercialization and development of MR contents in more various fields.

Acknowledgments. This research was supported by the MSIT (Ministry of Science and ICT), Korea, under the National Program for Excellence in SW supervised by the IITP (Institute for Information & communications Technology Promotion).

References

1. Gong, S.: Device and contents development for children applying mixed reality technology. J. Digit. Contents Soc. **17**(5), 339–348 (2016)
2. Yoon, K., Won, J., Park, J., Choe, J.: Design of tourism tour map App application using MR technology. Proc. Korean Soc. Comput. Inf. Conf. **25**(2), 146–148 (2017)
3. Kim, Y., Lee, Y., Nam, Y.: Mobile mixed reality storytelling as spatial paratexts. J. HCI Soc. Korea **10**(2), 57–64 (2015)
4. Hong, S., Shin, C.: Metaverse based mixed reality exhibition guide system. Mag. KIICE **16**(1), 31–37 (2015)
5. Lee, H.: A study on the 3D TV contents using mixed reality technique, Yonsei University graduate school of engineering Master's thesis, 13 (2003)

Smart Grid an Advanced Method for Saving Energy in Vietnam

Nhan Tran Van[1] and Hoanh Su Le[2](✉)

[1] Vietnam National University Ho Chi Minh City, Ho Chi Minh City, Vietnam
tranvannhan214@gmail.com

[2] Faculty of Information Systems, University of Economics and Law,
Vietnam National University Ho Chi Minh City, Ho Chi Minh City, Vietnam
lehoanhsu@gmail.com

Abstract. This study shows that in Vietnam, the problem of saving energy and enhancing the quality of electricity is largely purposes presently as well as in the future. Otherwise, Vietnam's power sector is one of the most rapidly expanding in Southeast Asia. These are the reasons this research aims to conduct a survey related to the smart power supply in accordance with the changes in both power output and demands by integrating the IoT technology.

1 Introduction

In Vietnam, the current grid system is mostly using the technology of the 1970s. However, with the advent of advanced power generation technologies, along with the problem of power shortages, power theft as well. Increase demand for electricity has led to the need for a modern electricity grid that meets all consumer requirements, that grid is called the smart grid.

The smart grid offers a number of improvements, such as increasing the stability, reliability and performance of the grid, minimizing power line losses. Modern technologies such as sensors and measurement, advanced electrical equipment are applied to develop the features of the grid.

There is no standard definition for smart grid, but it can be understood that the grid system uses advanced technology to increase the reliability and performance of the grid, from transmission to distribution. Smart Grids provide a better distribution infrastructure than conventional grids.

Power plants and power transmission networks are becoming increasingly old and outdated at Vietnam. That mean it is very difficult to manage and regulate the demand for electricity. The demand for electricity can be reduced but it can also increase continuously. One possible solution is to build more power lines, but then the existing systems will be wasted. Therefore, instead of using a temporary solution, a more reliable and long-term solution needs to be found. One solution that is considered optimal for a smart grid system is the use of smart meters.

There is no denying the benefits of smart grids, which allows users to optimize and save power consumption. While also reducing the monthly electricity bill, because of the ability to measure power consumption daily. As for electricity companies, smart

J. H. Park et al. (Eds.): PDCAT 2018, CCIS 931, pp. 322–328, 2019.
https://doi.org/10.1007/978-981-13-5907-1_34

grids will help them manage the grid better, improve system reliability, minimize power losses and protect the environment.

In Vietnam, the Government is aware of the inevitable trend of smart grid deployment to ensure national energy security, the efficient use of fossil fuels, the gradual exploitation of renewable energy sources, clear energy, environmentally friendly energy. Therefore, the Smart Grid Development Scheme was approved by the Prime Minister in Decision No. 1670/QD-TTg dated 8 November 2012. The overall objective of the project is "Develop smart grid with modern technology to improve power quality, power supply reliability; contributing to the management of electricity demand, promoting energy conservation and efficiency; To create conditions for raising labor productivity, reducing the need for investment in power source and grid development; To enhance the rational exploitation of energy resources, ensure national energy security, contribute to environmental protection and sustainable socio-economic development.

2 Smart Grid Trend in Vietnam

Over the last few decades Vietnam has made remarkable progress in reducing poverty and positioning its economy on a sustainable growth path [1]. As a consequence of robust economic growth, electricity demand in Vietnam grew at an average of 14% annually over the last decade [2]. With electricity consumption nearly matching generation in recent years and insufficient investment in new power plants, the median level of electricity usage in 2014 was 100 kWh per month per household [3], and the faction of households declaring unsatisfied electricity needs is below three percent. However, the electricity is becoming a heavier burden in Vietnamese households finances, and this affordability problem will increase if electricity prices are raised to finance the clean development of the energy system. The electricity grid is under constant strain by the growing economy. Realizing the large technical, institutional, and financial challenges posed by this level of expansion will be a key priority for Vietnam's grid system operators in the short term. In 2012, the Government of Vietnam (GoV) approved the smart grid development project in Vietnam which outlines a smart grid roadmap for Vietnam. The project is aimed at the integration of new monitoring, protection and control systems to improve grid reliability and make efficient use of infrastructure while facilitating future integration of scaled-up renewable energy options. The National Power Transmission corporation (NPT) has already started progressing some of the smart grid initiatives for transmission identified in the roadmap, such as, the deployment of Substation Automation System (SAS) and Wide Area Monitoring Systems (WAMS) as well as an information system for operation and supervision. To support GoV's efforts, the World Bank has closely engaged with NPT, the electricity regulator authority of Vietnam (ERAV) and the national load dispatch center (NLDC) to refine the existing smart grid roadmap on the basis of the lessons learned from the international experience with smart grid development. This report presents a technical analysis of Vietnam's existing smart grid roadmap. The cost-benefit analysis of HVDC reported incremental capital costs amounting to approximately \$13.3 million and an NPV of about \$23.5 million based on a 2,000 MW interconnection for 800 km of length [2].

Japan is one of the first countries to invest in smart grid research and development. Japan is also a top market for smart grid products and services globally, ranked 7th in the analysis conducted by the Department of Commerce in its 2017 Top Markets Report on smart grid opportunities [4].

Table 1 show the highlights trends to watch in 2017 for U.S. exporters and offers an expanded look at the top 50 markets for U.S. electricity transmission, distribution, and storage exports. And Vietnam is in the high-level rank (14). The overall smart grid rankings relative to the sub-sector rankings reflect variations in weighting among five categories [5]:

- *Smart Grid Market Growth Potential (30%):* Industry data and information on policies, regulations, and other local drivers of the smart grid technologies and services market.
- *Trade Factors and U.S. Competitiveness (30%):* Trade data and other information on exports of U.S. T&D equipment products and services in a given market.
- *Energy Storage Growth Potential (10%):* Energy storage deployment and information on renewable energy deployment and other drivers for the energy storage system market.
- *Key Economic and Energy Sector Investment Indicators (20%):* Broader economic data and power sector trends that impact investment in electricity infrastructure, and the development and growth of the smart grid in a given market.
- *Strength of Domestic Industry (10%):* Trade data and other information on the extent to which demand for smart grid technology and services will be met by the domestic industry – as opposed to international trade – in a given market.

Table 1. Overall smart grid rankings [5].

1	Mexico	**11**	China	**21**	Denmark	**31**	Nigeria	**41**	Costa Rica
2	Canada	**12**	Germany	**22**	Singapore	**32**	Italy	**42**	Morocco
3	UK	**13**	Ireland	**23**	Turkey	**33**	Israel	**43**	Ghana
4	India	**14**	Vietnam	**24**	Sweden	**34**	Thailand	**44**	Romania
5	Australia	**IS**	Malaysia	**25**	Kenya	**35**	Brazil	**45**	Peru
6	Korea	**16**	Philippines	**26**	Spain	**36**	Argentina	**46**	Bulgaria
7	Japan	**17**	Indonesia	**27**	Ethiopia	**37**	Colombia	**47**	Nicaragua
8	Egypt	**18**	France	**28**	Poland	**38**	South Africa	**48**	Mew Zealand
9	Saudi Arabia	**19**	Finland	**29**	Austria	**39**	Portugal	**49**	Russia
10	Chile	**20**	Netherlands	**30**	Belgium	**40**	Czech Republic	**50**	Kazakhstan

About developing a Smart Grid in Vietnam, Prime Minister Decision approving the project to develop the Smart Grid in Vietnam in order to improve the quality of power and reliability of power supply; contributing to the management of electricity demand, encouraging the use of energy saving and efficiency. The specific objective of this project is to build information and telecommunication infrastructure and to strengthen the monitoring and control system for the electric system and remote metering system.

At the same time, raising the capacity of forecasting additional electricity demand and planning electricity supply; Limiting the power shortage due to lack of power through peak load transfer during peak hours or emergencies. Implementation of technical solutions and management measures aimed at reducing power losses in power transmission and distribution systems from 9.23% in 2011 to 8% by 2015. In particular, the scheme will also enable customers to actively know and manage the detailed information on electricity use and purchase costs. The scheme guides the development of the Smart Grid in Vietnam in three phases [6].

Specifically, the first phase (from 2012–2016), implementation of the program to improve the efficiency of the operation of the power system, which will deploy applications to improve reliability, optimal operation of the transmission grid load, distribution grid, reduce power losses; Enhanced fault recording, power generation and power failure protection systems to ensure safe transmission on 500 kV power systems. During this phase, a number of testing and research programs will be implemented, promulgating technical standards for the Smart Grid [7, 8].

Phase 2 (2017–2022), continue to implement the program to increase the efficiency of operation of the power system, focusing on distribution grid; It is equipped with information technology - telecommunication infrastructure for distribution grid. At the same time, deploy the Smart Grid applications; development of technical regulations; Communication program for community.

Phase 3 (after 2022) will continue the program of equipping IT infrastructure for the distribution network, which will continue to implement the optimal operation tools from the transmission grid. to the electricity distribution grid; Continue to encourage the development of scattered power plants.

The Ministry of Industry and Trade is assigned to assume the prime responsibility for setting up a Steering Committee for Smart Grid in Vietnam headed by a leader of the Ministry of Industry and Trade as its head and the Electricity Regulatory Authority as its standing body. Organizing the construction and development of the Smart Grid in Vietnam in accordance with the approved roadmap.

The Vietnam Electricity Group shall work out specific plans for the programs, schemes and working groups for each period and submit them to the Ministry of Industry and Trade for ratification of specific targets for each program or project. Working Groups; Implement the Pilot Smart Grid projects in Vietnam in line with the approved Smart Grid development phases [9, 10].

2.1 Some of the Main Objectives of the Infrastructure Construction Project for Automatic Control of Electric Systems and Remote Measurement

By 2013: Complete set up of data collection system, SCADA control, remote metering system to all power plants with a capacity of more than 30 MW, substations of 110 kV up in the electrical system. By 2016: Utilizing all the functions of EMS in SCADA/EMS system at the National Power System Modernization Center and Regional Power System Modulation Centers. By 2022: SCADA/DMS systems for power utilities, remote metering systems are fully invested to all major power users.

2.2 Smart Grid is the Target of Vietnam Electricity Corporation (EVN) to Gradually Modernize the Grid System. Pioneer in this Field is the Electricity Corporation of Ho Chi Minh City (EVN HCMC)

Based on the project approved by the Prime Minister and the Electricity of Vietnam (EVN), the Electricity of Vietnam 1/2013. Deputy General Director of EVN HCMC - Head of Steering Committee for implementing the scheme, said that EVN HCMC has identified and developed Smart Grid in Ho Chi Minh City with a long-term goal to reduce losses. to improve the reliability of electricity supply, to ensure the beauty and safety of the electricity system throughout the city, contributing to the development of socio-economic development in Ho Chi Minh City. This is also one of the steps in the roadmap modernization of Ho Chi Minh City's electricity sector from now to 2020.

With solutions and roadmap implemented scientifically and rationally, the Corporation has achieved initial results. By the end of 2013, the reliability of electricity supply throughout the city has been raised. In particular, the SAIFI (number of power failures/year) decreased by 27.72%; SAIDI (total blackout) decreased 33.94% compared to 2012. Accordingly, the loss rate decreased to 5.3% thanks to the synchronous solutions such as: Build reasonable operation method The handling of non-load, overloaded transformers; balanced phase, reasonable stop; maintenance of counting and anti-electric hook-and-loop systems, electromechanical management, index management, regional losses [7].

2.3 Road Map for Deploying Smart Grid at EVN

In many areas of work to be implemented by the smart grid, EVN builds the smart grid development roadmap for the first phase with a 10-year vision through Decision 1795/QD-EVNCPC dated 24/4/2013. Approved Smart Grid Development Roadmap. The focus is on four components:

- *Measurement system for electricity business:*

EVN plans to invest in new electronic meters from 2013 onwards, replacing electromechanical meters with electronic meters with AMR Automatic Meter Reading (AMR) with many technological solutions; data collection, efficient use of data from the meter for customer care and management of production and business of EVNCPC. Step by step to implement the advanced Metering Infrastructure (AMI) – in order to be at 2022 to control the power demand of the customer as a key energy user. With the number of customers using electricity up to now about 3 million, the modernization of the metering system is a very necessary work and will be focused on EVN implementation in the coming time.

- *Component distribution grid automation to optimize system performance and increase power supply reliability:*

About automation grid 110 kV. From 2013, all new 110 kV TBAs will be built under the integrated computer control model. The existing 110 kV TBAs will be further improved to be able to be driven under the new model; These TBAs will be connected to the Operation Center.

Regarding to the automation of medium voltage electrical network: Including the construction of SCADA/DMS systems in the new and improved investment projects or replacement of equipment on the grid to be able to monitor and control the distance. In the period of 2013–2016, SCADA/DMS will be completed to invest in the previous plan (Thua Thien-Hue, Da Nang, Binh Dinh and Dak Lak) in order to monitor and control the remote be on the net. The next period from 2017 to 2022 will invest in new SCADA/DMS systems at the remaining scales, ensuring complete investment in SCADA/DMS systems by 2022.

- *Composition of specialized telecommunication systems and information technology infrastructure:*

+ Building specialized telecommunication systems:
EVN builds the telecommunication network with the transmission network from center to all terminal device. EVN conducts investment in OPGW fiber optic cables on 110 kV lines, creates additional redundant circuits and replaces equipment for transmission nodes, ensuring the construction of an inter-provincial transmission network at the end of 2013 as a basis for deployment. Inter-district transmission network will be completed in 2015. In parallel with the investment, EVN will expand telecommunication infrastructure exchange with partners to have strong and high-capacity private telecommunications systems.
+ Information technology infrastructure
Data Center ensures centralized processing of remote data collection, customer database with high security, reliability, security and redundancy.

Building customer care system - Call Center - to meet the information request for customers to use electricity.

- *Component of new and renewable energy sources:*

Research solutions for direct connection of new and renewable energy sources such as small hydropower, wind power, solar power, … scattered installation in households and production establishments. to the distribution grid; The ability to conduct two-way power purchase between the plan and customers. Initially implementing a pilot project to integrate wind and solar power sources as an experimental basis to integrate new and renewable energy sources into the system.

3 Conclusion

To sum up, this study described about Smart Grid trend in Vietnam. The metrics to monitor the Smart Grid initiatives from the technical, economic and regulatory point of view has been identified, analyzed and presented. The need for smart grid. Implementation of monitoring system to save energy and enhance the quality of electricity, which in turn let the consumers to save their electricity bills are the most important. That is the main thing was indicated in this study.

References

1. Huh, J.-H.: Smart Grid Test Bed Using OPNET and Power Line Communication. IGI Global, Pennsylvania (2017)
2. W.B. Group: Smart Grid to Enhance Power Transmission in Vietnam, Vietnam (2016)
3. Ha, D.M., Son, N.H.: Is Electricity Affordable and Reliable for all in Vietnam (2017)
4. M. o. I. a. T. a. I. o. E. V.: Co-organized by Japan Institute of Energy Economics, in Smart Grids Vietnam - Japan, Hanoi (2017)
5. I.T. Administration: Smart Grid Top Markets Report, Department of Commerce, US (2017)
6. E. new: Development Smart Grid at Viet Nam
7. V. E. C. a. H. C. M. c. (. HCMC): Smart grid development in Vietnam (2014)
8. Kabalci, E., Kabalci, Y., Develi, I.: Modelling and analysis of a power line communication system with QPSK modem for renewable smart frids. Int. J. Electr. Power Energy Syst. **34** (1), 19–28 (2012)
9. Wu, F.F., Moslehi, K., Bose, A.: Power system control centers: past, present, and future. Proc. IEEE **93**(11), 1890–1908 (2005)
10. Gungor, V.C., Lambert, F.C.: A survey on communication networks for electric system automation. Comput. Netw. **50**(7), 877–897 (2006)

Autonomous Flight Control Method of Drones for Enforcement of Traffic Law Violation

Jeonghoon Kwak[1], Sang-Geol Lee[2], and Yunsick Sung[1(✉)]

[1] Superintelligence Lab and Department of Multimedia Engineering, Dongguk University-Seoul, Seoul 04620, Republic of Korea
{jeonghoon, sung}@dongguk.edu

[2] Department of Electrical and Computer Engineering, Pusan National University, Busan 46241, Republic of Korea
leesg@pusan.ac.kr

Abstract. Recently, drones are used for monitoring in various fields. Especially, the pilots manually fly the drones in the sky in order to control the traffic law violation that occur at unspecified locations. However, in order to control traffic law violation by drone, it is necessary to fly autonomously and shooting traffic law violation rather than manually controlling the drones. This paper proposes autonomous control method of drones to crack down on traffic law violations. The pilot collects flight records to crack down on traffic law violations. The collected flight records generate a flight path for the autonomous flight of the drones. The generated flight path selects the optimal flight path for the drone to fly. The control signal is generated considering the obstacle and the flight path. The drones autonomously fly based on the control signal. It is possible to fly autonomously based on the proposed method by the drone and to crack down on traffic law violatios.

Keywords: Drone · Autonomous flight control · Traffic law violation

1 Introduction

Recently, a drone is being used to crack down on traffic law violations that occur at unspecified locations [1]. For autonomous flight of drones, various kinds of sensors such as Global Positioning System (GPS) and ultrasonic sensor, and camera images that can be mounted on drone are used to analyze and fly the surrounding environment [2, 3]. Some of the techniques that can be applied to autonomous flight of drones are deep learning based researches such as direct perception and end to end for autonomous driving of automobiles [4]. However, previous researches have focused only on avoidance of obstacles, or have focused on flying with a pre-set flight path plan. Research on autonomous flight is needed to automate tasks such as shooting vehicles that traffic law violations extensively with drones. In order to automate the drone, the autonomous flight control method is the most important core technology. In order to improve the existing drone, it is necessary to research a new approach to solve the optimal path planning problem and obstacle avoidance problem.

J. H. Park et al. (Eds.): PDCAT 2018, CCIS 931, pp. 329–334, 2019.
https://doi.org/10.1007/978-981-13-5907-1_35

This paper proposes a method for autonomous flight of the drone to control traffic law violation. The flight path for traffic law violations is generated by analyzing the flight records collected using the drone. A flight path that must be flown in order to enforce traffic law violations are planned by the generated flight path. In the process of flying to the flight path, a control signal is generated in consideration of information such as an obstacle. The drone fly to the flight location to control traffic law violations based on control signals.

In this paper, it is possible to intercept the vehicle violating the traffic law autonomously by using the drones. It can be applied not only to regulating the traffic law violations but also to surveillance and disaster relief delivery using the autonomous flight of the drone.

The structure of this paper is as follows: Sect. 2 describes related works. Section 3 details the proposed method for autonomous flight control of traffic law violations. Section 4 describes conclusions.

2 Related Works

This section analyzes the current situation of using drone to crack down on traffic law violations. And explains the methods for autonomous flight of drone.

2.1 Trends in the Enforcement of Traffic Law Violations Using Drones

Recently, drone has been flying to take pictures of traffic law violations in highways. It uses drone to control the speeding of the designated car, the speeding up of the road, the speed limit, and traffic accidents.

The Korean government takes the road by shooting a drone equipped with a camera in order to raise the awareness of safety of road traffic and compliance with traffic law violations. A driver of a car, and a violation of a designated car, and discovered a violation car as a driver of a highway bus by a drone equipped with a camera.

Drone of equipped with a camera can be used to prevent accidents as well as violating traffic law violations. By analyzing the images taken by the drone, accidental elements such as falling objects and sinkholes are prevented in advance.

2.2 Autonomous Flight Method Using Drone

The flight path that the drone should fly is set by designating the pilot based on the manually designated waypoint [5]. The drone autonomously fly based on the waypoints included in the flight path. The control signal is generated by the position difference of the waypoint at the current drone position. The pilot needs to set up the flight path in consideration of the environment.

The flight path for the flight is generated using the flight records of the drone [6]. The flight record includes the position of the drone and the direction of the camera mounted on the drone. The flight records are analyzed by K-Means algorithm to generate the K flight locations. The flight path is generated by connecting K flight locations. The flight path can be specified intuitively, and the drone can fly freely using the flight path. However, it is necessary to generate a flight path considering the environment in which the drone is flying.

Based on the images taken from the drone, the control signal is inferred with a deep neural network [4]. A deep neural network is learned by control signals for images and images taken from the drone as the pilot controls the drone. The captured image is input to the deep neural network to control the drone by inferring the control signal. However, there is a problem that is difficult to control with the flight path the pilot wants.

3 Autonomous Flight Control Method Using Drones

3.1 Overview

The process by which the drones generate control signals to autonomously fly is shown in Fig. 1. First, a flight path for traffic law violations is generated. Second, the flight path of the drone to fly is planned. Third, a control signal for controlling the drone is generated by using a flight path and a control signal for flying. The drone is controlled based on control signal for shoot of traffic law violations.

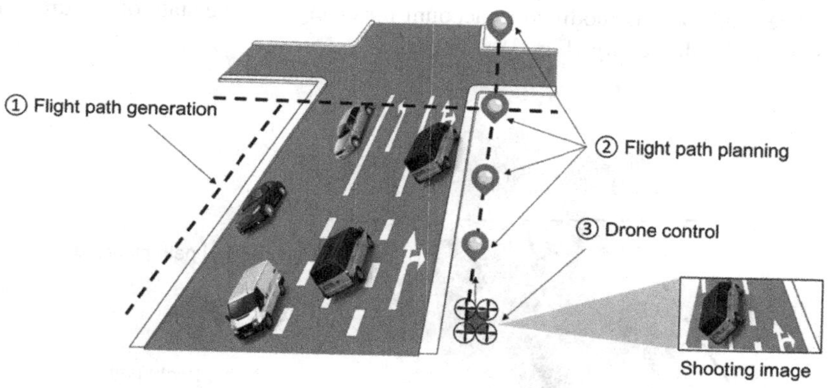

Fig. 1. The control signal generation process for the autonomous flight of the drone

3.2 Flight Path Generation

The flight path is generated by analyzing the flight records collected by the pilot flying the drone. The flight records are collected using GPS and sensing information measured by the sensors mounted on the drone. The flight state of the drone consists of GPS and sensing information that can be measured through the drone. Flight states are derived by analyzing the flight records. A flight path is generated by connecting flight states (Fig. 2).

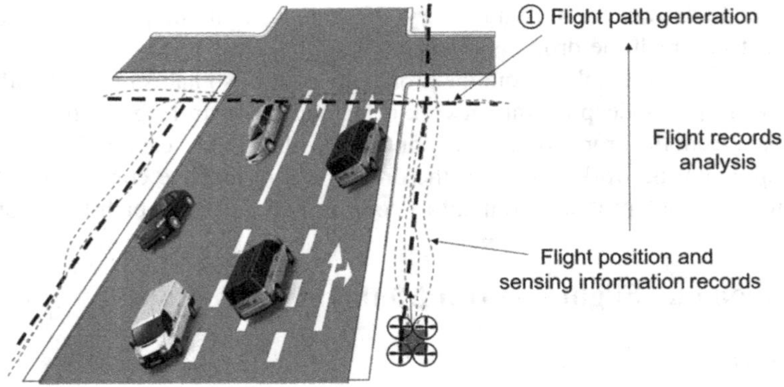

Fig. 2. Flight path generation process

3.3 Flight Path Planning

The optimal flight path is planned to the states of the drones using the generated flight path. The flight path of the drone is derived taking into account the current state of the drone. The flight path is modified to account for changes in the state of the drone that occurs when the drones fly (Fig. 3).

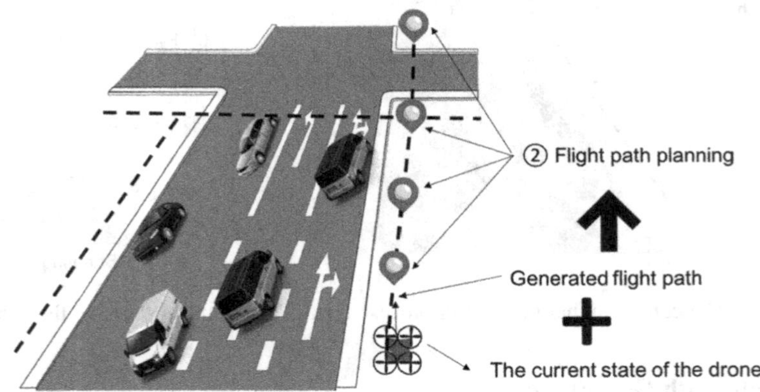

Fig. 3. Flight path planning process

3.4 Drone Control

The control signal is generated in consideration of the optimum flight path and the control signal based on obstacles. The drone fly based on the control signal. The drone fly by avoiding the flight path and obstacles specified by the pilot. The drone is able to fly and fly to shoot vehicles in violation of traffic regulations (Fig. 4).

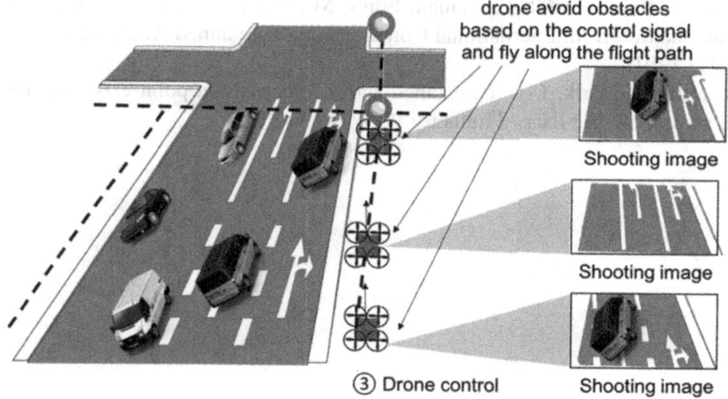

Fig. 4. Drone control signal generation and drone control process

4 Conclusion

This paper proposes a method of autonomous flight by drone to control traffic law violations. The flight path was generated using flight records to crack down on traffic law violations. The optimal flight path is calculated using the generated flight path. Control signal is generated by considering the inferred control signal for obstacle avoidance and optimal flight path. The drone was controlled based on the control signal.

It is possible to fly the drone considering the flight path for the autonomous flight and the state of the current drones. It is possible to control the flight in consideration of the flight path and obstacles to be performed in consideration of the image taken in the flight path and the drone.

Acknowledgments. This research was supported by Basic Science Research Program through the National Research Foundation of Korea (NRF) funded by the Ministry of Education (2018R1D1A1B07049990).

References

1. Ke, R., Wang, Y.H.: Motion-vector clustering for traffic speed detection from UAV video. In: IEEE First International Smart Cities Conference (ISC2), pp. 1–5 (2015)
2. Dentler, J., Kannan, S., Mendez, M.A.O., Voos, H.: A real-time model predictive position control with collision avoidance for commercial low-cost quadrotors. In: IEEE Conference on Control Applications (CCA), pp. 519–525 (2016)
3. Li, J., et al.: Real-time simultaneous localization and mapping for UAV: a survey. In: International Micro Air Vehicle Conference and Competition (IMAV), pp. 237–242 (2016)
4. Giusti, A., et al.: A machine learning approach to visual perception of forest trails for mobile robots. IEEE Robot. Autom. Lett. **1**(2), 661–667 (2016)

5. Santana, L.V., Brandão, A.S., Sarcinelli-Filho, M.: Outdoor waypoint navigation with the AR. Drone quadrotor. In: International Conference on Unmanned Aircraft Systems (ICUAS), pp. 303–311 (2015)
6. Kwak, J., Sung, Y., Park, J.H.: Unmanned aerial vehicle flight point classification algorithm based on symmetric big data. Symmetry **9**(1), 1–19 (2017)

Cost-Performance Comparison of Various Accelerator Implementation Platforms for Deep Convolutional Neural Network

Yechan Yu, HoJin Kim, Jinjoo Ha, Daewoo Kim, and Kang Yi(✉)

School of Computer Science and Electrical Engineering,
Handong Global University, Pohang, Republic of Korea
{21500429,21200213,21000770,21400067,yk}@handong.edu

Abstract. Due to the high accuracy, DCNN is a popular deep learning approach for object recognition and classification. But, the computing complexity of DCNN is too high for real-time application. Therefore, many acceleration methods like GPU and FPGAs are developed and competing with each other. The purpose of this paper is to assess the pros and the cons of many acceleration methods including GPGPU and FPGA-based approaches like Xilinx SDSoC, and Xilinx SDAccel. We will consider the installation cost (board price) as well as the operation cost (the energy consumption) and the speed of each acceleration method in the analysis.

Keywords: Reconfigurable high-performance computing · Heterogeneous computing systems · Intelligent computing and neural networks

1 Introduction

DCNN (Deep Convolutional Neural Network) is a deep learning adequate to object recognition based on vision data [1, 6]. Due to the large number of operations, computing accelerators such as GPGPU and FPGA are required. Many optimization techniques are suggested by past researches. In the previous research [2], we implemented DCNN for digit recognition and accelerated the recognizer by means of FPGA hardware accelerator [3, 4, 7]. Accelerator hardware platforms utilize GPGPU board and FPGA boards. In this paper, we will introduce optimization techniques and compare various DCNN acceleration methods in terms of cost and performance. The comparison includes the initial cost (hardware purchase and installation price) as well as the running cost which means the energy consumed by the accelerator hardware. By analyzing the energy consumption, the performance (the number of processed inputs per second) and the purchase price of the products, we can suggest a guideline such as the best performance-cost option for a specific period of continuous operations considering the life time of the hardware. These analysis results can give some resource investment advice for data center managers who should determine resource assignments and purchase new devices to maximize the performance and minimize the total cost of data center.

J. H. Park et al. (Eds.): PDCAT 2018, CCIS 931, pp. 335–344, 2019.
https://doi.org/10.1007/978-981-13-5907-1_36

2 Background

2.1 Accelerator System Development Environment

GPGPU Computing Model

General Purpose Graphics Processing Unit (GPGPU) is a widely used accelerator that supports parallel computing. The basic concept of GPGPU is utilizing the processing cores in GPU whose typical role is computing graphics. Since there is a huge number of processing units in GPU, it can be a powerful tool for SIMD parallelism. GPGPU computing model is shown in Fig. 1. There is one "Host", which can be an x86 processor, in connection with one or more GPGPU devices, and the host has control over the devices. A device has multiple "Multiprocessor", and each compute unit consists of L1 cache and shared memory and multiple "Core" [5]. The data transfer occurs between the main memory in the host and the global memory in the GPGPU through PCI-Express bus.

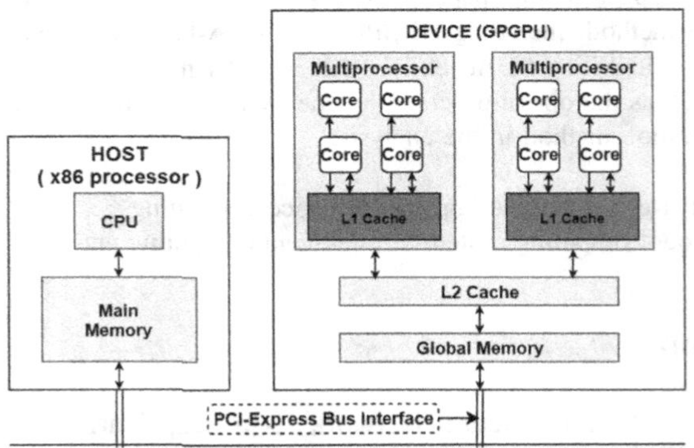

Fig. 1. GPGPU computing model

SDAccel Computing Model

SDAccel offers integrated environment of x86-baed processor and accelerator using Field Programmable Logic Device such as FPGA. The SDAccel Computing Model is shown in Fig. 2. Communication between processors and accelerator is performed via PCI-express bus. Host processor has a main memory and FPGA board has an off-Chip memory and an on-chip memory in FPGA. When we transfer data between Host processors and accelerator in SDAccel, host processor uses the main memory and FPGA uses the off-Chip memory like DRAM. But the memory access to the on-chip memory is faster than of the off-chip memory. To optimize computing on the FPGA, we need to minimize the traffic between the on-chip and off-chip memory. Unlike GPGPU, SDAccel has the advantage of hardware resource efficiency since

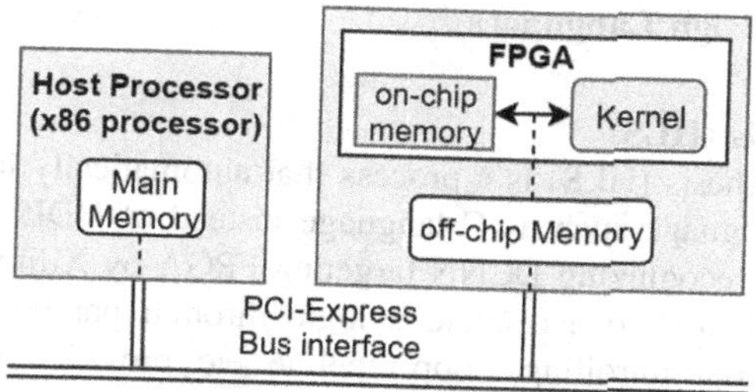

Fig. 2. SDAccel computing model

SDAccel allows optimal hardware architecture best fit to specific application design by the programmable logic circuit. The flexibility of SDAccel compared to the GPGPU makes the SDAccel approach very energy-efficient and cost-efficient alternative.

SDSoC Computing Model

SDSoC platform provided by Xilinx is used for the hardware and software codesign embedded environments where embedded processor in the PS (Processor System) unit utilizes the PL (Programmable) unit as the accelerator. SDSoC targets usually MPSoC devices from Xilinx including the Zynq-7000 SoC chip. One of the popular board with Zynq 7000 is the Zed board. This is a Standalone embedded system that has a processor and a Field Programmable Logic on a chip. The SDSoC Computing Model is shown in Fig. 3. The PS which runs software and the PL which implents hardware design communicate via the on-chip bus such as AXI bus. The detailed communication protocols are dealt with the Data mover that resides on the PL side.

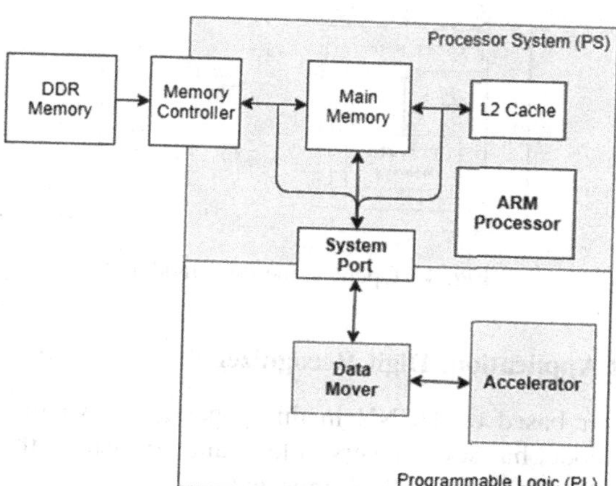

Fig. 3. SDSoC on-chip computing model

2.2 Accelerator Design Languages

High Level Synthesis (HLS)
The High Level Synthesis (HLS) is a process that automatically implement hardware from a high-level language such as C language instead of HDL. We can implement C/C++ code of digit recognizing DCNN targeting FPGA by Xilinx Vivado-HLS tool. In addition, it is possible to accelerate a logic through parallel logic optimization techniques such as Loop unrolling, Loop pipeline, etc. provided by HLS.

OpenCL
The accelerator optimization in heterogeneous computing environments targeting GPGPU or FPGA is commonly described in OpenCL. The OpenCL kernel program is written in OpenCL C which is based on C99 along with some additions. In OpenCL, there are four types of memories in a compute device: Global memory, Constant memory, Local memory and Private memory. Global memory and constant memory are accessible to all work items and belong to the off-chip memory. Local memory is shared among a work group and private memory is only visible to each work item. Local memory and private memory belong to the on-chip memory which has higher access speed than that of the off-chip memory. The memory model is shown in Fig. 4.

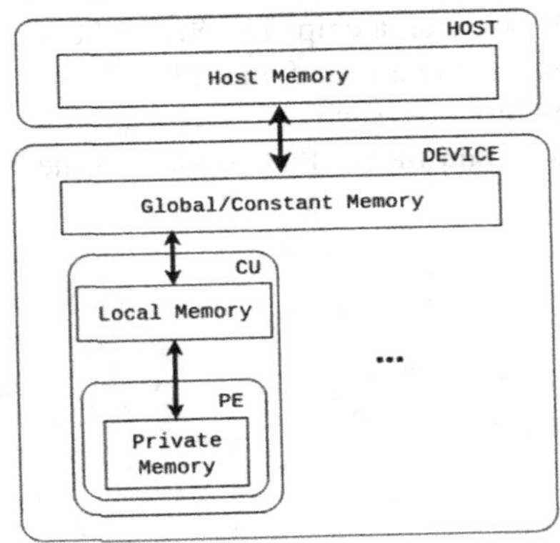

Fig. 4. OpenCL memory model

2.3 Our Target Application: Digit Recognizer

The digit recognizer based on DCNN in this paper is shown in Table 1. The digit recognizer of our model has seven layers in total and consists of three types of layers: Convolution, Max-Pooling, and Fully Connected.

Table 1. Specification of digit recognizer model

Layer number	Layer type	Input planes	Input plane size	Mask size	Output planes	Output plane size	# of operations
C1	Conv.**	1	1,024	25	32	784	1,999,300
P2	Max***	32	784	4	32	196	330,848
C3	Conv.	32	196	25	32	100	252,600
P4	Max	32	100	4	32	25	42,200
C5	Conv.	32	25	25	256	1	111,204
F6	Fully	256	1	1	256	1	197,888
F7	Fully*	256	1	1	10	1	7,710

*Fully Connected, **Convolutional, ***Max Pooling

3 Accelerator Design Optimization Techniques

3.1 Optimization Techniques for GPGPU in OpenCL

Use of On-chip Memory: As mentioned earlier, frequent access to the global memory causes poor performance. Thus, it is one of the key strategies of optimization to make use of the on-chip memory instead of the off-chip memory as much as possible. The on-chip memory in OpenCL refers to either local memory or private memory.

Optimal Number of Work Groups: A Device has its own number of Compute Units. As a work group is mapped to a CU, it can be said that the number of CUs and the number of work groups are closely related. From our experiment, we concluded that setting the number of work groups to multiples of the number of CUs yields a better performance.

Workloads vs. Memory Traffic: If work items in different work groups refer to the same address in the global memory simultaneously, global memory traffic gets heavy. In this case, we can reduce the global memory access by having less work items, which means a work item is given higher workload. The workload on a work item and the global memory access time is a trade-off. Reducing the work item size surely alleviates global memory traffic. However, it reduces the concurrency of computation as well. In that sense, we cannot say having less work items guarantees better performance. Therefore, it is important to figure out the optimal point where the combination of the two factors works best.

3.2 Common Optimization Technique for Both SDAccel and SDSoC

Memory Localization: The memory structure of SDAccel and SDSoC consists of on-chip memory and off-chip memory. Data communication between CPU and FPGA is handled by off-chip memory. In data communication, it is time-consuming to access this memory redundantly in the kernel. To resolve this problem, an on-chip memory such as a FIFO and a block RAM in the FPGA should be used during data transfer. Then the speed is increased, and it is called memory localization.

Data Type Conversion to Fixed Point: In FPGA, multiplication and division operations require a lot of hardware resources. For floating point data type, they use more time and resources than Fixed Point type. Therefore, it is one of the methods to optimize resources and time by converting DCNN's 32bit floating data type to Fixed Point type.

Loop Pipelining: Loop Pipelining technique uses a hardware instance to execute the next loop process before the end of the loop being processed in a sequential loop. This improves loop iteration throughput.

Loop Unrolling: The Loop Unrolling technique is a way to reduce the number of loop sequential iterations by processing loop operations in parallel on multiple hardware instances.

Array Partitioning: HLS has one read and one write port for one Array. If Loop Unrolling is applied, a problem occurs because of the multiple data accesses to one array. To resolve this problem, increase in the number of read and write ports by dividing one array is needed.

Loop Reordering: The Loop Reordering technique changes the order of nested loops. Because the optimizations of loops are applied from the inner loop, the nested loops need to be rearranged. The loops that optimizations are applied should be placed inside. Also, when choosing the loop to apply the optimization, it should be placed in the least accessible way considering the number of accesses of memory.

4 Experimental Results

4.1 Implementation Environments and Results

GPGPU Implementation

The implementation environment using GPGPU is as follows. The host code was written in C++, and the kernel code in OpenCL C with OpenCL 1.2 version. The host program was executed on Intel(R) Core(TM) i7-6700 CPU with 16 GB RAM, and the specification of the GPU on which the kernel ran is given in the Table 2.

In the result, running the DCNN on CPU took 3.14 ms while running on GPGPU took 1.3 ms, and 52 W of energy was consumed per sample.

Table 2. GPGPU specification

Product name	Max work group size	Local memory size	Max compute units
NVIDIA GeForce GTX 1060 6 GB	{1024, 1024, 64}	49152 Bytes	10

SDACCEL Implementation

The implementation environment using SDAccel is as follows. We used a board from COTS Technology mounted Kintex XCKU115 from Xilinx. SDAccel version v16.4 and Vivado HLS were used for hardware optimization. The host program was executed on Intel(R) Core(TM) i7-4770 CPU with 32 GB RAM, and the synthesis result of the FPGA by SDAccel is shown in the Table 3.

Table 3. FPGA resource utilization for our best results in SDAccel

Resource	FF	LUT	DSP	BRAM (18 KB)
Estimate	308,448	308,941	3850	1,391
Available	1,326,720	663,360	5520	6320
Utilization (%)	23.25	46.57	69.75	32.20

In the result, the execution time of the DCNN on FPGA took 2.40 ms per sample, and 22 W of energy was consumed.

SDSoC Implementation

The implementation environment using SDSoC is as follows. We used ZedBoard from Avent. SDSoC uses version v15.4 and Vivado HLS was used for hardware optimization. The host program was executed on dual-core ARM Cortex 9 in Zynq-7000 SoC, and the specification of the FPGA is shown in the Table 4.

Table 4. FPGA resource utilization for out best results in SDSoC

Resource	FF	LUT	DSP	BRAM (18 KB)
Estimate	23,431	13,522	132	228
Available	106,400	53,200	220	280
Utilization (%)	22.02	25.42	60.00	81.79

In the result, the execution time of DCNN using the PL of Zynq took 14 ms, and 2.4 W of energy was consumed per sample.

4.2 Performance and Cost Comparison

Table 5 shows the final performance and cost comparison. We measured execution time ad energy consumption. And, based on the electricity price in Korea (103.8 KRW/kWh on average), device purchasing price in the market (In the SDAccel case, the chip is assumed 50% off the list price) and cooling price in GPGPU [8]. The cost function is the sum of device price and electricity price by energy consumption per sample as shown in Eqs. 3 and 4. Equations 3 and 4 are driven through Eqs. 1 and 2. In terms of Cost/Sec and Cost/Sample, SDSoC has the best performance. However, GPGPU showed the best performance in terms of computing speed and device price.

Table 5. Each performance and cost

Type	sec/sample	W (J/sec)	Device cost (KRW)	Won/kWh	uWon/sample	uWon/sec
GPGPU	1.3 ms	52.0	343,860	103.8	3.90	3000
SDAccel	2.4 ms	22.0	2,750,000		1.52	634
SDSoC	14.0 ms	2.4	538,000		0.97	69.2

$$\frac{\mathrm{Cost_{run}}}{Sample}\left[\frac{\mathrm{Won}}{sample}\right] = \left[\frac{sec}{sample} \times \frac{J}{sec} \times \frac{kWh}{3600kJ} \times \frac{Won}{kWh}\right] \tag{1}$$

$$\frac{Cost_{run}}{time}\left[\frac{Won}{sec}\right] = \left[\frac{Won}{sample} \times \frac{sample}{sec}\right] \tag{2}$$

$$\mathrm{Cost_{tot}}\left(\mathrm{N_{Sample}}\right) = \mathrm{Cost_{Device}} + \frac{Cost_{run}}{Sample} \times N_{Sample} \tag{3}$$

$$\mathrm{Cost_{tot}}(\mathrm{time_{run}}) = \mathrm{Cost_{Device}} + \frac{Cost_{run}}{time} \times time_{run} \tag{4}$$

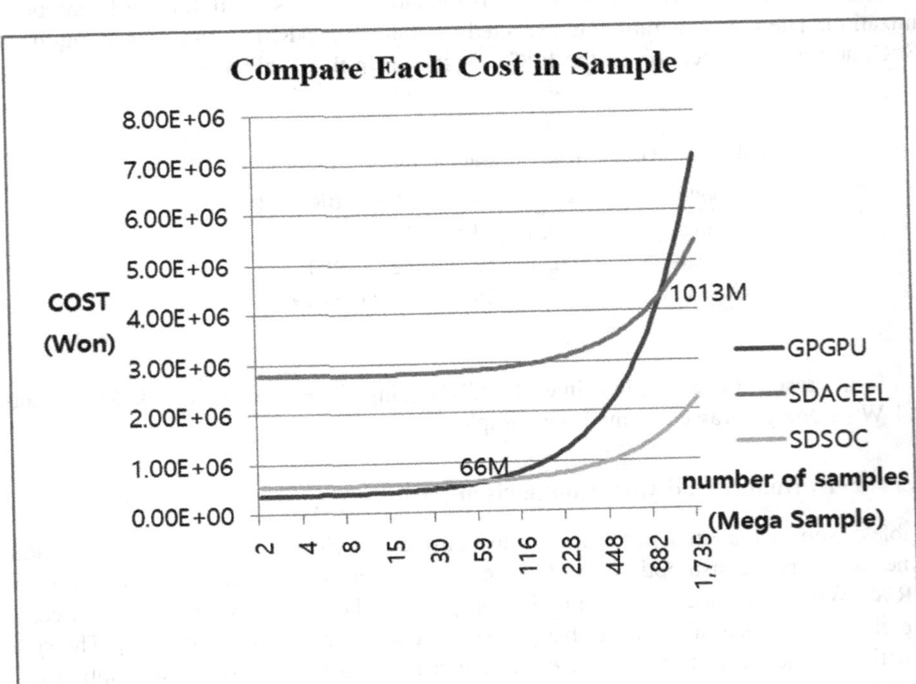

Fig. 5. Comparison of costs in sample

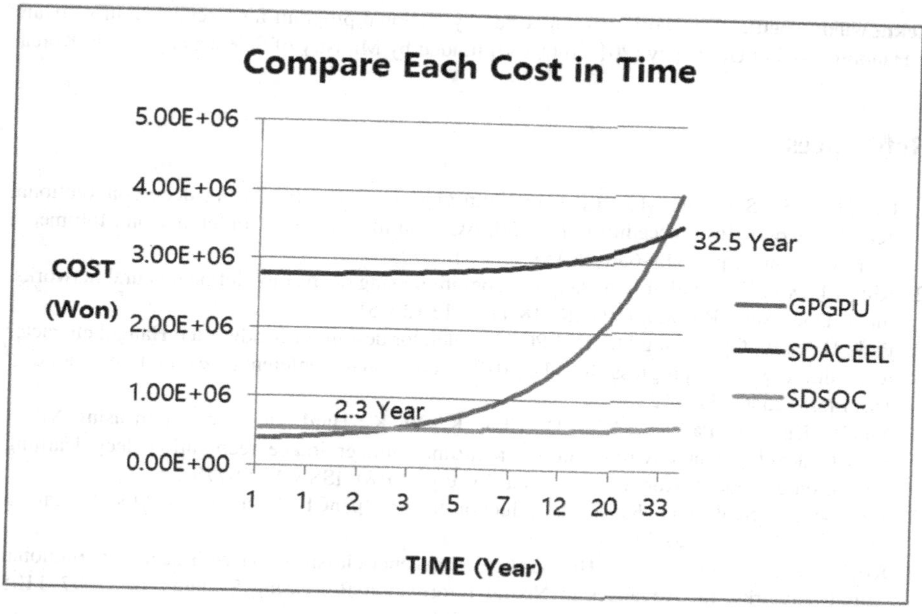

Fig. 6. Comparison of cost in time

GPGPU has the lowest device cost. But, SDSoC and SDAccel using FPGA are better than GPGPU in terms of energy-efficiency. Because of this, SDSoC and SDAccel have lower running cost. The results are shown in Figs. 5 and 6. Figure 5 shows the cost of each device according to the number of the processed samples. SDSoC is better than GPGPU when processing more than 66M samples, and SDAceel is better than GPGPU when processing more than 1,013M samples. Figure 6 also shows the cost of each approach over time. SDSoC is more efficient when used over 2.3 years than GPGPU, and SDAccel is more efficient than GPGPU when used over 32.5 years.

5 Conclusion

In this paper, we describe the optimization techniques that can be applied when implementing DCNN in various acceleration platforms and compared the platforms in terms of cost and performance. In order to assess accurately with real example, we implemented digit recognizer based on DCNN with GPGPU, SDAccel, and SDSoC and measured the execution time and power consumption of each cases. We also included the board cost of each platform as well as the energy consumption for the cost aspect consideration. Our experimental results assumed the electricity price in Korea shows that SDSoC is more attractive than any other approaches in terms of cost. The SDAccel shows moderate performance and energy consumption among three candidates. The problem with SDAccel is the high device price compared to low-end GPGPU. However, we have not yet considered the cooling cost, and we can think that it is most economic choice to use the FPGA in a country like Japan with higher electricity rates.

Acknowledgement. This work was supported by National program for Excellence in software at Handong Global University (2017-0-00130) funded by Ministry of Science and ICT in Korea.

References

1. Farfade, S.S., Saberian, M.J., Li, L.J.: Multi-view face detection using deep convolutional neural networks. In: Proceedings of the 5th ACM on International Conference on Multimedia Retrieval, China, pp. 643–650 (2015)
2. Kim, I.J., Xie, X.: Handwritten Hangul recognition using deep convolutional neural networks. Int. J. Doc. Anal. Recogn. (IJDAR) **18**(1), 1–13 (2015)
3. Park, H., et al.: Optimizing DCNN FPGA accelerator design for handwritten Hangul character recognition: work-in-progress. In: The 2017 International Conference (2017). https://doi.org/10.1145/3125501.3125522
4. Yu, Y., Kim, H., Park, Y., Kim, D., Choi, K., Yi, K.: Hardware acceleration using Xilinx SDSoC development environment for real-time number image recognition deep learning system on Zedboard, Korea Multimedia Society (2018). ISSN 1229-778X
5. Tompson, J., Schlachter, K.: An Introduction to the OpenCL Programming Model. Pearson Education, London (2012)
6. Krizhevsky, A., Sutskever, I., Hinton, G.E.: Imagenet classification with deep convolutional neural networks. In: Advances in Neural Information Processing Systems, pp. 1097–1105 (2012)
7. Sozzo, E.D., Baghdadi, R., Amarasinghe, S.P., Santambrogio, M.D.: A common backend for hardware acceleration on FPGA. In: IEEE International Conference on Computer (2017). https://doi.org/10.1109/ICCD.2017.75
8. Weijema, E.: How to calculate electrical costs for cooling and power consumption (2011). https://www.vmguru.com/2011/06/how-to-calculate-electrical-costs-for-cooling-and-power-consumption/

A Prediction Method of Solar Power Generator using Machine Learning Techniques

Jungseok Cho, Jeongdoo Lee, and Doosan Cho(✉)

Sunchon National University, Suncheon, EE, Korea
dscho@scnu.ac.kr

Abstract. We first purchased a solar power generator, installed it in the right place, and then collected power generation data for the next 12 months. First, we collected weather and power generation data to excel file. For the weather data, the data provided by meteorological agency was secured to ensure the reliability of the information. The collected data can be monitored anytime and anywhere using PC or mobile device. The user interface of these services is planned to be upgraded to be more usable after the accuracy of the service provided. The core development contents are collected data based power generation prediction technique. This is because system failure monitoring can be performed only when a power generation prediction system is configured. Our power generation prediction algorithm was developed based on machine learning. LSTM was used among linear regression and artificial neural network in machine learning. Experiments using the machine learning showed that the accuracy could be increased to 85% when compared with actual power generation.

1 Introduction

There is a need for a method that takes into consideration various factors in the power analysis of power generation facilities [1–4]. This is because there are days when the sunlight intensity is high and the relative humidity and precipitation probability are high, and sunlight intensity is high, and vice versa. The power output of the power plant not only has a complex relationship with the solar radiation intensity, but also has a complex relationship with each meteorological variable statistic. Each meteorological measurement also has a correlation, for example, the relative humidity and the precipitation probability are correlated. Of course, the correlation value does not remain constant according to the different weather factors, but the relative humidity, precipitation probability, and the electric current usually show a proportionally increasing relationship.

Table 1 shows the correlation coefficient for each meteorological measurement using the Pearson product-correlation coefficient, which is the covariance of the two variables divided by the product of the standard deviation. The higher the absolute value of the correlation coefficient, the stronger the correlation between the two meteorological indicators. A positive correlation indicates an increasing linear relationship, while a negative correlation indicates a decreasing linear relationship. The complex relationships between meteorological indicators and solar power plant outputs in this table provide insights into automated forecasting models using machine learning techniques.

J. H. Park et al. (Eds.): PDCAT 2018, CCIS 931, pp. 345–352, 2019.
https://doi.org/10.1007/978-981-13-5907-1_37

Table 1. Correlation between meteorological parameters

	Temperature	Cloudiness	Humidity
Temperature	1.0000	0.1955	0.7896
Cloudiness	0.1955	1.0000	0.4932
Humidity	0.7896	0.4932	1.0000

2 Prediction Modeling

Machine learning is a discipline that gives computers the ability to learn and intelligence. Artificial neural networks are machine learning that mimics the human brain structure. Conventional artificial neural networks use a sigmoid nonlinear function similar to neurons in the brain. Deep learning is used in various fields such as prediction, clustering, classification, recommendation, unmanned system, and game, as well as video, voice and natural language processing using various networks such as CNN (Convolution Neural Networks) and RNN (Recurrent Neural Networks). TensorFlow, an open platform that we use, is a machine learning system that Google has released as open source for developers, data scientists, and researchers. It is a library that uses dataflow graphs to perform operations. The nodes of the graph represent various operations such as differentials and the edges of the graphs connecting the nodes are multidimensional data arrays called tensors. In other words, the data stored in the multidimensional array is subjected to machine learning through various operations. Tensor flow supports various languages through Python, C++ and SWIG. It has rich expressiveness through dataflow graphs, computational structure, and automatic processing of differential calculations by defining target functions. Anyone can use it, but in order to use it efficiently, considerable effort is required in data processing/learning training/modeling. It is reported that at least one year of image recognition training should proceed with pre-processed data in order to achieve accuracy of 90% or more. We are also training to select data based on generation data of more than one year, and it is expected that it will take more than six months for the training learning process.

We show the observed and predicted meteorological indicator values as a time series that varies with weather patterns and seasons. As mentioned above, the solar radiation intensity varies with different meteorological methods, making it difficult to predict. The simplest form of linear regression analysis was chosen as the first step to develop an accurate prediction model. In order to generate each prediction model, training data (from February 16 to February 17, or February 17 to February 17, 2005) were used as input for the outputs of the weather data and power generation facilities. Machine learning technology predicts and outputs power generation output as well as four meteorological measurement indicators. To test the accuracy of the model first, the last two months of the input data set was used as the test suite. In March and April, we will continue to collect data and measure the accuracy and improve the prediction techniques to achieve the target accuracy. When the improvement of prediction technique enters into the stable state, accurate state judgment technique can be applied. Predictive models that use machine learning are generally more accurate as more training data is available.

The model we create is a simple function in the form below, which calculates the output value through weather statistics for several days of the year. You can also add time of day as an additional measure, but to make the experiment easier, the experiment should focus on the average output prediction. We plan to compare the accuracy of the initial predictive model (linear regression analysis) with the baseline and continue to improve.

Estimated average daily power generation value = F(Day, Temperature, Dew-Point, SkyCover, Precipitation, Humidity)

2.1 Linear Least Squares Regression

F represents a function to which various prediction techniques are applied. Inputs use values from 0 to 365 daily, temperature at Celsius, percentage between 0% and 100% of cloudiness, a precipitation probability of between 0% and 100%, and a percentage of humidity between 0% and 100%. All input data are normalized before applying the linear least regression method. To quantify the accuracy of each prediction model, we use the mean square root error (RMS-Error) between the predicted output and actual output. The RMS error is a well-known statistical value. The accuracy of the predicted value of the measured value by the time series model is expressed by the RMS error and 0 indicates the accurate prediction. Therefore, the closer the RMS-Error is to 0, the more accurate the prediction of the model.

First, the linear least squares regression method is applied to predict the output value. A linear least squares regression is a simple and commonly used technique for estimating the relationship between a dependent variable or a response variable (eg, output of a power plant) and a set of independent or predictive variables. The regression minimizes the sum of the squared differences between the observed output values and the output values predicted by the linear approximation of the predicted weather data. Applying linear least squares to 12 months of training data produces a predictive model with coeff of each measure.

$$\text{Output predicted value} = \text{coeff\_T} * \text{Temperature} - \text{coeff\_H} * \text{Humidity} - \text{coeff\_C} * \text{Cloud\_Index} + \text{coeff\_D} * \text{Dew Points}$$

Least squares regression is a method of finding a line that summarizes the relationship between two variables within the region of variable x. Let the line be the following Eq. (1).

$$\mathrm{Y} = \mathrm{a} + \mathrm{bx} \tag{1}$$

here,

$$\mathrm{b} = \mathrm{r} * \mathrm{SDy}/\mathrm{SDx}$$
$$\mathrm{a} = \mathrm{Y}' - \mathrm{bX}'$$

$$\text{Slope} = (\mathrm{N}\Sigma\mathrm{XY} - (\Sigma\mathrm{X})(\Sigma\mathrm{Y}))/(\mathrm{N}\Sigma\mathrm{X}\hat{\ }2 - (\Sigma\mathrm{X})\hat{\ }2)$$

Here, b is the slope of the regression line, a is the intersection of the vertical axis and the regression line on the coordinate plane, X 'is the average of the x values, Y is the average of the y values, SDx is the standard deviation of x, SDy is the standard deviation of y

To find the minimum-squared regression equation, proceed as follows.

Step 1:
Store the number of x values.
N = 55 (number of input values)
Step 2:
Computes XY, X^2 for the given values.
Step 3:
Calculate $\Sigma X, \Sigma Y, \Sigma XY$, and $\Sigma X^{\wedge}2$ for the given values.
Step 4:
Assign values to the slope formula given above.
Slope (b) $= (N\Sigma XY - (\Sigma X)(\Sigma Y))/(N\Sigma X^{\wedge}2 - (\Sigma X)$
Step 5:
Now re-assign to the intercept formula above.
Intercept (a) $= (\Sigma Y - b(\Sigma\ X))/N$
Step 6:
The calculated values are then substituted into the regression equation.

$$\text{Regression Equation}(y) = a + bx$$

Temperature variable and power generation linear regression equation = 812679.6037 + 6678.5199x
Meteorological variables and power generation linear regression equation = 943294.893 - 25094.6748x
Humidity variables and power generation linear regression equation = 866748.518 + 931.2996x
To calculate the predicted value of the y value for the variable x (21.4 humidity, 2.9 temperature, 0.0 cloud) = February 2, 2017, you can substitute the value in the above equation.

$$\begin{aligned}\text{Regression Equation (temperature y)} &= 812679.6037 + 6678.5199(2.9)\\ &= 832047.3114\end{aligned}$$

$$\text{Regression Equation} = 943294.893 - 25094.6748(0.0) = 943294.893$$

$$\begin{aligned}\text{Regression Equation (humidity y)} &= 866748.518 + 931.2996(21.4)\\ &= 886678.3294\end{aligned}$$

The actual power generation was 1011392, and the accuracy was 82.2675% (temperature), 93.2669% (cloud) and 87.6691 (humidity) respectively. We will finally develop a step by multiple regression analysis in the above single regression analysis and complete the prediction model by calculating coeff_T, coeff_H, coeff_C, and coeff_D. We will use the complete equation to obtain the output predicted value, which will provide a higher accuracy than the single equation (Fig. 1).

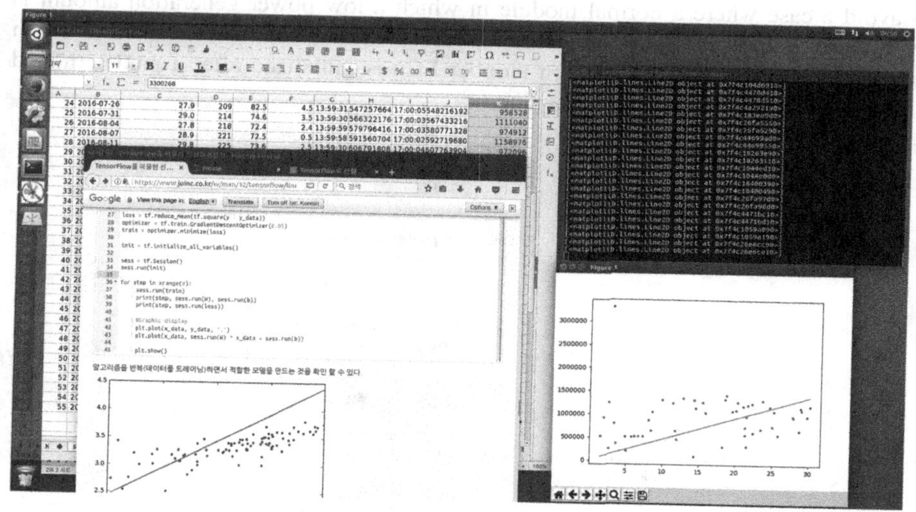

Fig. 1. Machine running with tensor flow (single regression analysis)

We validate the linear regression prediction model with a training data set (February 2016 to February 2017) and a test data set (March and April 2017) to verify the prediction accuracy. Here we observed an RMS error of 1.3085. The linear regression prediction model currently operates at an accuracy of 87.6% for 60 (day) tested values with 365 (day) trained values.

Implementation of Generation Status Diagnosis Algorithm: When the power generation forecasting algorithm has 85 ~ 90% accuracy, it is possible to capture abnormal development through suitable threshold value.

3 Experimental Results

Figure 2 shows the procedure for diagnosing the status of each module using the power generation prediction algorithm presented above. As shown in the figure, the power generation predicted value (Pp) is first calculated using the linear regression analysis method, and 0.8 pu of the estimated magnitude is set as the abnormal state determination reference value (Ps) of the module. The reference value is compared with the measured value (Pn) of each module actually outputted. If Pn is larger than Ps, it is judged as a normal module because it shows a normal output. On the other hand, if Pn is smaller than Ps, it indicates an abnormal state, so it is temporarily determined by the abnormal module to accumulate the determination number ECn and store it as an abnormal module in the warning table. If the determination number of the module temporarily determined as an abnormal state exceeds the predetermined number of times (10 times the current temporary setting - the subsequent optimization plan) by repeating the above-described process, the module is finally determined to be an abnormal module, Store it in a table and notify the administrator. In addition, in order to avoid a case where a normal module in which a low power generation amount is temporarily output due to the influence of clouds or shadows is determined to be an abnormal module, the determination number ECn is initialized after a predetermined time is determined to be a normal value.

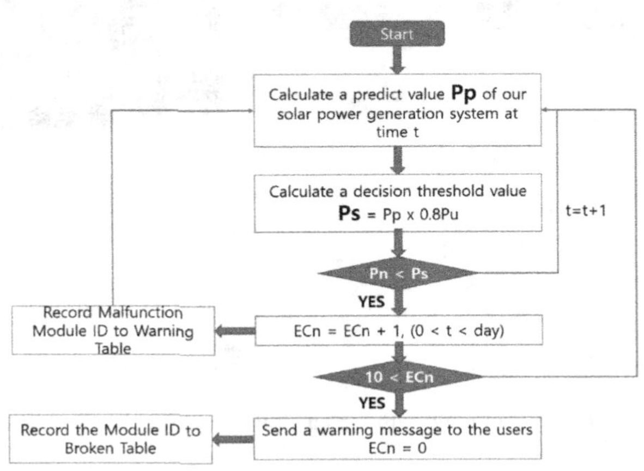

Fig. 2. State diagnostic algorithm flowchart

It is expected that it will take about one week to implement and implement the above-mentioned condition diagnosis algorithm in the system. Algorithms are already constructed, and the accuracy of prediction techniques using machine learning as a very simple and accurate technique will be the key to determining the accuracy of state determination.

Figure 3 shows the weighted linear regression technique using matlab and the results of the program execution. The x-axis represents the period from May 17 to January 18, and the y-axis represents the power generation in W units. The WLR prediction indicated by the orange line shows the predicted result of the program we implemented, and the blue test set shows the actual power generation. Since the influence of environmental variables such as humidity, solar radiation amount, air volume, air temperature, and the like has been formulated and predicted, the day of the week deviating from the formalized formula may increase in error. On average, the forecast results are at 89.5%. Figure 4 (left) shows only the prediction error. It can be seen that the error is high mainly in winter. When snow falls, most of the solar radiation is reflected and the generation amount is abnormally decreased. On the day when the solar radiation is good, it is difficult to make a formal prediction because the power generation amount can be obtained in summer. Figure 4 (right), and the numerical values are shown. It can be confirmed that the error value is greatly improved. If winter environment improvement devices, such as snow removal devices and pollutant cleaning devices, are available, this winter error increase part can be solved.

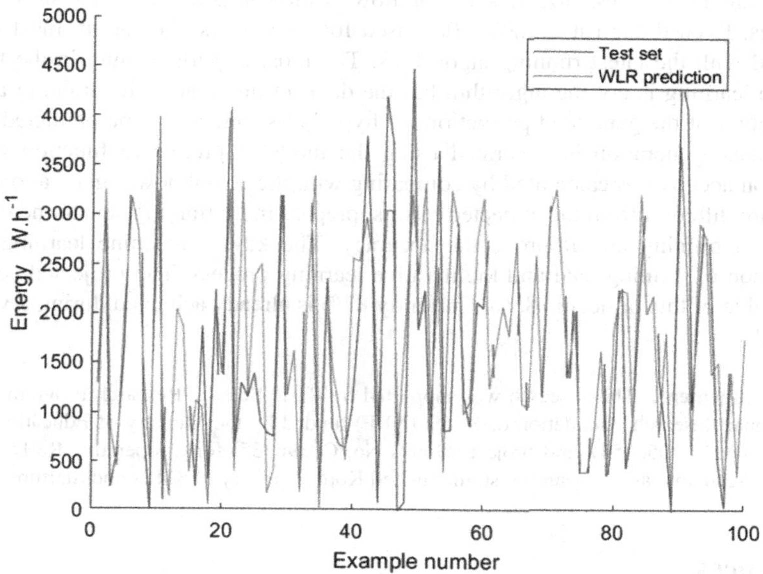

Fig. 3. Predicted results of weighted linear regression

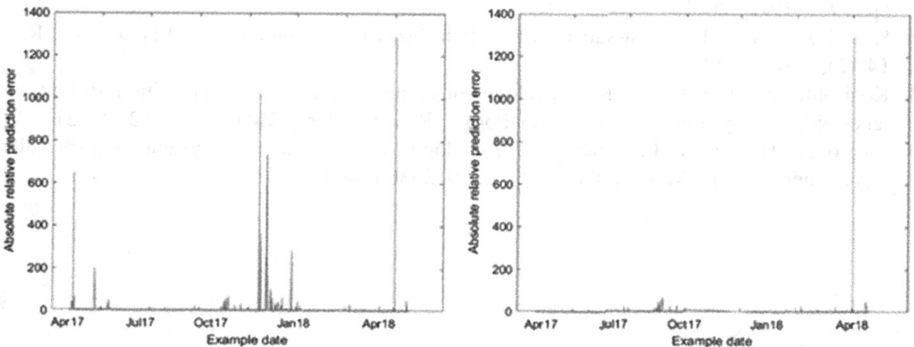

Fig. 4. (Left) Prediction error (R) Prediction error except winter data

4 Conclusion

At the core of the solar power generation prediction system is a machine learning algorithm called LSTM or linear regression. The prediction program is usually organized as the following. First prepare the training data set. We collected solar power generation for one year in database we built. We will use this as a training set. The running algorithm uses Google's tensor flow, which is primarily used in solar PV programs. Even though it is most often used for economics simulation, most of it is equipped with the latest running algorithms. The most important thing in the field of machine learning is not the algorithm but the data set preparation for training and the construction of the generated prediction (or hypothesis function). The predicted power of the latest generation is generated using the modeled predictive function and the prediction accuracy is calculated by comparing with the actual power generation. If the results are different from the expected results, prepare more training sets to increase the amount of training and improve the accuracy. The key to machine learning is the preparation of training data and the iterative learning process. The projected accuracy target value of this project is 85%. Currently 87% is already achieved during 2 years of this study.

Acknowledgement. This research was supported by Basic Science Research Program through the National Research Foundation of Korea (NRF) funded by the Ministry of Education (NRF-2018R1D1A1B07050054), and project (Grants No. C0396335) for Cooperative R&D between Industry, Academy, and Research Institute funded Korea Ministry of SMEs and Startups in 2016.

References

1. Choi, S.-H., Oh, D.-E.: The technical trend and prospect of platform integration for smart grid system. In: The Korean Institute of Electrical Engineers for the 42nd Summer Conference, pp. 1977–1978 (2011)
2. Kao, C., Chyu, C.-L.: Least-squares estimate in fuzzy regression analysis. Eur. J. Oper. Res. **148**(2), 426–435 (2003)
3. Koutroulis, E., Kalaitzakis, K.: Development of an integrated data acquisition system for renewable energy sources systems monitoring. Renew. Energy **28**(1), 139–152 (2003)
4. Forero, N., Hernandez, J., Gordillo, G.: Development of a monitoring system for a PV solar plant. Energy Conv. Manag. **47**(15–16), 2329–2336 (2006)

A Glyph Reduction Evaluation of Combined Jeongeum-Fonts

Ga-Yeon Kim and Jeong-Yong Byun(✉)

Department of Computer Engineering, Dongguk University,
123 Dongdae Ro, Gyeongju, Gyeongbuk 38066, Republic of Korea
{yeon, byunjy}@dongguk.ac.kr

Abstract. It is not easy to develop fonts for individual syllables because the scientific principles of Hunminjeongeum generate a huge number of syllables. So we need to develop syllable fonts combined with three phonemic glyphs. Here, the development of glyphs according to each syllable structure is also a great effort, so we think that it is important to reduce their number using structural properties of syllables and factor font drawing. Therefore, this paper aims to evaluate the reduction method of glyph number and its efficiency.

Keywords: Hunminjeongeum · Jeongeum · Haerye · Middle Initial · Final letters · Combined syllable font · Unicode · Grapheme Design target glyph

1 Introduction

King Sejong told that he newly made 28 letters in the preface of the book called Hunminjeoneum or Jeongeum that stands for correct sounds for the instruction of the people. However, the total number actually becomes not only 28 but also 45 [1, 2] because the set of last letters is equal to that of the initial letters of which the number is seventeen.

Scholar JungInji's preface of Jeongeum says "Everything in nature has sound and there are letters for them". In applying them for computer, it is not easy to develop a huge number of fonts for individual syllables because the scientific principles of Jeongeum can generate about 39.9 billion syllables [2] by the fives explanations such as the designs of letters explanation, initial sounds explanation, middle sounds explanation, final sounds explanation, and combining letters explanation. So we need to develop syllable fonts combining with three phonemic glyphs. Since the number of glyph sets is also large, developing it requires a great deal of time and effort. So we think that it is important to reduce their number using properties of syllabic structure and vector font drawing.

In this paper, we propose strategies to reduce the number of some middle letter glyphs as well as the initials using some properties of vector fonts and then will show some enhancement of reduction rate.

J. H. Park et al. (Eds.): PDCAT 2018, CCIS 931, pp. 353–358, 2019.
https://doi.org/10.1007/978-981-13-5907-1_38

2 Related Work

The most previous studies [2, 3] have been related on how to implement Jeongeum. First, it is fully to represent all syllables of Jeongeum using composite syllable fonts. The syllables consist of initial, middle, and final letters and each number is seventeen, eleven, and seventeen respectively.

We analyzed all syllables and identified that there are three factors that make syllable structure different. The first is the shape of middle letter. It has two types, vertical and horizontal. They can be repeated from 2 to 3 letters. Here, the same shapes are repeatedly written and mixed together. Second, the structure varies depending on the presence or absence of middle letters. Third, the number of first letters and final letters depends on how many. Format 1 stands for only one element glyphs of three graphemes such as first, middle, and final phonemic letter. Our study originally defined nine formats from format 1 to format 9.

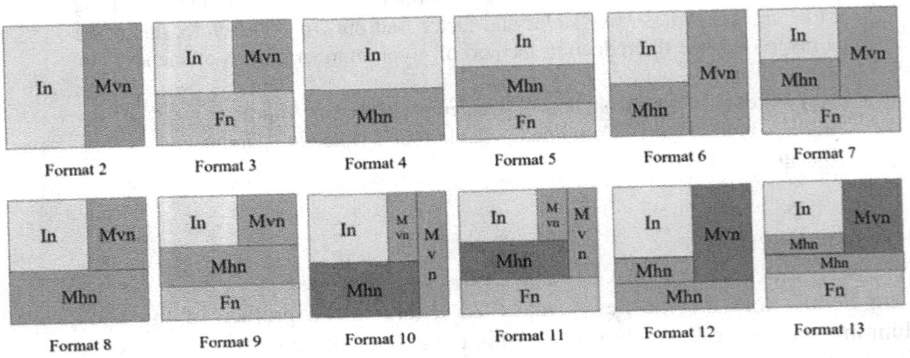

Fig. 1. Formats of syllable structure

3 Format Extension

As shown in Fig. 1, since the combination of some middle glyphs affects the shape of the middle glyphs, they need to be classified to design them more precisely. We analyzed some combination of middle letters and found two kinds such as vertical-horizontal-vertical and horizontal-vertical-horizontal. Four formats are extended from format 10 to format 13. Thus three types and thirteen formats were derived to produce a more balanced font for the full set of the syllables. Examples of syllables are shown in Table 1 as a result of developing the full set of glyphs. Second we have studied some reduction of initial letters by considering the letter shape according to properties of the formats and types of syllable. Some glyphs of initial letter reuse in case of the same shape and size. For instance, in the case of a format 3's initial letters, a single glyph is output three times to each position to complete a syllable.

Table 1. Example of formats

Format 2	Format 3	Format 4	Format 5	Format 6	Format 7
 ㅁㄴㅑ	 김	 소	 ㅎㅁㅡㅡㅂ	 ㅂㅅㅑ	 ㅎㅇㅡㅣㄱ
Format 8	**Format 9**	**Format 10**	**Format 11**	**Format 12**	**Format 13**
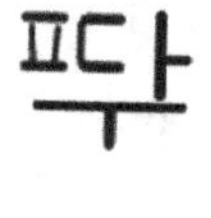 ㅍㄷㅜㅏ	 ㄷㄱㅣㅡㅇㄱ	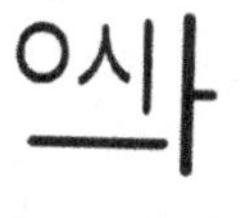 ㅇㅅㅣㅡㅏ	 ㅂㄷㅏㅂ	 ㅎㅁㅏㅗ	 ㅅㄷㅓㅜㅇ

We have newly developed MS IME that was actually implemented the principle of Hunminjeongeum. The four formats have shown much balanced fonts like between format 10 and format 13. Here, Fig. 2 shows an example of using only one set of glyphs repeatedly using the drawing characteristics of the vector font moving from left to right direction according to the complexity of the type. For example, two x and y-coordinates for first letter glyph are (0,510) and (612,786) in left case of Fig. 2. Middle case means that two glyphs are repeated twice with a set of glyphs of width 306 and right case says that three glyphs are repeated three times with a single set of glyphs of width 204. Here we reduced total three sets for two cases.

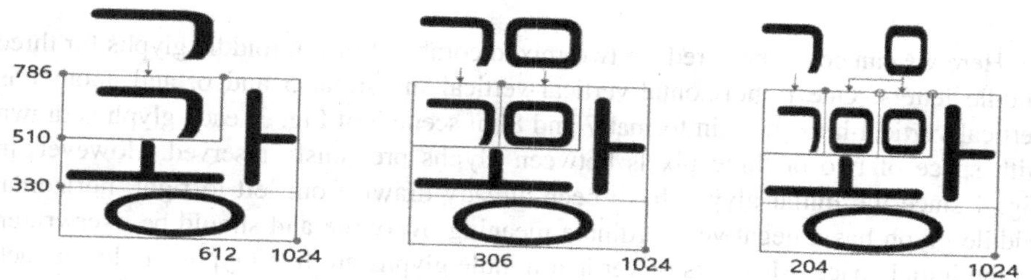

Fig. 2. Nature of vector font

4 Reduction of Middle Letters Glyphs

We consider the nature of vector fonts in order to reduce the number of glyphs for middle letters. As previous descried, shape of middle letter has two types such as vertical and horizontal. The vertical glyphs can be repeated from left to right like initial letters. Here horizontal glyphs are drawn from up to down but drawing direction of vector font is able to do so thus we will apply a reduction for just vertical glyphs but not horizontal. It invades the position of the already drawn glyph. Scene 4 of Fig. 3 shows vertical middle glyphs are repeated twice with a set of vertical glyphs. Figure 4 shows horizontal middle glyphs are overwritten with each different sets. Both figures shows initial glyphs are normally repeated with just one set.

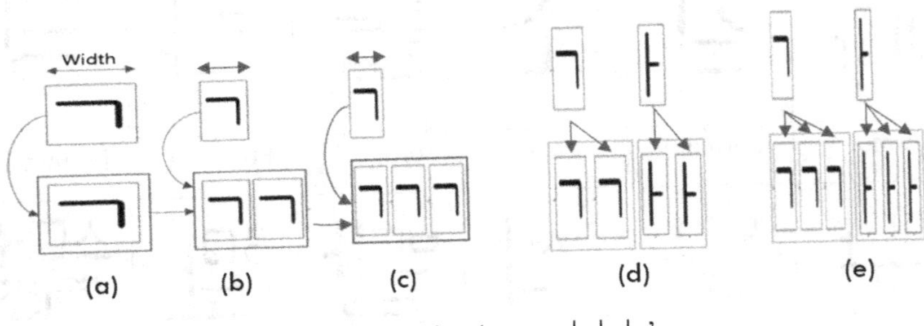

Fig. 3. Drawing 'ㄱㄱㄱㅏㅏㅏ'

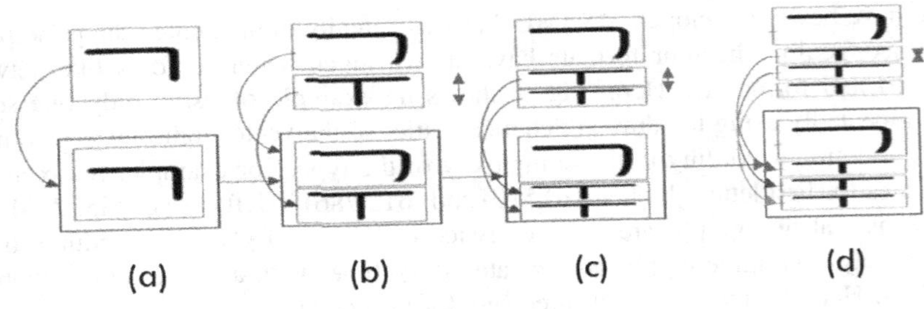

Fig. 4. Drawing 'ㄱㄱㄱㅜㅜㅜ'

Here we can consider to reduce two mixed combinations of middle glyphs for three middle letters. One is horizontal-vertical-vertical in format 5 and 6, and another is vertical-vertical-horizontal in format 7 and 8. In scene 3 of Fig. 3, each glyph is drawn with space of two or three pixels between glyphs previously reserved. However, in Fig. 4 since the initial glyphs have been already drawn from left to right, horizontal middle glyph has a negative coordinate meaning overwrite and should be overwritten on the initial space. Members of vertical middle glyphs are five glyphs of eleven such as {ㅏ, ㅑ, ㅓ, ㅕ, ㅣ}. Format 2 and 3 can reduce total three sets for type 2 and 3. Since format 4 and 5 have horizontal middle glyphs, the number is zero.

5 Implements and Evaluation

Figure 5 shows three kinds of syllable font. (a) is called a completed syllable font designed without considering the structure of syllable, and a font code is given to a glyph of one syllable. Due to a huge number of syllables in case of Jeongeum font, it is not relevant and not economical. (b) is to show a combined syllable font by two glyphs such as initial glyph and middle glyph without any reduction.

We assigned vector font set for glyphs to Unicode then each glyph has its own Unicode value. U+E511 means the reducing glyph font code for initial letter Giyuk of type 3 and format 2 and U+E731 is also the reducing glyph font code for middle letter A of type 3 and format. (c) of Fig. 5 is an example of glyph set reduction which shows drawing the combined triple Giyuk glyph ‘ㄱㄱㄱ’ by repeating a single initial glyph ‘ㄱ’. Here two sets of type 3 glyphs are reduced by using just one sets with the width of single glyph.

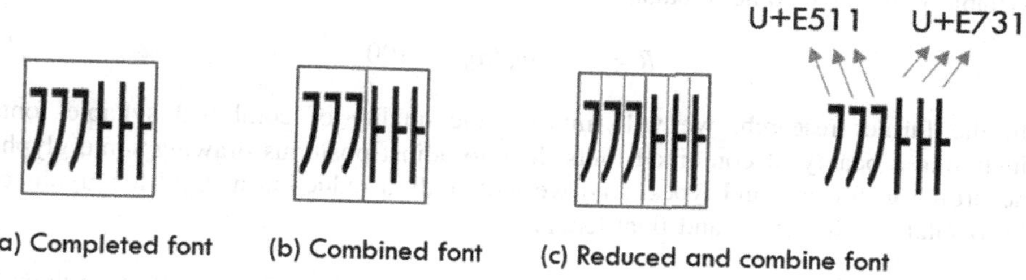

Fig. 5. Completed syllable and combined syllable

6 Conclusions

The number of glyphs can be reduced by repeating a single glyph for type 2 or type 3. In case of middle letters, six vertical shape glyphs are able to be reduced and five horizontal shape are not able to be reduced because their drawing direction is top to down. Generally speaking, that of vector font is left to right.

Table 2. Reduction evaluation

	Before Reduction				After Reduction			
Format	Init	Mid	Fin	Tot	Init	Mid	Fin	Tot
1	138	66	120	324	87	45	120	252
2	138	30	0	168	87	15	0	102
3	138	30	120	288	87	15	120	222
4	138	36	0	174	87	36	0	123
5	138	36	120	294	87	36	120	243
6	138	45	0	183	87	39	0	126
7	138	45	120	303	87	39	120	246
8	138	45	0	182	87	39	0	126
9	138	45	120	302	87	39	120	246
Tot	1,242	376	600	2,218	783	303	600	1,686
Reduction Rate : 24%								

The total number of design target glyphs before reducing is 2,218. This means that 2218 glyph fonts should be designed to represent the whole syllables of Hunmn-jeongeum. Table 2 shows a reduction evaluation result for the design target glyphs by applying a new design philosophy to reduce glyph sets having same width of type 2 or type 3.

Finally we were able to reduce 22.9% of the design target glyphs in generating combined Jeongeum syllable fonts that Hunmingeongeum wants to represent. In the below Eq. in (1), bg is the number of glyphs before reduction, and ag is the number of glyphs after reduction. Also, we show that a reduction of middle letter glyphs is economical in designing the least set of optimized glyphs. However the combined syllable fonts have some unbalanced shape.

$$R = (1-(bg/ag)) * 100 \tag{1}$$

In the further research, we will improve the quality of combined syllable fonts including a beauty of combined fonts. It is to adjust positions drawing some glyphs according to formats and types. And we will study a reduction method for glyphs of horizontal middle letters and final letters.

Acknowledgement. This research was supported by Basic Science Research Program through the National Research Foundation of Korea (NRF) funded by the Ministry of Education (NRF-2015R1D1A1A01060408).

References

1. Kang, S.H.: Hunminjeongeum Research. Sungkyunkwan University Publishing Division, Seoul (1991)
2. Byun, J., Lim, H.: A legislation principle of Hunminjeongeum creation and Hangul code system. In: Proceeding of 3rd Domestic Conference on Hangul and Korean Language Processing, KIISE, pp. 155–158 (1991)
3. Kim, G., Kim, H., Byun, J.: An economic evaluation of combined syllable fonts for Hunminjeongeum. In: Proceedings of 2017 Korean Computer Congress, KIISE, Ramada Jeju Hotel, Jeju Korea, pp. 1820–1821 (2017)
4. Byun, J., Hong, S., Kim, H.: A design scheme of combined syllable fonts for Hunminjeongeum. In: Park, J.J., Pan, Y., Yi, G., Loia, V. (eds.) Advances in Computer Science and Ubiquitous Computing. LNEE, vol. 421, pp. 1096–1102. Springer, Singapore (2017). https://doi.org/10.1007/978-981-10-3023-9_170
5. Byun, J., Lee, H.: A web input method for full set syllables defined by Hunminjeongeum. J. KIISE Comput. Pract. Lett. **19**(6), 371–375 (2013)
6. Hong, S.-B., Kim, H.-Y., Kim, G.-Y., Byun, J.-Y.: A reduction of first and middle letter Glyphs for Jeongeum-Font. In: Proceedings of the 4th International Conference on BigData Applications and Services 2017. Korea Bigdata Society and University of Uzbekistan, Tashkent, Uzbeikistan (2017)
7. Byun, J.: An improvement of hangul code for implementing Hunminjeongeum principle. KIISE Rev. **12**(8), 72–76 (1994)

Software Systems and Technologies

Software Requirements for an Ultra Large Scale System to Compute Multi Dimension Mean Failure Cost

Mouna Jouini[1(✉)], Latifa Ben Arfa Rabai[1], and Ridha Khedri[2]

[1] SMART Laboratory, Higher Institute of Management, Tunis, Tunisia
jouini.mouna@yahoo.fr, latifa.rabai@gmail.com
[2] Department of Computing and Software, McMaster University, Hamilton, ON, Canada
khedri@mcmaster.ca

Abstract. In previous work, we presented a quantitative cyber security risk assessment model that quantifies the security of a system in financial terms. Our model assesses the cost of the failure of an information system security with regards to threats dimensions. In this assessment, we consider that the threats world can be divided into several threats dimensions and perspectives. In this paper, we discuss the specification and design of an automated tool that manage and maintains information that pertains to estimating the security risk supported by our risk assessment model.

Keywords: Cyber security metrics · Security risk assessment · Security Multi dimension mean failure cost (M2FC) · Cloud computing Automated tool

1 Introduction

Information is a critical resource for all organizations since information supports service continuity and business and helps managers to make suitable and effective decisions. Securing assets and critical elements of organization like information, hardware, software, connectors… have become more and more important with the information technologies development.

Information is a curtail asset for organization that must be protected and managed with efficient ways. It lifts up the value of organization and its benefits. However, information is vulnerable because of a number and a wider variety of threats and vulnerabilities. We must thus protect the security of this information. Information security can described as to protect information from various threats to ensure business continuity, minimize risks and maximize profits and business opportunities [1, 13–16].

In order to ensure organizations to protect their information assets, we must use security strategies to reduce threats risks. NIST SP 800-30 defines risk as, 'a function of the likelihood of a given threat-source's exercising a particular potential vulnerability, and the resulting impact of that adverse event on the organization'. Risk occurs due to security threats and vulnerabilities. The aim of risk assessment is to identify and assess

J. H. Park et al. (Eds.): PDCAT 2018, CCIS 931, pp. 361–370, 2019.
https://doi.org/10.1007/978-981-13-5907-1_39

the risks in a particular environment [2–4, 13–16]. These assessments help identify these inherent business risks and provide measures, processes and controls to reduce the impact of these risks to business operations. The security risk assessment includes a threat assessment, vulnerability assessment and includes mitigation measures recommendations. It is used for decision-making, and for planning purposes and risk management.

There are two groups for security risk assessment as quantitative risk assessment and qualitative risk assessment. Quantitative models [6–8] are based on numerical values to estimate the security risk. However, qualitative approaches [5] can be taken by defining risk in more subjective terms. It does not use variable values to estimate risks but rather evaluate qualitatively the influence of each variable on the risk.

We deal in this article with quantitative risk assessment models to estimate and evaluate security breaches mainly because qualitative security risk assessment approaches present a subjective risk evaluation and give general and approximate results. In fact, we propose in this paper an automated tool that helps organizations to construct the information security or to establish a security risk assessment.

We have implemented, in this paper, a system that supports the computation of the Mean Failure Cost (M2FC) for each stakeholder. We used UML as a modeling language to design the tool and ASP.NET as technology for implementation.

The paper will be divided into four sections. Section 1 deals with the M2FC model presentation. Section 2 presents the specifications of our tool by stating the objectives of the application and operational requirements. Section 3 focus on modeling the tool. Section 4 describes the implementation with an illustration through a concrete example.

2 Motivation

With the rapid advent of new information technologies, it is very necessary for organizations to use automated tools to manage their assets and data. Automated tools allow organizations to work more efficiently and to maximize productivity. In fact, the use of tools lead to faster communication between organizations and customers (users of thee tool), electronic storage and the protection of information, working and communicate in secure environment, automated processing of information. These advantages improve company's efficiency by developing automated processes to take burden off your customers and in the whole organization.

For example, the maritime environment is so extensive and so complex. In fact, terrorist risks to merchant shipping present significant challenges to control and prevention due to the nature of the global maritime trade environment and present a big part of the spectrum of terrorist threats. Between 80% and 95% (depending on measure) of global trade is carried by ship [17, 18], with more than 1.2 million seafarers [17, 19] on 120,000 vessels in the global maritime fleet, [17, 18] making calls at over 2,800 ports in the world [17, 19]. To deal with this problem, security tool is used to allow port and facility managers, and vessel security officers to perform their own individualized security assessments. Specifically, tool uses of risk matrices to help identify the most likely risks to their shipping and port facilities, and develop security and training plans accordingly. It identifies for users how serious a risk is, based on the expected negative impact (destructiveness, cost) of an event, the vulnerability of a ship, port or facility to attack, and the associated probability of that event occurring. The tool allows ports to security threats events priorities (higher/ lower probability of risk and higher/ lower impact), and allows ports to test more realistic threats scenarios to be happened. Moreover, it lets users to understand the range of the possible hypothetical events that could occur on their ships.

In our case, our developed tool, which is called as the M2FC tool, support the computation of the Multidimensional Mean Failure Cost for each stakeholder (M2FC) metric for large systems. Our M2FC metric is developed in [9–12] and permit assessing security risk for each system's stakeholders with regards to security dimensions.

In addition, our tool resolves the complexity of calculation for large systems. In fact, ultra large systems recover software intensive systems with unprecedented amounts of hardware, lines of source code, numbers of users, and volumes of data… For instance, the Multidimensional Mean Failure Cost for each stakeholder (M2FC) allow security risks with regards to the diversification of security threats perspectives, security threats dimensions, system users…the developed tool is applied to example to a cloud computing system as a large scale system.

Moreover, our tool facilitating the risk decision process by consideration of the appropriate risk treatment option. It evaluates on financial terms for each system users the risk due to security threats breaches. It enables users to identify their assets and then evaluate their assets and their associated risks. The tool warns thus users for the degree of security of the system. It helps and guides as well decisions makers taking appropriate countermeasures for their system and managing the level of security risk.

3 A Multi Dimensional Mean Failure Cost (M2FC): A Measure of Cyber Security

The Multidimensional Mean Failure Cost model (M2FC) [9–12] presents the stakeholders' losses as result of security breaches with regards to threats perspectives and dimensions where a dimension can be defined as an elementary aspect or extent of this world while a perspective is a focused view of the threat world that encompasses some related dimensions. It estimates the security of a system in terms of the loss that each stakeholder incurs due to security breaches considering several dimensions within the threat world. For example, when considering the deployment dimension, we give the mean failure cost per site where a security breach occurs.

The M2FC process proceeds in four steps: Generation of stakes matrix, generation of Probabilities of failure requirements matrix, generation of probabilities of failure dimension matrix and generation of the threats probabilities occurrence vector.

The stakes matrix Vs represents the cost that stakeholder Si would lose if the system failed to meet the security requirement Rj. The probability of failure requirements Cs matrix represents the probability of failing requirement Rj due to a failure originating from element Dk. The probability of failure dimension matrix PFRs represents the probability that an element Ck fails once threat Tp has materialized. Finally, the threats probabilities occurrence vector Ps represents the probability that threats materialize during unitary period of operation.

Given the matrix of stakes Vs, the probabilities of failure requirements matrix PFRs, the impact matrix Cs and the threat vector Ps, the vector multi dimensional mean failure costs (M2FC) is defined by the following equation:

$$M(s, D) = Vs \circ PFRs \circ Cs \circ Ps \quad (1)$$

The M2FC considers the following criteria:

- The threat configuration under which the application operates.
- The impact that security breakdowns have when threats dimensions have failed.
- The dependency that exists between threats dimensions and security requirements.
- The costs that each stakeholder has when security requirements specification have failed.

Given the matrix of stakes Vs (derived collectively by the stakeholders), the probabilities of failure requirements matrix PFRs (derived by the systems architect), the probabilities of failure dimensions matrix Cs (derived by the security analyst) and the vector of threat Ps (derived by the security analyst), we can derive the multi dimensional mean failure costs vector (one entry per stakeholder) by the formula 2.

$$M^2FC = Vs \circ PFRs \circ Cs \circ Ps \quad (2)$$

4 General Characteristics

A stakeholder of a system designates a role more than a person or a group of persons. The following list presents the set of relevant stakeholders: The system administration team, the requirements engineers of a system, the users of a system, the architect of a system, the security analyst, and the Verification and Validation team (see Fig. 1).

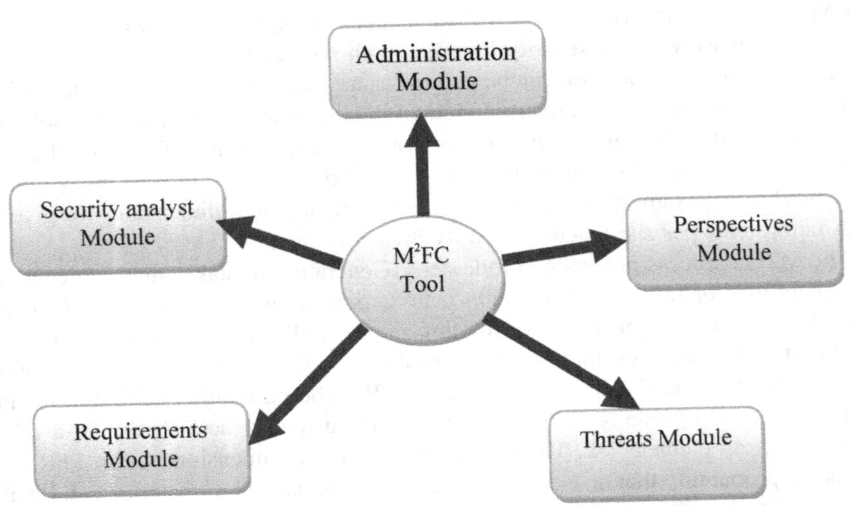

Fig. 1. General description of the M2FC tool functionalities.

The application should be able to:

- Calculate the mean cost of security failure for each stakeholder.
- Ensure communication between members of the same group messages or suggestions.
- Define the requirements, threats perspectives, threats dimensions (components, connectors, period of time …) and threats.
- Support the different classes of stakeholders to introduce their own data by providing details and empirical values.

Members of the application are grouped into 6 groups each of which has its own requirements and its own tasks. The common goal of these groups is to provide the data necessary to calculate the average cost of failure of each stakeholder.

The back office part is managed by the following system users: administrator, the engineer, the architect and the verification and the validation team as given in the following figure (Fig. 2).

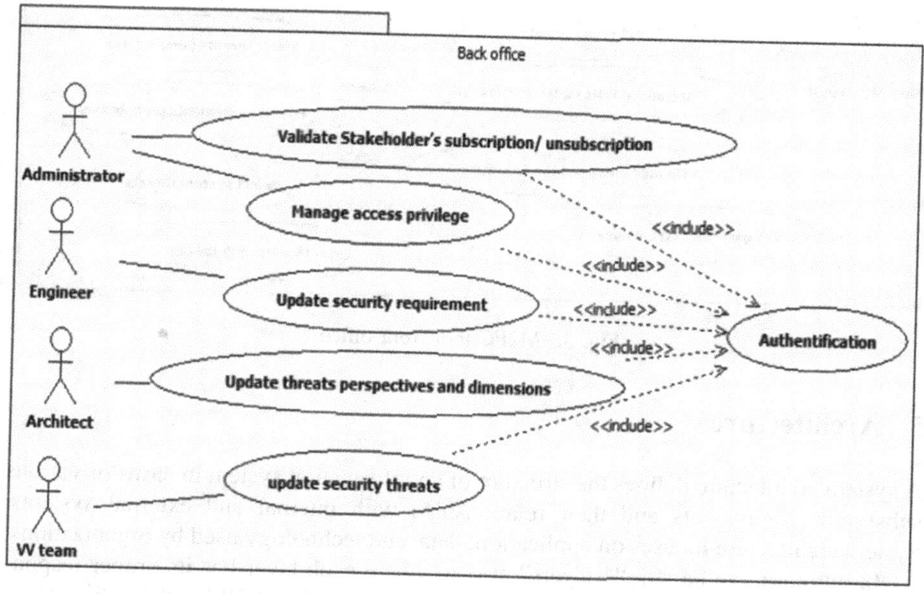

Fig. 2. M2FC tool back office.

The front office part is managed by the security analyst user of the system as shown in the following use case diagram (Fig. 3).

We represent in turn the system functionalities from a user/stakeholder perspective.

- The Engineers requirements should be able to create add or delete the list of system requirements.
- The system architect designs the security threats of system based on the requirements given by the system engineer.

- Given a security threats list, the Verification and Validation team establish which dimension element are probably to be affected by each security threat.
- The system administrator maintains information about global all system users. He specifies and manages login and access privileges for all users.
- The main mission of security team is to select the monetary unit of requirement cost, enter the cost for each security requirement in dollars/hour, Indentify the elements of the selected dimensions, selected threats perspectives and dimensions, determine probabilities of failure of security dimensions and of threats, and compute and consult the M2FC value.

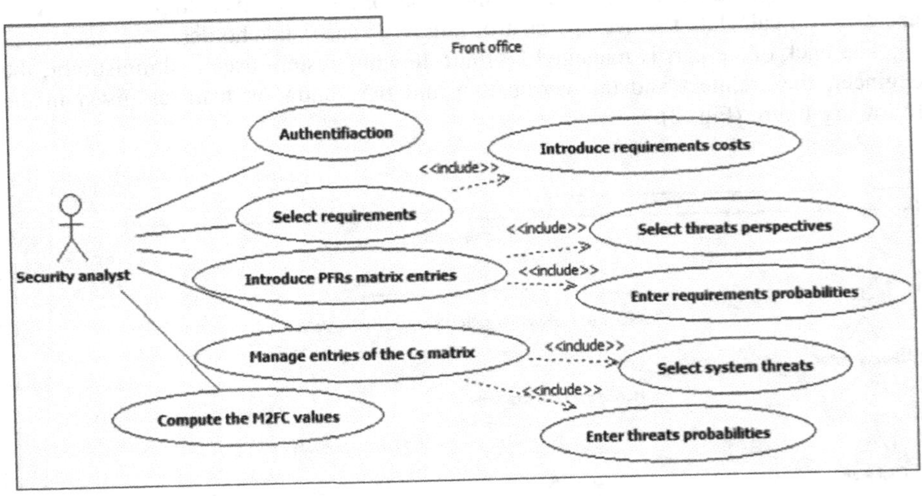

Fig. 3. M2FC tool front office.

5 Architecture

A system architecture defines the structure of an information system in terms of various subsystem components and their relationships with internal and external systems. System architecture focuses on application, data, and technology used by organizations.

In our case, we have a "3-layers" system where each layer has its proper responsibilities. We describe below the layered architecture of our ASP.NET Application.

The M2FC tool architecture is divided into three main layers. In fact, layers are the logical groupings of the software components that make up the application or service. They help to differentiate between the different kinds of tasks performed by the components, making it easier to create a design that supports reusability of components. It helps to support a strong separation of concerns that, in turn, supports flexibility and maintainability.

- Interface layer

The interface layer or the presentation layer provides physical interaction between the system and a human operator on a computer (a host). This layer can also be used to read input entered by users. For example the connection web page shows system parameters (login and password) entered by a system user to login to its space. The administration page displays system characteristics and the perspectives web page displays system perspectives and dimensions and probabilities entered per system requirement.

- Computation layer

This layer implements the core functionality of the system, and encapsulates the logic components. They generally consist of components located within the business layer, which may expose service interfaces that other users can use.

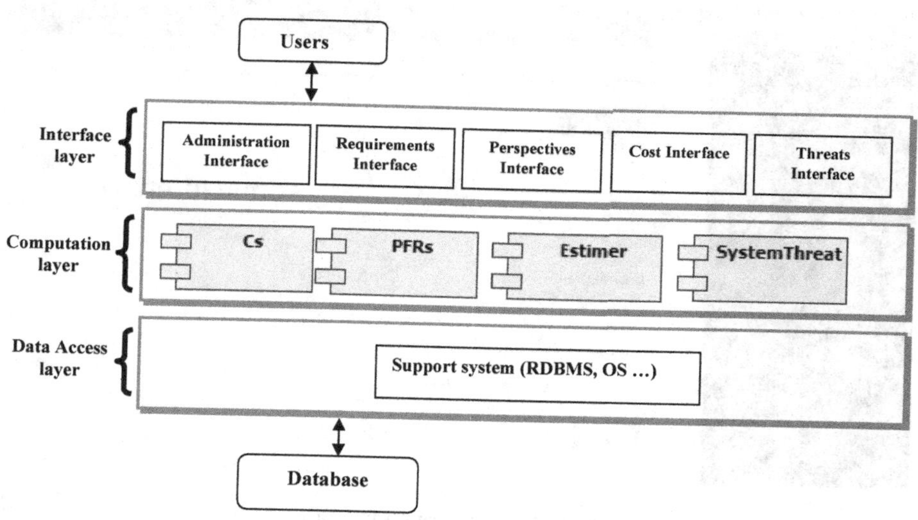

Fig. 4. M2FC tool architecture.

- Data Storage layer

This layer provides access to data that is hosted within the database of the system for example in performing fundamental operations like create, read, and update. Therefore we can here accessing data in whatever form it's stored. Our tool uses a relational database system (RDBS) to store all information about the organization, requirements, users security, perspectives, threats…

The three layers are the interface layer, the computation layer and the data layer as shown in Fig. 4.

6 Application of the Automated Tool for Computing the M2FC

We have applied this tool to the cloud computing system. We will illustrate an example of Cloud Computing system to evaluate the security in terms of loss incurred by each stakeholder.

We take into account the architectural perspective in which we consider the deployment sites dimension and the components (or architectural components) dimension. Our assessment varies according to the stakes that each stakeholder has in meeting each security requirement per system site. In this case, we opt for using the deployment dimension (i.e., sites dimension) as the leading dimension. For each site of the considered system, we have the lists of stakeholders, security requirements, components, and threats.

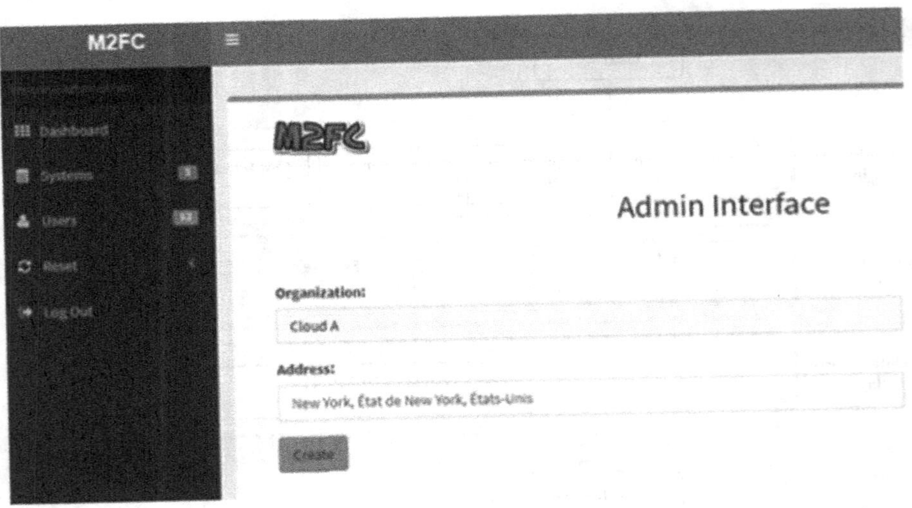

Fig. 5. System creation interface

We assume that the considered system is deployed on two sites. We have S = {S ite1, S ite2} and the component dimension D has the following elements which is equals to D = {Browser, Router/Firewall, Web server}. We recognize two stakeholders for this example namely: Gold user and Silver user. We consider as well the set R of requirements to include the three facets of security R = {Integrity, Availability}. We consider that we dealing with the set T = {Monitoring virtual machines from host (MVM), Communications between virtual machines and host (CBVH), Virtual Machine modification (VMm)} of threats.

Figure 5 shows the interface of a new organization (system) creation. Using our empirical data, we compute the vector of Multidimension Mean Failure Costs for the components of each site using our tool.

7 Conclusion

Security metrics are important indicators of how well security services are present in the information system and can be used to measure organization's security maturity level and provide effective decision-making support as what is the expected result if an organization invested in a specific security project.

In this paper we discuss the multi dimension man failure cost as a security model that estimate in financial terms security risk with regards to security threats dimensions. We have presented as well an evolving tool that supports the compute of the extended mean failure cost.

References

1. ISO/IEC 17799: (E), Information technology—Security techniques—Code of practice for information security management (2005)
2. Kwok, L., Dennis Longley, D.: Information security management and modeling. Inf. Manage. Comput. Secur. **7**, 30–40 (1999)
3. NIST SP 800-53: Information Security Handbook: A Guide for Managers (2006)
4. National Institute of Standards and Technology, Information Security - Guide for Conducting Risk Assessments (2002)
5. Stoneburner, G., Goguen, A., Feringa, A.: Risk management guide for information technology systems: Recommendations of the national institute of standards and technology. National Institute of Standards and Technology (NIST) Special Publication 800-30, U.S. Government Printing Office (2001)
6. Aissa, A.B., Abercrombie, R.K., Sheldon, F.T., Mili, A.: Quantifying security threats and their potential impacts: a case study. ISSE **6**(4), 269–281 (2010)
7. Mayer, N.: Model-based management of information system security risk. Ph.D. Thesis (2009)
8. Avizienis, A., Laprie, J.C., Randell, B.: Dependability and its threats: a taxonomy. In: IFIP Congress Topical Sessions, pp. 91–120 (2004)
9. Jouini, M., Ben Arfa Rabai, L.: A Security Risk Management Model for Cloud Computing Systems: Infrastructure as a Service. SpaCCS 2017, pp. 594–608 (2017)
10. Jouini, M., Ben Arfa Rabai, L., Khédri, R.: A multidimensional approach towards a quantitative assessment of security threats. ANT/SEIT 2015, pp. 507–514 (2015)
11. Jouini, M., Ben Arfa Rabai, L.: A security framework for secure cloud computing environments. IJCAC **6**(3), 32–44 (2016)
12. Jouini, M., Ben Arfa Rabai, L.: A multi-dimensional mean failure cost model to enhance security of cloud computing systems. IJERTCS **7**(2), 1–14 (2016)
13. Goettelmann, E., Dahman, K., Gateau, B., Dubois, E., Godart, C.: A security risk assessment model for business process deployment in the cloud, SCC. IEEE, pp. 307–314 (2014)
14. Zhao, X., Dai, M., Ren, S., Li, L., Duan, Z.: Risk assessment model of information security for transportation industry system based on risk matrix. Appl. Math. Inf. Sci. **3**, 1301–1306 (2014)
15. White, J.M.: Security Risk Assessment Managing Physical and Operational Security (2014)
16. Sun, L., Srivastava, R.P., Mock, T.J.: An information systems security risk assessment model under dempsterschafer theory of belief functions. J. Manage. Inf. Syst. **22**(4), 109–142 (2006)

17. Nincic, D.J., Bruce, C.: The utility of risk assessment tools in maritime security analysis. http://iamu-edu.org/wp-content/uploads/2014/07/28_TheUtilityofRisk.pdf
18. Daly, J.C.K.: Al-Qaeda and maritime terrorism, part I, The Terrorism Monitor, Jamestown Foundation (2003)
19. Richardson, M.: A time bomb for global trade: Maritime-related terrorism in an age of weapons of mass destruction, Viewpoints, Institute of South East Asian Studies (2004)

A Study of an Automated Software Effort Measurement Method

Yeong-Seok Seo(✉) and Hyun-Soo Jang

Department of Computer Engineering, Yeungnam University, 280 Daehak-Ro, Gyeongsan, Gyeongbuk 38541, Republic of Korea
ysseo@yu.ac.kr

Abstract. Software companies have adopted project management methodologies suitable for their organizations and have made significant efforts in successfully applying them to improve the quality of software. In particular, a technology that can measure and analyze software project data is essential for effective project management and productivity improvement. Of these software project data, software effort is the key metric to be measured, given its direct relation to process improvement and quality but also have general management interest. However, in practice, there have been many difficulties in actually measuring effort data because of problems in continuous and consistent measurement. Therefore, in this paper, we propose an automated software effort measurement method that can apply during the entire software development life cycle, to overcome these problems and to achieve improvement of effort measurement outcomes. Experiments are performed to evaluate the proposed method from the viewpoint of effort measurement accuracy. The results indicate that the proposed method shows a significant improvement compared to the existing methods.

Keywords: Effort · Measurement and analysis · Software project management
Software quality · Software tools and environments

1 Introduction

Software measurement and analysis can lay the foundation for effective software project management and can also have a significant impact on an organization's software development productivity [1]. Of the various software project data, the software development effort (note that "effort" is the amount of work units (time) required to complete any given task) is one of the most important indicators [2]. Based on the collected effort, we estimate the software development effort at the planning stage and perform systematic project management through the allocated effort by tasks. In order to conduct these activities, we should basically be able to measure effort accurately with less trouble. If accurate effort measurement is possible, then from an individual perspective, the effective planning and processing of each assigned task are possible, deficiencies in individual processes can be identified, and opportunities for improving these deficiencies can be provided when performing a software development project [3, 4]. From an organizational perspective, software project managers can easily

J. H. Park et al. (Eds.): PDCAT 2018, CCIS 931, pp. 371–380, 2019.
https://doi.org/10.1007/978-981-13-5907-1_40

identify and monitor the overall progress, and reliable effort estimation is possible through historical data, which will directly affect the quality of the final product [5, 6].

Many companies have been making or attempting effort measurement owing to its importance. However, the reliability of the collected effort is quite low because developers record their individual effort by directly measuring it through additional efforts other than development, or the effort collection process is not automated. Even with the existing effort measurement methods, developers should make further efforts or there are problems with accuracy. The existing effort measurement methods can be divided into a self-report method in which developers directly record the effort data and an automatic-report method using a software program. The self-report method has the possibility of fabrication because each developer has to manually enter his/her own effort or review it. Even for the automatic-report method, it is impossible to apply it to the whole stages of the software development cycle. Further, it is very difficult to collect accurate effort data because the method does not take into account "the time for thought" to solve problems and to find the necessary key information during development. Additional efforts may be required from developers to adjust this information, which may affect primary tasks assigned to them.

Therefore, in this paper, we propose a method that enables software developers to measure effort more accurately and automatically without additional efforts while performing a software project. This method, which uses Windows Application Programming Interface hooking (Windows API hooking) technology [7, 8], can collect events of working tools and input devices that developers actually use, and measure the actual work time by task, tool, and development phase by analyzing the collected data. In particular, the proposed method can measure effort by considering "the time for thought" to solve problems when developing a software product. In addition, this method can improve the effort measurement accuracy by measuring actual effort through extracting website information when using web browsers. Although web browsers are generally used for a software development, it is a difficult issue to identify the actual work time from them.

The remainder of this paper is organized as follows: Sect. 2 describes existing studies related to effort measurement. Section 3 introduces the effort measurement method proposed in this study. Section 4 shows the experimental results and analysis, and Sect. 5 concludes this paper and suggests future work.

2 Related Work

Research on software effort measurement is largely divided into self-report and automatic-report methods.

The most commonly known self-report methods are Personal Software Process (PSP) [3, 4] and Team Software Process (TSP) [5, 6]. Software developers record their daily effort to use in process improvement. However, there is a limitation in that actual effort must be recorded manually by developers, and thus it is very difficult to guarantee reliability and validity of the collected data. Although a variety of tools have been developed to support manual data collection, such as Process Dashboard [9],

PSP.NET [10], and WBPS [11], they still need to input data manually with additional efforts through a context switch between development and measurement activities.

There are several types of automatic-report methods. The first type is a chunk-based method [12–14]. This method divides 24 h per day into certain time intervals. When an event related to a software task occurs at a specific interval, that interval is measured as valid effort. Although the chunk-based method enables automated effort measurement, it is not very sophisticated since it measures even one software task event at a certain time interval as the entire work time. Further, additional reviews should be made periodically to check if the measured effort by developers is suitable. The review activity puts a lot of burden on developers. The second type of automatic-report method is an interval-based method [15, 16]. It first measures events related to software tasks (e.g., compilation in development support tools), and then measures the final effort by developing a discriminant to identify actual effort within the intervals between events. This method may be more automated than the chunk-based method, yet it can only be applied at the implementation stage where compilation is necessary but not the whole software development life cycle. The interval-based method can be used in combination with the self-report method. The interval-based method measures actual effort according to effort decision formulas. Here, if there is any additional work time recorded by developers, it can be included as work time. In this way, the final effort is drawn by combining the effort collected through the interval-based method and the effort recorded by an individual developer. The last type is a task-based method [17]. The method is developed as a software development task measurement system, called as TaskPit. It can record the time spent for each task such as programming, testing, and documentation, by binding between tasks and applications. In addition, it can record the amount of work and the amount of deliverables of each task. TaskPit is helpful for effort measurement and process improvement in a development task level. However, this mainly focuses on grasping how developers are working, rather than achieving accurate effort measurement by identifying and analyzing the actual work time in detail.

3 Personal Effort Measurement System (PEMS)

The proposed automated personal effort measurement method is named the personal effort measurement system (PEMS). PEMS enables the members participating in a software development project to measure effort, that is, actual work time, by analyzing working activities on each computer while working on their computers. To this end, this study applied API hooking provided by the Windows operating system (OS). The Windows OS is used by many people and is run by events as an event-driven OS. Here, hooking refers to a mechanism that can identify and monitor all events such as messages, mouse movements, and keystrokes occurring in programs such as operating systems and application software [7, 8]. If we create a function that acts as a hook for an application program running on the Windows operating system, we can extract the process names of the program or events of input/output (I/O) devices occurring in the program.

The overall approach of PEMS using these characteristics is shown in Fig. 1. In the following subsections, we present each step of PEMS in more detail.

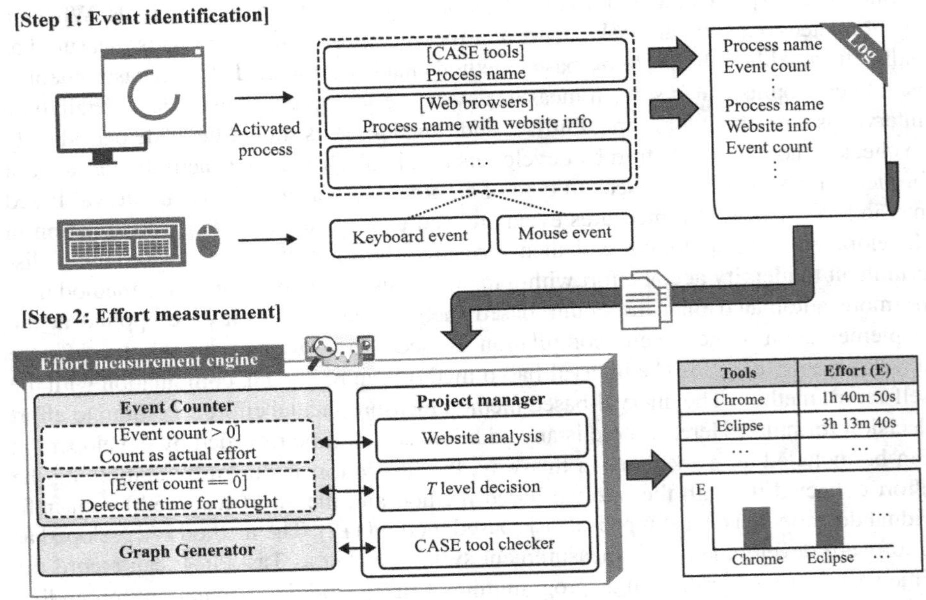

Fig. 1. Overall approach

3.1 Step 1: Event Identification

Step 1 is the stage of collecting the keyboard and mouse events generated from various software tools that software project participants (named as developers) use while working on a computer, and recording the information in a log file.

Developers use various software tools related to specific tasks assigned to them. Of the various tools that developers run on the computer for their tasks, PEMS identifies the process names of the tools that the developers are currently using. In other words, PEMS identifies the process names of the tools currently activated by the developers. Keyboard or mouse events are generated to activate and use the tools, and PEMS also identifies the counts of these events. These identified process names and event counts are recorded in the log file every second.

As shown in Fig. 1, process names and event counts can be collected as mentioned above to measure effort for the usage time of various Computer-Aided Software Engineering (CASE) tools that are used as general working tools at each stage of the software development process. Further, even for tools that each software development group wants to additionally include in effort measurement through discussion, project managers can input them into PEMS to collect process names and event counts. Tools that are not specified as working tools (tools not related to an assigned software project task) are recorded in the log file as NULL because they are not related to effort measurement.

For web browsers, unlike CASE tools, the title bar information of web browsers is recorded in the log file in addition to process names and event counts. Using web browsers can be an activity that is not related to an assigned task. However, there is a strong possibility that they can be used in web search activities relevant to the task in many ways. Thus, by collecting the titles of websites as well, PEMS was designed to enable identifying additional effort in the next step to detect the time related to actual work while using web browsers. In the case of web browsers, just like CASE tools, project managers can input work-related website title information to PEMS so that it records websites not related to work as NULL in the log file.

3.2 Step 2: Effort Measurement

Step 2 is a stage of providing the measured effort by minutely analyzing the log file obtained through Step 1.

As shown in Fig. 1, successive computations and algorithms are constructed in the form of an effort measurement engine to automate the analysis of the input log file. When the log file is inputted, Event Counter continuously analyzes keyboard and mouse event counts occurring for each process recorded in the log file every second, and derives the measured effort.

The entire pseudocode for effort measurement is shown in Table 1. First, the working tool of the corresponding process is determined to have performed a software project task if the event count of a specific process is greater than 0. Then, one second is assigned to actual effort for the working tool, because the log is recorded in every second, as described in Step 1. In the case of web browsers, although only websites related to the work can be recorded in the log file, it is also necessary to analyze further whether the search engines used for the tasks performed the search for the actual work. Note that the search words used in the general search engines are presented in the title bar of web browsers. Thus, by analyzing the title bar recorded in the log file, the execution time of web browsers is assigned to actual effort if the terms or names related to software project tasks are included in the title bar information.

Next, if event counts for a specific process are equal to 0, it can be determined that developers are usually taking a break without using the working tool of the corresponding process. However, it can also be determined that they are thinking to solve a problem while they are using the working tool for the process. We designate "time for thought" (T) to measure even these occurrences as actual effort. If the event count of the corresponding process is continually 0 for less than or equal to T, the time for the event counts is allocated to actual effort of the working tool of the corresponding process. If the event count of the corresponding process is continually 0 for more than to T, the time for the event counts exceeded T is classified as a break time rather than actual effort. When the developers return from a break and start working again, actual effort is measured again. In practice, T should be customized to obtain the best measurement results in software companies.

The final effort is derived by collecting all effort of the working tool of each process, as presented in Table 1. The derived effort is visualized in the form of graphs and tables by Graph Generator to easily identify effort by each tool, each task, and each development phase. In particular, the working tools that perform similar functions are grouped and represented together by CASE Tool Checker when it effort visualized by

Table 1. Pseudocode for effort measurement

Pseudocode

```
Require: a log file derived from Step 1
Ensure: effort from the log file

1: Map(process_name, Effort)
2: i
3: idx_time_i
4: cumulative_effort
5: event_count_i
6: T

7: idx_time_1 = 0;
8: CP = process_name_1;

9:for i = 1 to EOF do

10: if event_count_i > 0 then
11:    if CP == process_name_i then
12:        idx_time_(i+1) ← idx_time_i + 1;
13:    else
14:        Map(CP, cumulative_effort for CP + idx_time_i);
15:        idx_time_(i+1) ← 1;
16:        CP ← process_name_i;
17:    end if

18: else if (continually)event_count_i == 0 for ≤ T then
19:    idx_time_(i+1) ← idx_time_i + 1;

20: else if (continually)event_count_i == 0 for > T then
21:    Map(CP, cumulative_effort for CP + idx_time_i);
22:    idx_time_(i+1) ← 0;

23: else if i == EOF then
24:    Map(CP, cumulative_effort for CP + idx_time_(i+1));
25: end if

26:end for
```

(Comments)
1: // a Map data structure to save effort
2: // index of (process name, (website info), event count) in the log file
3: // temporary variable to calculate effort of the *i*th process name($process\_name_i$)
4: // total effort cumulated by idx_time_i (initial value is 0)
5: // the number of keyboard and mouse events for the *i*th process name
6: // the time for thought
8: // *CP* is a current process
9: // *EOF* is End Of File
11: // In case that the existing process name is continued
12: // 1 is one second
13: // In case that the existing process name is changed
18: // Identification of the time of thought
20: // Identification of a break time
24: // *i* in $idx\_time_{(i+1)}$ is an index value just before *EOF*

tasks and development stages. For example, using the category "Search", the measured effort from various search engines is represented by integration. This allows users to easily and intuitively understand the results of effort measurement at a glance, and to analyze these results from various angles.

4 Experiment

4.1 Experimental Design

In order to verify the performance of PEMS for software development tasks, we conducted a tool development project for algorithmic problem-solving using a hash structure. Experiments were performed on three participants who have one year work experience as software developers in a CMMI level 3 [18] software company. They are skilled in programming languages, such as Java (J2SE, J2EE, J2ME), C#/.NET, Python, HTML5. In our experiment, PEMS were compared with the self-report method, the chunk-based method, and the interval-based method, which are mentioned in Sect. 2. Task-based method could not applied because the specific algorithm for effort measurement was not expressive enough to implement and experiment it.

For the self-report method, apart from the three participants, three more self-report record monitoring personnel were assigned to supervise effort measurement recording of the participants. This is intended to reduce the possibility of bias mainly arises from self-reporting and to improve the reliability of effort measurement results.

For the existing methods, the chunk-based method that is the basis of this study and the interval-based method were implemented as program modules and activated with PEMS during our experiment. The time interval of the chunk-based method was set as one second to make fair comparison with PEMS.

4.2 Experimental Results

The experimental results for participants A, B, and C are shown in Fig. 2. The x-axis of the graph represents the tools each participant used while performing the project. Web browsers used in the search for solving problems while conducting tasks include Internet Explorer, Chrome, and Microsoft Edge. The y-axis of the graph represents effort expressed in seconds. The table below the graph describes the exact numerical values for each graph item. SR is the self-report method, IbM is the interval-based method, and CbM is the chunk-based method.

As shown in Fig. 2, PEMS shows a more accurate effort measurement than the existing methods based on SR. In addition, PEMS shows outstanding results of effort measurement for all tools represented on the x-axis, and the effort differences between PEMS and SR is minimal. Because PEMS can be affected by T, we also present the effort measurement results in the variation of T ($T = 10$ and $T = 15$), as provided in Fig. 2. The results show that the concept of T applied in this study affects the actual effort measurement and the measured effort increases gradually a little as T increases. We additionally analyzed the log file by setting from $T = 5$ to $T = 50$, and the best

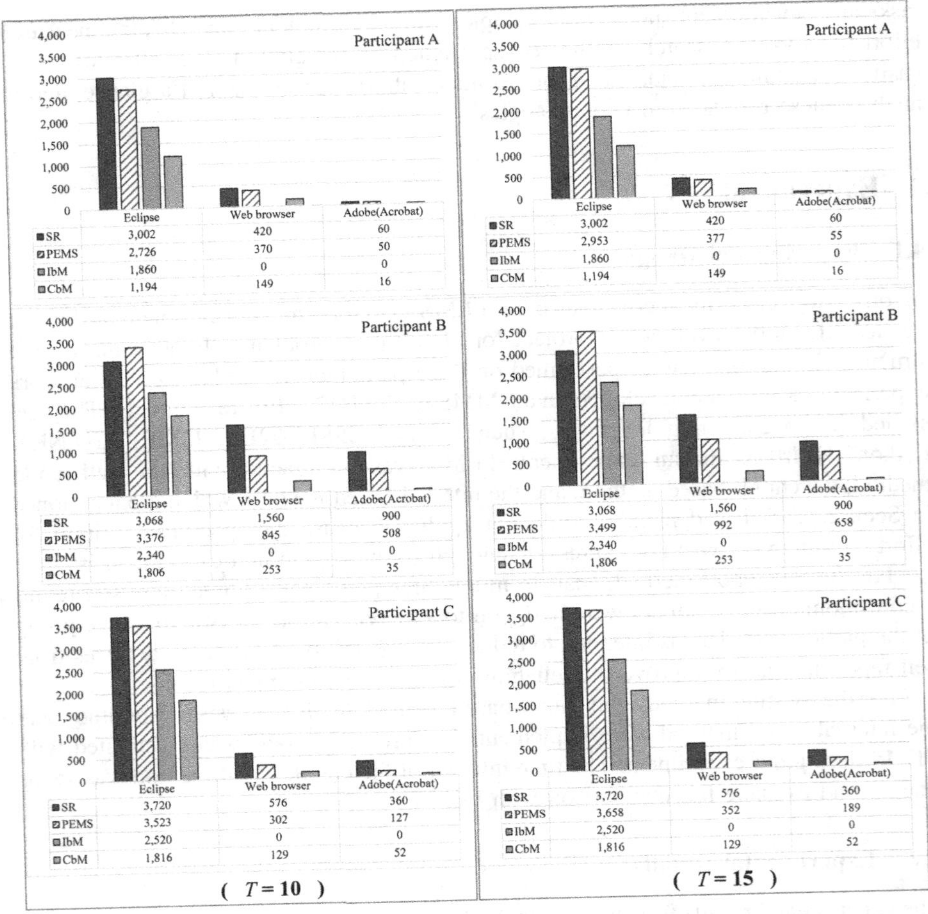

Fig. 2. Experimental results from three participants in the variation of *T*

T for Eclipse were 5 for the participant B (SR: 3,068, PEMS: 3,059) and 15 for the participant A (SR: 3,002, PEMS: 2,953) and C (SR: 3,720, PEMS: 3,658).

Note that the results of Web browser and Adobe (Acrobat) are zero in IbM. This is because IbM can be only applied for the implementation tools that include a compile function for source codes.

Table 2. Effort measurement accuracy from the participants (%, *T* = 15)

	Participant A			Participant B			Participant C		
	SR vs. PEMS	SR vs. IbM	SR vs. CbM	SR vs. PEMS	SR vs. IbM	SR vs. CbM	SR vs. PEMS	SR vs. IbM	SR vs. CbM
Eclipse	98.37	61.96	39.77	114.05	76.27	58.87	98.33	67.74	48.82
Web b.	89.76	0.00	35.48	63.59	0.00	16.22	61.11	0.00	22.40
Adobe	91.67	0.00	26.67	73.11	0.00	3.89	52.50	0.00	14.44

Table 2 presents the accuracy rates of effort measurement based on the numerical values represented in Fig. 2. The rate is derived by dividing the measured effort of each method by that of the self-report method. The best accuracy rate of PEMS is 98.37% for Eclipse of Participant A, which is 36.41% and 58.59% higher than the results of the existing methods.

5 Conclusion

Measuring effort accurately is one of the most challenging issues for software project management. Thus, this study proposed PEMS that could automatically measure effort without additional efforts from participants during the entire software development life cycle. Using API hooking, process names and keyboard/mouse events currently running on the computer were identified. Based on the log file that records them, a more accurate effort measurement was possible through Effort measurement engine of PEMS. In particular, we achieved a more accurate effort measurement by reflecting the time for thought in effort measurement and by analyzing the utilization of website information for software development tasks.

Although there were considerable encouraging results in this study, we will conduct further studies on the following issues. By actually applying the proposed method to small- and medium-sized companies, we will investigate the generalization of our results. In addition, we will develop a more accurate and automated effort measurement method on computers that use operating systems other than Windows.

Acknowledgement. This work was supported by the National Research Foundation of Korea (NRF) grant funded by the Korea government (MSIT) (No. NRF-2017R1C1B5018295).

References

1. Sommerville, I.: Software Engineering, 10th edn. Pearson, Boston (2016)
2. Singh, J.: Functional Software Size Measurement Methodology with Effort Estimation and Performance Indication, 1st edn. Wiley-IEEE Computer Society Press, Hoboken (2017)
3. Humphrey, W.S.: PSP(sm): A Self-Improvement Process for Software Engineers, 1st edn. Addison-Wesley Professional, Upper Saddle River (2005)
4. Blokdyk, G.: Personal Software Process: Best Practices Guide. CreateSpace Independent Publishing Platform, USA (2017)
5. Humphrey, W.S.: TSP: Leading a Development Team, 1st edn. Addison-Wesley Professional, Reading (2005)
6. Blokdyk, G.: Team software process: Fast Track. CreateSpace Independent Publishing Platform, USA (2017)
7. Richter, J.: Programming Applications for Microsoft Windows. Microsoft Press, Redmond (1999)
8. Yosifovich, P., Russinovich, M.E., Solomon, D.A., Ionescu, A.: Windows Internals, 7th edn. Microsoft Press, Redmond (2017)
9. Process Dashboard. https://www.processdash.com. Accessed 29 June 2018

10. Nasir, M.H.N.M., Yusof, A.M.: Automating a modified personal software process. Malays. J. Comput. Sci. **18**(2), 11–27 (2005)
11. Thisuk, S., Ramingwong, S.: WBPS: a new web based tool for personal software process. In: 11th International Conference on Electrical Engineering/Electronics, Computer, Telecommunications and Information Technology, pp. 1–6. IEEE, Nakhon Ratchasima (2014)
12. Johnson, P.M., Kou, H., Paulding, M., Zhang, Q., Kagawa, A., Yamashita, T.: Improving software development management through software project telemetry. IEEE Softw. **22**(4), 76–85 (2005)
13. Fauzi, S.S.M., Nasir, M.H.N.M., Ramli, N., Sahibuddin, S.: Software Process Improvement and Management: Approaches and Tools for Practical Development, 1st edn. IGI Global, Hershey (2011)
14. Artemev, V., et al.: An architecture for non-invasive software measurement. In: Petrenko, Alexander K., Voronkov, A. (eds.) PSI 2017. LNCS, vol. 10742, pp. 1–11. Springer, Cham (2018). https://doi.org/10.1007/978-3-319-74313-4_1
15. Hochstein, L., Basili, V.R., Zelkowitz, M.V., Hollingsworth, J.K., Carver, J.: Combining self-reported and automatic data to improve programming effort measurement. In: 10th European Software Engineering Conference Held Jointly with 13th ACM SIGSOFT International Symposium on Foundations of Software Engineering, pp. 356–365. ACM, New York (2005)
16. Hochstein, L., Basili, V.R., Vishkin, U., Gilbert, J.: A pilot study to compare programming effort for two parallel programming models. J. Syst. Softw. **81**(11), 1920–1930 (2008)
17. Suthipornopas, P., et al.: Industry application of software development task measurement system: TaskPit. IEICE Trans. Inf. Syst. **E100.D**(3), 462–472 (2017)
18. Chrissis, M.B., Konrad, M., Shrum, S.: CMMI for Development, 3rd edn. Addison-Wesley Professional, Amsterdam (2011)

A Generic Approach for the Verification of Static and Dynamic Behavioral Properties of SCDL/WS-BPEL Service-Component Architectures

Taoufik Sakka Rouis[1(✉)], Mohamed Tahar Bhiri[2], Mourad Kmimech[3], and Layth Sliman[4]

[1] LIPAH Laboratory, FST, University of Tunis El Manar, Tunis, Tunisia
srtaoufik@yahoo.fr
[2] MIRACL Laboratory, ISIMS, University of Sfax, Sfax, Tunisia
[3] UR-OASIS Laboratory, ENIT, University of Tunis El Manar, Tunis, Tunisia
[4] Efrei - École d'Ingénieur Généraliste en Informatique, Villejuif, France

Abstract. Web systems verification is a crucial activity throughout the systems development life cycle, especially in the phase of service-component architectural design. Indeed, this activity allows the detection and consequently the correction of errors early in Web systems development life cycle. In this paper, we discuss the behavioral verification problem on the SCDL/WS-BPEL service-component architectures. To do so, the Wright formal ADL and the Ada concurrent language were used as a target models. To achieve this, a set of systematic translation rules are proposed. This allows the verification of the standard behavioral properties using the Wr2fdr tool. In addition, using an Ada dynamic analysis tool, we could detect the potential behavioral properties such as the deadlock of an Ada concurrent program.

Keywords: Verification · Concurrent program · Model-Checker
Static behavioral properties · Dynamic behavioral properties
Service-Component architecture

1 Introduction

Today, we witness a growing interest in the domain of service-component architecture technologies. This interest is mainly motivated by the reduction of cost and development time of complex Web systems. In the context, the Service Component Definition Language (SCDL) and the Web Service Business Process Execution Language (WS-BPEL) are the standards de-facto used in the modeling and implementing of Service-Component Architecture (SCA). This SCDL language [14] is an XML based formatted language which allows expressing all the relations in an SCA architectural element. The WS-BPEL language (abbr. BPEL) [6] offers a standard based approach to build flexible business processes by the orchestrating and choreographing of multiple Web services. In addition, it aims to model the behavior of component processes by specifying both abstract and executable business processes. It also defines an interoperable

J. H. Park et al. (Eds.): PDCAT 2018, CCIS 931, pp. 381–389, 2019.
https://doi.org/10.1007/978-981-13-5907-1_41

integration model that should facilitate the expansion of automated process integration both within and between businesses.

For several years now, these SCDL and WS-BPEL technologies appear as the powerful complementary models for the development of service-component architectures. However, they lack a formal foundation for the specification and verification of their structural, behavioral and non-functional properties. As solutions for this problem, several works have been proposed to translate these source models into another which supports specific analyzers. For example, in our previous work [7] we proposed to map SCA to Acme/Armani for the verification of the structural and non-functional properties of SCA software architecture. For the verification of the WS-BPEL behavioral specifications, numerous works proposed to translate the WS-BPEL activities into a formal technique. For example, the works presented in [3], [16] and [12] propose, respectively, to translate the WS-BPEL activities into D-LOTOS, BPMN and PetriNet.

In this work, we target a formal verification of the behavioral properties of SCDL/WS-BPEL service-component architectures. To achieve this, the model transformation approach is used to translate the source architecture to an Ada concurrent program. In this study, the Wright formal ADL is used as an intermediate modeling language. The choice of these two languages in our verification approach is justified mainly by the following three factors:

- The Wright ADL defines eleven standard properties related to the consistency of software architecture among which four -assimilated to behavioral contracts- are automated by the Wr2fdr tool [15]. The latter contracts can be checked with the FDR2 model-checker.
- The semantics similarity of the Wright process and the Ada task favors the formalization of the Wright configuration by an Ada concurrent program.
- The presence of different analysis tools related to the detection of the dynamic and specific behavioral problems of an Ada concurrent program. For example, using FLAVERS, INCA or SPARK [11], we can detect the potential behavioural properties such as the deadlock of an Ada concurrent program. In addition, using an Ada dynamic analysis tool such as GNAT Programming Studio (GPS) or its extension GNATprove [10].

The remainder of this paper is structured as follows: Sect. 2 proposes an overview of the main related works. Section 3 deals with our systematic rules allowing the translation of SCDL/WS-BPEL source software architecture to the Wright target software architecture; Sect. 4 exhibits the translation rules of the Wright software architecture into an Ada concurrent program. An overview of the main results is discussed in Sect. 5. Finally, Sect. 6 provides a conclusion and possible future work.

2 Related Work

The approach shared by most of the existing works in the field of Web services architectures consistency verification is the use of techniques and general tools such as B, LTS, CSP and FSP. To do so, numerous works offer more or less systematic translations of source architecture to the target model. In this section, only the works related to the behavioral verification of the SCDL/WS-BPEL architectures are mentioned.

Yeung proposes in [18] to translate the WS-BPEL web service to CSP to verify the behavior properties of web services architecture. In this paper, formal verification can be carried out based on the notion of CSP trace-refinement and can take advantage of the FDR2 model checking. The authors of [4] propose to use the FSP formal language to check if a web service composition implemented in WS-BPEL satisfies a web service composition specification captured by Message Sequence Charts (MSCs). Both the WS-BPEL process and the MSC are translated to FSPs. Each FSP represents a finite labelled transition system. Using the LTSA model checking tool, this FSP target specification can check the safety and progress properties as well as properties expressed in the LTL logic. In [5], Foster et al. use this LTSA to check the compatibility of web service compositions in WS-BPEL. Since the semantics of Petri Nets is formally defined by mapping each WS-BPEL process to a Petri net, a formal model of WS-BPEL can be obtained. This approach has been followed in several works. For example, in [12] Verbeek et al. formalize some WS-BPEL activities used for the orchestration of Web services as a class of Petri Net called workflow nets. For this class of Petri nets, a verification tool named Wolfan has been developed. This tool can verify properties such as the termination of a workflow net and detection of nodes that can never be activated. The authors of [17] propose to map most of the basic and structured activities of WS-BPEL and the Web Service Choreography Interface (WSCI) to Coloured Petri Nets (CPN). In [8] Hamel et al. propose to use the Event-B method to check the structural and behavioural properties of an SCA component assembly. To achieve this, the B-Invariant and B-Event are profitably used to formalize the patterns proposed by Barros [1]. Using the ProB animator, Hamel et al. [8] validate their formal approach on an event-B specification.

3 Translation of SCDL/WS-BPEL to Wright

In this section we propose a set of rules allowing the translation of SCA software source architecture to a Wright target architecture. This allows the verification of standard behavioral properties supported by the Wr2fdr tool accompanying the Wright ADL.

Regarding the structural aspect, an SCA software architecture is generally described in an XML SCDL file. The latter expresses all the relations in a composite. In this context, a composite is an assembly of heterogeneous components. Each SCDL component is based on a common set of abstractions such as services, references and properties. In the context, services and references describe, respectively, what a component provides and what a component requires from its external environment. These services and references can be matched with bindings. Hence, each SCDL markup can be specified in Wright as follows:

- An SCDL composite can be translated to a Wright configuration;
- An SCDL component can be translated to a Wright component;

- An SCDL component's reference can be translated to a Wright port with the same name;
- An SCDL component's service can be translated to a Wright port with the same name;
- An SCDL wire connects two SCA components. Hence, we propose to translate an SCDL wire to a Wright connector that proposes two roles.

Concerning the WS-BPEL behavioral descriptions, we propose to translate each WS-BPEL process by a CSP process. In this translation, each primitive activity is translated to a CSP event. Since WS-BPEL provides three kinds of activities, we suggest translating each activity by a specific event as follows:

- An <invoke> activity is used to initialize an appeal of an operation Oper. This activity can be modeled in CSP by an initialized event as follows: _invokeOper
- A <receive> activity is used to wait for a message from an external operation Oper. This observed activity can be modeled in CSP by an observed event as follows: receiveOper
- A <reply> activity is used to initialize a response to an external operation Oper. This activity can be modeled in CSP by an initialized event as follows: _replyOper

In addition, WS-BPEL provides typical structured activities such as: <sequence>, <flow>, <terminate>, <if>, <switch>, <while>, <repeatUntil>, etc. These control structures can express a causal relationship between multiple invocations by means of control and data flow links. For the WS-BPEL control structures, we propose the following translation rules:

- The <sequence> construct is used in WS-BPEL wherever a series of activities needs to occur sequentially, although they may be contained one or more times within looping or concurrent construct activities. This <sequence> construct can be modeled in CSP by a set of events separated by the prefixing operator (->).
- Concurrency in WS-BPEL permits us to model the concurrent transitions in the message sequence charts. In WS-BPEL, this is specified using the <folw> construct. However, the concurrency in CSP is modeled by the parallel composition operator (| |). Hence, using the CSP parallel operator (| |), we can model the WS-BPEL flow activities by a set of concurrent processes.
- In WS-BPEL, the conditional branching introduces decision points to control the execution flow of a process. Each conditional structure such as <if> or <switch> can be modeled in CSP by the adequate choice operator:
  - ([ ]) deterministic choice operator: if the choice between these activities is an external choice. In other words, if these activities are observed (receive activity).
  - (| ~ |) nondeterministic choice operator: if the choice between these activities is an internal choice. In other words, if these activities are initialized (invoke or reply activity).
- In WS-BPEL, as in most programming languages, loops are used to repeat activities. Each looping structure such as <forEach>, <while> or <repeatUntil> can be modeled in CSP by a recurrent process as follows: P = … ->P.

4 Translation of Wright to Ada

The main structural concepts treated in this section are: configuration, component, connector, port, role, computation, glue, attachments, process, initialized event, observed event, successfully terminated event, prefixing operator, deterministic choice operator and nondeterministic choice operator. To achieve this, we proposed an Ada package called ArchWright allowing the representation in Ada of the main structural concepts coming from the Wright ADL. Hence, a Wright configuration can be translated to an Ada concurrent program using the ArchWright package. For traceability reasons, we keep the same identifiers used in the Wright specification.

The Wright ADL is one of the first approaches allowing the description of the behavioral aspect of architectural elements. Indeed, the behavior of a Wright component (respectively of a connector) is described locally through the ports (respectively roles) and, generally, through a computation (glue respectively) using a CSP process algebra. Hence, in this work, we propose to implement each:

- Wright configuration by an Ada concurrent program using the ArchWright package.
- Wright component by an Ada record compound with two fields: (1) Ports that represents the component's ports. It is modeled by an array of CSPTask, where the CSPTask is the task type proposed to implement in Ada the CSP process; (2) Computation that represents the component computation. This field can be modeled by a single CSPTask.
- Wright connector by an Ada record compound with two fields: (1) Roles that represents the connector roles. It is modeled by an array of CSPTask; (2) Glue that represents the glue of this connector. This field can be modeled by a single CSPTask.

Table 1 illustrates the principle of the translation of the main Wright architectural elements into Ada. For traceability reasons, we keep the same identifiers used in the Wright specification.

Regarding the CSP behavioral concepts, a simple CSP process can be compound with a set of observed and initialized events separate with the specific CSP prefixing operator. However the Ada concurrent language defines a powerful behavioral tasking model. An Ada task represents the basic element of each Ada concurrent program. It consists of two parts: task specification (declaration) and task body. This task specification can provide a set of services (called entries). These entries can have in, out or in out formal parameters. Each entry exported by an Ada task indicates the possibilities of an Ada rendezvous. Based on the similarity between the CSP process and the Ada task, we offer an intuitive correspondence provided below for translating a CSP process to an Ada task:

- A CSP process leads to an Ada task;
- A CSP event naturally corresponds to an Ada entry. In order to differentiate between an observed and an initialized event, we propose to use the same prefixed notation used in CSP: an observed event is denoted by (e) and an initialized event is denoted by (_e)
- The recursion operator can be translated by an Ada loop;

Table 1. Ada formalization of the Wright concepts

Wright	Ada

Component	`type CompWright (portsNumber : natural)`
	` is record`
	`    Ports: array (portsNumber) of CSPTask;`
	`    Computaion: CSPTask;`
	` end record;`

Connector	`type ConnectorWright (rolesNumber : natural)`
	` is record`
	`    Roles: array (rolesNumber) of CSPTask;`
	`    Glue: CSPTask;`
	` end record;`

Configuration	`with ArchWright;`
	`use ArchWright;`
	`procedure Configuration is`
	`  …`
	`begin`
	`  …`
	`end Configuration ;`

- The CSP prefixing operator (->) can be specified by the Ada sequential instruction;
- The CSP successfully terminated event (denoted by TICK or §) can be implemented with the Ada "terminate" instruction.
- The CSP internal choice operator (denoted by Π or $| \sim |$) allows the future evolution of a process to be defined as a choice between two sub-processes, but does not allow the environment any control over which one of the component processes will be selected. This internal choice can be implemented in Ada with a simple conditional structure (if). If we have multiple composite processes, the Ada conditional structure (case) can be used.
- The CSP external choice operator ([ ]) allows the future evolution of a process to be defined as a choice between two sub processes, and allows the environment to resolve the choice by communicating an initial event for one of the processes. This deterministic choice can be implemented with the Ada "select" instruction.

Table 2 illustrates these proposed rules allowing the translation of a CSP specification to Ada.

Each attachment of a component's port with a connector's role can be implemented in Ada by a sequence of entries call of the events specified in the interconnected port/role. The general form of the entries call (event) is specified as follows:

- If the port entry corresponds to an initialized event: we call this entry, then we call the similar entry of role (Component.port._event; Connector.role._event;).
- Else, if this entry corresponds to an observed event, we call this entry after the call of the similar entry in the role (Connector.role.event; Component.port.event;).

Table 2. Ada formalization of the CSP process

CSP	Ada
`P = _request -> result -> §`	**task** P **is** **entry** _request; **entry** result; **end** P ; **task body** P **is** **accept** _request; **accept** result; **end** P;
`P1 |~| P2`	**if** internCondition **then** P1 **else** P2 **end if**;
`P1 [ ] P2 [ ] TICK`	**select** P1; **or** P2; **or terminate** ; **end select**;

5 Discussions and Results Summary

The means to establish connections between software architecture and a concurrent language like Ada are limited. For example, Naumovich et al. [13] offer a manual translation of Wright into Ada without explanation rules. In our previous work [2] we established a set of simple rules allowing translating Wright software architecture into Ada. In this study, a significant improvement of our translation rules was proposed.

Our verification approach is validated on several uses cases available at our SourceForge repository [9]. The main advantage of our verification approach can be summary as follows:

- The first translation (SCDL/WS-BPEL to Wright): allows the verification of eleven standard properties related to the consistency of software architecture, among which four, assimilated to behavioral contracts, are automated by our Wr2fdr tool [15]. The latter contracts can be checked with the FDR2 model checker.
- The second translation (Wright to Ada): can be used in the verification of the specific behavioral properties of the source description using static and dynamic analysis tools associated with Ada such as test data generator and debugging. In addition, the refinement of the abstract architecture, allows step by step obtaining a coherent concrete architecture vis-à-vis the verified abstract architecture. The correction of each refinement step can be ensured by the static analysis tools associated with Ada: the refined concurrent program must keep the same properties checked on the abstract concurrent program.

6 Conclusion

In this paper, an approach for verifying the behavioral coherence of SCDL/WS-BPEL component-service architectures has been proposed. To achieve this, it has been proposed to map SCDL/WS-BPEL to the Wright ADL, thus allowing the checking of the standard properties supported by this ADL. As a second step, a set of exogenous translation rules allowing the translation from the Wright specification to an Ada concurrent program has been put forward. The choice of this Ada language is justified by the presence of many Ada analysis tools able to detect several error types. Versus the properties to be checked, we must choose the adequate Ada analysis tool. For example, FLAVERS and INCA tools promote the property-oriented trace, while the SPIN and SMV tools promote the property-oriented state.

Currently, we are extending this work by an automation of these translation rules using the Xtext, ATL and Xpand model transformation languages.

References

1. Barros, O.: Business process patterns and frameworks: reusing knowledge in process innovation. Bus. Process Manag. J. **13**(1), 47–69 (2007)
2. Bhiri, M.T., Fourati, F., Kmimech, M., Graiet, M.: Transformation exogène de Wright vers Ada. Technique et Science Informatiques **31**(7), 839–868 (2012)
3. Chama, I.E., Belala, N., Saïdouni, D.E.: Formalizing timed BPEL by D-LOTOS. IJERTCS J. **5**(2), 1–21 (2014)
4. Foster, H., Uchitel, S., Magee, J., Kramer, J.: Model-based verification of web service compositions. In: Proceedings of 18th IEEE International Conference on Automated Software Engineering, Montreal, Canada, pp. 152–163. IEEE, October 2003
5. Foster, H., Uchitel, S. Magee, J., Kramer, J., Magee, J.: Compatibility verification for web service choreography. In: Proceedings of the IEEE International Conference on Web Services, San Diego, CA, USA, pp. 738–741. IEEE, June 2004
6. Maâlej, A.J., Lahami, M., Krichen, M., Jmaïel, M.: Distributed and resource-aware load testing of WS-BPEL compositions. In: ICEIS (2), pp. 29–38 (2018)
7. Haddad, I., Kmimech, M., Sakka Rouis, T., Bhiri, M.T.: Towards a practical approach to check service component architecture. In: 11th International Conference on Semantics, Knowledge and Grid, pp. 65–72. IEEE (2015)
8. Hamel, L.M., Graiet, G.M., Kmimech, M.: Formal modeling for verifying SCA composion. In: RCIS Conference, pp. 193–204 (2015)
9. https://sourceforge.net/projects/SCA2WrightToAda (2018)
10. Hoang, D., Moy, Y., Wallenburg, A., Chapman, R.: SPARK 2014 and GNATprove - A competition report from builders of an industrial-strength verifying compiler. Int. J. Softw. Tools Technol. Transf. **17**(6), 695–707 (2015)
11. Maalej, Maroua, Taft, Tucker, Moy, Yannick: Safe dynamic memory management in ada and SPARK. In: Casimiro, António, Ferreira, Pedro M. (eds.) Ada-Europe 2018. LNCS, vol. 10873, pp. 37–52. Springer, Cham (2018). https://doi.org/10.1007/978-3-319-92432-8_3
12. Mateo, J.A., Ruiz, V.V., Macià, H., Díaz, G.: A coloured petri net approach to model and analyse stateful workflows based on WS-BPEL and WSRF. In: SEFM Workshops, pp. 389–404 (2014)
13. Naumovich, G., Avrunin, G.S., Clarke, L.A., Osterweil, L.J., Applying static analysis to software architectures'. In: ACM SIGSOFT 1997, Software Engineering Notes, vol. 22(6), pp. 77–93 (1997)
14. OASIS. Service Component Architecture Assembly Model Specification Version 1.1. Oasis, July 2017. https://www.oasis-open.org/standards

15. Sakka Rouis, T., Bhiri, M.T., Kmimech, M., Moussa, F.: Wr2Fdr Tool Maintenance for models Checking. In: SoMeT Conference, pp. 425–440 (2017)
16. Strobl, S., Zoffi, M., Bernhart, M., Grechenig, T.: A tiered approach towards an incremental BPEL to BPMN 2.0 Migration. In: ICSME Conference, pp. 563–567 (2016)
17. Yang, Y., Tan, Q., Xiao, Y., Liu, F., Yu, J.: Transform BPEL workflow into hierarchical CP-nets to make tool support for verification. In: APWeb Conference, pp. 275–284 (2006)
18. Yeung, W.L.: Mapping WS-CDL and BPEL into CSP for behavioural specification and verification of web services. In: ECOWS, IEEE Computer, pp. 297–305 (2006)

Pattern-Based Approaches for Business Process Improvement: A Literature Review

Nesrine Missaoui[1,2(✉)] and Sonia Ayachi Ghannouchi[2,3]

[1] Higher Institute of Computer Science and Communication Techniques of Sousse, University of Sousse, Sousse, Tunisia
missaouinesrine2015@gmail.com
[2] RIADI Laboratory, University of Manouba, Manouba, Tunisia
[3] Higher Institute of Management of Sousse, University of Sousse, Sousse, Tunisia
sonia.ayachi@isgs.rnu.tn

Abstract. The continuous evolutions of information technology, together with the various changes, have a significant impact on Business Processes (BP) and their performance. To effectively deal with these changes, several solutions that introduce a new way to better control a BP were proposed. Among them, emerges the concept of business process improvement (BPI) that focuses on continuous improvement and evolution of BPs by adopting a number of techniques. Nowadays, a new technique for modeling and executing a BP has regained a lot of attention. It is based on the concept of patterns defined as reusable solutions used for dealing with problems occurring in a certain context, concerning a given BP. Several studies were interested in this concept, by proposing a pattern approach for modeling, executing and improving a BP. For this, the aim of this paper is to identify and evaluate the different types of patterns proposed in literature by suggesting a systematic review of the various patterns proposed. The result of the review was analyzed using a number of criteria that enable us to identify the most suitable patterns to the area of process improvement and thus positioning our work in this area. This can later be iteratively corrected and completed in order to obtain a continuously improved set of BPIP.

Keywords: Business Process Improvement · Business Process Patterns BPI-patterns · Literature review · Evaluation · Criteria catalogue

1 Introduction

A business process (BP) is a sequence of activities that transform an input into an output. It helps companies define their objectives while linking the strategic and operational levels. In literature, several terms relating to the management and improvement of a BP were mentioned such as BP Redesign, BP Reengineering [1], Business Process Management (BPM) [2] and Business Process Improvement (BPI) [3]. Among these terms, BPM and BPI were identified as the most suitable approaches for managing and improving processes [3, 4]. They present a number of techniques and tools used for describing not only the situations before and after the

J. H. Park et al. (Eds.): PDCAT 2018, CCIS 931, pp. 390–400, 2019.
https://doi.org/10.1007/978-981-13-5907-1_42

improvement but also the actual act of improvement. Among the concepts that promote a better presentation of the improvement process is the pattern concept where several researchers have proposed a pattern-based approach for BPM [5, 7]. These patterns present a number of reusable models that help to solve problems and find solutions for a specific context. It can be applied in different domains such as Software development [6], Workflow Management [24, 25], BPM and BPI. Therefore, according to the domain and based on the different approaches proposed, a pattern can be classified into four types: Process Patterns (PP) allows defining examples that show the way to link activities and solve a common problem [7], Business Process Patterns (BPP) are a set of patterns used for modeling a BP [8], Process Improvement patterns (PIP) present the improvement of a particular aspect of a process and the fourth type is Business Process Improvement Patterns (BPIP) which is an approach that allows making significant changes to a BP. They can be applied during the entire lifecycle of a process promoting its continuous improvement [4]. Each of these types has its own characteristics as well as a specific structure that differentiates them. However, the common goal of these patterns is to improve the performance of a BP by proposing new ideas that facilitate the modeling or the execution of a process.

The presented work focuses on the improvement of a BP where several studies have been inspired by the concept of patterns and have introduced a number of approaches based on BPIP or Change patterns [4, 9, 10]. Nevertheless, these researchers have generally been limited to studying a set of patterns that have been previously applied and evaluated without presenting a review of the studies proposed in this area. Therefore, it is necessary to study additional types of patterns and examine their quality levels in order to propose an approach of pattern reuse promoting a better development and improvement of BPM projects.

The major contribution of the paper is presenting a study of the different approaches adopting this concept. The study is based on a literature review methodology facilitating the process of research and analysis of a large collection of patterns approaches. The results of the review comprise a set of studies defining various types of patterns. These patterns have been analyzed and evaluated using a criteria catalogue from which we have been able to select the most suitable approaches for presenting a BPI pattern which will help in creating a list of improvement patterns favoring greater process flexibility.

The remainder of this paper is structured as follows. Section 2 presents the research procedure for the literature review methodology and their results. Section 3 focuses on defining the criteria catalogue used for evaluating the results of the review. Section 4 summarizes the results of the evaluation. Section 5 provides a synthesis of the review and an overview of our BPI approach. Section 6 presents the conclusion and our future work.

2 SLR Research Methodology for Pattern Approaches

The aim of this paper is to identify a set of patterns that can be used during a BPI step, by establishing a systematic review of the different studies proposed. Indeed, in the field of business process engineering, two types of research methodologies have been presented and adopted which are the methodology of SLR (Systematic Literature Review) and the Design Science Research Methodology (DSRM). The SLR methodology [11] was chosen due to the fact that this methodology is more appropriate to the objective of this work which is the study of patterns that are already published in the literature. Specifically, the systematic review was conducted using the following procedure: (1) formulate the research(s) question(s), (2) present the research strategy, (3) define the inclusion and the exclusion criteria, (4) results analysis. The research questions (RQ) with which we establish the SLR are:

RQ1: "What type(s) of a pattern(s) are being tackled in the study?"
RQ2: "What is the main objective of the defined patterns?"
RQ3: "Which types of patterns can be applied in the field of BPI?"
RQ4: "What are the requirements/steps needed to apply the given pattern?"
RQ5: "How to ensure their reusability?"
RQ6: "How to ensure that the presented pattern has the required level of quality?"

To perform the research strategy, we have selected the following string research: ('Business Process Improvement' OR 'Process Improvement Patterns' OR 'Process Patterns' OR 'Business Process Patterns' OR 'Business Process Improvement Patterns' OR 'Patterns for improving business processes').

This string has been applied to the following data sources: ACM Digital Library, IEEE Explore Digital Library, Science Direct-Elsevier, Springer Link, BPTrends and Google Scholar. Overall, these libraries include the work of the most relevant conferences and journals in the domains of BPM and BPI. For the inclusion and exclusion criteria, we considered that a research work is included if and only if its title, abstract and content are related to the fields of BPM and BPI. Contrariwise, papers were excluded if they are not linked to this area. Papers with the same proposal are deleted and only the full version is included. There's no restriction to the date of publication, and we have considered proposals for which an implementation or evaluation has been established and/or those who have not presented an implementation of their approach. Our SLR research, summarized in Table 1, resulted in a total of 391 papers that were reviewed manually. According to the inclusion and exclusion criteria, 26 research papers have passed the first step of filtering and were further evaluated using a set of evaluation criteria.

Table 1. Results of the SLR before and after applying the criteria

Results	ACM	IEEE	Science Direct	Springer	BPTrends	Google Scholar	Total
Before	46	71	51	118	4	101	391
After	3	2	4	9	1	7	26
Selected papers	[14, 26, 29, 31]	[9, 23]	[10, 18]	[4, 7, 15–17, 19, 22, 27, 28, 30]	[3]	[6, 8, 13, 20, 21, 24, 25]	

3 Patterns Evaluation: Criteria Catalogue

To better understand the structure of the proposed patterns, it is recommended to define a set of criteria that facilitate their classification and definition. Several studies have attempted to present a number of criteria allowing the identification and presentation of patterns. Among these studies, Becker et al. [12] have presented a list of criteria to evaluate BPPs: General criteria, Representational criteria, and Feature criteria. In addition, other attributes describing the structure of a pattern were proposed by Falk et al. [4]. They include a number of attributes that address the problem, the context and the solution that a pattern is intended to solve or improve. Following the work of these two authors and to facilitate the analysis of the results obtained during the SLR, we have defined a criteria catalogue also considered as criteria of selection and analysis of patterns. The catalogue presents a combination of the works proposed by [12] and [4] to which we have added our own criterion that is the evaluation criteria. It regroups three classes:

(1) General criteria for a general description of the pattern. It contains a number of attributes presenting the name, the type of patterns, the source from which the pattern has been identified (approach origin), the objectives determining the reason for applying the pattern, the accessibility of the pattern, the origin of the proposed pattern and the domain to which the pattern has been defined and applied (application domain).

(2) Pattern presentation criteria for introducing detailed information about the structure of a pattern. It is presented by four criteria. The solution criteria define the steps of applying a pattern while being based on the notation, formalization and representation attributes (T for Textual, Gr for graphical representation). The context criteria help to determine the requirements needed to apply a given pattern based on the type of modeled views: Data (D), Control flow (CF), Message (M), Resource (R), General (G) and the abstraction level: L0 for presenting a design model and L1 for defining a pattern meta-model [4]. The adaptability defines the degree to which a pattern can be customized for a specific use case (St: static for creating or improving pattern models, DC: design choice, CP: configuration Point, F: Formal approach). And the guidelines criterion providing recommendations on the way to use and combine patterns [12].

(3) Evaluation criteria define a list of attributes estimating the feasibility of the patterns and assessing their quality levels. It contains attributes presenting the validation of a pattern by implementing the pattern using a tool support system or by applying it in a case study scenario. The effects attributes are related to the analysis of the performance of a BP after the application of the pattern: a pattern can be evaluated based on the KPI metric or by simply listing its benefits.

4 Results of the Review

The list of research papers that passes the primary selection step of the SLR has been analyzed and evaluated based on the criteria catalogue presented above. As illustrated in Table 2, 94% of the selected papers define an approach for modeling, managing or improving a BP based on PP (22%), BPP (33%), and BPIP (39%). We note that most studies propose a BPP or a BPIP approach since it facilitates the phases of modeling and improving a BP. Moreover, the proposed approaches are "result-oriented": 61% over 39% for PO, meaning that their objectives are generally having reliable results for detecting and correcting problems in a BP.

Table 2. Results of the evaluation: general criteria

S	Design Pattern	PP	BPP	BPIP	Name	Type*					App.O*		Obj*		Access*			Pattern. O*			App. Domain*							
						D	MM	Co	An	Mi	R	In	PO	RO	FR	L	CL	CS	LR	PM	SD	WM	Mod	P. Con	BPM	SPI	BPI	G
[3]				X	X	X		X			X			X		X			X								X	
[4]				X	X		X				X			X	X			X	X								X	
[6]	X				X	X					X		X		X			X			X							
[7]		X			X	X					X			X		X			X									X
[8]			X		X	X							X		X				X						X			
[9]				X	X	X					X		X			X			X						X		X	
[10]				X	X	X					X	X		X		X		X					X		X		X	
[13]		X		X	X		X		X		X	X		X	X			X	X						X		X	
[14]			X		X			X	X		X		X			X		X		X					X			
[15]			X		X		X			X	X			X		X				X					X			
[16]				X	X	X					X			X		X		X	X						X		X	
[17]			X	X	X	X					X			X		X			X		X						X	
[18]				X	X	X					X	X	X			X		X	X					X			X	
[19]		X	X		X	X					X	X		X		X		X	X						X			
[20]				X	X	X		X				X		X	X			X								X		
[21]			X		X	X					X			X		X		X	X				X					
[22]				X	X			X			X			X	X			X									X	
[23]				X	X	X					X	X	X		X				X								X	
[24]			X		X	X					X		X		X					X		X						
[25]		X	X		X	X							X		X				X			X						
[26]			X	X	X		X				X			X		X			X								X	
[27]	X				X		X				X			X		X			X				X					
[28]		X	X		X					X	X	X	X			X			X						X			
[29]		X			X		X				X		X			X			X				X		X			
[30]				X	X	X	X				X			X		X		X	X						X		X	
[31]		X	X		X	X					X			X		X		X				X						

(*) **Type**: D: design pattern for modeling new processes; MM: Meta-Model for presenting the structure of a pattern; Co: compliance pattern representing situations that a BP model must adhere to; An: anti pattern representing situations that should not occur in BP model; Mi: mining pattern used generally in workflow system. **App. O**: Research (R), Industry (In). **Access**: Free (FR), Limited (L), Closed (CL). **Pattern O**: Case study (CS), Literature review (LR), Process Mining (PM). **Objectives**: Procedure oriented (PO) representing the way to apply a pattern, Result oriented (RO) defining a number of building blocks to enhance software design. **App. Domain**: Software Development (SD), Modeling (Mod), Process configuration (P. Con), Software Process Improvement (SPI), General (G)*.

For the general criteria, 68% of the research papers have their origin from a scientific research that allowed the authors to propose a list of patterns applied in various domains. Furthermore, all patterns have a name that differentiates them from others where 51% of these patterns were mainly defined based on a literature review [3, 9] and 55% of them are design patterns that presented new ways of modeling a process. On the other hand, for the BPIP, 21% of the works proposed a meta-model facilitating the definition of a pattern e.g. [4, 13].

The pattern presentation criteria, presented in Table 3, most of the studies adopt the semi-formal (64%) methods by generally using BPMN (26%) and other types of notation (35%) as a modeling notation in order to have a textual and a graphical representation of a pattern (68%). In addition, the approaches are presented according to the two types of levels of abstraction where 60% of the studies propose a set of pattern models (L0) while 37% of them define a meta-model of patterns (L1) e.g. [4, 15].

Table 3. Results of the evaluation: pattern presentation + evaluation criteria

S	Pattern Presentation criteria																					Evaluation criteria			
	Solution									Context															
	Notation				Forma			Repres		Views					Level.abstr		Adaptability				Guidelines	Validation		Effect	
	UML	BPMN	G	Other	F	SF	I	T	Gr	D	CF	M	Re	G*	L0	L1	ST	DC	CP	F		Tool support	Case study	KPI	Benefits
[3]			X			X		X	X	X	X		X		X		X					-	-	X	
[4]	X					X		X			X	X	X			X	X				X		X	X	
[6]	X			X	X			X	X					X	X			X			X	X	X		X
[7]	X						X	X						X	X		X					-	-		X
[8]		X				X		X	X					X	X		X				X	X		-	-
[9]				X		X		X	X	X	X				X		X				X	X			X
[10]		X				X		X	X	X	X				X		X	X			X	X	X	-	-
[13]	X			X		X		X	X	X	X		X		X	X	X				X	X	X		X
[14]				X	X			X						X		X				X	X	X		-	-
[15]				X	X			X	X					X		X				X	X	X	X		X
[16]			X		X			X		X	X		X			X			X		X	X	X	X	
[17]		X				X		X	X	X	X				X		X				X	X	-	-	X
[18]		X				X		X	X		X	X			X		X	X			X	X	X		X
[19]		X				X		X	X		X				X			X			X	X	X		X
[20]			X				X	X						X	X		X				X	-	-		X
[21]		X				X		X	X	X	X				X		X					-	X		X
[22]			X	X		X		X		-	-	-	-	-	-	-	X				X	X	X	X	
[23]				X		X		X	X		X		X		X		X				X	X			X
[24]				X		X		X	X				X			X	X					X		-	-
[25]				X		X		X	X		X				X		X		X			X	X	-	-
[26]				X	X			X			X					X				X	X	-	-	-	-
[27]	X					X		X	X					X		X	X				X	-	-	X	X
[28]			X		X			X						X	X					X	X	X	X	-	-
[29]	X				X			X	X					X		X				X	X	X	X		X
[30]		X				X		X	X	X	X	X			X	X	X				X	X	X	X	
[31]		X				X		X	X	X	X				X		X				X	X	X	-	-

The use of these tools facilitates the modeling of a process according to different views. Therefore, the majority of the studies cover in their modeling two types of views: 34% for Control flow [9, 19] and 22% for data [13, 17]. For the adaptability criterion, 65% of the approaches propose a static pattern that can be customized to create new models or evaluate existing ones. In addition, all the studies include a step of validation for their approaches where 49% present an implementation of their patterns and 37% apply their approach on a real example of a process. For the effect attribute, 52% of the approaches present only the benefits of applying the proposed patterns, whereas 22% of them especially those defining a BPIP use KPI metrics for evaluating the proposed pattern. Based on these results, we can determine the importance of the concept of process improvement where the majority of the studied literature proposes an approach for improving a process. However, these studies lack detailed information about the improvement steps especially those who did not present a validation step for their proposed approaches. For this, to better positioning our work, a process for selecting the most suitable approaches to the BPI area is presented in Sect. 5 where a list of patterns has been selected and on which our BPI approach will be based on.

5 Synthesis and Solution Overview

The literature review together with the evaluation step facilitates the process of identification as well as the analysis of the different patterns proposed in literature. In fact, the three classes of criteria foster a better understanding of the concept of patterns which allow us to have a general view on the structure of a pattern and the way it is modeled and implemented. The main advantages of our review are that our SLR research not only focuses on presenting BPIP defined in literature but also other types of patterns used for modeling and executing a process. In addition, compared to the works of both Falk et al. and Becker et al., the criteria catalogue used during the analysis step is considered as a catalogue that can be applied for the analysis of all types of patterns. Whereas, those proposed by [4] were mainly used for presenting a meta-model for BPI pattern and the criteria defined by [12] were principally used for defining a BP pattern. However, during the evaluation phase, it was sometimes difficult to assess a pattern according to these criteria due to the fact that some of the studied approaches lack a detailed description of the defined pattern. Also, a major challenge we had to deal with is the selection of the appropriate research studies that are considered as most suitable for presenting an approach for managing or improving a BP, resulting in discussions about what could be defined as a BPIP. For this, we have considered that a BPIP is a pattern that helps to transform an "as-is" process model into a desired "to-be" process model, while taking into consideration the weaknesses detected during the analysis of the "as-is" process.

To present this idea, we proposed a process for identifying the approaches related to the concept of BPIP. The selection process was based on the evaluation results as well as the number of research questions detailed in Sect. 2. We have selected, from each criterion, a set of attributes that are considered as answers to the SLR research questions. For each attribute, we have chosen the most relevant values for defining a pattern. These values were selected by taking into consideration their usefulness to better present the framework of our work that focuses on improving a BP based on BPIP. Thus, the selected attributes represent the steps needed to define a BPI phase. Each value was scored as follows: (1) presents the value considered as essential for describing a pattern, (0.5) presents the value that is considered less suitable to choose. Table 4 presents the set of attributes selected and their values.

Table 4. Selected attributes and their values for the selection process

Research questions	Attributes + Values	
RQ1 & RQ3	Application domain	BPM/BPI (1) Mod (**0.5**) WM (**0.5**)
	Types	MM (1) D (1) An (**0.5**)
RQ2	Objectives	R-O (**0.5**) P-O (1)
	Pattern Origin	CS (1) LR (**0.5**)
RQ4	Notation	(1)
	Formalization	SF (1) F (**0.5**)
	Context	Views (1) Abstraction level (1)
	Guidelines	(1)
RQ5	Adaptability	St (1) DC (**0.5**)
RQ6	Validation	Case Study (1) Tool support (**0.5**)
	Effect	KPI **1**

According to these values and for each approach selected from the SLR research, we have added up the scores of the defined attributes. Table 5 present the results of the analysis where we find that the majority of studied papers concentrates on presenting the main objectives of the pattern (69% for RQ2), defining the way to apply it (92% for RQ4) and ensure its reusability (69% for RQ5).

Table 5. Results of the selection process

Sources	Research questions					
	RQ1 & RQ3	RQ2	RQ4	RQ5	RQ6	Total
[3]	2	1	4	1	1	9
[4]	2	1	5	1	2	11
[6]	1	2	4,5	0,5	1,5	9,5
[7]	1	1	3	1	0	6
[8]	2	1,5	5	1	0,5	10
[9]	2	1,5	5	1	0,5	10
[10]	2,5	1,5	5	1,5	1,5	12
[13]	2	2	5	1	1,5	11,5
[14]	1	2	4	0	0,5	7,5
[15]	2	0,5	4,5	0	1,5	8,5
[16]	2	2	4,5	0	2,5	11
[17]	2	1	5	1	0,5	9,5
[18]	2	2,5	5	1,5	1,5	12,5
[19]	2	2	5	0,5	1,5	11
[20]	1	1,5	4	1	0	7,5
[21]	1,5	2	4	1	1	9,5
[22]	1	1,5	3	1	2,5	9
[23]	2	1,5	5	1	0,5	10
[24]	1,5	1,5	4	1	0,5	8,5
[25]	1,5	1,5	4	1	1,5	9,5
[26]	2	1	4,5	0	0	7,5
[27]	1,5	1	5	1	1	9,5
[28]	1	1,5	4,5	0	1,5	8,5
[29]	2,5	1,5	3,5	0	1,5	9
[30]	3	2	5	1	2,5	13,5
[31]	1,5	1,5	5	1	1,5	10,5

Therefore, based on the results of the analysis, the studies with the highest score were chosen as the most relevant approaches for defining a BPIP where a list containing different types of patterns has been selected.

The list contains a multitude of patterns such as design patterns [6], change patterns [9, 10, 17, 18], improvement patterns [4, 16, 27, 30] and other types related to the modeling of a BP [29], the identification of weaknesses in a process [13], and the definition of time attributes in a BP [19]. This list is considered as a starting point for our future research on and it will be further analyzed in order to identify possible improvements and present new types of BPIP that ensure greater process flexibility and promote its continuous improvement.

6 Conclusion

The aim of this paper was to precisely conduct a literature review on the existing pattern approaches by adopting the SLR methodology to select, from the wide variety of patterns, those suitable to be used and implemented in the domain of BPI. To evaluate the results of the SLR, we have defined a catalogue with a set of criteria that were inspired by the works of Becker et al. and Falk et al. These criteria helped us in having a detailed description of the identified patterns and thus having a better understanding of the structure of a pattern. We have also presented an overview determining the list of studies on which our BPI approach will be based on. In addition, our study focuses on the continuous improvement of a process by proposing an approach for pattern reuse. More precisely, the approach will be applied to the lifecycle of a BP, taking as a starting point, a BP model and a pattern base containing a set of predefined patterns.

The global reuse approach of these patterns includes five steps: (a) Constitute the pattern database. (b) Preselect, from the database, a list of patterns to use. (c) Select from the predefined list the most appropriate patterns, to solve the problems detected during the analysis of a BP model. (d) Evaluate the effectiveness of the applied patterns and (e) Update the pattern database by proposing ideas for improving existing patterns and/or creating new patterns. This process will be applied on the BPM lifecycle where, for each phase of the cycle, a step(s) of pattern reuse will be defined, to improve the performance of the corresponding lifecycle phase and promotes the continuous improvement of a BP based on an iterative evaluation of the pattern database.

Our future works include two major parts: the first part is devoted to define a meta-model describing the structure of a pattern based on the analysis of previously proposed meta-models. Afterward, a list of BPIP will be created with the proposal of a guidance approach facilitating the selection of patterns. The second part focuses on the proposal of a pattern reuse approach ensuring that our list is applicable in various types of domains by developing a prototype allowing automatic selection of patterns adapted to a given case study. The prototype will be applied to different process examples which will help in validating our approach.

References

1. Grover, V., Jeong, S.R., Kettinger, W.J., Teng, J.T.C.: The implementation of business process reengineering. J. Manag. Inf. Syst. **12**(1), 109–144 (1995). https://doi.org/10.1080/07421222.1995.11518072
2. Dumas, M., Rosa, M.L., Mendling, J., Reijers, H.: Fundamentals of Business Process Management, 1st edn. Springer, Heidelberg (2013). https://doi.org/10.1007/978-3-642-33143-5
3. Forster, F.: The Idea Behind Business Process Improvement: Toward a Business Process Improvement Pattern Framework. BPTrends (2006)
4. Falk, T., Griesberger, P., Leist, S.: Patterns as an artifact for business process improvement - insights from a case study. In: vom Brocke, J., Hekkala, R., Ram, S., Rossi, M. (eds.) DESRIST 2013. LNCS, vol. 7939, pp. 88–104. Springer, Heidelberg (2013). https://doi.org/10.1007/978-3-642-38827-9_7

5. Van Hilst, M., Fernandez, E.B.: A pattern system of underlying theories for process improvement. In: Proceedings of the 17th Conference on Pattern Languages of Programs, New York, NY, USA, pp. 8:1–8:24 (2010). https://doi.org/10.1145/2493288.2493296
6. Gamma, E., Helm, R., Johnson, R., Vlissides, J.: Design Patterns: Elements of Reusable Object-Oriented Software. Addison-Wesley Professional, Upper Saddle River (1994)
7. Störrle, H.: Describing process patterns with UML. In: Ambriola, V. (ed.) EWSPT 2001. LNCS, vol. 2077, pp. 173–181. Springer, Heidelberg (2001). https://doi.org/10.1007/3-540-45752-6_14
8. Atwood, D.: BPM Process Patterns: Repeatable Design for BPM Process Models. BPTrends (2006)
9. Kim, D., Kim, M., Kim, H.: Dynamic business process management based on process change patterns. In: International Conference on Convergence Information Technology, Gyeongju, pp. 1154–1161 (2007). https://doi.org/10.1109/iccit.2007.91
10. Weber, B., Reichert, M., Rinderle Ma, S.: Change patterns and change support features - enhancing flexibility in process-aware information systems. Data Knowl. Eng. **66**(3), 438–466 (2008). https://doi.org/10.1016/j.datak.2008.05.001
11. Kitchenham, B., Charters, S.: Guidelines for performing systematic literature reviews. Software Engineering, Technical report, EBSE/EPIC, vol. 2 (2007)
12. Becker, M., Klingner, S.: A criteria catalogue for evaluating business process pattern approaches. In: Bider, I., et al. (eds.) BPMDS and EMMSAD 2014. LNBIP, vol. 175, pp. 257–271. Springer, Heidelberg (2014). https://doi.org/10.1007/978-3-662-43745-2_18
13. Höhenberger, S., Delfmann, P.: Supporting business process improvement through business process weakness pattern collections. In: Proceedings of the 12th International Conference Business Informatics, Osnabrück, pp. 378–392 (2015)
14. Awad, A., Barnawi, A., Elgammal, A., Elshawi, R., Almalaise, A., Sakr, S.: Runtime detection of business process compliance violations: an approach based on anti patterns. In: Proceedings of the 30th ACM Symposium on Applied Computing, NY, USA, pp. 1203–1210 (2015). https://doi.org/10.1145/2695664.2699488
15. Chapela-Campa, D., Mucientes, M., Lama, M.: Discovering infrequent behavioral patterns in process models. In: Carmona, J., Engels, G., Kumar, A. (eds.) BPM 2017. LNCS, vol. 10445, pp. 324–340. Springer, Cham (2017). https://doi.org/10.1007/978-3-319-65000-5_19
16. Niedermann, F., Radeschütz, S., Mitschang, B.: Business process optimization using formalized optimization patterns. In: Abramowicz, W. (ed.) BIS 2011. LNBIP, vol. 87, pp. 123–135. Springer, Heidelberg (2011). https://doi.org/10.1007/978-3-642-21863-7_11
17. Uronkarn, W., Senivongse, T.: Change patterns detection and traceability impact analysis of business process models. In: Yang, G.-C., Ao, S.-I., Huang, X., Castillo, O. (eds.) Transactions on Engineering Technologies, pp. 441–455. Springer, Dordrecht (2015). https://doi.org/10.1007/978-94-017-9588-3_33
18. Ayora, C., Torres, V., De la Vara, J.L., Pelechano, V.: Variability management in process families through change patterns. J. Inf. Softw. Technol. **74**, 86–104 (2016). https://doi.org/10.1016/j.infsof.2016.01.007
19. Lanz, A., Weber, B., Reichert, M.: Time patterns for process-aware information systems. J. Requir. Eng. **19**(2), 113–141 (2014). https://doi.org/10.1007/s00766-012-0162-3
20. Appleton, B.: Patterns for conducting process improvement. In: Conference on Pattern Languages of Program Design, PLoPD 1997 (1997)
21. Gschwind, T., Koehler, J., Wong, J.: Applying patterns during business process modeling. In: Dumas, M., Reichert, M., Shan, M.-C. (eds.) BPM 2008. LNCS, vol. 5240, pp. 4–19. Springer, Heidelberg (2008). https://doi.org/10.1007/978-3-540-85758-7_4

22. Lang, M., Wehner, B., Falk, T., Griesberger, P., Leist, S.: Evaluating business process improvement patterns by simulation: complete research. In: Twenty-Third European Conference on Information Systems (ECIS), Münster, Germany (2015). https://doi.org/10.18151/7217407
23. Pourshahid, A., Mussbacher, G., Amyot, D., Weiss, M.: Requirements for a modeling language to specify and match business process improvement patterns. In: 3rd International Workshop on Model-Driven Requirements Engineering, Brazil, pp 10–19. IEEE (2013). https://doi.org/10.1109/modre.2013.6597259
24. Russell, N., Hofstede, A.H.M., Edmond, D., Van Der Aalst, W.M.P.: Workflow Resource Patterns (2004). http://www.workflowpatterns.com/patterns/resource/
25. Russell, N., Ter Hofstede, A.H.M., Van Der Aalst, W.M.P., Mulyar, N.: Workflow control-flow patterns: A revised view (2006). http://www.workflowpatterns.com/documentation/
26. Rinderle-Ma, S., Reichert, M., Weber, B.: On the formal semantics of change patterns in process-aware information systems. In: Li, Q., Spaccapietra, S., Yu, E., Olivé, A. (eds.) ER 2008. LNCS, vol. 5231, pp. 279–293. Springer, Heidelberg (2008). https://doi.org/10.1007/978-3-540-87877-3_21
27. Cherfi, S.S.-S., Comyn-Wattiau, I., Akoka, J.: Quality patterns for conceptual modelling. In: Li, Q., Spaccapietra, S., Yu, E., Olivé, A. (eds.) ER 2008. LNCS, vol. 5231, pp. 142–153. Springer, Heidelberg (2008). https://doi.org/10.1007/978-3-540-87877-3_12
28. Smirnov, S., Weidlich, M., Mendling, J., Weske, M.: Action patterns in business process model repositories. Comput. Ind. **63**(2), 98–111 (2012). https://doi.org/10.1016/j.compind.2011.11.001
29. Tran, H.N., Coulette, B., Tran, D.T., Vu, M.H.: Automatic reuse of process patterns in process modeling. In: Proceedings of the ACM Symposium on Applied Computing, NY, USA, pp. 1431–1438 (2011). https://doi.org/10.1145/1982185.1982494
30. Yousfi, A., Saidi, R., Dey, A.K.: Variability patterns for business processes in BPMN. Inf. Syst. e-Business Manag. **14**(3), 443–467 (2016). https://doi.org/10.1007/s10257-015-0290-7
31. Zimmermann, B., Doehring, M.: Patterns for flexible BPMN workflows. In: Proceedings of the 16th European Conference on Pattern Languages of Programs: EuroPLoP 2011, Article No. 7 (2011). https://doi.org/10.1145/2396716.2396723

Security and Privacy

Research on Image Content Authentication Algorithm of Semi-fragile Watermarking for Adaptive Embedding Based on Characteristic of Image

Shan Yang, Liang Chen, and De Li(✉)

Department of Computer Science, Yanbian University, Yanji, China
379606352@qq.com, 371482788@qq.com,
leader1223@ybu.edu.cn

Abstract. In this paper, we propose a novel quantization-based semi-fragile watermarking algorithm for image authentication and tamper localization based on discrete wavelet transform (DWT). In this algorithm, the watermarking is generated by extracting image feature from the approximation subband in the wavelet domain. The linear congruential pseudo random number generator is used to control the watermark embedding location and realize adaptive embedding. The generation and embedding of watermark are carried out in the original image. The authentication are not require original image and any additional information, which improves the security and confidentiality of watermark. The scheme can resist the mild modification of digital image and able to detect the malicious modifications precisely. Experimental results show that the proposed algorithm can distinguish high quality JPEG compression, small amount of noise from shear attack, classification of intentional and incidental tampering, and has good localization ability.

Keywords: Semi-fragile watermarking · Content authentication
Discrete wavelet transform (DWT) · Adaptive embedding · Tamper location

1 Introduction

In the era of rapid development of information technology, multimedia data is convenient for storage, transmission and sharing, and enriches people's means of obtaining information. But at the same time, people can easily replicate and modify digital image, and multimedia data has a large amount of data, high redundancy, and the difficulty of tracking changes sources also make digital media security issues into a major problem in the informationization process, this will seriously hindered the development of information industry. In some applications, such as military documents, forensic evidence, news images or video, history literature, medical images, ticket anti-counterfeiting etc., the illegal tampering of digital content will have a serious impact on people or society [1]. So how to verify the authenticity, integrity and credibility of digital multimedia content in the complex network environment becomes an urgent problem.

J. H. Park et al. (Eds.): PDCAT 2018, CCIS 931, pp. 403–413, 2019.
https://doi.org/10.1007/978-981-13-5907-1_43

Digital watermarking technology, as an effective means of protecting intellectual property rights of digital media works, has received extensive attention from the society. Among them, semi-fragile watermarking technology can not only tolerate to some extent some common signal processing operations (such as adding noise, smooth filtering, lossy compression etc.), which contained on watermarking digital media, but also can make alarm reaction to malicious tampering with and have the ability to locate tamper with the region, so it is more important application value in the Internet age [2]. In order to improve the safety and reduce the computational complexity of the algorithm, Hu [3] proposed a fragile watermarking algorithm based on wavelet transform, the main innovation point of the algorithm lies in ascension format parametric integer wavelet coefficients. but this plan in resistance to JPEG compression cannot achieve a satisfactory effect. Zhao et al. [4] proposes a semi semi-fragile zero-watermark algorithm based on block compression perception, that the image is divided into several blocks. Then each image block is observed according to the compressed sensing theory, and the observed eigenvalues are preserved as semi-fragile zero-watermark information. Liu [5] proposed a semi-fragile watermarking algorithm based on Zernike moment. First, the original image was transformed by discrete wavelet transform, then the Zernike moment was taken as the image feature. Finally Zernike moment of low-frequency wavelet coefficient was taken as the watermark, which embedded into the DWT coefficient of the image.

This paper puts forward a semi-fragile watermarking algorithm based on wavelet domain, which can pass the authentication of common signal processing operations, and refuse to pass the authentication of malicious tampering and has strong localization ability. In the algorithm implementation process, first of all multilayer discrete wavelet transform is applied to the carrier image, and then the watermark is generated according to the image feature information of the carrier image. Embedded position determined randomly by random linear congruence generator, then inserts and updates the wavelet coefficients. When the watermark is extracted, the carrier image embedded with watermark information is firstly transformed by discrete wavelet transform. Then generate watermark information, and locate tamper areas. So the dependence on original information reduces and improves the security of watermark information. The experimental results show that this algorithm can realize the image content authentication function, and the location of malicious tamper. At the same time, it can withstand general non-malicious attacks such as JPEG compression and a small amount of noise.

2 Research on Related Technology

2.1 Image Authentication Technology Based on Digital Watermarking

Digital watermarks can be divided into fragile watermarks and semi-fragile watermarks, both of which are used to achieve image integrity authentication. Among them, the fragile watermark technology is very sensitive to any slight change of the detection object, while the semi-fragile watermarking technology allows general processing operations such as JPEG compression of images under the guarantee of the authenticity of image content. Both of these watermarking techniques can detect and locate tamper areas well.

2.2 Linear Congruent Generator

In order to ensure a good visual effect of carrier images and the invisibility of watermark information. For the selection of embedded locations, random embedding with the same probability is carried out in the intermediate frequency components of wavelet coefficients. In the random position determination, it is determined by the linear congruent pseudo random number generator.

Linear congruent pseudo-random number generator [6] is a uniformly distributed pseudo-random number generator. Its mathematical formula is:

$$x_{i+1} = (ax_i + c)(\text{mod}m) \quad (1)$$

There are three parameter multiplier a, increment c, and module m. Where m is a prime number, a is a prime root of m, $c \in [0, m)$. Suppose there is a minimum positive integer T, So that for any I there is a $x_{i+T} = x_i$. T is the period of the pseudo-random number. The values of the non-negative integers of T modules m are different in a period, So maximum period $T_{max} = m$. In theory, the bigger T is, the better the pseudo-random number is. In this paper, the value of m is selected as 2^{31-1}. When $c = 0$, the most general application of linear congruent pseudo random number can be obtained. It's also called the multiplication congruence generator.

The mathematical formula is as follows:

$$x_{i+1} = ax_i(\text{mod}m) \quad (2)$$

In this chapter, a pseudo-random number subject to uniform distribution is generated by the generator. Then the binary pseudo-random number is obtained, so that the embedding domain of the watermark is selected.

3 Semi-fragile Watermarking Based on Feature Recognition

Digital watermarking is an important way to ensure multimedia information security technology. It is mainly to verify the authenticity and integrity of digital multimedia works. Precision certification is sensitive to any minor modification and is suitable for using in situations where extreme precision is required [7]. However, in practical application, multimedia data may be changed by various attacks in network transmission, but small changes in data bits cause little visual change and make any significant difference. In the era of information sharing, network transmission may inevitably suffer from signal attacks such as JPEG compression. This attack will change a lot of data in the image, but will not affect the visual perception of the human eye. However, such processing operations will be rejected for accurate certification. In order to protect the integrity of information content and the inevitable routine signal processing operations, so the semi-fragile watermark has the two functions of fragile watermark and robust watermark.

3.1 Watermark Information Generation Algorithm

The advantage of wavelet transform is to provide rich feature information of image, and it has a great advantage in watermark information extraction. The coefficient component of the third layer wavelet decomposition $LL3(i, j)$, contains most of the energy of the image and can effectively characterize the image features. It has certain robustness before and after JPEG compression and has strong anti-interference ability. Therefore, it is an ideal choice to use it to extract image feature watermark.

The flow chart of watermark certification is shown in Fig. 1:

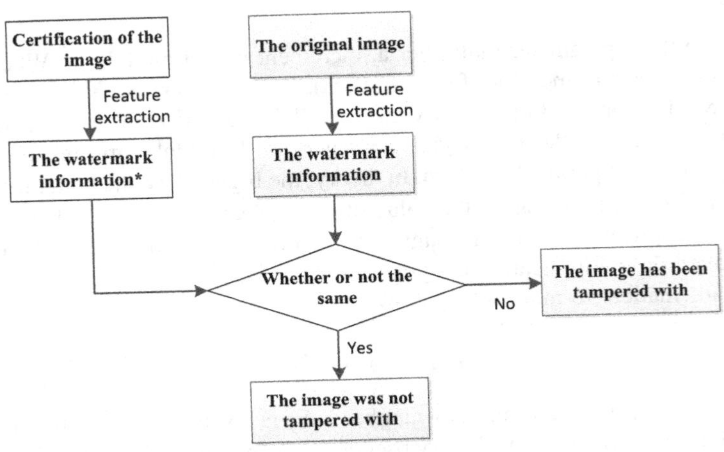

Fig. 1. Watermark certified flow chart

The watermark information generation algorithm is as follows:

Step 1: The carrier image is transformed by 3 - layer discrete wavelet transform.

Hypothesis: the carrier image I is a grayscale image of $M \times N$ size, and the 3 - layer wavelet transform is performed on I. The size of the wavelet coefficient Z is $M/8 \times N/8$. Where, $LL3(i, j)$ is the low frequency coefficient of three-layer wavelet decomposition.

Step 2: The calculation of watermark information.

Extract the binary highest bit of the coefficient component $LL3(i, j)$ as image binary watermark information, and generate W for embedded watermark of $M \times N$ size.

3.2 Watermark Information Embedding Algorithm

The process of watermark extraction is the inverse process of constructing zero-watermark. The algorithm is described as follows:

Step 1: Block processing of wavelet embedded domain.

The vertical components HL2 and horizontal components LH2 of the image are obtained by two-layer wavelet transform of the carrier image.

The image should be embedded with the watermark W with the size of $M/8 \times N/8$, and then segmented by HL2 and LH2. The size of each chunk is 2×2, so the two components of the wavelet coefficients are divided into $M/8 \times N/8$ pieces. To embed the size of $M/8 \times N/8$ for the watermark information W on and block processing, so the two components of the wavelet coefficients are divided into $M/8 \times N/8$ pieces. The position of each piece in the wavelet coefficient can be denoted as $B(i,j)$, among them: $(1 \leq i \leq 8M, 1 \leq j \leq 8N)$. We call it HL2 for $B_0(i,j)$, and LH2 for $B_1(i,j)$.

Step 2: Selection of wavelet embedded domain.

The linear congruence generator described in 1.2 is used to generate pseudo-random numbers that obey uniform distribution. After the initial value is determined, the 0,1 matrix $choice(i,j)$ of the binary value is generated.

Thus, the embedding domain determination principle is: at the time when $choice(i,j) = 0$, the corresponding watermark information is embedded into block $B_0(i,j)$; at the time when $choice(i,j) = 1$, The corresponding watermark information is embedded into block $B_1(i,j)$. In this way, the distribution of watermark information is more uniform and the invisibility of watermark is improved. When the initial value is uncertain, it is difficult to crack the watermark information even if you know the rules of watermark embedding. The security of watermark is guaranteed.

Step 3: Determine the quantification degree difference of watermark information embedding.

Each block $B_0(i,j)$ or $B_1(i,j)$ is embedded with a bit of watermark information. In order to ensure the invisibility of watermark embedding, the size of embedded watermark information can be adaptive embedding according to the size of wavelet coefficient. The definition parameter $weight(i,j)$ represents the maximum degree of quantization that each block can withstand $weight(i,j)$ is defined as:

$$weight(i,j) = mean(i,j) + 3 \times std(i,j) + 10 \times entropy(i,j) \tag{3}$$

Where $mean(i,j)$ represents the mean of $B(i,j)$, $std(i,j)$ represents the standard deviation of $B(i,j)$, and $entropy(i,j)$ represents the entropy of $B(i,j)$.

Step 4: Determine the quantitative step size of watermark information embedding.

Do the same thing for all the chunks, Find the matrix $weight_0(i,j)$ and $weight_1(i,j)$ that all the blocks of HL2 and LH2 can bear the maximum quantization degree. Determine the quantized step size of block $B(i,j)$, The definition formula is:

$$\Delta(i,j) = c + (d - c) \times \frac{weight(i,j) - w_{min}}{w_{max} - w_{min}} \tag{4}$$

Among them, w_{min} represents the minimum value of $weight(i,j)$ matrix and w_{max} represents the maximum value of $weight(i,j)$ matrix. C and d are the range of the selected quantitative step length, and c and d are selected according to the experiment.

Step 5: Watermark information embedding.

For block $B(i,j)$ to be embedded in the watermark, the adaptive quantization step is $\Delta(i,j)$, and x_i represents the wavelet coefficient within the block. $W(i,j)$ represents the watermark information to be embedded, and the watermark embedding process is the process of quantifying the absolute mean of the wavelet coefficient block. The absolute mean of the blocks $B(i,j)$ is defined as:

$$\bar{x} = \frac{1}{n}\sum\nolimits_{i-1}^{n} |x_i| \tag{5}$$

Define the formula: $(\bar{x} + \Delta(i,j)/2)/\Delta(i,j) = Q + r$. Among them, Q is the quotient and r is the remainder. Embed the watermark by modifying absolute mean $\bar{x}$, and set the modified absolute mean to $\overline{x'}$.

Watermark embedding rules are:

$$\begin{cases} \overline{x'} = \Delta(i,j) \times Q + \Delta(i,j)/2, Q\, and\, W\, have\, the\, same\, parity \\ \overline{x'} = \Delta(i,j) \times Q - \Delta(i,j)/2, Q\, and\, W\, have\, different\, parity \end{cases} \tag{6}$$

The embedding of watermark information $W(i,j)$ can be completed by updating the wavelet coefficients with the new absolute mean, so define the amount of changing in absolute mean as $g = \overline{x'} - \bar{x}$. The updating method of wavelet coefficient is:

$$\overline{x'} = x_i \times \left(1 + \frac{g}{\bar{x}}\right). \tag{7}$$

$\overline{x'}$ represents the updated coefficient, and the image with watermark is obtained through wavelet inverse transform.

3.3 Watermark Information Extraction Algorithm

When the watermark is extracted, two-layer wavelet transform is carried out for the watermark image I_W. The wavelet components HL2 and LH2 of the I_W can be obtained. Using the same algorithm as 3.2, the vertical detail sub-band and horizontal detail sub-band are segmented into blocks with a size of 2×2. The maximum watermark embedding degree $weight'(i,j)$ and the quantization step length $\Delta'(i,j)$ of each block are obtained. Random selection matrix $choice(i,j)$ is generated by using the same initial value conditions as the embedded algorithm. Define:

$$\overline{x_w}/\Delta'(i,j) = \tilde{Q} + r' \tag{8}$$

Among them, $\tilde{Q}$ is the quotient and r' is the remainder. When $\tilde{Q}$ is even, $W_m^*(i,j) = 0$, When $\tilde{Q}$ is odd, $W_m^*(i,j) = 1$. So get the extracted watermark W_m^*.

3.4 Image Authentication and Tamper Location

The watermark information W_m^* is generated by using image I_W containing watermark. After obtaining the two extracted watermark information W_m^* and W_0^*, because watermark is the authentication information of image, the tamper part of image is judged by comparing the difference between W_m^* and W_0^*. Tamper positioning matrix is defined as follows:

$$T = \left| W_0^* - W_m^* \right| \tag{9}$$

If $T = 0$, the image is not tampered with; if $T \neq 0$, then the point in the T matrix where is not 1, is considered to be the tamper area.

4 The Experimental Results and Analysis

4.1 Experimental Analysis of Semi-fragile Watermark Based on Image Authentication

The algorithm proposed in this paper, which extracts the features of the original image and inserts it into itself as watermark information to generate watermark images. The watermark information generated by the image under test is compared with the extracted watermark information.

(1) The algorithm uses PSNR (Peak Signal Noise Ratio) to accurately describe the change of image quality of the carrier before and after watermark embedding. Its definition is as follows:

$$PSNR = 10\,lg\frac{255^2}{\frac{1}{MN}\sum_{m=0}^{M-1}\sum_{n=0}^{N-1}[f(m,n) - g(m,n)]^2} \tag{10}$$

Normalize the NC value of the correlation function: NC value is often used to evaluate the robustness of hidden algorithms. Its definition is as follows:

$$NC(f,g) = \frac{\sum_{i=0}^{M-1}\sum_{j=0}^{N-1}(f(m,n) \times g(m,n))}{\sum_{i=0}^{M-1}\sum_{j=0}^{N-1}f(m,n)^2} \tag{11}$$

As can be seen from the above formula, the closer the NC value is to 1, the stronger the anti-attack capability of the algorithm is.

Among them: $f(m,n)$ is the pixel value of the original image; $g(m,n)$ is the modified pixel value. In general, if the value of PSNR is larger, the difference between watermark image and carrier image is smaller, so the algorithm has high transparency of watermark.

The 256 grayscale Lena images with the original carrier image size of 256×256 were tested. Without any attack, Fig. 2(a) is the original Lena image and Fig. 2(b) is the generated watermark image.

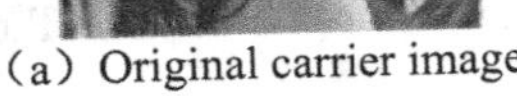

(a) Original carrier image

(b) Embedded watermark image of the carrier

Fig. 2. Host image and watermarked image

It can be concluded from the experiment that the two images are hard to be distinguished by the eye, indicating that the watermark information in the algorithm is highly invisible. By calculation, the PSNR of Lena graph is 50.55 dB. The comparison results of PSNR with other algorithms are shown in Table 1.

Table 1. Comparison of watermarked image's PSNR values of different algorithms

Algorithm	Image's PSNR values (dB)
This algorithm	50.55
The literature [8]	40.28
The literature [9]	41.04

(2) Judging whether the matrix is tampered according to the number of elements not equal to 1 in tamper matrix T [10]. Define BER as bit error rate Compression attacks: JPEG compression attacks with different compression factors for authentication images are treated separately.

Specific results are shown in Table 2.

Table 2. The change rate of image under different JPEG compression factors

JPEG compression factor	BER	PSNR	NC
30	23.8%	43.48	0.9992
40	21.7%	44.84	0.9959
50	4.2%	52.08	0.9980
60	2.6%	53.88	0.9984
70	2.3%	54.08	0.9985
80	2.3%	54.08	0.9985

(3) Noise attack: add different types of noise to the image, and the results are shown in the figure below (Table 3):

Table 3. The change rate of image under different JPEG noise attacks

Shear type	STREBGTH	BER	PSNR	NC
Salt and pepper noise	0.005	13.5%	38.8018	0.9987
Salt and pepper noise	0.01	19.8%	36.3102	0.9976
Gaussian noise	0.005	15.4%	29.4708	0.9997
Gaussian noise	0.01	23.6%	28.7037	0.9984

It can be concluded from the above experiments (2) (3) that JPEG compressed and slightly noisy images can be authenticated, BER $\leq 25\%$, so this algorithm allows images to be processed in a conventional way.

(4) Shear attack: it is a very common malicious tamper method, which randomly conducts shear experiments and tamper localization on Lena.

Table 4. Cropping experiment of Lena

Shear mode	1	2	3	4
The cut image				
Locate the image				
BER	11.36%	19.33%	8.66%	38.54%

As can be seen from Table 4, this algorithm can well indicate the tamper position of the image.

4.2 Algorithm Comparison and Experimental Results Analysis

We compared the experimental data of the algorithm in this paper with the experimental data in literature [11]. The comparison results are shown in Table 5:

Table 5. Watermark robustness test results contrast

The experimental data	The literature [11]		This algorithm	
Attack types The evaluation index	PSNR	NC	PSNR	NC
JPEG compression 30%	34.10	0.9379	43.48	0.9992
Salt and pepper noise 0.02	21.77	0.9824	29.34	0.9976
Gaussian noise 0.01	19.69	0.799	28.70	0.9984

As can be seen from the comparison results in Table 5, the algorithm in this paper is characterized by high robustness and high invisibility of the extracted watermark, strong resistance against various kinds of attacks and good image quality. The overall performance of the algorithm is superior to that of literature [11].

The experiment is carried out in the wavelet domain, which is combined with the compression standard of image, and has strong practicability. The image is transformed from the spatial domain to the transform domain, which can enhance the transparency of the watermark. This has good robustness to noise, filtering and JPEG compression.

The tamper location can be well realized in the experiment, but there are some errors in the experimental results. The extraction and generation of the printing information are provided by the image to be authenticated. It does not need to provide the original image and any additional information, which improves the security and practicability of the watermark.

5 Conclusions and Further Work

This paper presents a content authentication algorithm based on semi-fragile watermark. Make full use of the feature of multilayer multi-resolution of wavelet and combine image features to generate watermark information. The watermark information was embedded into the wavelet domain coefficients of the original carrier image. In the process of image verification, there is no need to provide the original carrier image. This allows for true blind extraction. The experimental results show that the algorithm is vulnerable to malicious attacks.

There are still many problems in this algorithm, that is, the image is not attacked and the extracted watermark and embedded watermark are not guaranteed to be the same. So on that basis, the further work is: firstly improve the algorithm of extracting and embedding watermarks to make them exactly the same; Secondly, it can reduce the error detection rate and improve the robustness.

Acknowledgments. 1. This research project was supported by the Education Department of Jilin Province ("Thirteen Five" Scientific Planning Project, Grant No. 249 [2016]).

2. This research project was supported by the National Natural Science Foundation of China (Grant No. 61262090).

References

1. Huang, S.: Image authentication algorithm based on semi fragile watermarking. Hunan University (2008)
2. Wang, X.Y., Yang, H.Y., Chen, L.K., Zhao, H.: Semi-fragile watermark embedding algorithm for image content authentication. Microcomput. Syst. **11**, 147–150 (2005)
3. Zhang, J.S., Yang, M., Zhou, L.X.: Semi-fragile watermark authentication algorithm for image content based on DWT. Sci. Technol. Bull. **33**(6), 192–195 (2017)
4. Zhao, C.H., Liu, W.: Image semi-fragile zero-watermark algorithm based on block compression perception. J. Autom. **38**(4), 609–617 (2012)
5. Liu, H., Lin, J., Huang, J.: Image authentication using content based watermark. In: IEEE International Symposium on Circuits and Systems, ISCAS 2005 (2005)
6. Wang, L.N., Guo, C., Li, P.: Experimental Course on Information Hiding Technology, pp. 22–23. Wuhan University Press, Hubei (2004)
7. Wang, B.B.: Research on digital watermarking algorithm of document image for content authentication. Shandong Normal University (2010)
8. Ma, X.M., Li, Y.S., Xie, J.L.: Experiment and analysis of point cloud de-noising using bilateral filtering method. Survey report, 89–115 (2017)
9. Qi, X.J., Xin, X.: A quantization-based semi-fragile watermarking scheme for image content authentication. J. Vis. Commun. Image **22**(2), 187–200 (2011)
10. Zhang, B.: Research on content authentication technology based on perception hash and digital watermark image. Beijing University of Posts and Telecommunications (2011)
11. Li, Z.W.: Research on dual digital watermarking algorithm based on wavelet transform domain. Anhui University of Technology (2016)

Autoencoder-Based on Anomaly Detection with Intrusion Scoring for Smart Factory Environments

Gimin Bae, Sunggyun Jang, Minseop Kim, and Inwhee Joe(✉)

Department of Computer and Software, Hanyang University, Wangsimni-ro, 222, Seoul 04763, Korea
{baegimin,mrjang28,minseop,iwjoe}@hanyang.ac.kr

Abstract. The industry 4.0 and Industrial IoT is leading new industrial revolution. Industrial IoT technologies make more reliable and sustainable products than traditional products in automation industry. Industrial IoT devices transfer data between one another. This concept is need for advanced connectivity and intelligent security services. We focus on the security threat in Industrial IoT. The general security systems enable to detect normal security threat. However, it is not easy to detect anomaly threat or network intrusion or new hacking methods. In the paper, we propose autoencoder (AE) using the deep learning based anomaly detection with invasion scoring for the smart factory environments. We have analysis F-Score and accuracy between the Density Based Spatial Clustering of Applications with Noise (DBSCAN) and the autoencoder using the KDD data set. We have used real data from Korea steel companies and the collected data is general data such as temperature, stream flow, the shocks of machines, and etc. Finally, experiments show that the proposed autoencoder model is better than DBSCAN.

Keywords: Anomaly detection · Intrusion detection · Scoring
Autoencoder · DBSCAN · Smart factory · Industrial IoT

1 Introduction

Cyber security is the set of technologies and processes designed to protect computers, networks, programs, and data from attack, unauthorized access, change, or destruction [1]. Industrial cyber attack give loss of production or damage to machines. In the face of a number of well documented cyber related incidents such as the Slammer Worm infiltration of an Ohio Nuclear plant and several power utilities [2]. Nowadays, the new threats exist in Industrial IoT. For example, the traditional manual control system by operating worker is only internal security breach that is to control direct physical access to the equipment. However, the threats are becoming increasingly external security incident because while internet connection, the system is exposed. The external attacks have always been higher than internal which that the externally generated incidents account for 70% of all events and internally generated incidents account for 5% of all events) [2]. There are a lot of attack methods and anomaly intrusions in the cyber world.

J. H. Park et al. (Eds.): PDCAT 2018, CCIS 931, pp. 414–423, 2019.
https://doi.org/10.1007/978-981-13-5907-1_44

Especially, In Industrial IoT environment, there are not only simple attack but also critical attacks such as stopping the factory. In more detail, Industrial IoT attack damage causes huge problems like all machines in the factory are stopped by the attack or broken the machines by the attack for a long time. It must be emphasized that maintain security operation of the Industrial IoT applications that connect machine. Therefore, it is necessary to analyze the attack behavior based on the real attack data and actual collected data.

Typically there are huge amounts of data and variety of data including system log, network log, network packet data, and sensing data from machine in the Industrial IoT systems. The data are the constituent elements of training data for deep learning in order to detect anomaly intrusions or many attacks. With that in mind, a more sophisticated approach was needed to solve the problems below mentioned.

We focus on scoring model for security solution using machine learning including artificial intelligence. The implementation of deep learning modeling can be divided into three parts: (1) preprocessing, (2) deep learning model and (3) evaluation.

2 Related Work and Background

The basic idea of automated intrusion detection is often acknowledged in 1980 by Anderson's How to Improve Computer Security Auditing and Monitoring [3]. This work has expanded the research into misuse detection of mainframe systems. In the 1990s, IDS technology proposed a new approach to address the increasing and sophisticated nature of network attacks and defined this as anomaly detection.

In 2009, Chandola et al. grouped the anomaly detection techniques into several categories based on their approach and compared the pros and cons. It is also available as a guideline for evaluating the effectiveness of anomaly detection techniques in specific domains [4]. Anomaly detection techniques also apply many machine learning algorithms. In 2002 Mukkamala et al. compared anomaly detection methods using support vector machines and neural networks [5]. In 2009, Tsai et al. studied how to apply machine learning to intrusion detection methods [6]. In 2015, An and Cho conducted an anomaly detection study using the VAE algorithm [7]. In 2016, Buczak et al. investigated methods of machine learning and data mining to support intrusion detection. A discussion of the problem for using ML and DM methods is presented and provides some recommendations for using the given method [1]. In 2018, Shone et al. proposed a network intrusion detection technology using asymmetric autoencoder [8]. Intrusion detection or anomaly detection has been actively studied from the past to the present.

2.1 Anomaly Detection

Anomaly Detection is an Intrusion Detection System (IDS) analysis method that collects and learns accumulated normal data, completes the profile of normal type usage, and then detects patterns with different forms as intrusion. It has the advantage of being able to detect the known intrusion and reduce the error (False Negative) that the actual intrusion is not an intrusion.

Anomalies are a pattern of undefined data in normal behavior. Assuming that N1 and N2 are normal categories of behavior after observing the behavior, the points P1 and P2, P3 which are not included in the normal category, fall outside the normal categories N1 and N2. It is anomaly (see Fig. 1) [4].

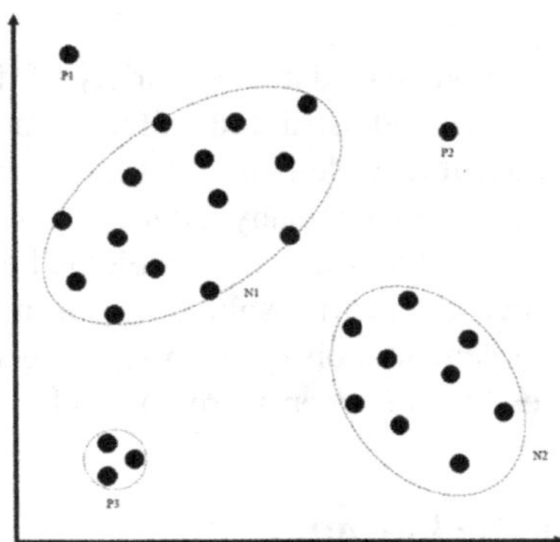

Fig. 1. Anomalies and normal behavior

2.2 Autoencoder

The existing neural network used the labeled data as the learning data given both the input value x and the target value y of the sample data. Learning to use these data is called supervised learning. On the other hand, the learning method of finding the characteristics of data with only the input value x of data is called unsupervised learning. Autoencoder [9] is a neural network based on unsupervised learning and output value $\dot{x}$ to be approximated to the input value x. The autoencoder consists of an encoder and a decoder. Basic Stacked autoencoder with three hidden layers (see Fig. 2).

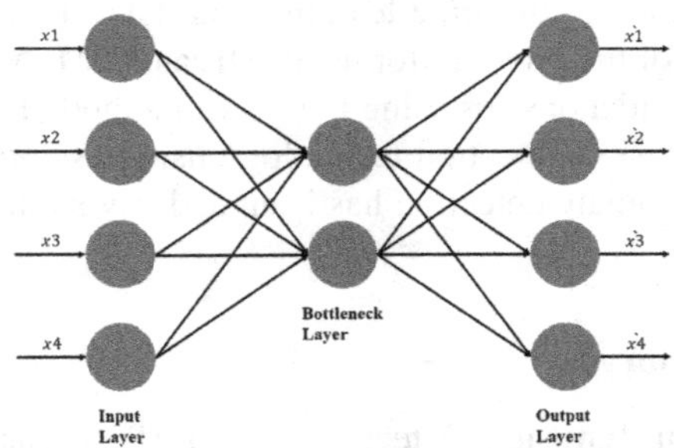

Fig. 2. Basic stacked autoencoder

3 Proposal Approach

In this paper, we propose an autoencoder based anomaly detection algorithm based on the data collected in the Smart Factory environment, which can potentially detect attacks that may affect production quality or prewarning signs of such attacks. The proposed anomaly detection algorithm is based on the assumption that normal data is formed into the same cluster with similar density, and the data (outliers) that do not form the cluster and the cluster formed in this way mean an ideal value.

In this paper compare the performance of the anomaly detection model using the previously proposed density-based clustering algorithm [1] and the newly proposed semi-supervised based autoencoder algorithm to detect the anomaly in smart factory environment.

Anomaly detection using the previously proposed density-based clustering algorithm [1] is not suitable for smart factory environments. The first is the speed of execution. In a Smart Factory environment, the data collected per day ranges from tens of gigabytes to hundreds of gigabytes, depending on the factory environment. For density-based clustering algorithms, the larger the data, the longer it takes for the autoencoder. Second, it is a cluster of anomaly data. Data collected in real life is not labeled. Therefore, it is difficult to confirm whether the formed clusters are normal or anomaly when using the density-based clustering algorithm. However, in the case of autoencoder using semi-supervised learning method, it is easy to notice if there is anomaly data.

In addition, in the case of the autoencoder based on the proposed map learning, there is an advantage that labels can be classified as accumulation of learning is accumulated. Finally, management of the learning model. For density-based clustering algorithms, no learning is saved. In other words, when new data is added, we need to re-cluster. However, in the case of the auto encoder algorithm, we can use neural networks to store the weights and models.

We propose anomaly detection algorithm is based on a model that is learned only by normal data. When new data is received, it is checked based on the difference of the output values restored by the decoder. To do this, you need to set the difference between the normal output and the anomaly output value and find the optimal threshold. The data collected is based on data collected by steel companies in Korea. Assuming that the collected data is normal data, it generates anomaly data by modulating normal data. Also we use KDD data sets and scenario-based data to verify the model. We constructed the model using a Python-based tensorflow library [15] with three encoder and decoder layers (consisting of 64, 32, and 16 neurons) and a bottleneck layer (consisting of 1 neuron). In addition, the mean square error was used for the loss function, and the model evaluation used the Confusion matrix (F-score, Accuracy) and the MAPE modified MMAPE [14].

3.1 Proposed Evaluation Method

Maximum Mean Absolute Percentage Error. In the conventional method of distinguishing normal and anomaly from anomaly detection based on the autoencoder, the difference between the input value and the output value is obtained as in the expression.

The difference between the values obtained is checked to see if it exceeds the threshold, and if it exceeds, it is classified as anomaly. Equation (1) shows the MAPE formula. A_t is the actual value and P_t is the predicted value. MAPE subtracts the predicted value from the actual value, divides it by the actual value, and then divides by n. Then multiply by 100 to represent the percentage. Although the concept of MAPE sounds simple and persuasive, there are two problems in actual use. (1) If the actual value is less than 1, MAPE can have a value close to infinity. (2) If the actual value is 0, the MAPE value itself can not be calculated. In this paper, we use the MMAPE scoring function based on the MAPE. The equation for MMAPE is equal to Eq. (2).

$$\mathrm{MAPE} = \frac{100}{n}\sum_{t=1}^{n}\left|\frac{A_{tn} - P_{tn}}{A_{tn}}\right| \tag{1}$$

$$\mathrm{MMAPE} = \begin{cases} if\ A_t == 0 & \frac{100}{n}\sum_{t=1}^{n}|A_{tn} - P_{tn}| \\ else & \frac{100}{n}\sum_{t=1}^{n}\left|\frac{A_{tn}-P_{tn}}{A_{max}}\right| \end{cases} \tag{2}$$

4 Experiments

4.1 Dataset

The data set used for learning and verification is based on a certain period (10 s) of the system, network, and sensor data collected from Korean steel companies and uses the average value of the collected values during the relevant period. The index of the integrated log contains time information. Sensor data are sensed values (flow, temperature, humidity, weight, etc.) in each process. Network data includes source IP, source PORT, bytes transferred from source to destination, connection count information from source to destination. The destination port, the bytes transferred from the destination to the source, and the number of connections from the destination to the source. In the case of the host data, the data collected from the smart factory system has the number of CPU status information processes and memory status information. Labels are not used for learning and are added to calculate the detection accuracy of the model. When the label is 0, it means normal data. When the label is 1, it means anomaly.

4.2 Preprocessing

In this paper, we normalize the data through the feature scaling method in preprocessing. Feature scaling is a method used to standardize the range of independent variables or data features. Also called data normalization, the method is performed in the data preprocessing step. Since the range of values between data that are not normalized greatly varies, some machine learning algorithms cause problems that normal learning can not be achieve. Another reason for applying feature scaling is that it converges much faster than if no feature scaling is applied in the gradient descent. The main methods used in feature scaling are as follows.

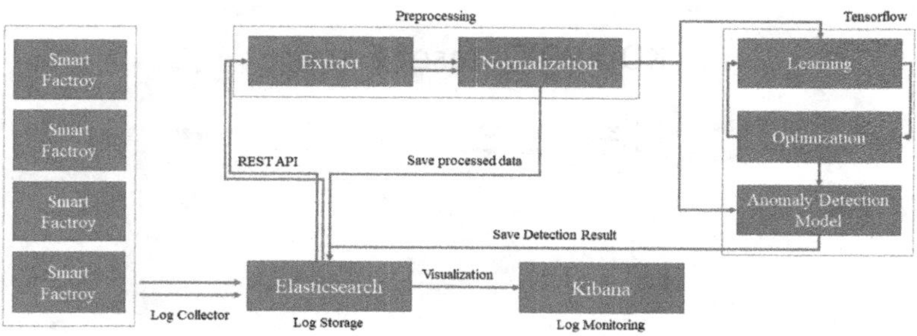

Fig. 3. Data collection and learning process

4.3 KDD Cup '99 Dataset Evaluation

The KDD Cup '99 (KDD) [10] data set was created by processing the tcpdump portion of the DARPA IDS evaluation data set from MIT Lincoln Lab [11]. KDD is primarily used when evaluating the performance of models in anomaly detection studies. Numerous intrusion detection studies use the KDD data set, and this paper also uses benchmarking data for the comparative evaluation of the proposed model. The experiment is done as follows. The size of the dataset is 1 K, 10 K, 100 K. Each data set has 5 attack rates (0%, 5%, 10%, 25%, 50%). Attack types such as DoS, PROBE, R2L, and U2L are randomly sampled and all are marked as anomaly [12, 13]. DBSCAN and auto encoder can set various parameters. We chose the best performance while increasing or decreasing the parameters. The results of the experiments are as follows (see Fig. 4).

4.4 Industrial IoT Dataset Evaluation

Based on the scenario-based verification data, we evaluated the performance of the model. The scenario consists of three types. In Scenario 1, sensor data, network data, and system data were modulated. In attack scenario 2, sensor data and network data were modulated. In Scenario 3, sensor data and system data are modulated.

Scenario 1. This scenario assumes that a hacker seizes control PC in a smart factory environment to cause massive damage to the factory, sends malicious commands like a normal command, and leaves internal data to the outside. There are also pre-signs that hackers take control PC in Smart Factory environment. The number of data for this scenario is 8640, and anomaly is 864, which is 10%. In the case of sensor data, the anomaly modulates the temperature, flow rate, and heating temperature of the cooling water. In the case of network data, the destination IP is specified differently, and the number and size of packets are modulated. We modified the system data based on the traces of hackers.

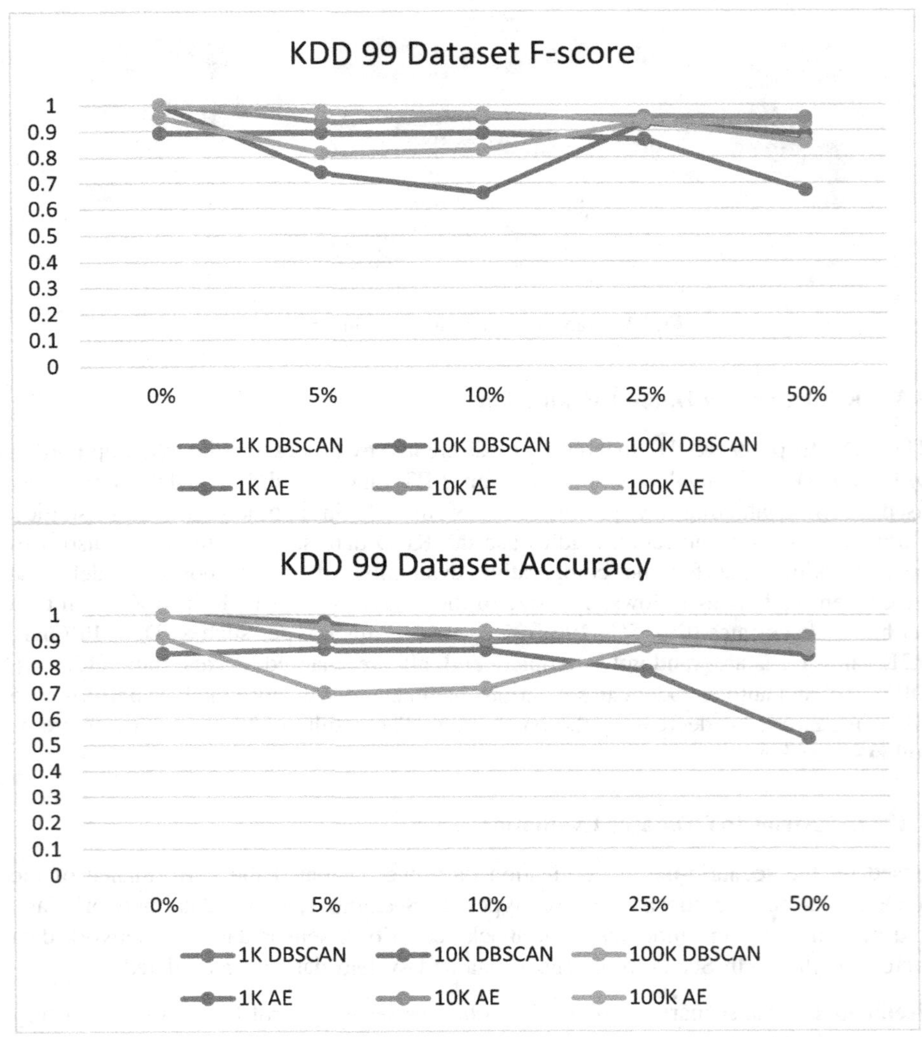

Fig. 4. Results by attack data each ratio in 1 K, 10 K, 100 K KDD 99 Dataset

Scenario 2. In Scenario 2, a hacker changes the values of sensors in the Smart Factory environment by turning malicious commands into normal commands using infected control PCs. There are preliminary indications in this scenario, and the number of data is 8640. The anomaly is 2592, which is 30%. The anomaly modulates the temperature of the upper and lower heating zones and the speed of the motor. In the case of system data, the trace of hacker's intrusion is altered, and the relevant part is information such as the number of remote accesses and PLC command information. In this scenario, the network data has not been tampered with.

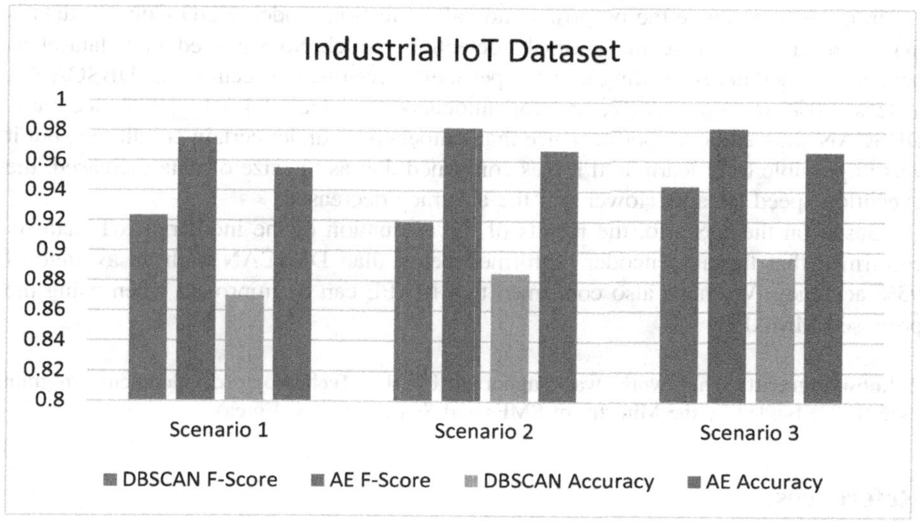

Fig. 5. Results by scenario based in industrial IoT dataset

Scenario 3. Scenario 3 is a scenario where a hacker leaks data from a smart factory environment. There are preliminary indications in this scenario, and the number of data is 8640. The anomaly is 432, which is 5%. In the case of network data, the anomaly has been tampered with the hacker's intrusion, and this part is information such as source IP, source PORT, destination IP, destination port, and packet size. In this scenario, the system data has not been tampered with.

5 Conclusion

Industrial IoT is steadily growing, and security threats to industrial IoT are also increasing. In this paper proposed the anomaly detection with Intrusion scoring for the smart factory environments. Also utilizes integrated industrial process data collected from the industrial IoT platform.

We propose anomaly detection algorithm is based on a model that is learned only by normal data. When new data is received, it is checked based on the difference of the output values restored by the decoder. To do this, you need to set the difference between the normal output and the anomaly output value and find the optimal threshold. The data collected is based on data collected by steel companies in Korea. Assuming that the collected data is normal data, it generates anomaly data by modulating normal data. Also we use KDD data sets and scenario-based data to verify the model. We constructed the model using a Python-based tensorflow library [15] with three encoder and decoder layers (consisting of 64, 32, and 16 neurons) and a bottleneck layer (consisting of 1 neuron). In addition, the mean square error was used for the loss function, and the model evaluation used the Confusion matrix (F-score, Accuracy) and the MAPE modified MMAPE.

In order to evaluate the proposed anomaly detection model, KDD Cup '99 data is mainly used to evaluate the anomaly detection model. So we used that dataset as benchmarking data. According to the experimental results, the accuracy of DBSCAN is at least 70% and up to 99%, and for autoencoder at least 84% to 100% accuracy. DBSCAN showed better performance than autoencoder under certain conditions, but it was impossible to re-learn, and it was confirmed that as the size of data increased, the execution speed became slower and the accuracy decreased.

Based on the scenario, the results of the evaluation of the industrial IoT data set confirmed that the autoencoder performed better than DBSCAN with an average of 95% accuracy. We have also confirmed that MAPE can be improved when using the proposed MMAPE.

Acknowledgment. This work was supported by the Technology development Program (S2521883) funded by the Ministry of SMEs and Startups (MSS, Korea).

References

1. Buczak, A.L., Guven, E.: A survey of data mining and machine learning methods for cyber security intrusion detection. IEEE Commun. Surv. Tutor. **18**(2), 1153–1176 (2016)
2. Byres, E., Lowe, J.: The myths and facts behind cyber security risks for industrial control systems. In: Proceedings of the VDE Kongress, vol. 116 (2004)
3. Anderson, J.P.: Computer security threat monitoring and surveillance. Technical Report, James P. Anderson Company (1980)
4. Chandola, V., Banerjee, A., Kumar, V.: Anomaly detection: a survey. ACM Comput. Surv. **41**(3), 15 (2009)
5. Mukkamala, S., Janoski, G., Sung, A.: Intrusion detection using neural networks and support vector machines. In: 2002 Proceedings of the 2002 International Joint Conference on Neural Networks, IJCNN 2002, vol. 2. IEEE (2002)
6. Tsai, C.-F., et al.: Intrusion detection by machine learning: a review. Expert Syst. Appl. **36** (10), 11994–12000 (2009)
7. An, J., Cho, S.: Variational autoencoder based anomaly detection using reconstruction probability. SNU Data Mining Center, Technical Report (2015)
8. Shone, N., et al.: A deep learning approach to network intrusion detection. IEEE Trans. Emerg. Top. Comput. Intell. **2**(1), 41–50 (2018)
9. Hinton, G.E., Salakhutdinov, R.R.: Reducing the dimensionality of data with neural networks. Science **313**(5786), 504–507 (2006)
10. KDD Cup 1999. http://kdd.ics.uci.edu/databases/kddcup99/kddcup99.html. Accessed October 2007
11. Lippmann, R., Haines, Joshua W., Fried, David J., Korba, J., Das, K.: Analysis and results of the 1999 DARPA off-line intrusion detection evaluation. In: Debar, H., Mé, L., Wu, Felix (eds.) RAID 2000. LNCS, vol. 1907, pp. 162–182. Springer, Heidelberg (2000). https://doi.org/10.1007/3-540-39945-3_11
12. Olusola, A.A., Oladele, A.S., Abosede, D.O.: Analysis of KDD '99 intrusion detection dataset for selection of relevance features. In: Proceedings of the World Congress on Engineering and Computer Science, vol. 1. (2010)

13. Özgür, A., Erdem, H.: A review of KDD99 dataset usage in intrusion detection and machine learning between 2010 and 2015. PeerJ PrePrints (2016)
14. Fawcett, T.: An introduction to ROC analysis. Pattern Recogn. Lett. **27**(8), 861–874 (2006)
15. Abadi, M., et al.: TensorFlow: large-scale machine learning on heterogeneous systems, 2015. tensorflow.org. Accessed 21 Nov 2016

Blockchain and Smart-Contracts Modeled in a *SwarmESB* Ecosystem

Ioana Stănescu, Lenuţa Alboaie, and Andrei Panu(✉)

Faculty of Computer Science,
Alexandru Ioan Cuza University of Iasi, Iaşi, Romania
{stanescu.ioana, adria, andrei.panu}@info.uaic.ro

Abstract. Blockchain has emerged as a trusted and secure distributed ledger for transactions while also being decentralised, distributed and its legitimacy not guaranteed by a trusted authority. Since the appearance of Bitcoin, Blockchain has known many implementations based on P2P architectures. This paper presents how the blockchain and smart contracts technologies can be integrated into the SwarmESB ecosystem. SwarmESB is a framework that helps building distributed applications, which benefit from privacy and scalability features. Our proposal will present the flexibility in building not only microservices based applications, but also decentralised applications employing blockchain and smart-contracts by modeling a sample Dapp.

Keywords: Blockchain · SwarmESB · Smart-contracts
Decentralised applications

1 Introduction

In the last years, Blockchain caught a lot of interest because of its initial application as distributed ledger for cryptocurrency (Bitcoin) which proposed a new model for distributed, decentralised applications, which can store information in a secure manner and make it publicly available without risking tampering with the data [1]. Blockchain, together with the concept of smart-contracts [2], opened the path to a new era for web applications, which lead to decentralised applications (Dapps) [3].

Regarding web applications, they have widely transformed in time, evolving from monolithic architectures, to service oriented architectures (SOA), to microservices and serverless. With such a diversity of architectures, which can grow in complexity, a need to ease dealing with concurrency, scalability and integration between components has developed.

The SwarmESB framework was developed in order to provide a method to integrate distributed processes for composing applications while being highly scalable and having privacy embedded in its design [4].

First, we will give an overview of how SwarmESB works and how it can model a distributed application. Then we will present how the blockchain technology can rest on top of the SwarmESB model, how adapters manage the distributed ledger and how smart contracts enhance transactions widening the application's functionality. Lastly, we will present a Dapp sample that depicts the integration between blockchain, smart-contracts and SwarmESB concepts.

J. H. Park et al. (Eds.): PDCAT 2018, CCIS 931, pp. 424–432, 2019.
https://doi.org/10.1007/978-981-13-5907-1_45

2 SwarmESB Overview

SwarmESB was developed as a new approach on communication and composability mechanisms for services in cloud systems [5]. It is a framework, which combines the message passing architecture of the Enterprise Service Bus (ESB) with choreographies in order to easily achieve integration between multiple distributed processes. It relies on four basic concepts: messages, swarms (choreographies), adapters and swarm clients [4].

Messages, which are found at the lowest level of the SwarmESB architecture, are sent using pub/sub mechanisms and are responsible with calling the adapters. A group of messages composes the swarms, each message from a swarm is related to a phase of the swarming process which is executed in the context of an adapter. Swarms (choreographies) at functional level are composed of messages. Each step of a choreography can be executed on a different process, without losing the context of the choreography. At structural level, a swarm, which is described in Swarm DSL (language based on javascript), contains four types of constructions: meta (metadata), vars (variables), constructor functions and phases.

Adapters are the long running processes in SwarmESB architecture and represent the component, which provides the actual services. They are responsible with the processing of data or external calls. Adapters having the same functionality can be associated into groups. As it can be seen in Fig. 1, between the client and the adapters lays the swarms layer which is responsible with all the logic triggered by a client's request and, through phases, makes calls to adapters.

Swarm clients are the component with which the client application communicates with the swarm based system through WebSockets. The client application can execute a swarm and listen for events sent during the swarm execution. The mechanism is similar with calling an API, which integrates multiple services, the difference being that by swarming privacy can be achieved [6, 7].

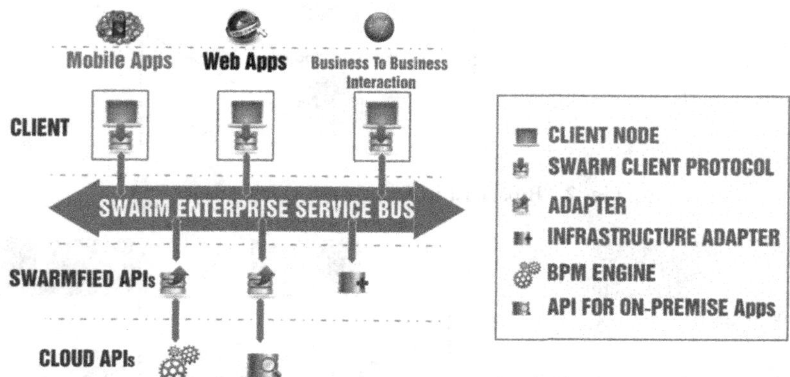

Fig. 1. SwarmESB architecture [5]

2.1 SwarmESB Enhanced for Blockchain Integration

In order to support building Dapps, a SwarmESB based architecture should at least allow creating P2P like systems and execution of smart-contracts. Considering the SwarmESB architecture, where adapters are the long running processes, which expose services, they would be used as miners and form a P2P network, while swarm choreographies will be used as smart-contracts.

Since adapters cannot directly communicate within a group, the communication will be provided by using swarm choreographies for internal calls in the network. From the point of view of the communication mechanism, swarms need to allow both broadcast and unicast. In the figures below (Figs. 2 and 3), a communication flow model is represented using swarm and does a ping-pong communication, both broadcast and unicast. As it can be seen in Fig. 2, from the constructor "startBroadcast", the "pingBroadcast" phase is called which executes on every available adapter which sends back "pong" to the calling adapter. Broadcasting in SwarmESB means executing a phase on top of every adapter from within a group (each swarm phase corresponds to a group of adapters), which means that even the calling adapter would receive its own "ping". In the following Fig. 3 is an example of direct communication, which executes the "pingDirect" phase on AdapterC who will respond back to AdapterA.

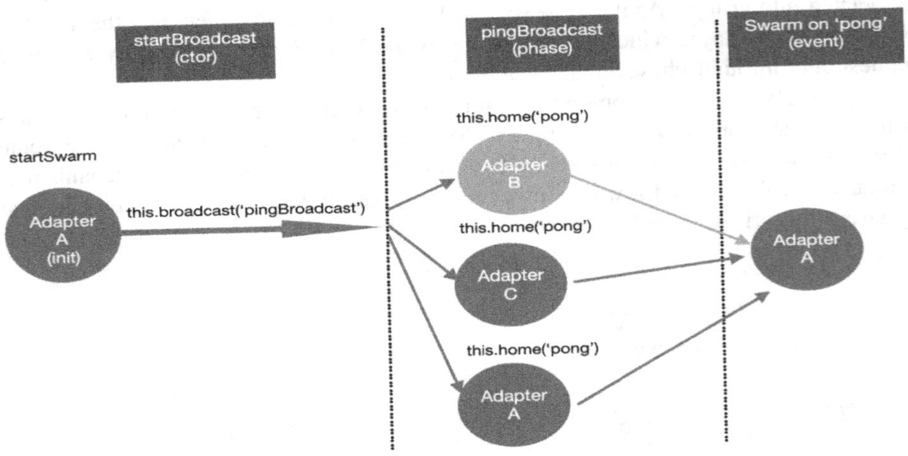

Fig. 2. Broadcast in SwarmESB (ping-pong)

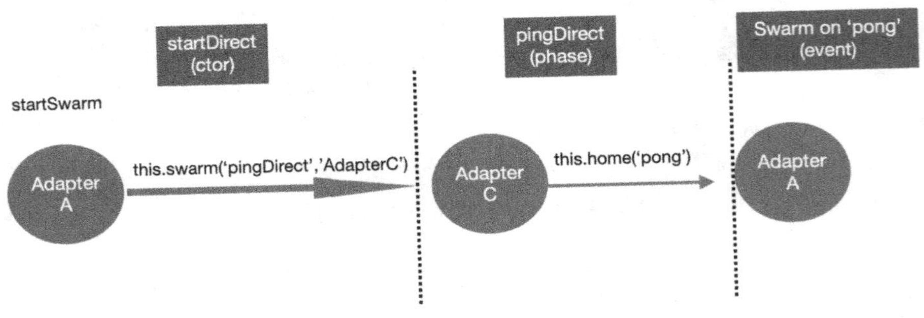

Fig. 3. Direct call in SwarmESB (ping-pong)

SwarmESB smart-contracts, as previously mentioned, are implemented using swarm choreographies. A smart-contract, in its basic form, should have a state, be called from the miners and generate transactions called "internal transactions". By using swarms (which are not long running processes), the state cannot be held by the contract, this is an issue that can be solved by keeping a contract's state at miners' level. Another disadvantage is that by not being a long running process swarm contracts cannot issue events or transactions based on a timer or external event, they can only execute based on a transaction.

By providing needed mechanisms for communication and support for smart-contracts, SwarmESB offers a suitable environment for building blockchain based applications. In the following chapter, we will present a detailed model of a decentralised application built on top of SwarmESB.

3 Blockchain Application Model Using SwarmESB

In our context, the miners are represented by the adapters and the smart-contracts by swarm choreographies. As miners, adapters should support a basic communication protocol composed of:

- broadcast its own identity when joining the network;
- broadcast when a new block is mined;
- require the chain from other peers when local version of the blockchain is unsynced;
- receive transactions from client applications.

The communication will be done through swarms choreographies, and the miners should implement the corresponding methods for the protocol such as: receive newly mined blocks, transactions or peers announcements (register them and respond back), sending current chain and transactions in the pool when requested. The communication scenarios are graphically represented in Figs. 4, 5, 6 and 7.

When a new transaction is to be sent to the miners network, the client executes a swarm calling the "addTransaction" constructor and passing the generated transaction. Then, the constructor broadcasts the transaction to the miners via the "sendTransaction" phase, at that point every miner should receive the transaction and add it to their pool of transactions in order to be mined.

Whenever a new block is mined, the miner will broadcast it to the network by calling the "ctor" with the parameters: "broadcast", its name, "addBlock" (phase name) and the block (params). The constructor then broadcasts a message to the "addBlock" phase in order to make calls to all the other adapters to send them the newly mined block. The miner who emitted the new block will not wait for any responses.

When a miner has an unsynced chain (it can no longer generate valid blocks for the blockchain), it will call the ctor constructor in order to request the current chain and transactions from a specific miner, to do so it passes the destination's name, its own identity (to identify itself) and "update". From the swarm the update phase will be

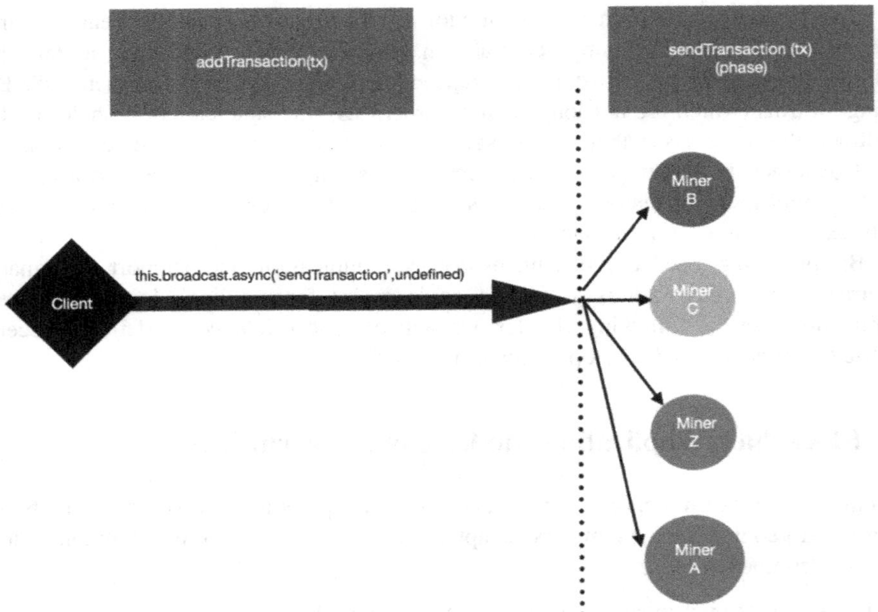

Fig. 4. Transaction broadcast from client to miners network

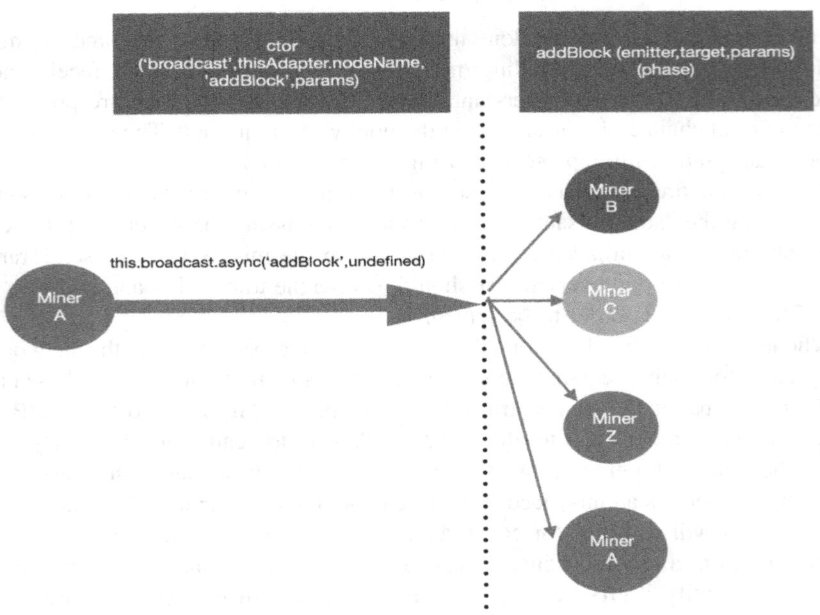

Fig. 5. Miner broadcasts mined block

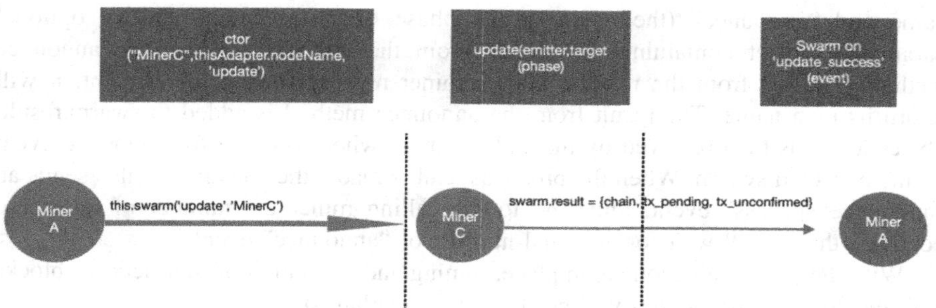

Fig. 6. Miner requests update from another miner (direct call)

executed calling the update method from the destination miner, which will return an object composed of the current chain, its pool of pending transactions and the pool of transactions, which await confirmation. The returned result is then added to swarm. result and passed to the calling miner through the update_success event.

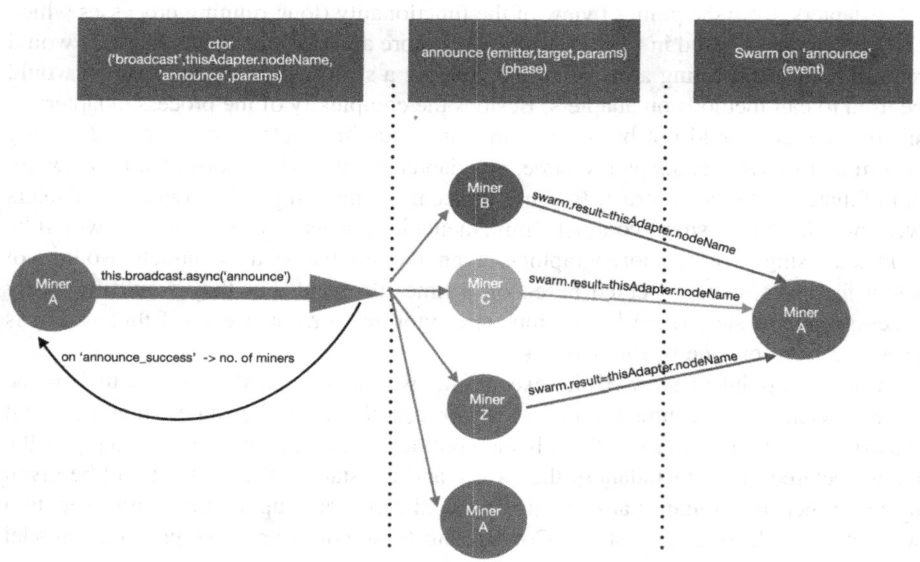

Fig. 7. Miner announcing itself in the network

When a new miner joins the network, it will call the constructor "ctor" when executing the swarm choreography used for internal communication, giving as parameter: "broadcast" (the message is meant to be broadcasted to all the peers), its name and “announce” (the name of the phase to be executed) and an optional parameters object containing the chain. From the announce phase, the announce method is called from the miners, after a miner registers the announcement, it will return its own name. The result from the announce method is added to swarm.result, object which is then received by the calling miner when listening for announce event on the executed swarm. When the broadcast call is made, the constructor also sends an “announce_success” event, returning to the calling miner the number of registered peers, so that it will wait for an equal number of “announce” events.

With the communication set in place, mining and validation of transactions, blocks and the chain are trivial and will not be further explained.

3.1 Smart-Contracts

Regarding smart-contracts, which are a necessary part of a decentralised application based on blockchain, when creating a model, the following facts were taken into consideration: a contract should have (receive) it’s previous state, should have access to the chain, receive transaction parameters and should be called from miners (adapters). In the SwarmESB environment, two possible ways to model smart-contracts were available: swarm choreographies and adapters.

Adapters, from the point of view of the functionality (long running processes which can execute tasks based in timers), would be a more appropriate choice, but they would require both instantiating an adapter and creating a swarm choreography which would be used to call methods on adapters. Besides the complexity of the process, adapters as smart-contracts would not be secure, since methods in adapters can be called by any swarm that knows the adapter’s name, an adapter’s state can be corrupted by a swarm not affiliated with the contract. Because of security and complexity concerns, adapters were not chosen for smart-contracts implementation, instead, smart-contracts would be modeled using swarm choreographies, even though the smart-contracts would not allow for tasks to be executed based on a timer or event, but they would be easily accessible, their state (held by the miner) cannot get corrupted even if that swarm is executed from outside of the network.

Form the point of view of the execution, swarms are called based on their name which means that a contract’s address is not calculated by the miners nor the client (based on the transaction which adds the contract in the system), but it actually is the name declared in the metadata of the swarm and the state of the contract will be given by the miner as parameter at execution time (the contract, upon successful execution will return back its current state). Considering these constraints, we propose a model described as follows:

```
var contract =
{
  meta: {
   name: "sampleContract"
  },
  vars: {
   state:null,
   method:null,
   params:null,
   chain:null,
   transaction:null
  },
  ctor: function (transaction,state,chain) {
   this.method = transaction.params.method;
   this.params = transaction.params;
   this.chain = chain;
   this.state = state;
   this.transaction = transaction;
   this.swarm(this.method); //method is the phase name
   //this.swarm calls the phase given as parameter
  },
  <phase>:{
   node: "All", // node is the adapters group, "All"
means that it can be executed on any adapter
   code: {...} // code to be executed in the current
phase
  }
}
```

The constructor ('ctor') is responsible with instantiating the variables that will be further used in the called phase. In SwarmESB in order to execute a phase, a constructor is called and it is responsible with dispatching the request to the corresponding phase.

As for how a contract is called, the miner will execute the corresponding swarm using a swarm client and will wait for a response. In order not to lock the block mining or validating process, the response from a contract should be awaited for a limited time. In case it does not finish in a timeframe, its result can be set to error or timeout, based on internal logic of miners.

4 Conclusions

SwarmESB, from its core architecture, proves itself capable to support not only applications based on microservices, but also decentralised applications based on blockchain and smart-contracts. Even though there are mechanisms that cannot be implemented in the same way as in other blockchain implementations (e.g. smart-contracts are not long running processes and the state it's passed at each swarm initialisation), using SwarmESB also brings its advantages by going one step forward to offer privacy in blockchain applications and by giving more power to smart-contracts, which, through the swarm choreography, can implement complex flows in a less constraining environment.

Acknowledgements. The dissemination of this work is partly funded by the European Union's Horizon 2020 research and innovation programme under grant agreement No 692178. It is also partially supported by the Private Sky Project, under the POC-A1-A1.2.3-G-2015 Programme (Grant Agreement no. P 40 371).

References

1. Nakamoto, S.: Bitcoin: A Peer-to-Peer Electronic Cash System (2008)
2. Szabo, N.: Smart contracts (1994). http://www.fon.hum.uva.nl/rob/Courses/InformationInSpeech/CDROM/Literature/LOTwinterschool2006/szabo.best.vwh.net/smart.contracts.html
3. Ethereum Whitepaper. https://github.com/ethereum/wiki/wiki/White-Paper#applications
4. Alboaie, L., Alboaie, S., Barbu, T.: Extending swarm communication to unify choreography and long-lived processes. In: Proceedings of the 23rd International Conference on Information Systems Development, ISD 2014, Croatia, 2–4 September (2014)
5. Alboaie, L., Alboaie, S., Panu, A.: Swarm communication – a messaging pattern proposal for dynamic scalability in cloud. In: Proceedings of the 15th IEEE International Conference on High Performance Computing and Communications – IEEE HPCC, China, pp. 1930–1937 (2013). https://doi.org/10.1109/hpcc.and.euc.2013.277
6. Alboaie, S., Cosovan, D.: Private data system enabling self-sovereign storage managed by executable choreographies. In: Chen, L.Y., Reiser, H.P. (eds.) DAIS 2017. LNCS, vol. 10320, pp. 83–98. Springer, Cham (2017). https://doi.org/10.1007/978-3-319-59665-5_6
7. Alboaie, S., Alboaie, L., Vaida, M.-F., Olariu, C.: Executable choreographies applied in OPERANDO. Comput. Sci. J. Mold. **24**, 417–436 (2016)

PINUS: Indoor Weighted Centroid Localization with Crowdsourced Calibration

Jehn-Ruey Jiang[1(✉)], Hanas Subakti[1], Ching-Chih Chen[1], and Kazuya Sakai[2]

[1] National Central University, Taoyuan City, Taiwan
jrjiang@csie.ncu.edu.tw
[2] Tokyo Metropolitan University, Hachioji, Tokyo, Japan

Abstract. PINUS, an indoor weighted centroid localization (WCL) method with crowdsourced calibration, is proposed in this paper. It relies on crowdsourcing to do the calibration for WCL to improve localization accuracy without the device diversity problem. Smartphones and Bluetooth Low Energy (BLE) beacon devices are applied to realize PINUS for the sake of design validation and performance evaluation.

Keywords: Bluetooth · Beacon · Calibration · Crowdsourcing
Device diversity · Indoor localization · Weighted centroid localization

1 Introduction

This paper proposes PINUS, an indoor weighted centroid localization method with crowdsourced calibration. Localization is the process of determining the location or position of a specific object. Some systems, such as the Global Positioning System (GPS) [1] and BeiDou Navigation Satellite System (BDS) [2], can provide world-wide accurate position information. However, they cannot be applied to indoor environments, since they require the line of sight (LoS) property and depend on satellite signals directly received for localization. This motivates many researchers to design and develop indoor localization systems (ILSs) or indoor positioning systems (IPSs) [3, 4] for specific applications. Typical ILS applications include visitor guidance and navigation in museum/building/shopping mall/station/airport, person/animal/asset/repository item tracking, industrial automated guided vehicle (AGV) steering, etc.

PINUS is based on the weighted centroid localization (WCL) method [5]. A target device (or target, for short), whose position is to be determined, receives signals from several anchor devices (or anchors, for short), whose positions are already known. An anchor periodically broadcasts beacon signals. The target can estimate the distance between itself and every anchor according to the received signal strength indicator (RSSI) values. It is well known that distance estimation based on RSSI values is not accurate. PINUS does not apply the estimated distance to calculate the target position directly. Instead, the distance is used for deriving the relative weight associated with every anchor for calculating the weighted composition of anchor positions, which in turn is assumed to be the target's position. Since signal RSSI values are not stable due

J. H. Park et al. (Eds.): PDCAT 2018, CCIS 931, pp. 433–443, 2019.
https://doi.org/10.1007/978-981-13-5907-1_46

to the multi-path effect, interference of signals, and/or different battery power levels of anchors, it is thus necessary to occasionally calibrate the distance estimation for deriving proper weights. PINUS performs distance estimation calibration on the basis of the Friis equation [6], and utilizes the crowdsourcing concept for devices to share calibration information for improving WCL localization accuracy. This accounts for the name of PINUS, which can be regarded as "PIN US," standing for the meaning to pin or locate a bunch of cooperative targets altogether. It is worth mentioning that PINUS has no crowdsourcing device diversity problem, since WCL depends on relative weights for position estimation.

In this study, a PINUS anchor is realized by a Bluetooth Low Energy (BLE) beacon device [7], and a target is realized by a smartphone for validating PINUS design and evaluating PINUS performance. Certainly, other devices, such as WiFi AP, Arduino and Raspberry Pi, can also be applied to realize PINUS. An Android App is developed for performing experiments. In the experiments, four anchors are deployed on the corners of a 5 m × 8 m area, and four arbitrary testing points within the area are selected for measuring localization errors. As will be shown later, distance estimation calibration, either derived by the target itself or by other devices, can evidently reduce localization errors.

The rest of this paper is organized as follows. Section 2 introduces some background knowledge and related work. In Sect. 3, the distance estimation calibration based on the Friis equation is described. Crowdsourced calibration is also elaborated in this section. PINUS implementation and experimental results are shown in Sect. 4. And finally, some concluding remarks are drawn in Sect. 5.

2 Related Work

2.1 Weighted Centroid Localization

There are many variants of the weighted centroid method. This paper adopts the one proposed by Blumenthal et al. [5], which is described below. On receiving a signal from an anchor, a target estimates the distance from itself to the anchor on the basis of the Friis equation [6], as shown below.

$$P_r = P_t G_t G_r \left(\frac{\lambda}{4\pi d}\right)^2 \tag{1}$$

In Eq. (1), P_r is the power received by the receiver (i.e., target), P_t is the transmission power of the transmitter (i.e., anchor), G_t is the gain of the transmitter antenna, G_r is the gain of the receiver antenna, λ is the wavelength, and d is the distance between the transmitter and the receiver.

In many cases, the received signal strength is defined as the RSSI value, which is of the unit of dBm and is 10 times the logarithmic value of the ratio of the received power to the reference power P_{ref}, usually taken as 1 mW. We have the following equation for RSSI.

$$\mathrm{RSSI} = 10 \times \log \frac{P_r}{P_{ref}} \tag{2}$$

To put the terms $P_r, P_t, G_r,$ and G_t in the Friis equation in the decibel (dB) unit (with P_r and P_t in dBm, and G_r and G_t in dBi), we have the following equation.

$$P_r = P_t + G_t + G_r + 20 \log\left(\frac{\lambda}{4\pi d}\right) \tag{3}$$

It is usually that P_t, G_t, G_r, and λ are fixed and known in advance. Therefore, a target can determine the distance d between itself and the anchor according to the RSSI value of the anchor's signal received.

On receiving signals from n anchors, a target can determine its position or location $L = (L_x, L_y)$ according to the following equation, where (L_x, L_y) is the coordinate of location L, and n is usually taken as an integer at least 3.

$$L = \frac{\sum_{i=1}^{n} w_i L_i}{\sum_{i=1}^{n} w_i} \tag{4}$$

Note that Li in Eq. (4) is the coordinate of the location of anchor i. Also note that $w_i = \frac{1}{d_i}$ in Eq. (4), where d_i is the estimated distance between the target and anchor i. As suggested in [5], we may set $w_i = \frac{1}{d_i^g}$ with g being a large value (say 3) to weight longer distance marginally lower for the case that anchors have long transmission range. However, in this paper g is set as 1 and thus $w_i = \frac{1}{d_i}$.

2.2 Crowdsourced Indoor Localization

Crowdsourcing is adopted by many indoor localization methods, most of which are based on fingerprinting [8]. In general, fingerprinting localization consists of two phases: offline site survey, and online positioning. In the site survey phase, a fingerprint (e.g., the RSSI values of WiFi APs) for every reference point (RP) is obtained for constructing the fingerprinting map or fingerprint database. In the positioning phase, the observed fingerprint of a target (e.g., smartphone) is used as the key to find out k nearest fingerprints of k RPs, where $k \geq 1$. The target's position can then be estimated with the k RP positions according to different rules.

The site survey phase of traditional fingerprinting localization is very labor- and time-consuming. Crowdsourcing can relieve the burden of site survey by allowing common users to participate in fingerprinting. According to [8], there are two types of fingerprinting crowdsourcing: active and passive. In active fingerprinting crowdsourcing [9–11], participants are requested to annotate fingerprints with RP location information, such as the location coordinate, recognized location on the floor plan, semantic labels of locations (e.g., room numbers), etc. In passive fingerprinting crowdsourcing [12–14], measurements of the participant device's sensors, such as the accelerometer, gyroscope, magnetometer, and barometer, can be used to infer location and/or participant movement trajectory information. Floor plan constrains can be utilized to further correct the information. Therefore, fingerprinting annotation can be done without user intervention.

A challenge of fingerprinting crowdsourcing is the device diversity problem. Since crowds own devices with diverse brands, models, and hardware specifications, inconsistency of fingerprint measurements is likely to happen. Inter-device calibration [15, 16] can be used to solve the problem. It is assumed that the fingerprints measured by two different devices at a same place have linear relation. The linear relation can in advance be derived and stored for all supported devices. Thus, supported devices can rely on the linear relation and each other's fingerprints for performing positioning.

3 The Proposed Method

In this section, we introduce the proposed PINUS method, whose basic idea is WCL with crowdsourced calibration. Below, we first describe the calibration for WCL and then the crowdsourced calibration for WCL.

3.1 Calibration for WCL

As mentioned earlier, when the WCL method is applied, a target can estimate the distance between itself and an anchor according to the RSSI value of the anchor's signal received. This is under the assumption that the four values P_t, G_t, G_r, and λ in the Friis equation are known. However, the values are usually not known and just set as default values when the WCL method is applied. Moreover, even the values are ever measured, they may change over time. Therefore, they need to be calibrated occasionally.

Below we describe how to achieve the calibration for WCL. It is assumed that there are points with known positions, which are called calibration points (CPs) in this paper. When a target stands on a CP and receives signals from one or more anchors, it can perform the calibration. Since the positions of the CP and anchors are known, the target can calculate the distance between itself and every anchor. Assume the target can receive the signal of a specific anchor with power P_r (represented as RSSI in the unit of dBm) and the distance between the target and the anchor is d, we can derive the following equation on the basis of Eq. (3).

$$P_r = P_t + G_t + G_r + 20(\log\lambda - \log 4 - \log\pi - \log d) \tag{5}$$

Note that all terms in Eq. (5) are of the unit of dBm or dBi. Let the term $(P_t + G_t + G_r + 20(\log\lambda - \log 4 - \log\pi))$ be C. We have the following equation.

$$P_r = C - 20\log d \tag{6}$$

Since P_r and d are known, then C can be calculated as $P_r + 20\log d$ and regarded as the calibration parameter associated with the anchor. Similarly, a target can derive the calibration parameter associated with every anchor and store it for further use. After the calibration, when a target receives a signal from an anchor, it can use P_r and the calibration parameter C associated with the anchor to derive the distance between itself and the anchor as follows.

$$d = 10^{\frac{C - P_r}{20}} \tag{7}$$

Similarly, when receiving signals from some anchors, a target can derive the distance between itself and every anchor based on Eq. (7). Afterwards, it can derive the weight associated with every anchor and easily derive its estimated position on the basis of Eq. (4).

3.2 Crowdsourced Calibration for WCL

The afore-mentioned calibration for WCL should be done for every anchor to take effect. It is obviously a labor- and time-consuming task. Therefore, crowdsourcing can be utilized to help complete comprehensive calibration. Such a manner is called crowdsourced calibration in this paper.

Crowdsourced calibration can be active or passive. In active crowdsourced calibration, a participant standing at a CP inputs the CP's coordinate to complete the calibration for all anchors whose signals can be received by the participant device. By the CP's coordinate, the distance between the participant and every anchor can be calculated, and then, along with the received power P_r, the calibration parameter can be calculated for every anchor. A floor plan with CP marks may be helpful for participants to input CP coordinate. Moreover, coordinate-embedded QR codes near CPs may automate the input of CP coordinates; they even may automate the calibration process when participants scan them.

In passive crowdsourced calibration, the sensors in the participant's device can be used to derive location information or movement trajectory of the participant to allow the participant to do the calibration implicitly. However, since the derived location information or the trajectory may have biases, it may not be suitable for the calibration.

The calibration parameters associated with anchors can be uploaded to a server for future sharing. The parameters can also be stored in the participant device's local memory to be exchanged with other participants through opportunistic communications. Note that the participant device ID should also be uploaded or exchanged along with those parameters. To keep the privacy of the participant, pseudo device ID can be used. The purpose of an ID is to indicate that some calibration parameters are derived by a same device. The ID is very important, because a target should always use calibration parameters provided by a same device. The reason for such a requirement will be explained later.

Now we discuss the aspect of device diversity. The proposed crowdsourced calibration in practice has no device diversity problem. To be more precise, a device A can directly use calibration parameters provided by another device B. Without loss of generality, assume that the receiver module of A is more sensitive than that of B by a factor f. That is to say, if B measures the received power as P, then A measures the received power as $f \times P$ under the same condition. Therefore, if B derives a calibration parameter of C, then A derives a calibration parameter of $C + 10 \log f$ (in the unit of dBm). Since B's calibration parameter is less than A's by $10 \log f$, the distance estimation by A will be shorter under B's parameter. To be more precise, if the distance between A and an anchor is really d, A will estimate the distance to be $d \times f^{-\frac{1}{2}}$ (i.e., $d/\sqrt{f}$) under B's calibration parameter according to Eq. (7). Now, assume A receives signals from n anchors and estimates the distance between itself and each anchor all under B's calibration parameters. Every distances estimated by A will be $f^{-\frac{1}{2}}$ times of the original distance. Now we consider Eq. (4). Since $f^{-\frac{1}{2}}$ appears both in the numerator and the denominator of Eq. (4), the final estimated position of the target will be the same no matter A's calibration parameters or B's calibration parameters are used. This also accounts for the reason why PINUS requires a target to use calibration parameters provided by a same device when estimating its position.

4 Experiments

4.1 Implementation

PINUS is realized by an Android App as a target and a BLE beacon device [7] as an anchor. Figure 1 shows the screenshots of the App. A BLE beacon device can periodically broadcast beacons or BLE advertising packets to be received by smart devices nearby. Seekcy beacon, as shown in Fig. 2, is adopted as the BLE beacon device in this paper. It is powered by a button battery to operate for several months. It can transmit beacons with the time interval of 20 ms to 10 s, and with the transmission range of several tens of meters. The packet can be up to 47 bytes in length, including 4 bytes of the access address, and 2 to 39 bytes of advertising data.

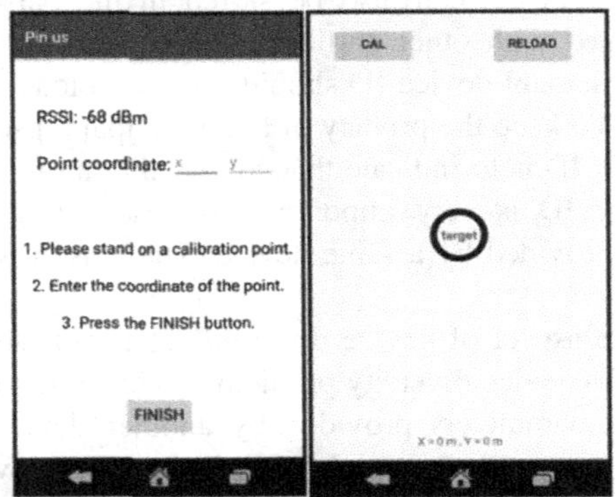

Fig. 1. Screenshots of the PINUS App.

Fig. 2. The Seekcy BLE beacon device.

4.2 Experiment Environment Setting

Experiments are conducted in a 5 m × 8 m area at the lobby of Engineering Building V, National Central University, with four anchors deployed on the area corners. Figure 3 shows the picture of the experimental environment. In the experiment, two types of smartphones are used as the target: device A is Sony Xperia XZ Premium, and device B is the Xiaomi Redmi 3S.

A coordinate system is set up for the 5 m × 8 m area, as shown in Fig. 4, and four testing point are chosen for evaluating the localization errors. For every testing point, 30 continuous samples are taken for the error evaluation.

Fig. 3. The picture of the experimental environment.

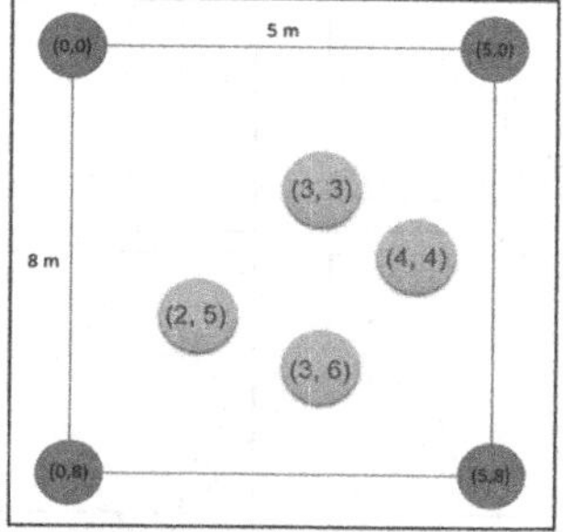

Fig. 4. The coordinates of anchors and testing points in the experimental area coordinate system.

4.3 Experimental Results

There are four sets of experiments. The first two sets adopt device A as the target, while the last two sets, device B. The first and the third sets of experiments apply the WCL method without calibration; they use default calibration parameters for position estimation for every anchor. In the second set of experiments, device A runs WCL with calibration under A's calibration parameters for all anchors. Specially, in the fourth set of experiments, device B runs WCL with calibration under A's calibration parameters for all anchors. Note that the fourth set of experiments are for the purpose of validating that the proposed method has no device diversity problem. That is, when a device uses another device's calibration parameters for all anchors, the localization error does not be influenced. Figures 5, 6, 7 and 8 show the results of 30 samples for four testing points for the four sets of experiments. Figures 9 and 10 show statistics of the experimental results, including the error mean, max and min of the 30 samples.

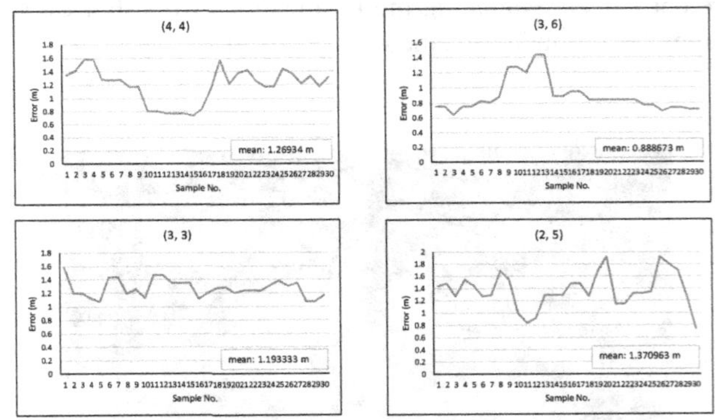

Fig. 5. The experimental results for device A without calibration.

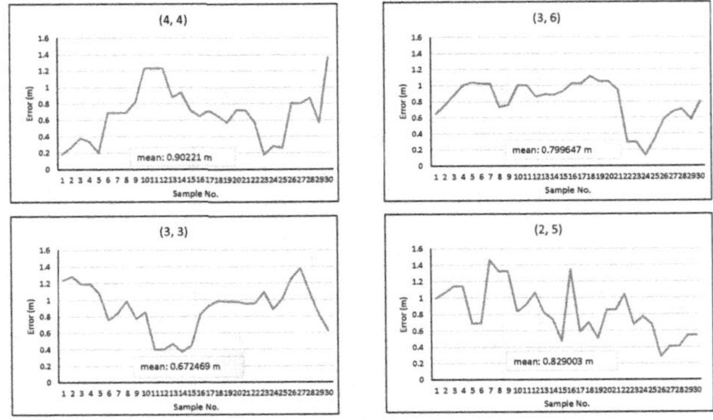

Fig. 6. The experimental results for device A with calibration.

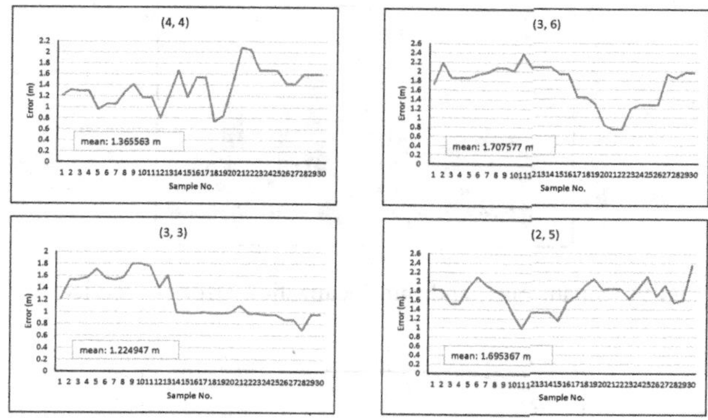

Fig. 7. The experimental results for device B without calibration.

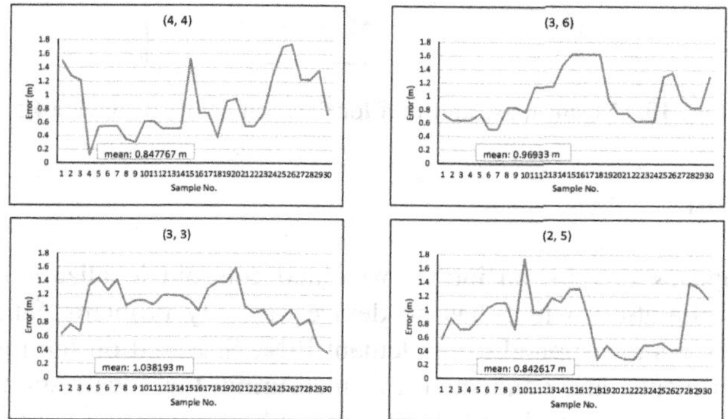

Fig. 8. The experimental results for device B with calibration using device A's calibration information.

By Figs. 9 and 10, we can observe that calibration for WCL significantly reduces the localization error no matter if a device uses another device's calibration information. By the figures, we can also observe that device B has much larger errors when calibration is not applied. However, when B applies A's calibration information to perform calibration, B's errors are reduced significantly and are similar to A's errors. This validates that PINUS has no device diversity problem.

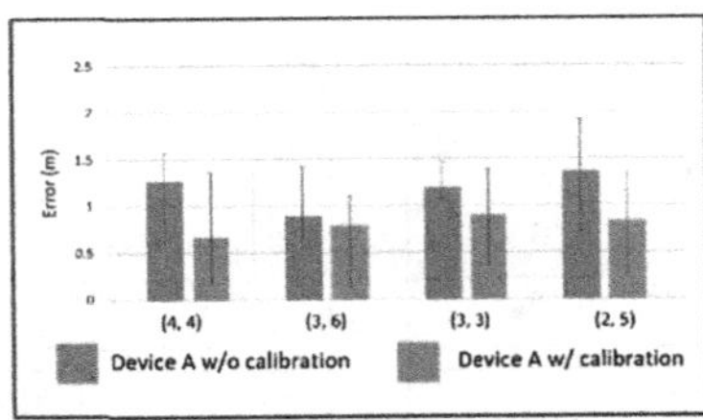

Fig. 9. Mean, max, and min localization errors of device A.

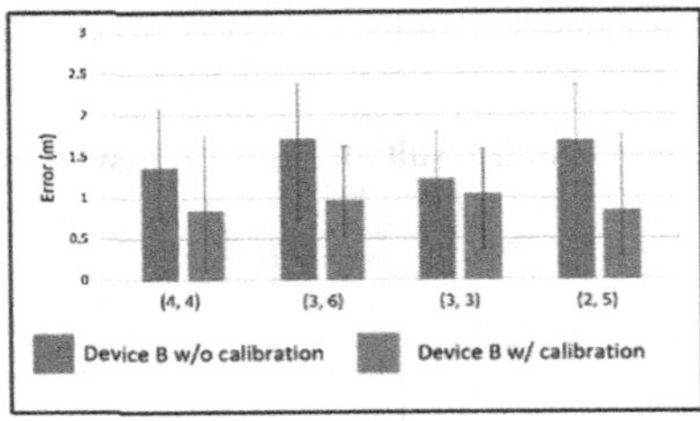

Fig. 10. Mean, max, and min localization errors of device B.

5 Conclusion

This paper proposes PINUS, an indoor weighted centroid localization method with crowdsourced calibration, which has no device diversity problem. Smartphones and BLE beacon devices are applied to implement PINUS. Based on the implementation, extensive experiments are conducted for measuring the localization errors. It is observed that the errors are reduced significantly when calibration is applied. They are reduced dramatically even when a device uses another device's calibration parameters for all anchors involved in the localization process.

In the future, we plan to apply PINUS to large-scale environments using different anchors, such as WiFi APs. More sophisticated calibration methods will be developed to consider various fading situations of radio signals. We also plan to apply filter technologies, such as the Kalman filter and the particle filter, to eliminate noises from received signal series in order to make the distance estimation more stable, which in turn makes position estimation more stable and more accurate.

Acknowledgment. This work was supported in part by the Ministry of Science and Technology (MOST), Taiwan, under grant numbers 106-2218-E-008-003-, and 107-2918-I-008-002-. Special thanks go to Tokyo Metropolitan University, Japan, for providing international cooperation research opportunity in 2018.

References

1. Xu, G., Xu, Y.: GPS: Theory. Algorithms and Applications. Springer, Heidelberg (2016). https://doi.org/10.1007/978-3-662-50367-6
2. Li, X., Zhang, X., Ren, X., Fritsche, M., Wickert, J., Schuh, H.: Precise positioning with current multi-constellation global navigation satellite systems: GPS, GLONASS, Galileo and BeiDou. Sci. Rep. **5** (2015). Article 8328
3. Xiao, J., Zhou, Z., Yi, Y., Ni, L.M.: A survey on wireless indoor localization from the device perspective. ACM Comput. Surv. **49** (2016). Article 25
4. Yassin, A., et al.: Recent advances in indoor localization: a survey on theoretical approaches and applications. IEEE Commun. Surv. Tutor. **19**, 1327–1346 (2016)
5. Blumenthal, J., Grossmann, R., Golatowski, F., Timmermann, D.: Weighted centroid localization in Zigbee-based sensor networks. In: Proceedings of IEEE International Symposium on Intelligent Signal Processing, pp. 1–6 (2007)
6. Balanis, C.: Antenna Theory Analysis and Design, 3rd edn, pp. 94–105. Wiley, Hoboken (2015)
7. Bluetooth Beacons. http://bluetoothbeacons.com/. Accessed 25 June 2018
8. Wang, B., Chen, Q., Yang, L.T., Chao, H.C.: Indoor smartphone localization via fingerprint crowdsourcing: challenges and approaches. IEEE Wirel. Commun. **23**, 82–89 (2016)
9. Park, J.G., Curtis, D., Teller, S., Ledlie, J.: Implications of device diversity for organic localization. In: Proceedings of IEEE INFOCOM, Shanghai, pp. 3182–3190 (2011)
10. Yang, S., Dessai, P., Verma, M., Gerla, M.: FreeLoc: calibration-free crowdsourced indoor localization. In: Proceedings of IEEE INFOCOM, pp. 2481–2489 (2013)
11. Zhang, C., Subbu, K.P., Luo, J., Wu, J.: GROPING: geomagnetism and crowdsensing powered indoor navigation. IEEE Trans. Mob. Comput. **14**, 387–400 (2015)
12. Rai, A., Chintalapudi, K.K., Padmanabhan, V.N., Sen, R.: Zee: zero-effort crowdsourcing for indoor localization. In: Proceedings of 18th Annual International Conference Mobile Computing and Networking, pp. 293–304 (2012)
13. Wu, C., Yang, Z., Liu, Y.: Smartphones based crowdsourcing for indoor localization. IEEE Trans. Mobile Computing **14**, 444–457 (2015)
14. Luo, C., Hong, H., Chan, M.C.: Piloc: a self-calibrating participatory indoor localization system. In: Proceedings of 13th International Symposium on Information Processing in Sensor Networks, pp. 143–153 (2014)
15. Ledlie, J., et al.: Mole: a scalable, user-generated WiFi positioning engine. J. Locat. Based Serv. **6**, 55–80 (2012)
16. Lee, M., Jung, S.H., Lee, S., Han, D.: Elekspot: a platform for urban place recognition via crowdsourcing. In: Proceedings of 2012 IEEE/IPSJ 12th International Symposium on Applications and the Internet, pp. 190–195 (2012)

A Study on Application Method for Automation Solution Using Blockchain dApp Platform

Seong-Kyu Kim[1,2], Hyun-Taek Kwon[2], Young-Kun Kim[2], Yong-Pil Park[2], Dae-Won Keum[2], and Ung-Mo Kim[3(✉)]

[1] School of Electronic and Electrical Computer Engineering, Sungkyunkwan University, Seoul, Republic of Korea
guitara7@skku.edu, guitara77@gmail.com

[2] Geobluelab, Seoul, Republic of Korea
t01046702069@gmail.com, 3ksoft@gmail.com, yongpilo@gmail.com, xgoldya@gmail.com

[3] Electronic and Electrical Computer Engineering, Sungkyunkwan University, Seoul, Republic of Korea
ukim@skku.edu

Abstract. The GOB Chain is a universal Blockchain platform designed to play the role of a perfect Blockchain hub for every Blockchain. First, it connects every Blockchain in order to facilitate the exchange of data between Blockchains based on W3C standard meta data, thus guaranteeing inter-operability and use, as well as searches of transaction information. In addition, it links the Blockchain technology to various existing legacy systems so that it can be immediately applied to the business in service. It allows the GOB chain to apply the Blockchain technology to existing businesses at the sites of the Fourth Industrial Revolution, thus enabling new productivity and competitiveness and improving profitability to a remarkable extent. The GOB Chain cooperates with all Blockchain technologies, companies, and researchers around the world to overcome the limitations of the existing Blockchain technology, and implements the Blockchain-platform ecology chain and the eco system, making our lives more convenient while revolutionizing every industry. In addition, it automates all block-chain systems and suggests a system that anyone can easily provide, operate, and maintain block-chain services.

Keywords: Blockchain · IoT · KYD · M2M · Whitechain Authentification · BoT · IIoT · SaaS · PssS · IaaS · BMS BaaS · Smart contract · Rainbowchain · dApp

1 Introduction

The Blockchain technology guarantees security and transparency by improving the existing way of providing information from the centralized system to connect and share the blocks, which constitute the transaction distribution note on the decentralized P2P

J. H. Park et al. (Eds.): PDCAT 2018, CCIS 931, pp. 444–458, 2019.
https://doi.org/10.1007/978-981-13-5907-1_47

network, thus making it impossible to perform manipulation or falsification [1–3]. At present, many companies are issuing digital encryption currency and exchanging issued coins or tokens through the market with legal currency and having them approved as their own values in order to implement a new capital market and continue to grow [4, 5]. Moreover, many global IT companies are conducting R&D on and investing in Blockchain technology as one of the key technologies of the Fourth Industrial Revolution and the technology has already been adopted by some enterprises.

According to the Sandtander report issued by IBM, Blockchain technology is expected to reduce the infrastructure costs of banks by around 20 billion dollars each year by 2022 [6–9]. Likewise, as Blockchain technology becomes the center of attention in the digital encryption currency market and the Fourth Industrial Revolution, it is reported to have reached the 3rd generation. For instance, at present, we cannot exchange the information or data generated in each Blockchain with other Blockchains. As mentioned here, we must resolve the urgent issue of data exchange between Blockchain platforms, and if a technology capable of resolving this problem should emerge, it will secure its position as the most critical technology in the future development of Blockchain technology. To answer this question, the creation of the fourth-generation Blockchain technology, called the GOB chain platform, was mandated. The GOB Chain is a universal Blockchain platform designed to play the role of a perfect Blockchain hub for every Blockchain [10–14].

First, it connects every Blockchain in order to facilitate the exchange of data between Blockchains based on W3C standard meta data, thus guaranteeing interoperability and use, as well as searches of transaction information. In addition, it links the Blockchain technology to various existing legacy systems so that it can be immediately applied to the business in service. It allows the GOB chain to apply the Blockchain technology to existing businesses at the sites of the Fourth Industrial Revolution, thus enabling new productivity and competitiveness and improving profitability to a remarkable extent. The GOB Chain cooperates with all Blockchain technologies, companies, and researchers around the world to overcome the limitations of the existing Blockchain technology, and implements the Blockchain-platform ecology chain and the eco system, making our lives more convenient while revolutionizing every industry.

2 Background Knowledge

2.1 Blockchain

Blockchain can substitute the current centralized ledger system to the distributed ledger due to the public key algorithm and has encryption technology and the low cost by distribution processing structure. Blockchain technology is the most threat to the payment settlement intermediary system [1, 2].

It is because P2P financial transaction is possible with internet without the intervention of the financial institutions or the trusted third party (TTP). Blockchain is the infrastructure security technology that keeps 'Bitcoin' which is the most widely used and activated among virtual money. Blockchain in Bitcoin is a kind of distributed digital ledger that saves transfer history of Bitcoin value which is issued periodically [15–18]. This ledger is made of encrypted technology to prevent forgery. Digital ledger to record transaction of encrypted money publicly to transfer the ownership of Bitcoin is also started from Genesis Block. And the chain where individual block contains information of the previous block and Blockchain that means all information of transaction are started from the storage and management by distribution into various nodes (Fig. 1) [7].

Blockchain can be explained in three forms including Public, Private and Consortium (Fig. 2). As for Public Blockchain, if all participants verify the transaction details, it may be helpful for integrity, but it may cause privacy problem. And, there is information such as internal information and trade secret that must be hidden in the capital market. Private Blockchain is a method to fill the realistic need in the actual operation by securing high privacy. And Consortium is Blockchain Half Centralized Blockchain that is composed of consortium by various institutions. As pre-selected nodes have rights, the N number of institutions operate one node each and transaction takes place if there is consent of nodes in each institution. The right of reading records of Blockchain is granted to all participants or to institutions only to disclose to specific personnel through API [19–22].

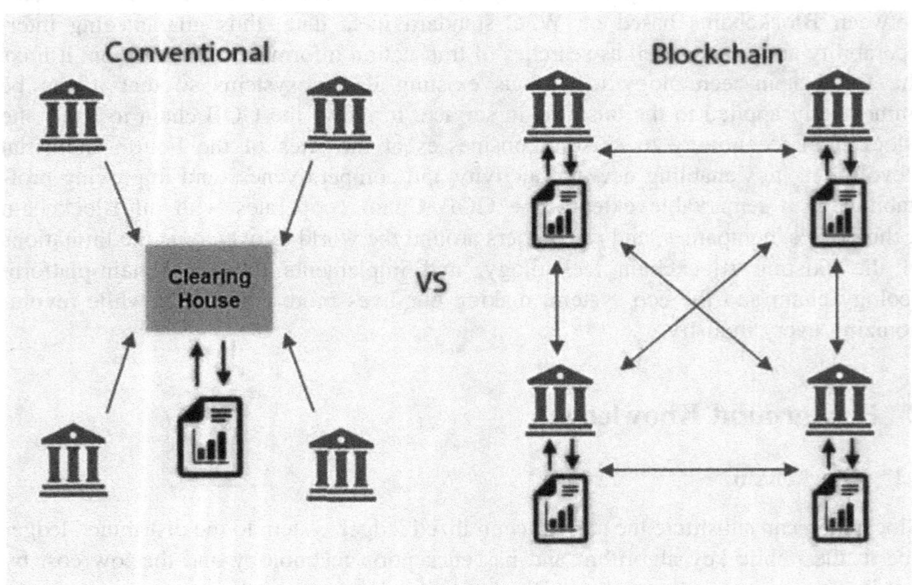

Fig. 1. Diagram of blockchain

Property	Public blockchain	Consortium blockchain	Private Blockchain
Consensus determination	All miners	Selected set of nodes	One organization
Read permission	Public	Could be public or restricted	Could be public or restricted
Immutability	Nearly impossible to tamper	Could be tampered	Could be tampered
Efficiency	Low	High	High
Centralized	No	Partial	Yes
Consensus process	Permissionless	Permissioned	Permissioned
Consensus Core algorithm	PoW, PoS	DPoS, PBFT, Raft Ripple, Tendermint(DPoS+PBFT)	DPoS, PBFT, Raft Ripple, Tendermint(DPoS+PBFT)

Fig. 2. Type of blockchain

However, we also present an automated Blockchain technique overcoming the difficulties of maintenance and difficulty in implementing the existing Blockchain.

2.2 Blockchain Using XML Language

XML stands for EXtensible Markup Language and is a Markup Language like HTML (Hyper Text Markup Language). It is a markup language like HTML (Hyper Text Markup Language). However, the precise meaning is not a markup language, but a language for defining a markup language. Unlike HTML, you can define tags and describe data [23–25].

XML is a standard for exchanging data electronically. In February 1998, the world wide web consortium (W3C) released the XML 1.0 standard as a way to overcome the limitations of existing HTML on the Internet and address the complexity of SGML. XML is a standardized text format markup language designed to transmit structured documents on the web. It is a subset of SGML, a standard that is much simpler than SGML and can be used directly on the Internet. Thus, XML uses a text-based approach and a tag-based approach to avoid data ambiguity and represent unique structures within the data, providing a mechanism for flexibility and extensibility to handle highly structured DB records and unstructured documents [26–31]. XML is an Internet standard that allows users to effectively express what they are trying to express using data. The connection of a huge network via the Internet is an innovative paradigm that can bundle everything from individuals, enterprises and companies to Internet [32–37]. However, these paradigms have the following problems. First, the problem of integration of heterogeneous application systems, second, the problem of data integration in different formats, the layout of web pages according to the third situation, and the files that can be interconnected can be retrieved from the part of information search, electronic commerce, And to accept and utilize it. And the fundamental countermeasures to solve problems have focused on the problem of automating information exchange. In addition, the use of MetaData as a solution to the problem and the establishment and use of the standard Shared Context have been suggested [38–42]. And the Hyper Text Markup Language (HTML) provides various conveniences for display, but it does not provide a standards-based approach to data processing. Like the current HTML page, based on a standard representation that makes it easy to move structured data on the Web, XML, the data standard, is used in a variety of ways in the future. In addition, a language describing a language is called a meta language (upper language), and a meta language is called a technology, and a language is called a target language (sub language). For example, there is a language called BASIC [42–45]. The language is divided into semantics that have a grammatical structure such as the principle of word organization, the principle of constant values for numbers and strings, the principle of statements, and the interpretation of the grammar [45–47]. At this time, the actual target language is the language BASIC, and the language that expresses the grammar of the BASIC language is called the META language. In other words, the language created to express the composition of the target language is the meta language (Fig. 3).

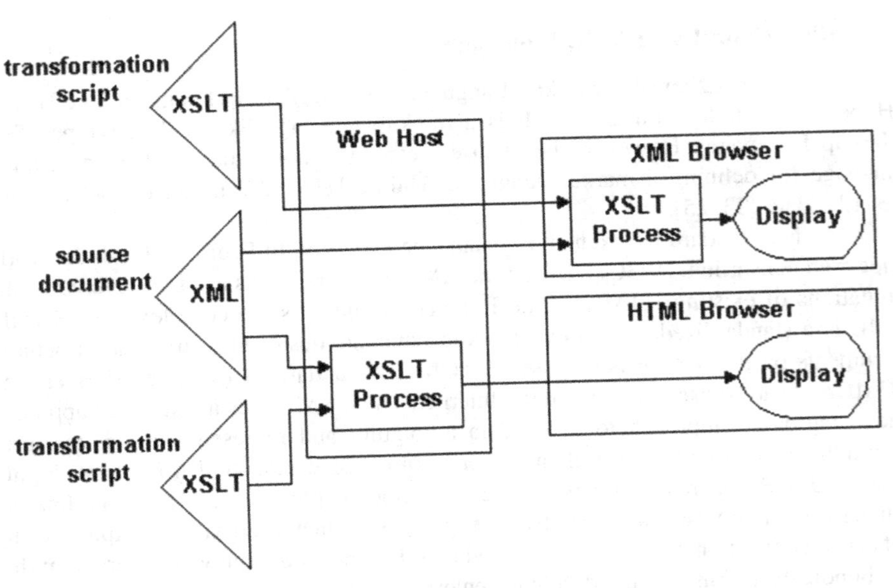

Fig. 3. Structure of XML

3 Design and Implementation of Automated Blockchain dApp

3.1 Issue Raising

At present, the Blockchain is available on various platforms such as Bitcoin [1], Ethereum, Hyperledger and EOS and serves a unique purpose. As well as its limited processing speed per second and slow speed of consent, the public-oriented Blockchain technology has the following technological limitations.

First, it is unable to process unstructured data such as photos, voice, and video.

Second, there is no data standard for exchange between Blockchains.

Third, both critical and personal types of information are exposed to Blockchains.

Fourth, there is only data in smart contact, but there is no user-friendly UI type view.

The GOB Chain can resolve these four problems and provide diverse services.

Also, the existing Blockchain is considered to be inferior in technology and less in SW. It is only about two to three years that the Blockchain started to be applied to real cases. Even the best known block-chain platforms, Hyper Leisure and Etherium, are considered inexperienced. This can lead to unexpected problems during the introduction of both technologies. Recent research shows that quantum computing can ultimately break this distributed book algorithm. So a quantum computer, which is a high performance computer (HPC), can be used to explore the block-chain algorithm. It takes 8 months to decompose a 129-digit number using 1600 high-performance computers connected in parallel, but it is dominant that a quantum computer can be solved in a matter of hours, and a Blockchain can be drilled. In this case there is a problem in that it is also possible to manipulate the distributed chain of the Blockchain.

3.2 Technology Propose Methodology

The GOB chain's platform architecture is available as a single integrated architecture based on the Blockchain technology, and can provide an end-to-end service from the time of initial data generation to the time of the final service (Fig. 4).

In layer 1, it implements the foundation of the Blockchain, BaaS (Blockchain as a Service), in order to provide every service in a Blockchain.

In layer 2, the W3C standard XML is used to standardize the data format, support the XML-Standard in diverse industries, and support the repository that simultaneously supports the SQL & NoSQl DB.

In layer 3, a middleware layer, the following functions are performed: Data Management Processing, Web Service, EAI, Configuration, APIs, Query Processing, etc.

In layer 4, various services are provided including dashboards, reporting, searches, analysis, dApps and community.

In layer 5, a user can use various devices and browsers to provide an HTML5-based view for every service provided by the GOB chain in order to provide diverse services.

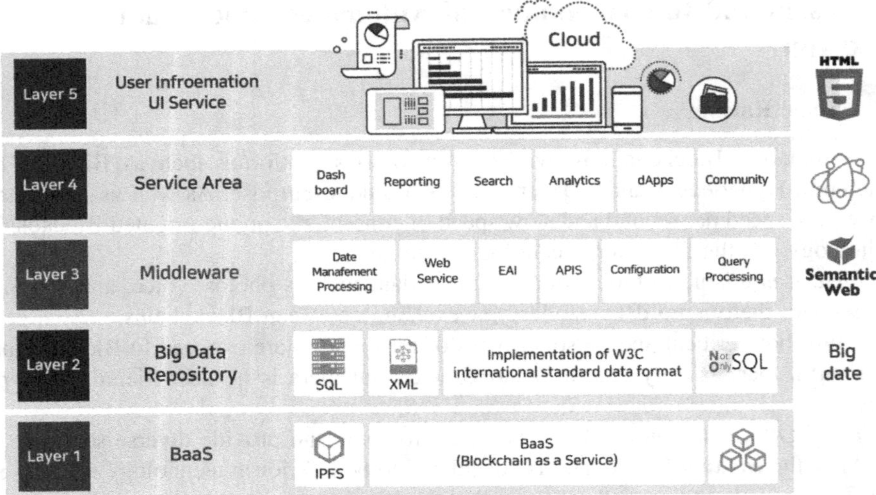

Fig. 4. Platform architecture of the GOB chain

The GOB Universal Blockchain HUB allows you to search for diverse types of transaction data generated in various Blockchains and exchange them between various Blockchains and supports links to the existing legacy systems. It collects the transaction data generated by various Blockchains and creates a common format called "Super Data Set" and saves it in the XML repository. The Blockchain mapper creates mapping data for the Super Data Set Transformations and saves them in the mapping repository. The form repository provides the web form template data required for various UIs. It also provides Search, EAI, Report, Dash Board, and Alarm services (Fig. 5).

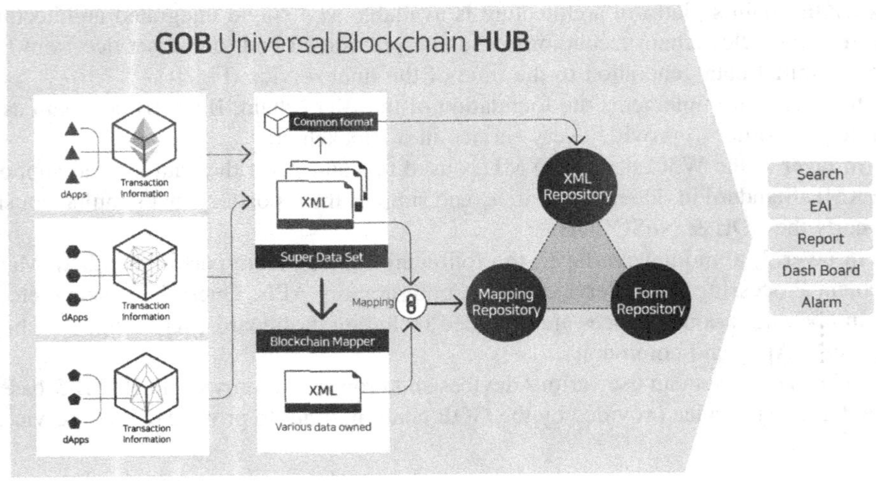

Fig. 5. Universal Blockchain HUB service concepts

4 Provides Automation Blockchain Architecture

4.1 Legacy System Data Exchange

While the current Blockchain technology has difficulty in supporting data exchange between different types of Blockchain platforms, the GOB platform provides a technology and a service capable of exchanging data between various Blockchain platforms. Also Various Blockchain platforms use the GOB Universal Blockchain HUB to exchange data between platforms. Various Blockchain transaction information is collected from the GOB Universal Blockchain HUB, and the Super Data Set is used to make a link to various legacy systems for the purpsose of data exchange. Connected in parallel, but it is dominant that a quantum computer can be solved in a matter of hours, and a Blockchain can be drilled. In this case, there is a problem in that it is also possible to manipulate the distributed chain of the Blockchain.

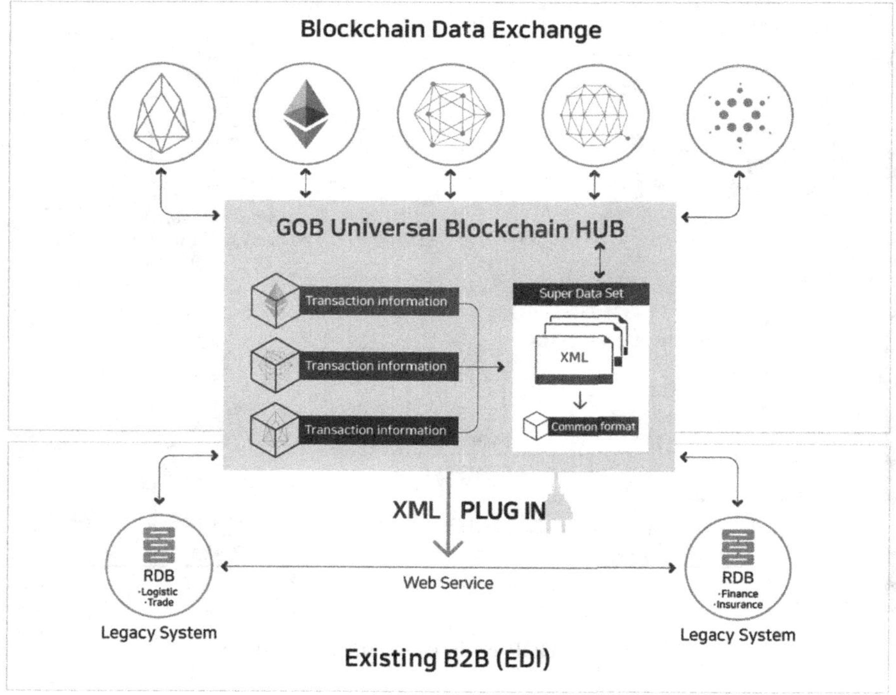

Fig. 6. Concepts of Blockchain data exchange

To ensure flawless operation in harsh environments, a certification exceeding the IP67-Class Dust- and Water-Proofing was acquired as shown in (Fig. 6) shows the test bed experiment on an actual ship for the water- and dust-proofing performance.

4.2 Decentralized Service

The GOB Chain Platform decentralizes Data, View and even GVM (Geoblue Virtual Machine) to provide the perfect decentralized service. By using the W3C standard format, Blockchain data are created as XML (logic data) and XSL (View + Logic), and are then distributed and saved in the IPFS. Also Using the browsers installed on the devices of end users, data and views are integrated and saved in the IPFS, from which it is possible to receive the distributed service immediately (Fig. 7).

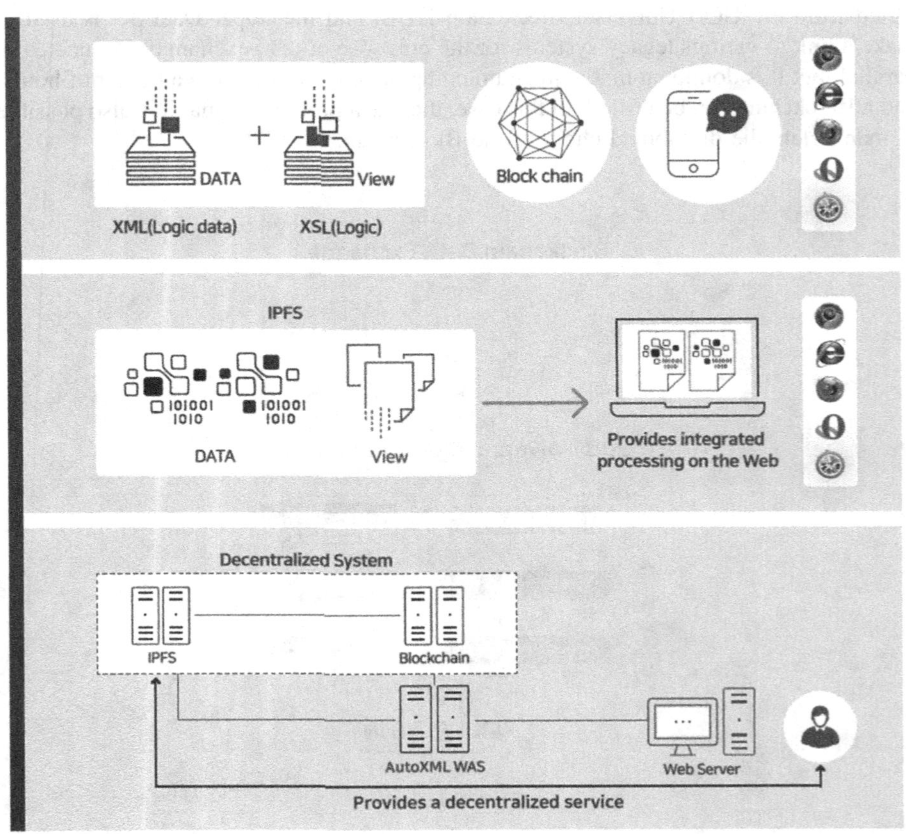

Fig. 7. The GOB chain's decentralized service

4.3 Transaction Virtual Data Set Service

The TDVS (Transaction Virtual Data Set) is a service that provides customized data via various Blockchain platforms. Transaction data are collected from various Blockchains via the GOB Universal Blockchain HUB. The transaction data collected from every

Blockchain are classified and saved in the AutoXML-based common format called "Super Data Set". To meet the diverse requests of individual users, the AutoXML data are customized with the TVDS, and the thus created indexing XML is called "Summary XML". The TVDS's individual summary XML can continue to support the 2nd to nth summary following the 1st summary; therefore, unlike the existing DB-oriented indexing, it can provide a very distinct N-dimensional summary service. If a user wishes to immediately check the basis data for a report provided in the individual dashboard, then the TVDS can provide the user with a web form that provides a view of the W3C standard XML metadata basis. It provides a customized dashboard service for each user, and saves data within a specific timeframe, thus allowing the user to reproduce the target analysis results (Fig. 8).

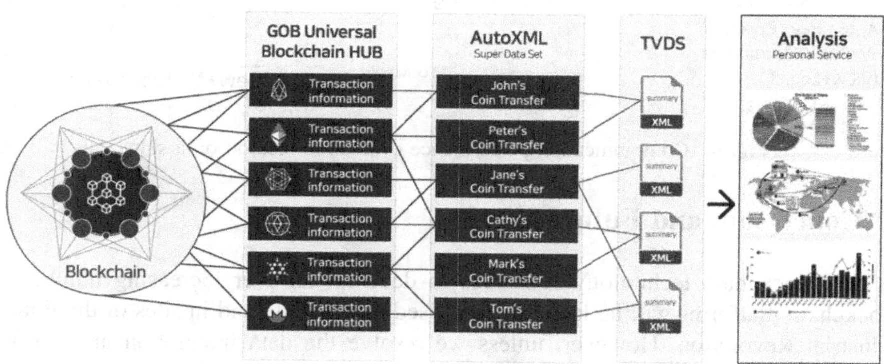

Fig. 8. GOB chai's TVDS service process

4.4 GVM (Geoblue Virtual Machine)

The smart contract created by AutoXML (XML + XSL) includes programming codes. To run them, the run-time environment – which is called AVM (AutoXML Virtual Machine) – is required.

AVM (AutoXML Virtual Machine) supports XLT through XML and XSL, which means that it can be used in every language that supports the W3C standard formats.

The smart contract used in the GOB platform openly saves data (XML) and views (XSL) in every distribution file system node (IPFS).

Unlike the existing method, this method can resolve the issue with performance degradation by performing calculation in the user's node in real time.

- Characteristics of XSLT used in AVM (AutoXML Virtual Machine).
- It can convert an XML document to a new XML structure.
- It can convert an XML document to various outputs.
- XSLT has every function (Turing complete) of the turing machine.

When manufacturing the dApp, XStyler can be used to generate a data schema (XML) and UI, and the deliverables (XML + XSL) can be combined with Blockchains for use. For example, the user can prepare a contract as follows: XML has the contract conditions (date, amount, return, etc.). XSL contains a logic that verifies the wiring details from the Blockchain or document's values and checks the wiring details to verify whether they match the contract conditions of XML (Fig. 9).

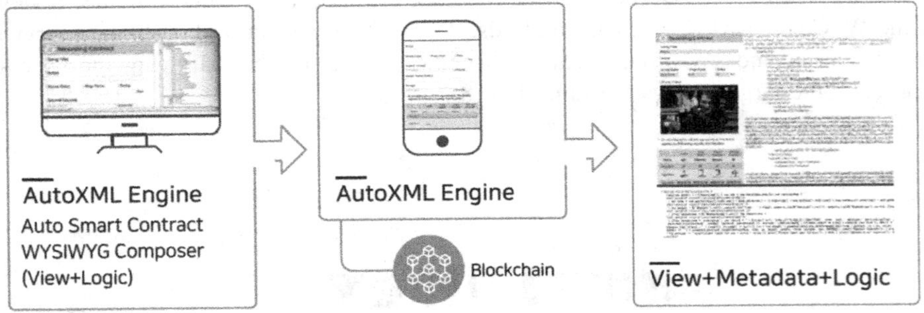

Fig. 9. Communication performance evaluation method on a ship

5 Conclusion and Future Work

As the Blockchain technology continues to develop, an ever increasing number of Blockchain platforms will be launched and used in daily life and in sites of the Fourth Industrial Revolution. However, unless we resolve the data integration and service issues of the current Blockchain platform, we will not be able to deal with the immediate limitations of the Blockchain industry. The GOB chain can resolve all of these issues and ultimately provide an ecology in which it will be possible to use all Blockchain platforms as a single platform. With the growing diversification of Blockchain-related activities, the role of the GOB Chain Platform will be increasingly emphasized and magnified. The GOB chain platform will then secure its position as the key global Blockchain platform that is as verifiable and transparent as numerous governments, individual and organizations around the world, including the UN, require it to be. In addition, it will be possible to exchange and sell GOB Platform for the GOB platforms of the future in linkage with various types of Blockchain Token, thereby securing its position as the dominant Token. Furthermore, it will be used in diverse payments and evolve into a highly reliable Platform that can be trusted as the key hub Token for value exchanges. As the platform supports the BaaS (Blockchain as a Service), the role of the GOB will be increasingly significant in the future; and as users and partners consistently execute eco-partner policies, diverse programs in each area of new start-ups will be promoted, producing even greater effects. With all these efforts, the technology will become a GOB platform that supports the mutual growth of all Blockchains by securing its position as the best Blockchain platform in the world. In the future, we will automate the application of Blockchains in real life and provide models applicable to existing systems (Figs. 10 and 11).

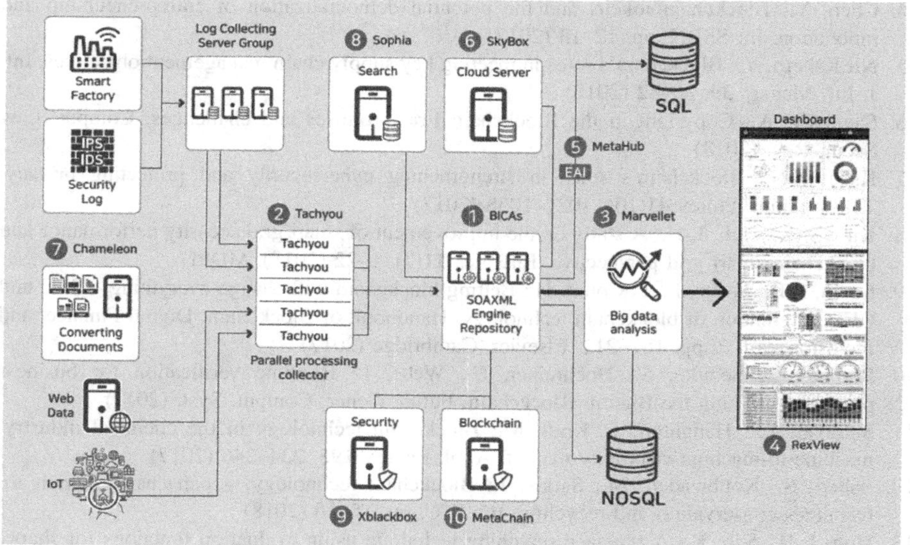

Fig. 10. Diagram of the GOB business system

Solution Name	Description
1 BICAs	Big data engine: collect and analyze IoT and various typical and atypical big data
2 Tachyou	Collecting engine: collect various data by classifying and arraying after applying various collecting rules
3 Marvellet	Big data analysis engine: analyze various business domain knowledge-based big data
4 RexView	HTML5-based business intelligence solution: provide dashboard/report
5 MetaHub	EAI/ESB solution: support to exchange between all various database
6 SkyBox	Cloud solution: provide all data and content service with cloud as the center
7 Chameleon	Document conversion solution: convert all MS-Office, PDF, HWP, etc. files into web documents
8 Sophia	Intelligent search engine: machine training meta data-based search and intelligent search
9 Xblackbox	Data security solution: provide data security with mata data-based tag security
10 MetaChain	Blockchain platform: meta data-based Blockchain platform

Fig. 11. The GOB business system is supported as shown in the following solution map

References

1. Nakamoto, S.: Bitcoin: a peer-to-peer electronic cash system, pp. 1–9 (2008)
2. Huh, J.-H., Seo, K.: Blockchain-based mobile fingerprint verification and automatic log-in platform for future computing. J. Supercomput. 1–17 (2018)
3. Huh, J.-H., Otgonchimeg, S., Seo, K.: Advanced metering infrastructure design and test bed experiment using intelligent agents: focusing on the PLC network base technology for Smart Grid system. J. Supercomput. **72**(5), 1862–1877 (2016)

4. Chen, Y.: Blockchain tokens and the potential democratization of entrepreneurship and innovation. In: SSRN, pp. 12–13 (2017)
5. Nir Kshetri, Y.: Blockchain's roles in meeting key supply chain management objectives. Int. J. Inf. Manag. **39**, 80–82 (2018)
6. Savelyev, A.: Copyright in the Blockchain Era: promises and challenges. Comput. Law Secur. Rev. (2018)
7. Kshetri, N.: Blockchain's roles in strengthening cybersecurity and protecting privacy. Telecommun. Policy **41**(10), 1027–1038 (2017)
8. Kim, S.-K., Huh, J.-H.: A study on the improvement of smart grid security performance and blockchain smart grid perspective. Energies **11**(7), 1–22 (2018). MDPI
9. Levin, R.B., Waltz, P., LaCount, H.: Betting blockchain will change everything – SEC and CFTC regulation of blockchain technology. Handbook of Blockchain, Digital Finance, and Inclusion, vol. 2, pp. 187–212. Elsevier, Cambridge (2017)
10. Prybila, C., Schulte, S., Hochreiner, C., Webe, I.: Runtime verification for business processes utilizing the Bitcoin Blockchain. Future Gener. Comput. Syst. (2017)
11. Sikorski, J.J., Haughton, J., Kraft, M.: Blockchain technology in the chemical industry: machine-to-machine electricity market. Appl. Energy **195**, 234–246 (2017)
12. Saberi, S., Kouhizadeh, M., Sarkis, J.: Blockchain technology: a panacea or pariah for resources conservation and recycling. In: TTC, pp. 15–16 (2018)
13. Huh, J.-H., Seo, K.: A typeface searching technique using evaluation functions for shapes and positions of alphabets used in ancient books for image searching. Int. J. Hybrid Inf. Technol. **9**(9), 283–292 (2016). SERSC
14. Huh, J.-H.: Server operation and virtualization to save energy and cost in future sustainable computing. Sustainability **10**(6), 1–20 (2018). MDPI
15. Qin, B., et al.: Cecoin: a decentralized PKI mitigating MitM attacks. Future Gener. Comput. Syst. (2017)
16. Wang, H., He, D., Ji, Y.: Designated-verifier proof of assets for bitcoin exchange using elliptic curve cryptography. In: TTC, pp. 21–24 (2017)
17. Löbbe, S., Hackbarth, A.: The Transformation of the German Electricity Sector and the Emergence of New Business Models in Distributed Energy Systems, pp. 287–318. Elsevier, Cambridge (May 2017). Chapter 15
18. Huh, J.-H.: Smart grid test bed using OPNET and power line communication. In: Advances in Computer and Electrical Engineering, pp. 1–425. IGI Global, USA (2017)
19. Dorri, A., Kanhere, S.S., Jurdak, R.: Towards an optimized blockchain for IoT. In: Proceedings of the Second International Conference on Internet-of-Things Design and Implementation, pp. 173–178. ACM (2017)
20. Pop, C., Cioara, T., Antal, M., Anghel, I., Salomie, I., Bertoncini, M.: Blockchain based decentralized management of demand response programs in smart energy grids. Sensors **18**, 162 (2018). MDPI
21. Huh, J.-H.: PLC-based design of monitoring system for ICT-integrated vertical fish farm. Hum.-centric Comput. Inf. Sci. **7**(20), 1–19 (2017)
22. Dorri, A., Kanhere, S.S., Jurdak, R., Gauravaram, P.: Blockchain for IoT security and privacy: the case study of a smart home. In: Proceedings of the 2017 IEEE International Conference on Pervasive Computing and Communications Workshops (PerComWorkshops), Kona, HI, USA, 13–17 March 2017 (2017)
23. Huh, J.-H., Kim, T.-J.: A location-based mobile health care facility search system for senior citizens. J. Supercomput. 1–18 (2018)
24. Underwood, S.: Blockchain beyond bitcoin. Commun. ACM **59**(11), 15–17 (2016)

25. Leiding, B., Memarmoshrefi, P., Hogrefe, D.: Self-managed and blockchain-based vehicular ad-hoc networks. In: Proceedings of the 2016 ACM International Joint Conference on Pervasive and Ubiquitous Computing: Adjunct, pp. 137–140. ACM (2016)
26. Pass, R., Shi, E.: Fruitchains: a fair blockchain. In: Proceedings of the ACM Symposium on Principles of Distributed Computing, pp. 315–324. ACM (2017)
27. Ngu, H.C.V., Huh, J.-H.: B+-tree construction on massive data with Hadoop. Cluster Comput. 1–11 (2017)
28. Karame, G.: On the security and scalability of bitcoin's blockchain. In: Proceedings of the 2016 ACM SIGSAC Conference on Computer and Communications Security, pp. 1861–1862. ACM (2016)
29. Huh, J.-H., Seo, K.: Hybrid advanced metering infrastructure design for micro grid using the game theory model. Int. J. Softw. Eng. Appl. **9**(9), 257–268 (2015). SERSC
30. Kiayias, A., Koutsoupias, E., Kyropoulou, M., Tselekounis, Y.: Blockchain mining games. In: Proceedings of the 2016 ACM Conference on Economics and Computation, pp. 365–382. ACM (2016)
31. Hori, M., Ohashi, M.: Adaptive Identity authentication of blockchain system-the collaborative cloud educational system. In: EdMedia+Innovate Learning, Association for the Advancement of Computing in Education (AACE), pp. 1339–1346 (2018)
32. Lee, S., Huh, J.-H.: An effective security measures for nuclear power plant using big data analysis approach. J. Supercomput. 1–28 (2018)
33. Florescu, D., Kossmann, D.: Storing and querying XML data using an RDMBS. Special Issue IEEE Data Eng. Bull. **1060**(22), 27–34 (1999). IEEE
34. Shanmugasundaram, J., et al.: Relational databases for querying XML documents: limitations and opportunities. In: Proceedings of VLDB, Edinburgh, Scotland (1999)
35. Huh, J.-H., Lee, D., Seo, K.: Implementation of graphic based network intrusion detection system for server operation. Int. J. Secur. Appl. **9**(2), 37–48 (2015). SERSC
36. Lin, I.C., Liao, T.C.: A survey of blockchain security issues and challenges. I. J. Netw. Secur. **19**(5), 653–659 (2017)
37. Eom, S., Huh, J.-H.: Group signature with restrictive linkability: minimizing privacy exposure in ubiquitous environment. J. Ambient Intell. Human. Comput. 1–11 (2018)
38. Huh, J.-H.: Implementation of lightweight intrusion detection model for security of smart green house and vertical farm. Int. J. Distrib. Sens. Netw. **14**(4), 1–11 (2018)
39. Kshetri, N.: Blockchain's roles in strengthening cybersecurity and protecting privacy. Telecommun. Policy **41**, 1027–1038 (2017)
40. Jabbar, K., Bjørn, P.: Growing the Blockchain information infrastructure. In: Proceedings of the 2017 CHI Conference on Human Factors in Computing Systems, pp. 6487–6498. ACM (2017)
41. Due.com: How blockchain improves security and transaction times. Nasdaq (2017). http://www.nasdaq.com/article/how-blockchain-improves-securityand-transaction-times-cm771339. Accessed 1 Aug 2018
42. Higgins, S.: Hours after launch, OpenBazaar sees first drug listings. CoinDesk (2016). http://www.coindesk.com/drugs-contraband-openbazaar/. Accessed 1 Aug 2018
43. Young, J.: Hackers eye e-commerce platforms, bitcoin-based OpenBazaar to capitalize. The Cointelegraph (2016). https://cointelegraph.com/news/hackerseye-e-commerce-platforms-bitcoin-based-openbazaar-to-capitalize. Accessed 1 Aug 2018
44. Mainelli, M.: Blockchain will help us prove our identities in a digital world. Harvard Business Review (2017). https://hbr.org/2017/03/blockchain-willhelp-us-prove-our-identities-in-a-digital-world. Accessed 1 Aug 2018

45. Bray, T., Paoli, J., Sperberg-McQueen, C.M., Maler, E., Yergeau, F.: Extensible Markup Language (XML). World Wide Web J. **2**(4), 27–66 (1997)
46. Deutsch, A., Fernandez, M., Suciu, D.: Storing semistructured data with STORED. In: Proceedings of ACM SIGMOD. ACM, Philadelphia (1999)
47. Florescu, D., Kossmann, D.: A performance evaluation of alternative mapping schemes for storing XML data in a relational database. Technical report, INRIA, France (1999)

Towards a Two Factor Authentication Method Using Zero-Knowledge Protocol in Online Banking Services

Manish Singh(✉) and Yichen Han(✉)

Wellington Institute of Technology, Lower Hutt, New Zealand
manish.singh@weltec.ac.nz, yichen.han@it.weltec.ac.nz

Abstract. The main objective of our work is to explore the applicability of cryptographic authentication techniques in two factor/step authentication techniques for online banking systems. In particular, we are investigating zero-knowledge protocol as the second step authentication in the aforementioned systems. Many of the existing two-factor authentication schemes involves the third party in their authentication scheme and/or send user information such as passwords over the network. We have proposed a model which utilizes zero-knowledge proof for second step authentication. The proposed system does not involve the third party or require user passwords to be sent over the network. We also have analyzed and discussed some of the security aspects such as key logging, shoulder surfing and eavesdropping which existing one-factor username password-based systems are not immune to.

Keywords: Zero-knowledge proof · Two-factor authentication
Online banking

1 Introduction

With the widespread use of Internet, an increasing number of people are using online banking service, which not only saves their time to wait in a queue but also let them operate their accounts at any time at everywhere. The outbreak of the "PRISM" incident in the United States [1], coupled with a lot of incidents caused by password leaks, has aroused our concerns about security problems on the Internet. Recently, in the UK, fraudsters infect customer's computers and spy on the emails sent between them and contractors. Fraudsters were able to get bank details for payment sent by the contractor. Using the contractor's email they send a new email to the customer with new bank details stating the former one was incorrect. Customers paid the money and fraudster immediately took the money out from the account. The concerned bank refused to pay back citing the mistake of customers [5].

One-factor cryptographic authentication, the username-password method, is still popular in many applications. However, with the increasing number of online crimes and skill level of malicious users, one factor authentication is not enough. Many application has adopted two factor or two step authentication schemes for an added layer of security. For example, Gmail uses two-factor authentication method,

J. H. Park et al. (Eds.): PDCAT 2018, CCIS 931, pp. 459–472, 2019.
https://doi.org/10.1007/978-981-13-5907-1_48

the username-password and verification code sent to users' mobile phones whenever the email is accessed from an unknown location or if the user wants to change the password [3].

Banks in different countries have adopted two factor authentication for online banking services they provide. Almost all of them use username-password based authentication scheme for first step. However they differ in the second step authentication method. Most of these second factor (step) authentication methods would rely on a third party, the communication company to send a text message or an outsourcing company to produce and maintain a device such as the USB key. Which could be the major source of information (passwords and all) leak. In order to minimize the transfer of sensitive information such as password, across internet we are focusing on a method that does not utilize third parties. Our work will investigate the applicability of zero-knowledge proof in such authentication system. In the first phase, we will propose a two factor authentication scheme using zero-knowledge proof for the second step. The second phase will investigate more on the applicability of the design. This is work in progress.

In the next section, we list existing cryptographic authentication technologies and applications which different banks are using across different countries and compare the advantages and disadvantages of these methods. After that, we introduce zero-knowledge proof and the related works on which our work is based on. In the subsequent sections, we give an overview of our proposed model, discussion on how it can minimize the weaknesses.

2 Background

We looked at the official websites of main banks in three countries, New Zealand, China and the United States, to find the authentication methods of their online banking services. We give summary of these methods outlining their advantages and disadvantages.

Authentication Methods of Online Banking Services in China

The authentication methods of most banks in China is similar, therefore, we make a detailed introduction of ICBC (Industrial and Commercial Bank of China) to be as a representative of banks in China [6]. In general, with this bank, transactions under 50,000 RMB is authorized by sending a code through text message to the customer's mobile device. For authorizing Transfers over the 50,000 RMB and interbank transactions USB key or One-time password device is used. Customers can choose from using either USB Key or OTP Device.

USB Key comes with a CPU, memorizer, and Chip Operation System installed. The memorizer contains user's secret key and digital certificate. A cryptographic algorithm, internally installed is used identify and verify user identity. User after entering the information on the related (transfer) page, plugs the USB Key to the computer. The user then has a fixed time to press a button in order to confirm the amount displayed as well as the transaction. USB key has the control chip of the digital certificate is designed to be only written-in, and all the algorithm made by the digital certificate is working inside the USB Key. The server as well as the device can only read the result. Thus, the secrecy of the secret key is guaranteed. Even in the situation where the installed key of USB Key is cut out by someone and the Trojan starts an illegal request, as long as the physical pressing operation is not activated, the process will be cancelled in a limited time. However, the USB device requires the user to download and install corresponding drive, which may not be compatible with all kinds of computer systems. Also, the bank keeps changing USB keys regularly and user needs to be updated with the new version. It may also create confusion and inconvenience as there are too many versions for users to choose from.

As the name suggests, **One-time password (OTP device)** can only be used once. This method uses two factor authentication, the static PIN code, which is set by the user, and the dynamic password, which is random and verified by the back-end server. Once transaction information is entered, the user enters the correct PIN code, thereafter, a sequence of numbers appears the webpage. The user would then enter these numbers into the device. After verifying the numbers, the OTP Device will display another sequence of numbers, and the user then type these numbers on the webpage. With all the numbers correct, the user can continue the process. In the incorrect PIN entered 10 times the device shuts down. OTP device is a classical cryptography system, which means there is just one key, the key to encrypt is the same as the key to decode.

The pin code is entered in the device not in the computer so anyone hacks into server cannot have access to it. Compare to the USB key OTP device is user friendly because it doesn't require to install any drive. However, the cost of the OTP device is high. In china it is over 100RMB. If some malicious user somehow has access to the server, there is a possibility to calculate all the possible dynamic passwords.

Authentication Methods of Online Banking Service in the United States

In the USA some banks allow limited amount of money when it comes to online transfers. They do not have additional authentication mechanisms other than the one factor username/password methods for these limited online money transfers. For example, Chase Bank sets a limit of $2000 [7]. There will be an email to inform the user if a transaction is made. If users want to transfer more money, they should go to the bank physically and do wire transfer with proven identities.

Limited amount of online banking ensures property security to some extent. Banks could regulate transactions of large amount money more easily. However, in the absence of additional authentication method, if someone obtains username/password they can carry out the money transfer. The user would know via email but it is very difficult to recover money once this happens. It is also inconvenient for users to for the bank and wait in the long queue every time they want to make a wire transfer. The cost of doing wire transfer is also not cheap. In 2016, according to 15 large banks in US, the average fee for a domestic wire transfer was $25. Incoming domestic wire fees were almost between $0 and $15 [8].

Authentication Methods of Online Banking Service in New Zealand

ANZ, the largest bank in New Zealand, serves more than 1 million people and supports over 30% of all home loans. In addition to username and password to login into the online banking interface, ANZ uses an authentication tool is called **OnlineCode** [9]. For any transfer over $1000 the online interface will send a unique OnlineCode through text message to user's mobile phone. The user is then required to enter the code into the online banking page. Users has an option of using Online Code for any other transactions [10]. ANZ also uses mobile internet banking where the first authentication happens by the means of user chosen PIN-code. For second factor it uses same OnlineCode method. This two-factor authentication method is easy to use and convenient for the customers. They customers of the bank do not require install or download any additional software on the computer or mobile phone. However, this method involves a third party, the telecommunication company, to send the message. If something goes wrong to the communication. On the event of telecommunication network failure there is no way to complete the transaction. The bank and its customer has to wait.

The banks of New Zealand BNZ uses online banking authentication tool called **NET Guard** [11] in addition to traditional username/password authentication. NetGuard card has unique serial numbers with a grid of numbers and letters. The customer are re required to enter three co-ordinates that is randomly selected from their NetGuard card to authenticate a transaction. This service is optional at the time of login. Users may choose to log into Internet Banking without using their NetGuard card each time. When the customer chooses to login without the card, they still can use NetGuard card to complete following tasks: make a one-off payment, update personal details, view online statements, and open up accounts. For business customers, BNZ gives a small digital NetGuard token after registering for Internet Banking for Business (IB4B) or Client Fund Services (CFS). After activation and PIN-setting, the token could produce a one-time code each time the customer login. For mobile banking, Mobile NetGuard technology is utilized on both BNZ's personal and business mobile banking applications. Mobile NetGuard gets the customer's information through the customer's initial login and automatically memorizes these details, which means that no codes or co-ordinates from card or token are required to enter after the first login. NetGuard Mobile is allowed on a couple of devices at a time for personal banking while for business banking it can be only used on one mobile device. If someone tries to activate on another iOS or Android device, it will be disabled on the first device [12].

NetGuard does provide an additional layer of security along with username and password. The official website of BNZ has a very detailed description and the introduction of NetGuard. However, we believe the cost of producing cards as well as tokens are high. The bank needs to replace the card or token, if the customer loses one. It involves a third party to produce and maintain.

Westpac, the second largest bank in New Zealand uses **Online Guardian Challenge Service** as its second factor authentication. The methods involves sending challenge questions, or rarely a one-time verification code sending to the customer's mobile number. The service is used when something happens that is not usual, such as, if users log on from a different country, reset the password, or send a large amount of money to some new accounts. Challenge security questions are mandatory for users using online banking, but registering mobile number is optional for standard online banking functions. Users has an option of changing the challenge questions and registered mobile number through online banking if they pass correctly a challenge [13].

Online Guardian Challenge Service is simple and easy to use. However, this service is not used regularly to authenticate all transactions. This irregular occurrence of second factor authentication cannot provide enough protection. Again a third party is involved in sending verification code through text message when required.

Another New Zealand bank, Kiwi bank, uses a method called **KeepSafe** which is a question-answer based authentication method. Customers are required to choose or create at least three sets of questions and answers. Every time a customer logs in, they have to answer one of the stored questions. The answer is displayed as empty boxes, two of which will be highlighted. The customer needs to select a couple of correct characters to fill in the two highlighted boxes. For example, if the correct answer is 'NewZealand', the characters to insert could be 'Z' and then 'L'. Users can only use the mouse or touchpad depending on the device they are using to select the characters, instead of typing them in from the keyboard [14].

KeepSafe is quick and extremely easy to use. However, as the answers are needed to be remembered, the customer may choose questions related to their daily life. Which makes it vulnerable to a form of social engineering as close friends and family members might know the answer. Also, after a successful log in there is no other authentication method to carry out other banking services.

Zero-Knowledge Proof

Zero knowledge proof was proposed by Shafi Goldwasser, Silvio Micali and Charles Rackoff in 1989 [4]. It is a special kind of interactive proof system. It involves a method using which the prover can prove to the verifier that a statement is true but do not reveal any secret information, just conveying the fact of the truth of that statement. The process involves two or more parties. The challenge is that prover should prove that it knows the fact or possesses a specific thing without sending the actual information secret information [15]. The general interactive proof system has two main properties, completeness and soundness. In addition to those two properties, zero-knowledge proof has one more property, zero knowledge.

Completeness: An honest prover could convince the honest verifier who correctly follows the protocol that the statement is true if the statement is truly reliable.

Soundness: If the statement shows false information, it is almost impossible for any cheating provers to successfully persuade the honest verifier to believe the truth of the statement, except with some small probability that the cheating prover should success.

Zero-knowledge: If the statement is true, the prover aims to make the verifier believe in the truth of the statement without revealing any information about the statement in the entire process. In other words, for the verifier, having the idea of the statement, which is not the secret, is sufficient to imagine a scenario showing that the prover knows the secret. This is formalized by showing that every cheating verifier has some simulator that, given only the statement to be proved (and no access to the prover), can produce a transcript that "looks like" an interaction between the honest prover and the cheating verifier.

Related Work

The authors in [17] proposed a protocol that ensures security when users log onto a system from an untrusted device utilizing zero knowledge proof, where the server can verify user identity without knowing the password and only the user knows the password. Since it is a two-factor protocol, it is based on the password and a trusted device which must be a mobile device, a laptop or a mobile phone. They have described a process to setup the trusted device. After selecting a valid username, the user then selects a password. The trusted device will generate a secret key using the password. Meanwhile, the trusted device randomly generates a secret key for itself. The two secret keys then generate two publics respectively. After that, the two public keys and the username are sent to the server. Then the user wants to log in from an untrusted device. The user uses the trusted device to send a message to the server indicating the login attempt which only has life time of a minute. Next, the trusted device requires the user to enter the password and it computes the secret key from the password. After that the trusted device runs several rounds of a zero-knowledge proof with the server. After the server is convinced that the trusted device knows the password, it generates a token encrypted by the public key of the trusted device. Then the trusted device decrypts the token and shows it on the screen. The user then enters the username and token on the untrusted device. This is a half-way login. At this point, the trusted device will run another few rounds of zero-knowledge proof to result in another encrypted token showing on the screen. The user should enter that token to the untrusted device. This time, the server not only verifies the correct of token but also ensures that the untrusted device is the only device requesting to log-in. Our proposed model utilizes the protocol described above to suit our needs. Since our research focuses on the authentication methods for online banking service, the trusted device and untrusted device assumption is not suitable. In our design, we single software that can run zero-knowledge proof to the server.

As explained above in the background section, banks in different countries are using different second factor authentication methods, some of which are costly, relies on third parties having access to secret information, less convenient and prone to attacks like social engineering as well as shoulder surfing. Sending verification text methods relies on third party. Methods such as OTP, USB key and NETGuard involves high cost. Question answer based method are vulnerable to social engineering and shoulder surfing. US wire transfer methods are not convenient for the customers. Almost all the method explained above involves exchanging some sort of secret user information.

We propose a two factor authentication method which utilizes zero knowledge protocol as the second factor authentication. Although we have explained the method for online money transfer specifically it can easily apply to other banking transactions also. It does not involve any third party having access user's secret information.

In the following section, we explain an interface of a software that bank users can install to their device and verify their identifications without compromising their password to the server.

3 Proposed Model

The software (hereafter referred as Application) Application is designed for money transferring and password changing. If users don't have a device with the application, they can log in to the online banking account using any device, but can just view the amount, and cannot transfer money as well as change the password. The two factors ensure that only with both factors authenticated, the user can do money transaction or password changing.

Characters and Assumptions

User: Online banking service user who wants to send money or change important private information. We assume the user has already successfully login the online banking service by username and standard password.

Username: The username is unique and there is no same username for two users. It is used for both online banking service logon and the Application logon.

Standard password: The password the user uses to login to the online banking account.

App Password: The App password is the one user collects to log in to the Application, which should be not as the same as the standard password.

Application: The two-factor zero-knowledge proof authentication application is shorted for the Application in the following context. Users should download the Application on computer or mobile phone. We assume that one user could only install the Application on one particular device.

Device: Our Application is assumed to have versions both for computer and mobile phone so that users can use for most situation. The device is one particular device where the Application is installed. The user must have this device in order to do the transaction through online banking service. We assume the device is secured.

Server: In our scheme, all interaction of the services are done through communication with the server.

App Password Setup Process

This process takes place after the user has already login the online banking through website or mobile phone and wants to use the Application for the first time. The user can download the installation package from the official website of the bank.

After the device has the Application installed, the user enters his username. The server then verifies whether the username is correct. If the screen shows the username doesn't exist, that means two situations. One is the user enters the wrong username, the other is he hasn't logged in online banking yet. Then the server checks whether this username has already linked with two public keys in the database. This step is shown in

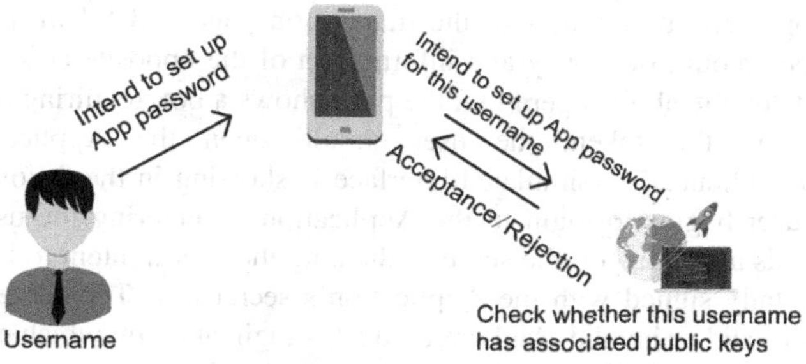

Fig. 1. App password set up

Fig. 1. If it has no associated public keys, the log in intent is accepted and the user can continue the App password setup process. If this username has two associated public keys, the intent is rejected. This implies that it is not the first time the user is trying to log in. User has already set up the App password and could login to the Application directly.

Then, the user enters the standard password. If the pair of username and standard password is correct, the user's identity for the first part is verified and he can continue to the next step. If it is wrong, the user must have entered the wrong standard password and the process is broken.

In the next step, the user collects a password for the Application, which is called the App password. The Application uses that password to generate a secret key. This can be done by using standard technique such as in [2, 16]. The Application then randomly generates a secret key for itself. Two public keys associated with the two secret keys are computed respectively and sent along with the username to the server. This step is shown in Fig. 2. It is important to mention that once the above process is set up, the Application erases all references to the App password and to the two public keys, just saves the two secret keys.

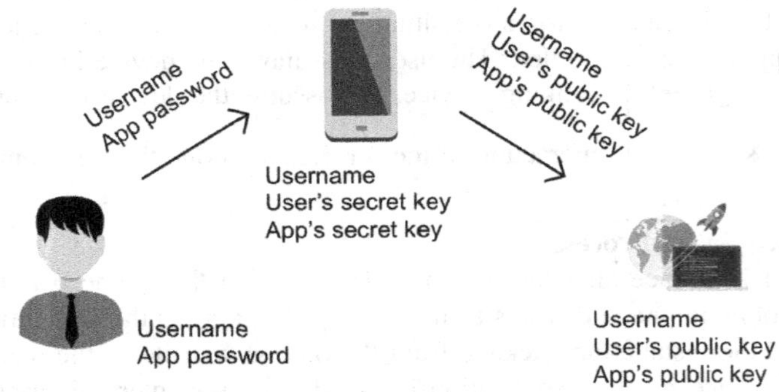

Fig. 2. Generating and sending public keys to the server

Transfer Process

After having set up the App password, the user may want to send money. He begins to login to the online banking service by username and standard password. After he successfully login, the user jumps to the transaction page and fill in the transaction information, the amount of money and information of the opposite account. Once the user has completed the above operation, the page shows a box requiring him to enter a token. To obtain the token, the user should open the Application on his computer/mobile phone. The simulated interface is showing in the following Fig. 3.

Then, the user begins to login to the Application by entering the username. The Application sends a message to the server indicating the user's intent to log in and this message is digitally signed with the Application's secret key. The server verifies the signature, either valid or invalid. And it records this login attempt, which expires within a minute if not continued. The above step is shown in Fig. 4.

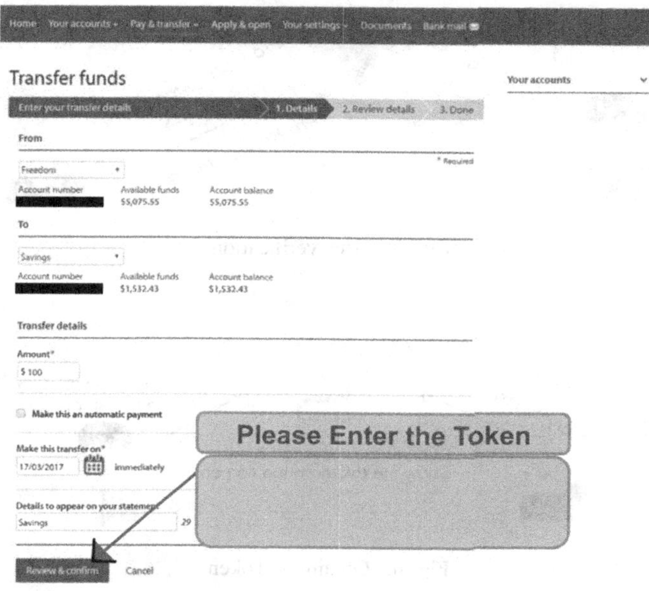

Fig. 3 Online interface asking Token

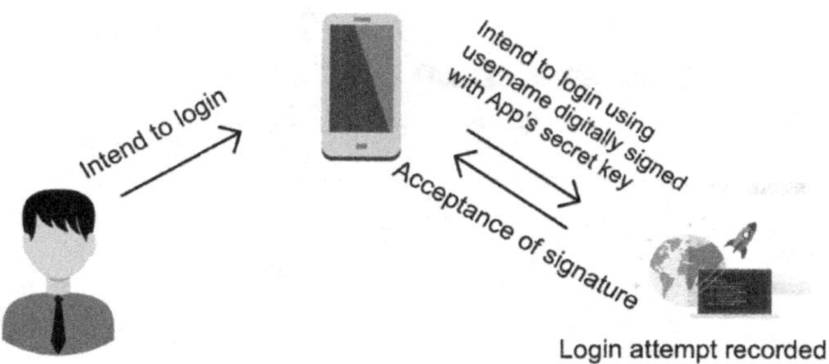

Fig. 4. Log in process

Next, the Application requires the user to enter the App password. The Application then uses the password to compute the user's secret key. Then, it communicates with the server of several rounds of zero-knowledge protocol to prove the Application knows the user's secret key, as shown in Fig. 5. The zero knowledge protocol could be a challenge response protocol as described in [18] or using discrete algorithm as described in [19]. Every message is signed with the Application's secret key.

When server is convinced that the Application currently knows the password, it generates a random token, which is encrypted by the Application's public key, and send the token to the Application. This token expires if it remains unused for two minutes. The Application decrypts the token with the user's secret key and shows it on the screen. The user then enters the token into the box on the transaction page. These steps are

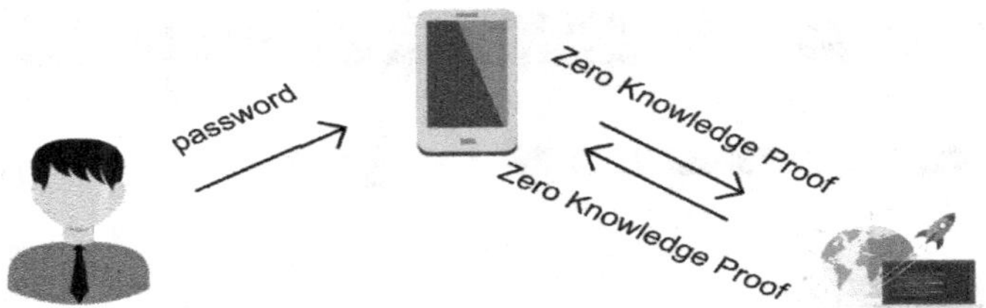

Fig. 5. User verification

Fig. 6. Obtaining Token

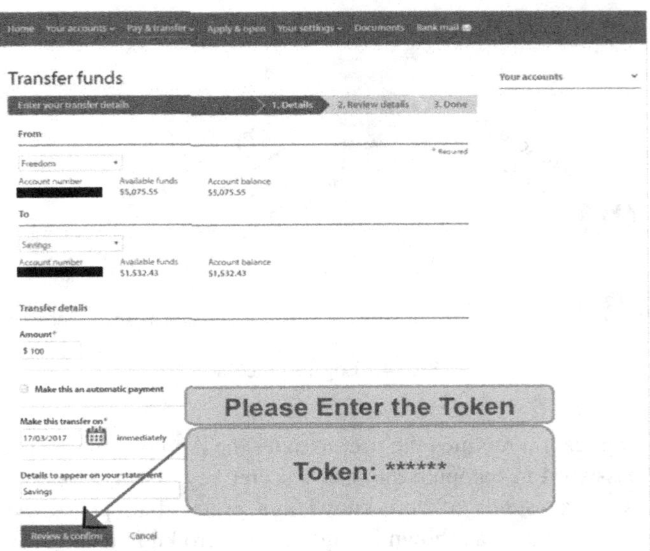

Fig. 7. User entering Token

shown in Fig. 6. And the simulated interface of the transaction page is shown in Fig. 7. The server checks whether the token is matched with the one it generated before. If it is matched, the transaction process is successfully completed. If it is not matched, the process is unsuccessful.

Reset the Standard Password

When the user wants to reset the standard password, the process is similar to the above process. He begins to login to the online banking service by username and standard password. After he successfully login, he jumps to the password resetting page and fill in the new password. Then the page shows a box requiring him to enter a token. The simulated interface is shown in Fig. 8. To obtain the token, the user should open the Application on his computer/mobile phone. The simulated interface is showing in the following figure. The following processes are login to the application and get a token, similar to the money transferring process.

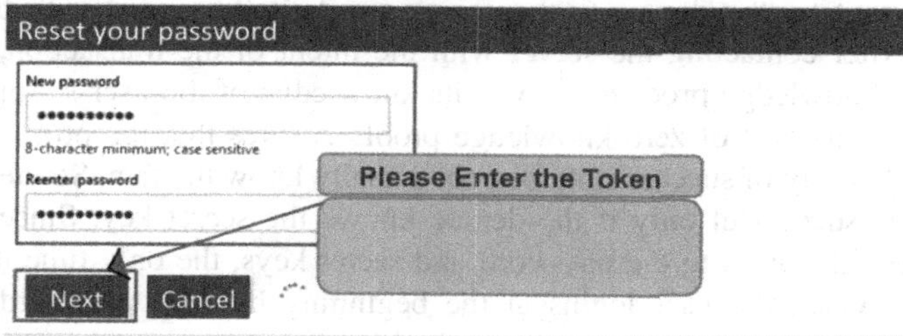

Fig. 8. Password reset requiring Token

In the above assumption part, we assume one user can only install the application for one unique device to ensure safety. If the user lose the device or wants to change the device, he should go to the bank himself with proved identification to delete the former username-public keys records in the server. Then he could return to the App password setting process after he has installed the Application for a new device. Although it may cause inconvenience for users, we have to add this process to ensure the safety since the secret keys are saved in the device. More than two devices could increase the potential to leak the secret keys.

4 Discussion

Generally, password based systems require users to send passwords to the server either in plaintext or encrypted form and the hashed passwords are saved in the server. Those passwords have always the risk of getting leaked either from server or from the communication network. In our proposed scheme, we avoid this by not sending the password to the server. The server only receives the public keys. A successful login to do the money transfer in the proposed scheme requires the Application's secret key stored in the device. One of the ways that a device can be compromised is using a keylogger. To achieve this, someone must install the hardware or software when the user is not in control of the device for enough time. Based on our assumption of device being secured and owned by the user, it is hard for the adversary to install hardware or software keylogger on user's device. Thus, we can conclude that only the user with device having the Application which has the knowledge of the secret key can do a successful money transaction.

At the moment in this scheme, the Application can only be installed on one device, and the server saves one pair of public keys in the database. If an adversary shoulder surfs and is able to see login credentials, it is not possible to carry out the money transaction without the device with Application. If someone steals the device the user can immediately call the bank to lock the Application. It is therefore difficult to realize compromised shoulder surfing with this scheme.

The device sends the server a pair public keys and the username. With the use of public key cryptography and Zero knowledge proof it is obvious that the security of the scheme does not rely on keeping these three data to be secret. During a transfer process, all messages sent by the device to the server are digitally signed with the device's private key. After contacting the server with the intent of the transaction, the device applies a zero-knowledge proof to convey its knowledge of the user's App password. The soundness property of zero-knowledge proofs ensures that the prover has only a negligible probability of success if it does not actually know the fact. So the transaction process can be successful only if the device knows the secret key. Since the device always deletes references to the password and secret keys, the only time it knows the secret keys is when the user logins at the beginning by App password. Thus, the knowledge of app password is required which is not sent over the network. If an adversary gets hold of the token and tries to use it before the user can terminate the login session and report the bank. Since it can be used only once it is of no use if it someone gets it by watching while it is getting used.

5 Conclusion and Future Work

In our project, we proposed a two-factor authentication application using zero-knowledge proof to improve the safety of online banking service while not depending on third-party suppliers. Methods involving text message or token would rely not only on banks but also on a third party company. Question-based methods appear relatively simple but have potential threatens since closed relatives or friends might also know the answer. Our scheme depends on the zero-knowledge proof, which has the zero-knowledge property to not reveal the password to the server while managing to proof the identification of the user. After making relevant definition and assumption, we design the working process with simulated interfaces, including setting up app password, making money transaction and resetting the standard password. The scheme requires both the possession of a device and the knowledge of the password. The scheme can is secure in some compromised situations, such as a keylogger, shoulder surfing and eavesdropping under unsecured network.

As mentioned above the scheme allows Application to be installed on only one device. However, in the real situation, most people have more than one device. Users may want to login to the application using both mobile phone and computer. In the future work, we would find ways to let users install the application on more than one device without decreasing the security and keeping the two factors authentication method utilizing the zero-knowledge proof. Other important work is to measure the time required to carry out rounds for zero-knowledge proof as well as the optimum number of rounds.

References

1. Sottek, T.C., Kopfstein, J.: Everything you need to know about PRISM. https://www.theverge.com/2013/7/17/4517480/nsa-spying-prism-surveillance-cheat-sheet. Accessed 17 July 2013
2. Tamimi, A.A., Al-Allaf, O.N.A., Alia, M.A.: Cryptography based authentication methods. In: Proceedings of the World Congress on Engineering and Computer Science, pp. 199–204 (2014)
3. Shah, N.: Google official blog. https://googleblog.blogspot.co.nz/2011/02/advanced-sign-in-security-for-your.html. Accessed 7 Dec 2011
4. Fiat, A., Shamir, A., Feige, U.: Zero-knowledge proofs of identity. J. Cryptol. **1**, 77–94 (1988)
5. Christie, S.: FRIDAY AFTERNOON FRAUD How hackers can pose as your email contacts to take your cash… and the banks will NOT refund you. The Sun. https://www.thesun.co.uk/living/2699976/how-criminals-can-hack-into-emails-and-trick-you-into-transferring-cash-and-banks-have-no-way-to-stop-it/. Accessed 25 Jan 2017
6. Touryalai, H.: World's 100 Biggest Banks: China's ICBC #1, No U.S. Banks in Top 5. https://www.forbes.com/sites/halahtouryalai/2014/02/12/worlds-100-biggest-banks-chinas-icbc-1-no-u-s-banks-in-top-5/#1c75f0ce22ab. Accessed 12 Feb 2014
7. Chase (2017). https://www.chase.com/online/digital/online-banking.html
8. Tierney, S.: Wire Transfers: What Banks Charge. https://www.nerdwallet.com/blog/banking/wire-transfers-what-banks-charge/. Accessed 8 Sept 2017
9. real banks: Banks around the World (2017). https://www.relbanks.com/oceania/new-zealand/anz-new-zealand
10. ANZ: OnlineCode (2017). https://www.anz.co.nz/banking-with-anz/banking-safely/online-code/
11. Relbanks: Banks around the world (2017). https://www.relbanks.com/oceania/new-zealand/bank-of-new-zealand
12. BNZ: Help & Support (2017). https://www.bnz.co.nz/support/banking/privacy-and-security/netguard
13. Westpac: Westpac Online Guardian (2017). https://www.westpac.co.nz/branch-mobile-online/safety-and-security-online/westpac-online-guardian/
14. KIWIBANK (2017). https://www.kiwibank.co.nz/about-us/security/how-we-protect-you/KeepSafe/
15. Micali, S., Rackof, C., Goldwasser, S.: The knowledge complexity of interactive proof systems. SIAM J. Comput. **18**, 186–208 (1989)
16. Goldreich, O., Lindell, Y.: Session-key generation using human passwords only. In: Kilian, J. (ed.) CRYPTO 2001. LNCS, vol. 2139, pp. 408–432. Springer, Heidelberg (2001). https://doi.org/10.1007/3-540-44647-8_24

17. Nyguyen, Q., Rudoy, M., Srinivasan, A.: Two factor zero knowledge proof authentication system. (n.d.) (2014)
18. Datta, N.: Zero knowledge password authentication protocol. In: Patnaik, S., Tripathy, P., Naik, S. (eds.) New Paradigms in Internet Computing. Advances in Intelligent Systems and Computing, vol. 203, pp. 71–79. Springer, Heidelberg (2013). https://doi.org/10.1007/978-3-642-35461-8_7
19. Chaum, D., Evertse, J.-H., van de Graaf, J., Peralta, R.: Demonstrating possession of a discrete logarithm without revealing it. In: Odlyzko, A.M. (ed.) CRYPTO 1986. LNCS, vol. 263, pp. 200–212. Springer, Heidelberg (1987). https://doi.org/10.1007/3-540-47721-7_14

Security Vulnerability Analysis of Wi-Fi Connection Hijacking on the Linux-Based Robot Operating System for Drone Systems

Jinyeong Kang and Inwhee Joe(✉)

Department of Computer and Software, Hanyang University, Seoul, South Korea
{achieve365, iwjoe}@hanyang.ac.kr

Abstract. In this paper we describe the security vulnerabilities of the Erle-Copter quadcopters. Due to the fact that it is promoted as a toy with low acquisition costs, it may end up being used by many individuals which make it a target for harmful attacks. In addition, the video stream of the drone could be of interest for a potential attacker due to its ability of revealing confidential information. Therefore, we perform a security threat analysis on this particular family of drones. We set the focus mainly on obvious security vulnerabilities like the unencrypted Wi-Fi connection or the user management of the GNU/Linux operating system which runs on the drone. We show how the drone can be hacked in order to hijack the Erle-Copter. Our aim is to sensitize the end-user of Erle-Copters by describing the security vulnerabilities and to show how the Erle-Copter can be secured from unauthorized access. We provide instructions to secure the drones Wi-Fi connection and its operation with the official smartphone app or third party PC software.

1 Introduction

The Erle-Copter is a remote controlled quadrocopter from the Spain company Erle-Robotics. The drone is controlled by the user via an app which is currently available for iOS and Android. Unofficial software is available for bada, Symbian and Windows Phone. The second version of the Erle-Copter was unveiled at CES Las Vegas in 2012. This drone offers an upgraded camera and more sensors. It was first presented at the International Consumer Electronics Show in Las Vegas in the year 2015 (winning the CES Innovations Award for Electronic Gaming Hardware). The drone consists of plastic and foam material and measures about 30 cm. The connection to the drone is realized using Wi-Fi and the drone is then controllable from smartphones and tablets using iOS or Android operating systems, which allows the transmission of the video stream from both cameras. The drone features include object tracking and the compatibility for AR-Game applications.

Quadrocopter and other remote controlled drones gain more and more in popularity and are getting cheaper in price, for this reason the Erle-Copter which is a popular device at low cost is used in this investigation and tested for its security vulnerabilities. In this paper the focus will be set on how to hijack the Erle-Copter and how to secure them.

J. H. Park et al. (Eds.): PDCAT 2018, CCIS 931, pp. 473–482, 2019.
https://doi.org/10.1007/978-981-13-5907-1_49

Even though the Erle-Copter is not a professional drone, its features turn it into an interesting object for research and gaming purposes. It is easily controlled through the mobile application where the user gains access to the various sensors. The available API and documentation allows to create own applications and projects. Bonus features include the AR.Drone academy [1] for SHARING flights, movies and pictures and augmented reality games (which are only supported under iOS). After powering up the drone, it will set up an open Wi-Fi Access Point.

This paper is structured as follows: Sect. 2 introduces the Erle-Copter and its technical aspects. In Sect. 3 the security vulnerabilities of the Erle-Copter using different attacking scenarios are described. A way of securing the drone is explained in Sect. 4. Section 5 will introduce possible future work and finally, in a summary is given.

2 Technical Specifications of the Erle-Copter

The Erle-Copter uses an OMAP 3630 CPU. This processor is based upon a 32 bit ARM Cortex A8 and runs with 1 GHz, it also uses a PowerVR SGX530 GPU with a frequency of 800 MHz on the System on a Chip (SoC) constructed by Texas Instruments. The drone memory includes 1-GB DDR2-RAM, running with 200 MHz and 128 MB NAND-Flash. The drone runs the Kernel Linux uclibc 2.6.32.9-gbb4d210 using uclibc as c-library. This allows a lightweight and fast reacting system. The quadrocopter supports the WLAN standards b/g/n for Video and flight data. For navigation, an Inertial Navigation System (INS) is used. The vertical QVGA (320×240) camera is used for ground speed measurement and video stream (Fig. 1).

Fig. 1. The Erle-Copter

A flight recorder is to be released by Erle-Copter, offering added GPS support, improved stability in flight and flights after set courses, a coming-home safe return function, review flights in 3D and sharing flights with in-flight data with other Erle-Copter Academy users [2]. If one part of the drone gets broken, every single part of the drone can also be bought from the Erle-Robotics store and the user can then replace them individually. The app offers an easy to use interface, which makes the control of the drone relatively easy, compared to more professional RC controlled drones. Most of the controlling aspects are handled automatically, like the start- and landing phases, calibration and position holding. That way, the user only needs to issue 'high level' commands, and can totally forget about the complex handling required by other drones.

2.1 Sensors and Inertial Navigation System (INS)

The Erle-Copter is equipped with multiple sensors, which are placed under the hull. This includes a miniaturized Inertial Navigation System (INS) which measures in six degrees of freedom. This allows the drone to perform yaw, pitch and roll movement. Other sensors perform flight stabilization. The height control is realized through the usage of a pressure sensor and ultrasonic sensors.

2.2 Monitoring

Occupy-Wall-Street-protester Tim Pool configured one of the first generation Erle-Copter, which he described as Occucopter, for the purpose of monitoring the police by the citizens. As soon as the police notice the controller of the drone, they might interrupt it but the control automatically switches to the next person [3, 4]. Using a 3rd or 4th generation mobile communication standard (like UMTS/LTE) it is also possible to control such drones from a much larger distance than just the normal Wi-Fi range. A proof of concept has already been shown by Alcatel-Lucent Bell Labs Acceleration Program and Parrot R&D [5], piloting the Erle-Copter from a 1000 m (3280 ft) distance with a smartphone. This increase in control distance will increase the potential privacy threat.

3 Security Vulnerabilities of the Erle-Copter

We investigated the security vulnerabilities of the Erle-Copter running in the standard out of the box configuration using firmware version 2.2.9. The following interesting details regarding the security have been found.

3.1 Port Scan

After connecting the battery to the drone, it automatically boots up and is ready to use within seconds. It will check the motors and set up an unencrypted Wi-Fi hotspot named erle-robotics-frambuesa followed by a random number with 6 digits. After connecting to this Wi-Fi hotspot, a port scan using the Nmap software on the drones IP has been performed. This scan already shows two interesting open well-known system ports, port 21 (FTP) and port 23 (Telnet). The other ports are related to the drone operation. Table 1 will show the usage of all ports.

Table 1. Open ports running on the Erle-Copter

Port	Explanation
21(TCP)	FTP Server which serves video and image files recorded by the drone
23(TCP)	Telnet Server offering a root shell
5553(TCP)	VIDEO: The H264-720p frames of the camera are available here if the phone application is recording
5554(UCP)	NAVDATA: Current telemetry data (status, speed, rotor speed) is sent to the client here (15 cmds/s demomode, 200 cmds/s full/debug mode)
5555(TDP)	VIDEO: The video stream of the drone is available to clients here
5556(UDP)	ATCMD: The drone is controlled in the form of AT commands. These control commands are sent periodically to the drone (30 cmds/s)
5559(TCP)	CONTROL port: Some critical data (ROS to ROS), such as configurations are transferred here

3.2 Port 5559

The connection to the Control bridge automatically running is not password protected, allowing anonymous access to the/data subdirectory of the drone. After connecting an anonymous bridge to the ROS in the drone, it is automounted to/services/. This allows direct access to the connected services and possibly gain access to confidential data or place malicious files on it (Fig. 2).

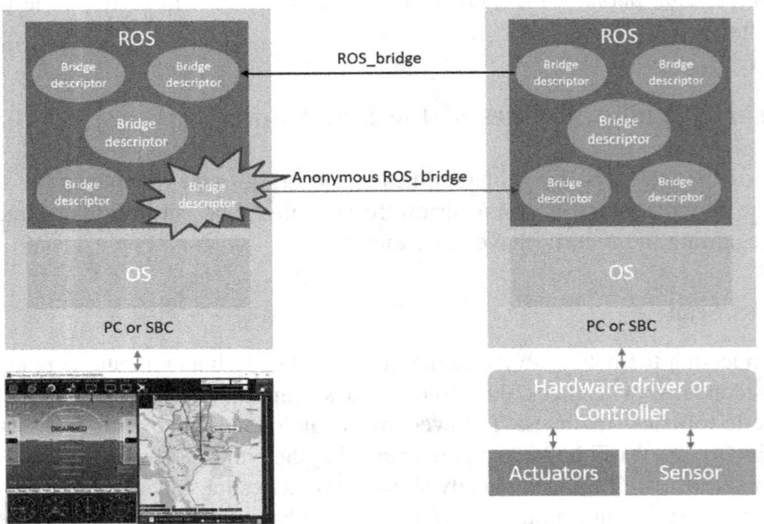

Fig. 2. Message communication of Erle-Copter using 5559 port

In the next step one could connect to the Telnet port which leads to a root shell. The root account is not password protected which gives an attacker free access to the entire drone operating system. This will allow an attacker to perform malicious actions like changing important config files or in the worst case scenario, wipe the filesystem to make the drone unusable. After investigating the filesystem, the following interesting shell scripts have been found:

3.3 Filesystem Backdoor

Using these files, an attacker has even more options to cause harm. This could include automatic moving of malicious files to a connected USB device every time a device is connected, changing the configuration of the reset-script. It also prevents initialization script execution by preventing drone pairing for unauthorized users. The shellscript/bin/reset_config.sh shows that, after using the reset button below the batteries, only the config file/data/config.ini will be reset. All other files remain untouched. This allows an attacker to always regain access to the drone, even after the user might use the reset button if he notice something is wrong with his drone (Table 2).

Table 2. Interesting shellscripts and Linux files on the Elre-Copter

File path	Explanation
/bin/check_update.sh	Update script
/bin/init_gpios.sh	Initialisation of GPIO ports used for connecting the navigation board to the SoC
/bin/pairing_setup.sh	Shell script for pairing using the Smartphone app
/bin/parallel-stream.sh	Camera streams
/bin/reset_config.sh	Reset config.ini while keeping total flighttime value
/bin/Wifi_setup.sh	Start of Wi-Fi connection and other services

3.4 Combination of Attacks

A combination of attacks - like the virus-copter [7] or the SkyJack-Hack, can be used by a malicious attacker to take over the drone entirely. This hack actively searches for hotspots generated by the Erle-Copter. This is done by "seeking out any wireless connections from MAC addresses owned by the Parrot company" (refer to [8]). If the hack finds a drone nearby, it disconnects the real user by deauthenticating him, then initiates a new connection to it while the drone is waiting for the new connection from the real owner. Additional files will be transferred to the drone to control it from the attacker's machine (Fig. 3).

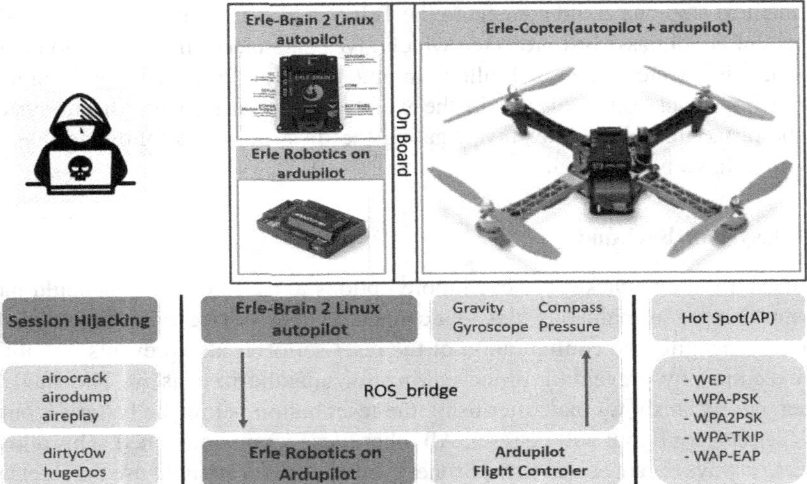

Fig. 3. Combination attacks scenario

4 Hacking Test

The objective is to capture the WPA/WPA2 authentication handshake and then use aircrack-ng to crack the pre-shared key. This can be done either actively or passively. "Actively" means you will accelerate the process by deauthenticating an existing wireless client. "Passively" means you simply wait for a wireless client to authenticate to the WPA/WPA2 network. The advantage of passive is that you don't actually need hijacking capability and thus the Linux version of aircrack-ng can be used (Fig. 4).

Now at this point, aircrack-ng will start attempting to crack the pre-shared key. Depending on the speed of your CPU and the size of the dictionary, this could take a long time, even days (Fig. 5).

Here is what successfully cracking the pre-shared key looks like (Figs. 6 and 7):

Here are the basic steps we went through:

1. Start the wireless interface in monitor mode on the Erle-Brain autopilot AP channel.
2. Start airodump-ng on Erle-Brain autopilot AP channel with filter for bssid to collect authentication handshake.
3. Use aireplay-ng to deauthenticate the wireless client.
4. Run aircrack-ng to crack the pre-shared key using the authentication handshake.
5. Penetration and Conquer of ROS control system using 5559 port.

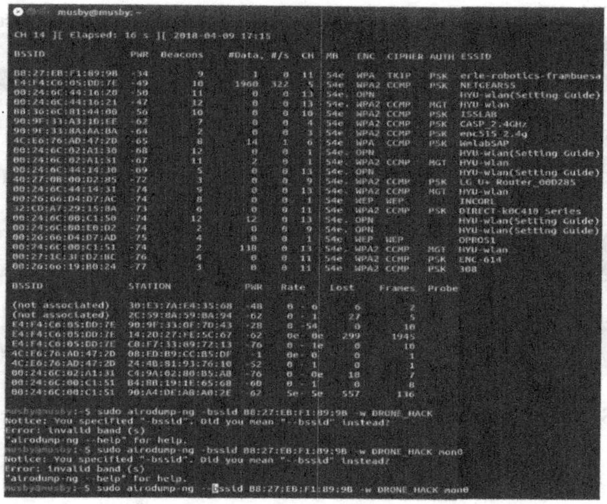

```
musby@musby: ~

CH 14 ][ Elapsed: 16 s ][ 2018-04-09 17:15

BSSID              PWR  Beacons    #Data, #/s  CH  MB   ENC  CIPHER AUTH ESSID

B8:27:EB:F1:89:9B  -34        9        1    0  11  54e  WPA  TKIP   PSK  erle-robotics-frambuesa
E4:F4:C6:05:DD:7E  -49       10     1960  322   5  54e  WPA2 CCMP   PSK  NETGEAR55
00:24:6C:44:16:20  -50       11        0    0  13  54e. OPN              HYU-wlan(Setting Guide)
00:24:6C:44:16:21  -47       12        0    0  13  54e. WPA2 CCMP   MGT  HYU-wlan
88:36:6C:81:44:00  -56       10        0    0  10  54e  WPA2 CCMP   PSK  ISSLAB
90:9F:33:A3:1D:EE  -62        7        0    0   4  54e  WPA2 CCMP   PSK  CASP_2.4GHz
90:9F:33:8A:AA:BA  -64        2        0    0   3  54e  WPA2 CCMP   PSK  enc515_2.4g
4C:E6:76:AD:47:2D  -65        8       14    1   6  54e  WPA  CCMP   PSK  WmlabSAP
00:24:6C:02:A1:30  -68       12        0    0   1  54e. OPN              HYU-wlan(Setting Guide)
00:24:6C:02:A1:31  -67       11        2    0   1  54e. WPA2 CCMP   MGT  HYU-wlan
00:24:6C:44:14:30  -69        5        0    0  13  54e. OPN              HYU-wlan(Setting Guide)
40:27:0B:00:D2:85  -72        3        0    0   9  54e. WPA2 CCMP   PSK  LG U+ Router_00D285
00:24:6C:44:14:31  -74        9        0    0  13  54e. WPA2 CCMP   MGT  HYU-wlan
00:26:66:D4:D7:AC  -74        8        0    0   1  54e  WEP  WEP         INCORL
32:CD:A7:29:15:8A  -73        6        0    0  11  54e  WPA2 CCMP   PSK  DIRECT-k0C410 Series
00:24:6C:00:C1:50  -74       12       12    0  13  54e. OPN              HYU-wlan(Setting Guide)
00:24:6C:00:E0:D2  -74        2        0    0   9  54e. OPN              HYU-wlan(Setting Guide)
00:26:66:D4:D7:AD  -75        4        0    0   1  54e  WEP  WEP         OPROS1
00:24:6C:00:C1:51  -74        2      138    0  13  54e. WPA2 CCMP   MGT  HYU-wlan
00:27:1C:3F:D2:BC  -76        4        0    0  11  54e  WPA2 CCMP   PSK  ENC-614
00:26:66:19:B0:24  -77        3        0    0  11  54e  WPA2 CCMP   PSK  308

BSSID              STATION            PWR   Rate    Lost    Frames  Probe

(not associated)   30:E3:7A:E4:35:68  -48    0 - 6      6       2
(not associated)   2C:59:8A:59:8A:94  -62    0 - 1     27       5
E4:F4:C6:05:DD:7E  90:9F:33:0F:7D:43  -28    0 -54      0      10
E4:F4:C6:05:DD:7E  14:2D:27:FE:5C:67  -62   0e- 0e    299    1945
E4:F4:C6:05:DD:7E  C8:F7:33:89:72:13  -76    0 - 1e     0      10
4C:E6:76:AD:47:2D  08:ED:B9:CC:85:DF   -1   0e- 0       0       1
4C:E6:76:AD:47:2D  24:4B:81:93:76:10  -52    0 - 1      0       1
00:24:6C:02:A1:31  C4:9A:02:80:B5:A8  -76    0 - 0e    18       7
00:24:6C:00:C1:51  84:8B:19:1E:65:68  -60    0 - 1      0       8
00:24:6C:00:C1:51  90:A4:DE:A8:A0:2E  -62   5e- 5e    557     136

musby@musby:~$ sudo airodump-ng -bssid B8:27:EB:F1:89:9B -w DRONE_HACK
Notice: You specified "-bssid". Did you mean "--bssid" instead?
Error: invalid band (s)
"airodump-ng --help" for help.
musby@musby:~$ sudo airodump-ng -bssid B8:27:EB:F1:89:9B -w DRONE_HACK mon0
Notice: You specified "-bssid". Did you mean "--bssid" instead?
Error: invalid band (s)
"airodump-ng --help" for help.
musby@musby:~$ sudo airodump-ng --bssid B8:27:EB:F1:89:9B -w DRONE_HACK mon0
```

Fig. 4. Start airodump-ng to collect authentication handshake

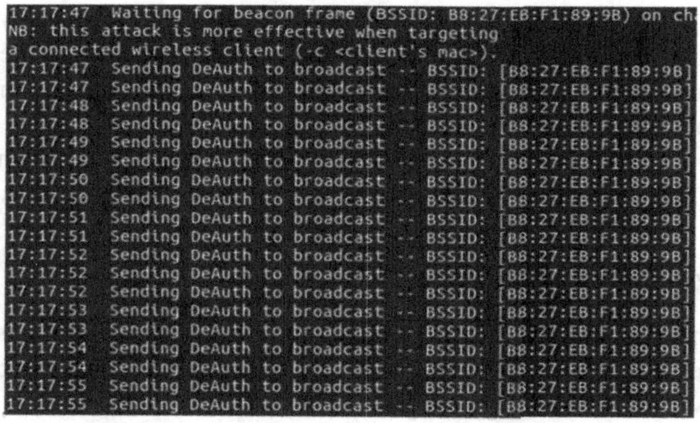

```
17:17:47  Waiting for beacon frame (BSSID: B8:27:EB:F1:89:9B) on ch
NB: this attack is more effective when targeting
a connected wireless client (-c <client's mac>).
17:17:47  Sending DeAuth to broadcast -- BSSID: [B8:27:EB:F1:89:9B]
17:17:47  Sending DeAuth to broadcast -- BSSID: [B8:27:EB:F1:89:9B]
17:17:48  Sending DeAuth to broadcast -- BSSID: [B8:27:EB:F1:89:9B]
17:17:48  Sending DeAuth to broadcast -- BSSID: [B8:27:EB:F1:89:9B]
17:17:49  Sending DeAuth to broadcast -- BSSID: [B8:27:EB:F1:89:9B]
17:17:49  Sending DeAuth to broadcast -- BSSID: [B8:27:EB:F1:89:9B]
17:17:50  Sending DeAuth to broadcast -- BSSID: [B8:27:EB:F1:89:9B]
17:17:50  Sending DeAuth to broadcast -- BSSID: [B8:27:EB:F1:89:9B]
17:17:51  Sending DeAuth to broadcast -- BSSID: [B8:27:EB:F1:89:9B]
17:17:51  Sending DeAuth to broadcast -- BSSID: [B8:27:EB:F1:89:9B]
17:17:52  Sending DeAuth to broadcast -- BSSID: [B8:27:EB:F1:89:9B]
17:17:52  Sending DeAuth to broadcast -- BSSID: [B8:27:EB:F1:89:9B]
17:17:52  Sending DeAuth to broadcast -- BSSID: [B8:27:EB:F1:89:9B]
17:17:53  Sending DeAuth to broadcast -- BSSID: [B8:27:EB:F1:89:9B]
17:17:53  Sending DeAuth to broadcast -- BSSID: [B8:27:EB:F1:89:9B]
17:17:54  Sending DeAuth to broadcast -- BSSID: [B8:27:EB:F1:89:9B]
17:17:54  Sending DeAuth to broadcast -- BSSID: [B8:27:EB:F1:89:9B]
17:17:55  Sending DeAuth to broadcast -- BSSID: [B8:27:EB:F1:89:9B]
17:17:55  Sending DeAuth to broadcast -- BSSID: [B8:27:EB:F1:89:9B]
```

Fig. 5. Use aireplay-ng to deauthentication the wireless client

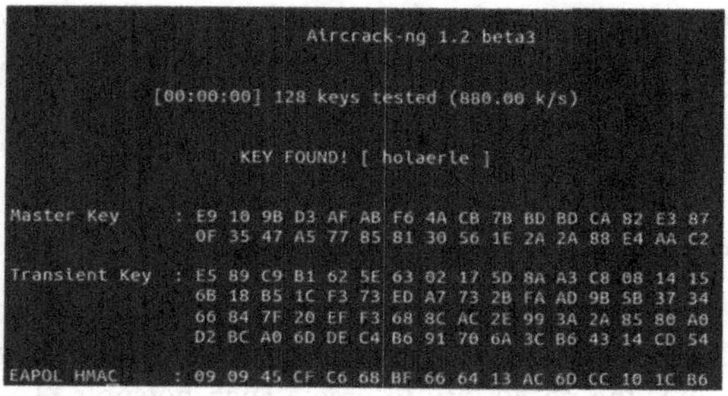

```
                          Aircrack-ng 1.2 beta3

             [00:00:00] 128 keys tested (880.00 k/s)

                     KEY FOUND! [ holaerle ]

Master Key     : E9 10 9B D3 AF AB F6 4A CB 7B BD BD CA 82 E3 87
                 0F 35 47 A5 77 85 81 30 56 1E 2A 2A 88 E4 AA C2

Transient Key  : E5 B9 C9 B1 62 5E 63 02 17 5D 8A A3 C8 08 14 15
                 6B 18 B5 1C F3 73 ED A7 73 2B FA AD 9B 5B 37 34
                 66 84 7F 20 EF F3 68 8C AC 2E 99 3A 2A 85 80 A0
                 D2 BC A0 6D DE C4 B6 91 70 6A 3C B6 43 14 CD 54

EAPOL HMAC     : 09 09 45 CF C6 68 BF 66 64 13 AC 6D CC 10 1C B6
```

Fig. 6. Run aircrack-ng to crack the pre-shared key

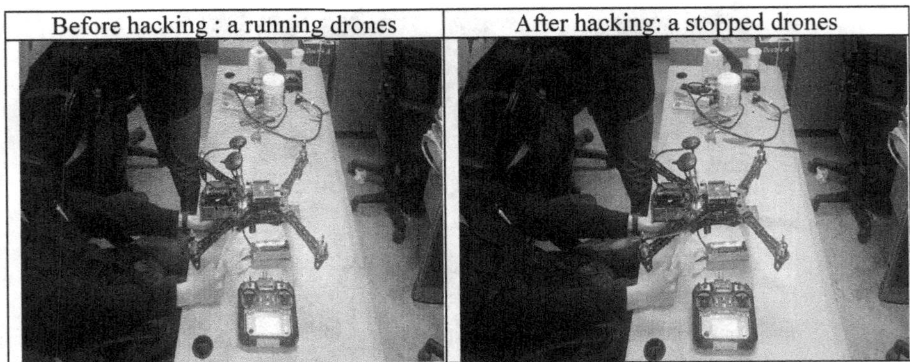

Before hacking : a running drones
After hacking: a stopped drones

Fig. 7. Control System conquer of Erle-Copter through ROS-bridge between autopilot and ardupilot

5 Securing the Erle-Copter

Secure operation of drones in general is an important factor [9, 10]. The only security feature currently active is the MAC address filter. This system can easily be tricked by spoofing the MAC address of the owner's device when connecting to the drone. To get the needed address the attacker simply monitors and reads out the network traffic while the victim is operating the drone. We will show how one can secure the Erle-Copter in such a way that an attack is almost not possible. Our method follows a similar way as in [11] and [12]. This is realized by cross compiling the needed programs.

5.1 Securing the Wi-Fi Connection

As previous investigation of the first version of the Elre-Copter has shown, it is possible to eavesdrop the videosteam of the drone (refer to [6]). This is possible due to the encrypted Wi-Fi network hijacking set up by the drone on start. In this chapter, a way of securing this network connection is shown. Our approach to a more secure concept that can be used to control the drone is that the drone is just a client and the control device acts as the access point. We will show how the wpa_supplicant can be used on the Erle-Copter to let it connect to an encrypted Wi-Fi network. This tool is a WPA Supplicant for Linux, Mac OS X, and Windows systems and supports WPA and WPA2 (IEEE 802.1X [13]). To use the wpa_supplicant on the drone, we first have to cross-compile it to run on the ARM architecture.

5.2 Using the Secure Network

As stated before, we are going to use an external (secured) access point in order to use encrypted Wi-Fi. This hotspot can be generated by the control device as current mobile phones and tablets offer the possibility to create such hotspots. To keep the app running, the drone must maintain its original IP address (192.168.1.1). This is the only IP address that must remain free if the user does not want to use a third party app to control the drone. The following method of starting the secure connection requires a third device to initiate the connection from the drone to the secured network. First, the drone should start up normally after connecting the battery. Now the secure hotspot can be started on the control device that will be later used to control the drone. The third device must now connect to the unsecure drone network which is called ardrone2_ followed by a random number. Still on the third device a telnet connection must be initiated to the drone. After issuing the commands:

```
$ wpa_passphrase $ESSID $PASSWORD > /etc/wpa_supplicant.conf
$ ifconfig ath0 $ADDRESS
$ iwconfig ath0 essid $ESSID
$ wpa_supplicant –B –Dwext –i ath0 –c /etc/wpa_supplicant.conf
```

via telnet. $ESSID must be replaced with the name of the secure network and the password must be set. $ADDRESS will be the new address the drone uses in the new network. To keep the original app running, the user should set this to 192.168.1.1. The drone will close the hotspot and connect to the control device on which the secured hotspot is running. By keeping the initiation of the connection to the secure network manual, we ensure the normal operation is still possible after reconnecting the battery, so even in case of a misconfiguration no special operations are required.

5.3 Benefits of Using an External Network

Since the new (secured) network can be connected to the internet, different new possibilities are possible. This includes the operation of the drone over the internet. The videostream could also be transmitted over the internet like a flying webcam which can be controlled from far away. The only problem is the reduced operating time of the drone due to the low battery time.

6 Conclusions

In this paper we have shown that the Erle-Copter can be secured in the way that it uses encryption instead of an unencrypted Wi-Fi to operate. Several attacking scenarios that a malicious attacker could use to disrupt normal operation have been shown. Using the option to use an external network connection secures the drone from such attacks. Additionally, the risk of showing the contents of a connected USB device is not given anymore.

Future work will include adding scripts to the drone or second device to remove the need for a third device to initiate the connection to the network. The security could be improved even more with a randomly generated passphrase which is given to the drone operator by touching a NFC device on the drone with the smartphone app. An app for tablets and smartphones that is allowing the drone to use a custom IP to operate would be of interest too. If this custom app is given, a way of automatically securing the drone by applying our method to it would be the next thing to develop.

Acknowledgment. This work was supported by the Technology development Program (S2521883) funded by the Ministry of SMEs and Startups (MSS, Korea).

References

1. Demgen, A.: AR.Drone-Academy: Soziales Netzwerk für Drohnen-Flieger verfügbar, August 2012. http://www.netzwelt.de/news/93209-ar-drone-academy-soziales-netzwerk-drohnen-flieger-verfuegbar.html
2. AR.Drone 2.0 flight-recorder. http://ardrone2.parrot.com/apps/flight-recorder/
3. Sharkey, N., Knuckey, S.: OWS Fights Back Against Police Surveillance by Launching "Occucopter" Citizen Drone, December 2011. http://www.alternet.org/story/153542/ows_fights_back_against_police_surveillance_by_launching_occucopter_citizen_drone
4. Wagstaff, K.: Occupy Wall Street's New Drone: 'The Occucopter', December 2011. http://techland.time.com/2011/12/21/occupy-wall-streets-new-drone-the-occucopter/
5. Méchaly, A.: One flew over the cornfield, October 2012. http://www2.alcatel-lucent.com/blogs/corporate/2012/10/one-flew-over-the-cornfield/
6. Samland, F., Fruth, J., Hildebrandt, M., Hoppe, T., Dittmann, J.: Erle-Copter: security threat analysis and exemplary attack to track persons. In: Casasent, D.P. (eds.) SPIE Proceedings, Intelligent Robots and Computer Vision XXIX: Algorithms and Techniques, Juha Röning; 83010G, vol. 8301 (2012)
7. Ackerman, E.: AR Drone that infects other drones with virus wins DroneGames, December 2012. http://spectrum.ieee.org/automaton/robotics/diy/ar-drone-that-infects-other-drones-with-virus-wins-dronegames
8. Kamkar, S.: SkyJack, December 2013. https://github.com/samyk/skyjack
9. Hacking Drones and the Dangers It Presents, July 2012. http://www.npr.org/2012/07/08/156459939/hacking-drones-and-the-dangers-it-presents
10. Drone hack explained: Professor details UAV hijacking, July 2012. http://rt.com/usa/news/texas-professor-drone-hacking-249/
11. node-cross-compiler. https://github.com/felixge/node-cross-compiler/
12. ardrone-wpa2. https://github.com/daraosn/ardrone-wpa2
13. IEEE_802.11. http://en.wikipedia.org/wiki/IEEE_802.11

Author Index

Printed in the United States
By Bookmasters